BIOENERGY

BIOENERGY:

Principles and Applications

Edited by

YEBO LI

Department of Food, Agricultural, and Biological Engineering
Ohio State University
USA

SAMIR KUMAR KHANAL

Department of Molecular Biosciences and Bioengineering
University of Hawai'i
USA

WILEY Blackwell

Published by John Wiley & Sons, Inc., Hoboken, New Jersey
Published simultaneously in Canada

For general information on our other products and services or for technical support, please contact our Customer Care Department within the United States at (800) 762-2974, outside the United States at (317) 572-3993 or fax (317) 572-4002.

Wiley also publishes its books in a variety of electronic formats. Some content that appears in print may not be available in electronic formats. For more information about Wiley products, visit our web site at www.wiley.com.

Library of Congress Cataloging-in-Publication data applied for

ISBN: 9781118568316

Cover image: Gettyimages/BanksPhotos

Set in 10.5/12.5pt AGaramond by SPi Global, Pondicherry, India

Printed in the United States of America

10 9 8 7 6 5 4 3 2 1

Contents

9   OILSEED-BASED FEEDSTOCKS, 127

Chengci Chen and Marisol Berti

10   LIGNOCELLULOSE-BASED FEEDSTOCKS, 143

Sudhagar Mani

11 ALGAE-BASED FEEDSTOCKS, 170

Xumeng Ge, Johnathon P. Sheets, Yebo Li, and Sudhagar Mani

SECTION III   BIOLOGICAL CONVERSION TECHNOLOGIES, 199

12 PRETREATMENT OF LIGNOCELLULOSIC FEEDSTOCKS, 201

Chang Geun Yoo and Xuejun Pan

18   BIOGAS PRODUCTION AND APPLICATIONS, 338

Samir Kumar Khanal and Yebo Li

19   MICROBIAL FUEL CELLS, 361

Hongjian Lin, Hong Liu, Jun Zhu, and Venkataramana Gadhamshetty

SECTION IV   THERMAL CONVERSION TECHNOLOGIES, 385

20   COMBUSTION FOR HEAT AND POWER, 387

Sushil Adhikari, Avanti Kulkarni, and Nourredine Abdoulmoumine

21   GASIFICATION, 407

Sushil Adhikari and Nourredine Abdoulmoumine

22   PYROLYSIS, 423

Manuel Garcia-Perez

SECTION V   BIOBASED REFINERY, 439

23   SUGAR-BASED BIOREFINERY, 441

Samir Kumar Khanal and Saoharit Nitayavardhana

SECTION VI   BIOENERGY SYSTEM ANALYSIS, 505

27 TECHNO-ECONOMIC ASSESSMENT, 507

Ganti S. Murthy

28 LIFE-CYCLE ASSESSMENT, 521

Ganti S. Murthy

List of Contributors

Nourredine Abdoulmoumine
Department of Biosystems Engineering, Auburn University, Auburn, AL, USA

Sushil Adhikari
Department of Biosystems Engineering, Auburn University, Auburn, AL, USA

Hasan K. Atiyeh
Biosystems and Agricultural Engineering, Oklahoma State University, Stillwater, OK, USA

Marisol Berti
Department of Plant Sciences, North Dakota State University, Fargo, ND, USA

Jeffrey M. Bielicki
Department of Civil, Environmental, and Geodetic Engineering, and The John Glenn School of Public Affairs, Columbus, OH, USA

Chengci Chen
Central Agricultural Research Center, Montana State University, MT, USA

Rakshit Devappa
Biorefining Research Institute, Lakehead University, Thunder Bay, ON, Canada

Thaddeus Chukwuemeka Ezeji
Department of Animal Sciences, The Ohio State University/Ohio Agricultural Research and Development Center (OARDC), Wooster, OH, USA

Venkataramana Gadhamshetty
Civil and Environmental Engineering, South Dakota School of Mines and Technology, Rapid City, SD, USA

Manuel Garcia-Perez
Biological Systems Engineering, Washington State University, Pullman, WA, USA

Xumeng Ge
Department of Food, Agricultural, and Biological Engineering, The Ohio State University/Ohio Agricultural Research and Development Center, Wooster, OH, USA

Scott C. Geleynse
School of Chemical Engineering and Bioengineering, Washington State University, Richland, WA, USA

J. H. Van Gerpen
Department of Biological and Agricultural Engineering, University of Idaho, Moscow, ID, USA

B. Brian He
Department of Biological and Agricultural Engineering, University of Idaho, Moscow, ID, USA

Lian He
School of Engineering and Applied Science, Washington University in St. Louis, St. Louis, MO, USA

Gal Hochman
Department of Agriculture, Food, and Resource Economics, and Rutgers Energy Institute, State University of New Jersey, New Brunswick, NJ, USA

David Hodge
Department of Chemical Engineering and Material Sciences, and Department of Biosystems and Agricultural Engineering, Michigan State University, East Lansing, MI, USA

Sami Kumar Khanal
School of Environment and Natural Resources, The Ohio State University, Wooster, OH, USA

Samir Kumar Khanal
Department of Molecular Biosciences and Bioengineering, University of Hawaii at Manoa, Honolulu, HI, USA

Avanti Kulkarni
Department of Biosystems Engineering, Auburn University, Auburn, AL, USA

Yebo Li
Department of Food, Agricultural, and Biological Engineering, The Ohio State University/Ohio Agricultural Research and Development Center, Wooster, OH, USA

Wei Liao
Department of Chemical Engineering and Material Sciences, and Department of Biosystems and Agricultural Engineering, Michigan State University, East Lansing, MI, USA

Hongjian Lin
Department of Biosystems and Bioproducts, University of Minnesota, St. Paul, MN, USA

Hong Liu
Department of Biological and Ecological Engineering, Oregon State University, OR, USA

Xiaolan Luo
Department of Food, Agricultural, and Biological Engineering, The Ohio State University/Ohio Agricultural Research and Development Center, Wooster, OH, USA

Sudhagar Mani
BioChemical Engineering Program, College of Engineering, University of Georgia, Athens, GA, USA

Armando G. McDonald
Department of Biological and Agricultural Engineering, University of Idaho, Moscow, ID, USA

Matthew J. Morra
Department of Biological and Agricultural Engineering, University of Idaho, Moscow, ID, USA

Ganti S. Murthy
Biological and Ecological Engineering, Oregon State University, Corvallis, OR, USA

Saoharit Nitayavardhana
Department of Environmental Engineering, Faculty of Engineering, Chiang Mai University, Chiang Mai, Thailand

Michael Paice
Michael Paice & Associates, Richmond, VA, USA

Xuejun Pan
Department of Biological Systems Engineering, University of Wisconsin-Madison, Madison, WI, USA

Ajay Shah
Department of Food, Agricultural, and Biological Engineering, The Ohio State University, Wooster, OH, USA

Johnathon P. Sheets
Department of Food, Agricultural, and Biological Engineering, The Ohio State University/Ohio Agricultural Research and Development Center, Wooster, OH, USA

Devin Takara
Department of Molecular Biosciences and Bioengineering, University of Hawaii at Manoa, Honolulu, HI, USA

Yinjie J. Tang
School of Engineering and Applied Science, Washington University in St. Louis, St. Louis, MO, USA

Victor Ujor
Department of Animal Sciences, The Ohio State University/Ohio Agricultural Research and Development Center (OARDC), Wooster, OH, USA

Arul M. Varman
School of Engineering and Applied Science, Washington University in St. Louis, 1 Brookings Drive, St. Louis, MO, USA

Mark R. Wilkins
Biosystems and Agricultural Engineering, Oklahoma State University, Stillwater, OK, USA

Chang Geun Yoo
Department of Biological Systems Engineering, University of Wisconsin-Madison, Madison, WI, USA

Xiao Zhang
School of Chemical Engineering and Bioengineering, Washington State University, Richland, WA, USA

Jun Zhu
Department of Biological and Agricultural Engineering, University of Arkansas, Fayetteville, AR, USA

Preface

Our modern society depends on energy for almost everything. Energy can be considered as a basic need of today's society. We need energy for home appliances used for preparing and storing food, lighting our homes and streets, heating/cooling our homes/offices, and powering our rapidly proliferating entertainment gadgets. Our vast transportation networks (air, land, and sea) and various modes of communication essentially depend on energy, as do all industrial processes that supply commodities for our daily needs. Energy also has vast socio-economic implications in the rural areas of developing countries, where women and children often spend as much as 4–6 hours collecting firewood for cooking. Thus, energy is an inevitable part of the growth, prosperity, and well-being of our society.

Global energy consumption is expected to increase by nearly 56% by 2040, due mostly to increased demand from emerging nations such as China and India. In recent years there has been a significant shift in the dynamics of energy consumption. Non-OECD (Organization for Economic Cooperation and Development) countries, which account for 90% of global population growth and 70% of economic outputs, are expected to have over 85% growth in energy consumption from 2010 to 2040. Energy consumption in non-OECD developing Asian countries, especially China and India, will increase by 112% from 2010 to 2040. The rest of the non-OECD countries are also projected to show strong growth in energy consumption during that period: for example, by 76% in the Middle East, by 85% in Africa, and by 62% in Central and South America. It is forecast that China's energy consumption will grow by as much as twice that of the USA between 2010 and 2040.

Currently, over 85% of total energy consumption is met through the use of non-renewable sources such as petroleum, coal, natural gas, and nuclear energy. Our heavy dependence on these rapidly depleting non-renewable energy sources has several irreparable consequences such as impacts on economic development, national security, and local and global environments, especially through climate change. Thus, we must act quickly and decisively to develop sustainable, affordable, and environmentally friendly energy sources. Bioenergy derived from renewable bioresources, such as biomass (energy crops, agri- and forest residues, algae, and biowastes) is considered to be the most promising alternative.

With the growing interest in bioenergy, there is a need to prepare a new cadre of the workforce in this emerging field who could lead the research, development, and implementation efforts of bioenergy technology. There are many reference books available on this subject. However, there is no comprehensive book that could be used as a textbook at an undergraduate level. This 29-chapter textbook is one of the very first that covers comprehensively both the fundamental and application aspects of bioenergy. The chapters are organized in such a way that each preceding chapter builds up a foundation for the following one. Every effort has been taken to maintain consistency throughout the book, even though the chapters were contributed by different authors. We strove to maintain clarity in explaining the concepts, and textboxes have been provided throughout the book to further clarify the concepts/terminology. At the end of each chapter, exercise problems have been provided, which instructors can use as an assignment for the class. A solution manual is also available.

The textbook is divided into six sections. Section 1 consists of seven chapters focusing mainly on the fundamental aspects of bioenergy. Section II comprises four chapters covering different bioenergy feedstocks. Section III consists of eight chapters focusing on various biological conversion technologies. Section IV has three chapters about thermochemical conversion technologies. The four chapters in Section V cover various aspects of biorefineries. Finally, Section VI comprises three chapters focusing on bioenergy system analysis. This organization will help students easily grasp the content presented in the textbook.

The editors, especially Samir Kumar Khanal (SKK), drew inspiration in preparing this textbook from King Bhumibol Adulyadej's (Thailand) self-sufficiency initiatives. SKK is also particularly thankful to Andrew G. Hashimoto (Professor Emeritus, University of Hawai'i), Ju-Chang (Howard) Huang (Chair Professor Emeritus, Hong Kong University of Science and Technology, Hong Kong), Heinz Eckhardt (RheinMain University, Wiesbaden, Germany), Akhilendra Bhusan Gupta (Malaviya National Institute of Technology, Jaipur, India), Shihwu Sung (University of Hilo, Hawaii), Chongrak Polprasert (Professor Emeritus, Asian Institute of Technology, Bangkok, Thailand), Dulal Borthakur (University of Hawaii), and Kenneth Grace (Associate Dean for Research, University of Hawaii) for their supports and encouragements. We sincerely hope that this textbook will be valuable especially to undergraduate students and instructors. The book will be equally useful to graduate students, decision makers, practicing professionals, and others interested in bioenergy.

We gratefully acknowledge the hard work and patience of all the authors who have contributed to this textbook. The views or opinions expressed in each chapter are those of the authors and should not be construed as opinions of the organizations for which they work. Special thanks go to SKK's former and current graduate students at the University of Hawai'i at Mānoa (UHM), Saoharit Nitayavardhana, Devin Takara, Pradeep Munasinghe, Surendra K.C., Sumeth Wongkiew, Duc Nguyen, Chayanon Sawatdeenarunat, and Edward Drielak, for reviewing some of the chapters and helping with solution manual. Furthermore, we are highly indebted to Justin Jeffrey, Editorial Director at Wiley, for his relentless support for our textbook project and Shummy Metilda, Production Editor at Wiley, for enforcing the publication deadline. Last but not least, we extend our sincere gratitude, love, and appreciation to our family members for their support through the years. Finally, we would like to salute the people of developing countries for their rational use of energy in their daily activities, sacrificing their comfort for the rest of the world.

– Yebo Li and Samir Kumar Khanal

Acknowledgments

The following individuals helped us in reviewing the book chapters.

Dulal Borthakur, University of Hawai'i at Manoa, Honolulu, HI, USA

Thomas Canam, Illinois Institute of Technology, Chicago, IL, USA

Shaoqing Cui, Ohio State University, Columbus, OH, USA

Xumeng Ge, Ohio State University, Columbus, OH, USA

Lee Jakeway, Hawaiian Commercial and Sugar Company, Puunene, HI, USA

Harold Keener, Ohio State University, Columbus, OH, USA

Long Lin, Ohio State University, Columbus, OH, USA

Xiaolan Luo, Ohio State University, Columbus, OH, USA

Ned Mast, Green Arrow Engineering, LLC, Wooster, OH, USA

Venkata Mohan, CSIR-Indian Institute of Chemical Technology (CSIR-IICT), Hyderabad, India

Sue Nokes, University of Kentucky, Lexington, KY, USA

Deepak Pant, Flemish Institute for Technological Research (VITO), Antwerp, Belgium

Stephen Park, Southern Illinois University, Carbondale, IL, USA

Ratanachat (Siam) Racharaks, Yale University, New Haven, CT, USA

Troy Runge, University of Wisconsin, Madison, WI, USA

Ajay Shah, Ohio State University, Columbus, OH, USA

Vijay Singh, University of Illinois at Urbana Champaign, Champaign, IL, USA

Juliana Vasco-Correa, Ohio State University, Columbus, OH, USA

Caixia Wan, University of Missouri, Columbia, MO, USA

Mary Wicks, Ohio State University, Columbus, OH, USA

Fuqing Xu, Ohio State University, Columbus, OH, USA

Liangcheng Yang, Illinois State University, Normal, IL, USA

Julia Yao, University of Kentucky, Lexington, KY, USA

SKK is particularly thankful to the following colleagues who supported/hosted him during the preparation of the textbook:

Hyeun-Jong Bae, National Chonnam University, Gwangju, South Korea

Piyarat Boonsawang, Prince of Songkla University, Songkhla, Thailand

GuangHao Chen, Hong Kong University of Science and Technology, Hong Kong

Wen-Hsing (Albert) Chen, Ilan National University, Yilan City, Taiwan

Berhanu Demessie, Addis Ababa Institute of Technology, Addis Ababa, Ethiopia

Akhilendra Bhusan Gupta, Malaviya National Institute of Technology, Jaipur, India

Zhen Hu, Shandong University, Jinan, China

JaeWoo Lee, Korea University, Seoul, South Korea

Po-Heng (Henry) Lee, Hong Kong Polytechnic University, Hong Kong

Xie (Sally) Li, Tongji University, Shanghai, China

Hui Lu, Sun Yat Sen University, Guangzhou, China

Hans Oechsner, Universität Hohenheim, Stuttgart, Germany

Juan Camilo Acevedo Paez, Universidad de Santander, Cucuta, Colombia

Ashok Pandey, Center of Innovative and Applied Bioprocessing, Punjab, India

About the Companion Website

This book is accompanied by a companion website:

www.wiley.com/go/Li/Bioenergy

The website includes:

- Figures
- Tables

Scan this QR code to visit the companion website

Bioenergy Fundamentals

Introduction to Bioenergy

Samir Kumar Khanal and Yebo Li

What is included in this chapter?

This chapter provides an introduction to non-renewable and renewable energy resources. Different forms of non-renewable and renewable energy and their current demand/consumption are discussed. An overview of bioenergy, its merits and demerits, and current status are also presented.

1.1 Energy

Our modern society depends on energy for nearly everything, including our basic needs that we often take for granted (e.g., to supply drinking water, produce food, and even provide air in some cases). Whether we admit it or not, we are addicted to energy in order to power appliances, light our homes, streets, and offices, and, perhaps more importantly, power the advanced technological gadgets we keep in our pockets. Many of the things we commonly overlook, like our vast transportation networks, are heavily reliant on an abundant and consistent supply of energy. Yet energy, in the form of electricity and fuel, is not as ubiquitous worldwide as it is in the USA and Western Europe. In many rural areas of developing countries, energy is derived from burning wood and local biomass resources, and the ability to secure energy consistently has significant socio-economic implications affecting the quality of life for local communities. In all cases, in both developing and developed nations, energy is essential for the growth, prosperity, and well-being of society.

Sustainability (i.e., meeting the needs of present generations without compromising the needs of those in the future) is another key issue of great concern caused by the rapidly growing global population and the corresponding increase in energy demand. As of October 2011, the world population reached 7.0 billion, and the United Nations projects that the population will continue to grow to 10.1 billion by the end of the twenty-first century (UNFPA, 2011). Standards of living are also on the rise, particularly in developing countries, a fact that is expected to contribute significantly to

---

Bioenergy: Principles and Applications, First Edition. Edited by Yebo Li and Samir Kumar Khanal.
Companion website: www.wiley.com/go/Li/Bioenergy

increased energy consumption and the stress already being placed on our diminishing non-renewable resources (i.e., fossil fuels), which currently meet over 85% of primary energy demands.

Discussions of energy are often associated with countries' gross domestic product (GDP), and per capita energy consumption is frequently viewed as an index of development. For example, developing countries have very low per capita energy consumption, typically less than 1.0 metric ton of oil equivalent per year (toe/year) compared to over 4.0 toe/year for developed countries (IEA, 2010a). For example, developing countries such as Nepal, India, Kenya, and Ghana have per capita total primary energy supply (TPES; inclusive of total energy consumed for exports, imports, etc.) of just 0.34, 0.54, 0.47, and 0.41 toe/year, respectively. In comparison, the world's average per capita TPES is 1.83 toe/year (IEA, 2010a). GDP growth and increases in electricity demands are linearly correlated with a coefficient of approximately 1 (i.e., every 1% increase in GDP is associated with a 1% increase in electricity demand; IEA, 2010b).

In 2010, global energy consumption was around 524 quadrillion Btu (QBtu – 13,205 million metric tons of oil equivalent or Mtoe). This value is projected to increase by 56% to 820 QBtu (20,664 Mtoe) by 2040 based on the International Energy Outlook 2013 (EIA, 2013). Currently, over 85% of total energy consumption is met through the use of non-renewable sources such as petroleum, coal, natural gas, and nuclear energy. China recently overtook the USA with total primary energy consumption of 2,550 Mtoe (101.2 QBtu; EIA, 2013).

According to the International Energy Outlook 2013 (EIA, 2013), the dynamics of energy consumption are changing dramatically. Non-OECD (Organization for Economic Cooperation and Development) countries, which account for 90% of global population growth and 70% of economic outputs, are expected to have over 85% growth in energy consumption from 2010 to 2040. Energy consumption in non-OECD developing Asian countries, especially China and India, will increase by 112% from 2010 to 2040. The rest of the non-OECD countries are also projected to show a strong growth in energy consumption during the projected period: for example, by 76% in the Middle East, by 85% in Africa, and by 62% in Central and South America. It is projected that China's energy consumption will grow by as much as twice that of the USA between 2010 and 2040. Because of its large population, however, on a per capita basis, China's energy use will still be half that of the USA. Another country of interest is India, which currently ranks as the third largest consumer (615 Mtoe or 24.4 QBtu) of energy. Despite the unprecedented rapid development of China seen over the last decade, the growth rate and energy consumption of other countries, such as India, Brazil, Indonesia, and the Middle Eastern nations, are expected to be even higher.

Many developing countries continue to struggle to provide sufficient energy to address the basic needs of all citizens. In India, nearly 840 million people lack proper access to modern energy services. Based on a 2009 estimate, over 1.3 billion people in sub-Saharan Africa and developing Asian countries do not have access to electricity, and nearly 2.7 billion people in these countries, 40% of the world's population, still rely on traditional biomass for cooking. Perhaps not surprisingly, 84% of people without electricity live in remote rural areas of these countries (IEA, 2011b). In these situations, bioenergy, especially biogas (which is covered in Chapters 17 and 18), may be a viable and attractive option for supplying cheap and consistent energy to rural populations.

Throughout this chapter, there is a fundamental, resounding question: What major energy resources can be implemented to meet the rising energy demands of rapidly growing populations? The answer is not easy for many reasons, including the uncertainty of various factors such as the availability of non-renewable resources; threats to the environment, such as climate change; geopolitics and energy security; changing governmental policies and regulations in light of emission and safety concerns, especially for nuclear power plants; and the unpredictable cost of fossil fuels. One thing, however, is certain: our future energy portfolio will be extremely diverse and an increasing

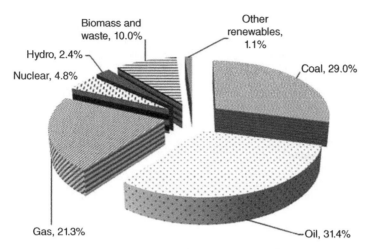

FIG. 1.1   World total primary energy supply by fuel type in 2012. Source: Data from IEA, 2014.

share of our energy will come from renewable sources, which have the highest annual growth rate of 2.5% among all energy resources (EIA, 2013).

1.2 Non-renewable Energy

Non-renewable energy is derived from materials or resources that cannot be replaced during a human lifetime and that, by definition, are available only in limited (finite) reserves. The world TPES by fuel type in 2012 is shown in Figure 1.1. Petroleum, natural gas, coal, and nuclear energy, which are considered to be conventional non-renewable energy resources, are discussed in detail in the rest of this section.

Petroleum, the primary source of *transportation fuel*, is a *fossil fuel* that accounts for nearly 31% of the world's TPES based on 2012 data. Table 1.1 shows the major crude oil consumption by nations and regions in 2011. The total world liquid fuel consumption in 2011 was 87.3 million barrels (bbl) per day, with the USA accounting for nearly 22% of total demand (EIA, 2012). It is interesting to note that the USA imported nearly 45% of the petroleum (crude oil and petroleum products) that it consumed in 2011 (EIA, 2012).

Coal is estimated to be the world's most abundant fossil fuel. Globally, it is the second most heavily used resource after petroleum, and accounted for nearly 29% of the TPES in 2012. Coal plays a critical role in electricity generation, comprising over 40% of the global electricity output in 2010 (IEA, 2011b). China is the largest consumer of coal, with a demand of 2,653 million metric tons coal equivalent (Mtce) or 73.5 QBtu in 2010, over 48% of the total global demand. The USA and India are the second and third largest consumers of coal, respectively. Table 1.2 compares the major coal-consuming regions of the world.

Coal is considered to be one of the highest polluting energy sources in terms of greenhouse gas (GHG) emissions. Carbon dioxide (CO_2) released from coal-fired power plants is a major concern, and significant attention is required to improve the efficiency of these plants and develop technologies for carbon capture and storage (CCS). In China and India, where coal is abundant and

Table 1.1 Major petroleum oil consumers in the world in 2011

Countries/regions	Petroleum oil consumption (million barrels per day)
USA	18.34
Europe	15.08
China	8.92
Japan	4.48
Africa	3.51
India	3.43
Saudi Arabia	2.99
Brazil	2.79
Russia	2.73
Germany	2.42
Canada	2.24
South Korea	2.23
Iran	2.03
France	1.82

Source: Data from http://www.eia.gov/

Table 1.2 Annual coal consumption by major economies in the world in 2010

Countries/regions	Coal consumption, Mtce (QBtu)
China	2648 (73.5)
USA	750 (20.8)
Europe	465 (12.9)
India	453 (12.6)
Japan	174 (4.8)
Russia	171 (4.7)
Africa	157 (4.4)
South Korea	107 (3.0)

Source: Data from http://www.eia.gov/

electricity supply is limited, coal consumption is expected to grow by as much as 70% and 190%, respectively, by 2035 (IEA, 2011b). The 450 Scenario in the World Energy Outlook 2011 aims to limit atmospheric CO_2 equivalent (CO_2eq) to around 450 parts per million (ppm), and will require the share of coal used for electricity generation to decline from 40% in 2009 to 15% in 2035. To meet this goal, power generation will need to originate from alternative resources such as renewables, nuclear, and natural gas, which emit significantly less CO_2.

Natural gas is another important component of the world's current energy portfolio, accounting for almost 21% of the TPES in 2014. Based on the New Policies Scenario (World Energy Outlook 2011), global gas demands in 2009 were 3,076 billion m^3 and are projected to increase to 4,750 billion m^3 by 2035, nearly a 54% increase. Table 1.3 shows the major consumers of natural gas in 2010. The USA is the largest with a total annual demand of 673 billion m^3. The second largest consumer is Russia, where natural gas comprises over 50% of total primary energy use (IEA, 2011b). Ranked third is Iran, followed by China. China is believed to play a key role in the future growth of natural gas demand due to its rapid economic growth, new energy policies (favoring natural gas as a less polluting resource), and access to new resources via pipelines from Russia.

Table 1.3 Natural gas consumption by major economies in the world in 2010

Countries/regions	Natural gas consumption billion m^3 (ft^3)
USA	673 (23,775)
Europe	584 (20,638)
Russia	424 (14,961)
Iran	145 (5,106)
China	107 (3,768)
Japan	105 (3,718)
India	64 (2,277)
South Korea	43 (1,515)
Africa	101 (3,554)

Source: Data from http://www.eia.gov/

Natural gas is primarily used for power generation and currently comprises nearly 21% of global electricity production (IEA, 2011b). The second largest consumption of natural gas is in buildings, typically for space and water heating, and accounted for nearly 24% of the total natural gas consumed in 2009. In industry, 17% of the total gas consumed is used to produce steam and heat for the production of goods and commodities. In addition, there is a growing market in the transportation sector for use as fuel; that is, *compressed natural gas (CNG)* or *liquefied natural gas (LNG)*. Presently, 70% of the world's natural gas–powered vehicles are concentrated in five countries: Pakistan, Argentina, Iran, Brazil, and India.

The overall growth in natural gas demands is due to policy intervention in favor of natural gas over other fossil fuels that require stricter emission regulations to protect the environment. Natural gas–fired power plants emit the lowest level of CO_2 compared to all fossil-fuel plants, and produce almost half that of coal-fired plants. Moreover, there are fewer safety concerns related to natural gas. Limitations in nuclear power production due to safety concerns are likely to lead to additional growth in the demand for natural gas for power generation.

Nuclear energy is primarily used for power generation and supplied 2,461 terawatt-hours (TWh) of electricity (about 4.8% of the total global primary energy supply) in 2014. In France, nearly 75% of the total power produced domestically originates from nuclear facilities, followed by Ukraine (48%), South Korea (31%), and Japan (27%). In terms of total nuclear-based electricity generation, the USA leads the world with 790.2 billion kilowatt-hours (kWh), followed by France (421.1 billion kWh), Russia (161.7 billion kWh), Japan (156.2 billion kWh), and South Korea (147.7 billion kWh; http://www.nei.org/). In 2011, there were 441 nuclear plants operating in 30 countries with a gross installed capacity of 393 gigawatts (GW; 374 GW net), and nearly 83% could be found in OECD countries. Non-OECD countries, however, are leading recent efforts to build new nuclear plants, representing 55 of the planned 67 new facilities under construction. China reportedly had 28 new nuclear power plants under construction in 2010 (IEA, 2011b). Two important considerations that favor nuclear power generation are its ability to provide electricity at a relatively low price; and its potential to curb GHG emissions. Thus, if climate change becomes a serious tangible threat and renewables are unable to meet energy demands, nuclear energy will likely become the primary option for immediate deployment.

The safety of nuclear power plants has been a longstanding issue after two major nuclear power plant accidents, at Three Mile Island in the USA (1979) and Chernobyl in Ukraine (1986). More recently, following the March 11, 2011 tsunami in Japan, significant damage to the Fukushima Daiichi nuclear plant resulted in the release of radioactive materials into the environment. The incident resulted in significant policy debates in many countries over the risk and safety of nuclear power plants, whose long-term role in supplying energy to the world is in question. The Fukushima incident has caused significant public concern about the safety of nuclear power and, as a result, Germany has decided to phase out its nuclear power facilities before the end of their economic life. Similarly, Switzerland has decided to phase out nuclear power generation by 2034 and France, a major nuclear power producer, plans to invest in more renewable sources of energy. Significant delays in obtaining permits and rising costs of construction to meet additional safety measures, coupled with public concerns, are expected to be major hurdles to developing nuclear energy in the future. In spite of this, non-OECD countries such as China, India, and Russia plan to continue the expansion of their nuclear power capacity by 109 GW, 41 GW, and 28 GW, respectively, from 2010 to 2035 (EIA, 2013).

Unconventional non-renewable resources include oil shale and natural gas hydrates in marine sediments, as well as natural gas from coal beds (coal bed methane), low-permeability reservoirs (tight gas), and shale formations (shale gas, discussed further shortly). Unconventional proven reserves account for a significant share of the total reserves in the USA and Canada. Oil shale is a sedimentary rock that contains solid bituminous materials known as kerogen, which release petroleum-like liquids during pyrolysis, a process in which rock is heated under limited oxygen conditions.

While extending throughout the continental USA, shale, a fine-grained sedimentary rock composed of clay, is considered to be an "unconventional reservoir," as it has not been historically exploited for energy (see Figure 1.2). **Shale gas** (a gaseous resource in contrast to shale oil) had been virtually untapped and ignored until the 1973 energy crisis, when new recovery technologies were developed by the US Department of Energy and commercially tested by Mitchell Energy and Development Corporation in the 1980s and 1990s. The recent boom in shale gas production is a result of two main technologies developed during this period: horizontal wells and hydraulic fracturing. Compared to conventional vertical wells, horizontal wells offer the ability to replace four or more vertical wells due to greater reservoir exposure and to reduce surface disturbance from construction. To overcome the low permeability of the shale and to make gas production economically viable, horizontal drilling is used in conjunction with hydraulic fracturing. Hydraulic fracturing involves the use of an aqueous fracturing fluid at high pressure to create fissures and interconnected cracks, which increase the permeability of the shale and allow greater flow rates of gas into the well.

As the fluid injected during the hydraulic fracturing treatment needs to be recovered and disposed of before the gas can flow out of the shale, treatment of the hydraulic fracturing fluid poses a substantial challenge. The aqueous fluid that returns to the surface after pressure is relieved is called "flowback" and constitutes 10–40% of the total fracturing fluid. The flowback period depends on the geology of the formation and may last several weeks. The created brine solution may be composed of salts, metals, oils, greases, soluble organic compounds, and radioactivity from radioactive underground sources (Gregory et al., 2011). The most common method of disposal is by injecting the wastewater underground. The wastewater is normally confined to a designated Class II–type well that is regulated under the Underground Injection Control program to minimize possible pollution and contamination. The disposal well is placed thousands of feet underground in porous rock that is separated from groundwater by an encasement of impermeable rock. The nature of the well's geological requirements limits the use of this disposal method to only particular shale wells. If a Class II well is not available locally, the wastewater may be transported or piped to a suitable location, thus increasing the costs. Other treatment options for shale gas wastewater include freeze–thaw/ evaporation, ion exchange, capacitive desalination, electrodialysis reversal, and artificial wetlands. These treatment options, however, suffer from limitations, including restriction to colder climates, high costs, reliance on additional chemical treatments, and the limited salinity tolerance of plant species.

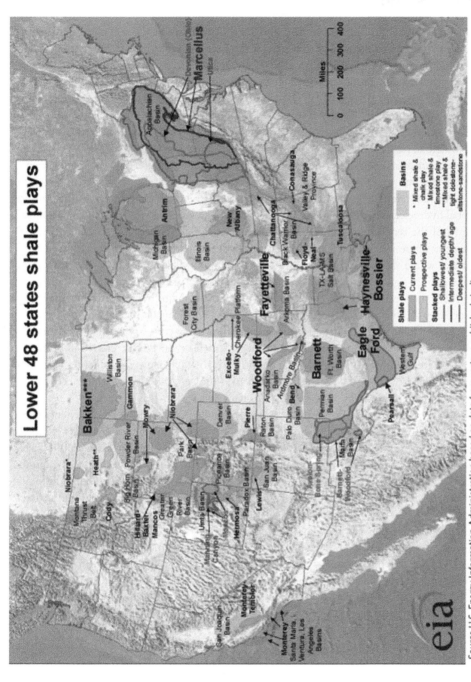

Source U.S. Energy Information Administration based on data from various published studies.
Upeate: May 9, 2011

FIG. 1.2 US shale plays. Source: Data from US Energy Information Administration.

1.3 Renewable Energy

Renewable energy is derived from resources that are available on a renewable basis and inexhaustible in the foreseeable future. Examples of these types of resources are wind, solar, geothermal, ocean (tidal), hydro, and biomass. Although only about 10% (1,452 Mtoe or 57.6 QBtu) of total energy use was met by renewables in 2010, the share of renewables is anticipated to reach 15% (i.e., 3,098 Mtoe or 123 QBtu) by 2040 (EIA, 2013). The growth in electricity generation from renewables is expected to originate from four primary resources: wind, hydro, biomass, and solar photovoltaics. The European Union (EU) is likely to lead this growth. Renewable resources have significant and growing potential, particularly since the technology for harnessing renewable energy is developing rapidly and production costs have been decreasing. Some of the major renewable energies are discussed in this section.

Globally, **hydroelectricity** is currently the largest source of renewable electricity generation. In 2009, net global hydroelectricity generation was 3,145 billion kWh (3,145 TWh), which accounted for almost 84% of the renewable electricity generated, and 17% of the total electricity generated (from all sources). There is significant potential for growth in hydroelectricity generation in non-OECD countries where the water resources are relatively underutilized. However, it is important to note that there is also significant natural fluctuation in energy output depending on rainfall, glacier melt, and other weather-related conditions. Table 1.4 presents the major hydroelectricity-generating countries as of 2010. China is the largest, followed by Brazil and Canada.

The major issues facing hydroelectric projects are environmental concerns created by damming rivers and the large capital investment required. Smaller hydroelectric projects, referred to as micro- and pico-projects, are more viable in developing countries as they have lower installation costs and can be privately owned by co-operatives. Micro- and pico-projects can also provide electricity in remote rural areas that are not connected to the national power grid.

Wind energy, another important resource, uses wind turbines to convert the kinetic energy of air flow into electricity. Both onshore and offshore wind energy sources are being aggressively explored by many countries. The net wind energy electricity generation was 328 billion kWh in 2009, which accounted for 7% of the total renewable net electricity generated. The USA produces the largest amount of electricity from wind, followed by Spain, Germany, China, and India. Denmark, however, has the largest penetration of wind-derived energy into its national grid (nearly 20% of its electricity generation; IEA, 2011a).

The **solar** energy system converts sunlight into energy that can be used to heat water or to generate electricity directly using photovoltaic (PV) technology. Currently, total solar PV electricity generation is about 20 billion kWh worldwide, and is expected to reach 740 billion kWh in 2035 (IEA, 2011b), corresponding to an average growth rate of 15% per year. Significant growth in PV

Table 1.4 Major hydroelectricity-producing nations in 2010

Country	Net hydroelectricity generation (billion kWh)
China	715
Brazil	400
Canada	350
USA	260
Russia	165
Norway	115
India	110

Source: Data from http://www.eia.gov/

electricity is anticipated to occur in the EU due to strong governmental support. Countries such as the USA, China, and India are also likely to have strong growth and may meet and exceed the EU in total electricity generation from PV.

Natural heat located close to the Earth's surface created by geological hot spots is referred to as a **geothermal** resource. This geothermal heat can be converted into electricity. Net global geothermal electricity generation in 2010 was 64 billion kWh. The USA is presently the largest producer of geothermal electricity (16 billion kWh), followed by the Philippines (9 billion kWh), Indonesia (8.5 billion kWh), and New Zealand (6.6 billion kWh). In Iceland, nearly 26% of net electricity generation originates from geothermal resources (EIA, 2012).

The lunar- and solar-induced tidal motions of oceans can be captured using wave energy converters (WECs) for producing electricity. For example, Canada and Norway have several small tide mills underwater for electricity generation. The first large-scale **tidal energy**-based power plant was built on the Rance river in Brittany, France. The Rance tidal power plant has an installed capacity of 240 megawatts (MW) and has been in operation since 1966. Presently, the world's largest tidal power plant is the Shihwa Lake tidal power plant in South Korea, with a generating capacity of 254 MW. This plant has been in operation since 2011.

Bioenergy is generated from feedstocks of biological origin, also known as biorenewable resources. Examples of bio-based feedstocks are crops rich in starch (corn, cassava, sorghum) and sugar (sugarcane, sugarbeet, sweet sorghum); oil-rich seeds (soybean, rapeseed, canola, palm fruits); lignocellulosic biomass (agricultural residues, forest residues, dedicated energy crops); and organic wastes (animal manure, industrial waste, agri-waste, food waste, sewage sludge). These feedstocks can be used to produce energy/energy carriers in three forms: liquid (e.g., ethanol, butanol, and biodiesel); gas (e.g., hydrogen and methane); and solid (e.g., fuel wood and biobriquettes).

Technologies for producing ethanol from sugar and starch-rich crops, biodiesel from seed oil and waste (cooking) oil, and methane from biomass are already available commercially in many developed countries. In 2012, total global bioethanol and biodiesel production was about 64 million gallons/day (242 million liters/day) and 14 million gallons/day (53 million liters/day), respectively (EIA, 2012). The USA is the leading biofuel-producing nation with a total capacity of 37 million gallons/day (140 million liters/day), followed by Brazil with 22 million gallons/day (84 million liters/day). Europe, on the other hand, is paving the way for anaerobic biotechnology in the form of biogas production. Biogas technology is currently being implemented for the production of electricity and heat using combined heat and power (CHP) units, upgraded to natural gas quality for injection into gas pipelines, and compressed for use in CNG engines for transportation. Europe alone produced 3,000 MW of electricity from biogas in 2007 and nearly half of that came from Germany, a global leader in biogas technology.

1.4 Why Renewable Energy?

There are several factors driving the rapid growth in renewable energy. Some of the motivation for this is discussed here. Renewable energy development policies in the USA emphasize the importance of energy independence, whereas in Europe the aim is in relation to climate change.

1.4.1 Energy Insecurity

The OPEC (Organization of the Petroleum Exporting Countries) oil embargo of 1973 against the USA significantly affected the US economy because of soaring petroleum prices. For the first time in

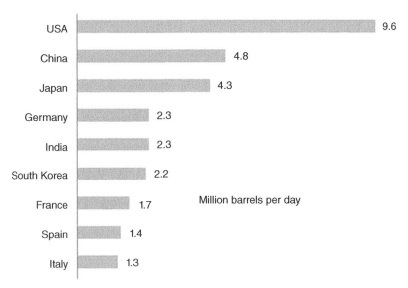

FIG. 1.3   Top 10 oil-importing nations. Source: Data from EIA, 2011.

its history, the USA realized its vulnerability caused by an insatiable dependence on petroleum imports. The economies of many other developed countries also strongly rely on the continual availability of energy at an affordable price. Many nations currently meet their energy demands through imports from geo-politically unstable countries. For example, in 2011 the USA imported nearly 45% of its liquid fuels (net imports of 8.4 million barrels/day) to meet its transportation needs, and Japan imported nearly 84% of its total energy needs. China also depends heavily on imported oil and brought in nearly 52% of its consumption from overseas in 2010. Figure 1.3 shows the top 10 oil-importing nations. It is evident that the development of renewable energy could significantly reduce the world's reliance on imports and the risk of energy insecurity.

1.4.2 Depletion of Energy Resources Reserves

The economies of the OECD countries have primarily been built on energy. These countries account for about 43% of total global energy demand, but contain less than 20% of the world's population. Moreover, there is a growing demand for energy in non-OECD countries, particularly emerging economies such as China, India, Brazil, Russia, and Indonesia, where people have only recently begun to enjoy the comforts of their rising affluence. The International Energy Outlook 2013 projects that the energy demand in non-OECD countries will increase by nearly 85% of current levels by 2040 (EIA, 2013). Most (>85%) of our energy needs today are met through the exploitation of non-renewable energy resources, the reserves of which are known to be limited. Table 1.5 illustrates proven reserves of non-renewable energy resources correlated with current consumption rates. It is apparent from the table that known energy reserves may be exhausted within one to two generations. In addition, as reserves begin to empty, oil becomes increasingly difficult to extract until market prices rise enough to make extraction economically viable. The result is higher consumer fuel prices. Thus, renewable energy resources are and will become critical to buffer increasing global energy demands and associated cost increases.

Table 1.5 Global proven reserves of non-renewable energy
resources

Energy resources	Years proven reserves to last
Petroleum oil	48
Coal	150
Natural gas	48
Uranium	90

Source: Data from IEA, 2011b; OECD NEA, 2010.

1.4.3 Concern about Climate Change

The use of fossil-derived energy is a major source of GHG emissions, especially CO_2. Energy-derived CO_2 emissions are considered to be a significant contributor to global climate change. Among the various types of fossil fuels, coal releases the most CO_2 per unit of energy, followed by petroleum. Natural gas (also derived from fossil fuel) is considered to be the cleanest form of non-renewable energy in terms of emissions of CO_2 and other pollutants. Globally, China is the largest emitter of energy-related CO_2 with estimated emissions of 7.5 giga metric tons (Gt) in 2010, followed by the USA (5.4 Gt) and India (1.6 Gt). Coal is the biggest contributor to total emissions due to its extensive use in China and India. Energy-related CO_2 emissions worldwide were estimated at 30.5 Gt in 2010, and the World Energy Outlook 2011 projects that emissions could be as high as 36.4 Gt by 2035, leading to a long-term increase in the Earth's temperature by more than 3.5 °C. The effects of global warming are likely to be devastating, such as rising sea levels that cause the large-scale dislocation of human settlements, along with erratic rainfall patterns, drought, flood, and heat waves (IEA, 2011b). Some researchers believe that we may already be witnessing the beginning of some of these severe and unpredictable weather events.

The 450 Scenario (World Economic Outlook 2011) aims at reducing the release of energy-related CO_2 to 21.6 Gt by 2035 through strong policy interventions. The goal of the 450 Scenario is to limit the long-term atmospheric CO_2 concentration to 450 CO_2eq. This change is expected to cap the global average temperature increase to 2 °C. The projected CO_2 emissions reduction will likely shift the share of fossil energy in the total global energy mix from 81% in 2009 to 62% in 2035. Renewable energy resources are anticipated to play a significant role in supplying carbon-neutral energy and tackling global climate change.

1.5 Bioenergy

Bioenergy also has the potential to contribute a portion of future energy needs; however, there are many challenges that need to be overcome. This section discusses the current status, benefits, and limitations of bioenergy. It is important to note here that presently there is no formal definition of bioenergy and biofuel, but for convenience and clarity, the following definitions of bioenergy and biofuel are used in this book.

- **Bioenergy** is a generic term that includes solid, liquid, and gaseous fuels of biological origin. The energy, which is in the form of heat and power (i.e., electricity) for stationary applications or as a liquid biofuel for transportation, can be collectively termed bioenergy.

- **Biofuel** is a more specific term that refers to any liquid fuel derived from biological materials with sufficient energy density and combustion characteristics to make it a suitable transportation fuel. Common examples of biofuels are ethanol, biodiesel, and biomass-derived Fischer–Tropsch fuels. Methane and hydrogen are gaseous fuels that can be compressed for use in jet propulsion and may also be considered biofuels.

1.5.1 Current Status of Bioenergy Production

The technologies for sugar and starch-derived ethanol, and seed oil and waste oil–derived biodiesel are commercially available. Feedstocks that produce these types of fuels are known as first-generation feedstocks as a result of early research into these renewable resources. Current commercial facilities and processing are expected to continue to dominate biofuel production in the foreseeable future. Currently, the USA and Brazil are the major biofuel-producing countries.

The *second- and third-generation feedstocks*, which consist of non-food feedstocks such as *lignocellulosic biomass, Jatropha*, and *microalgae*, can also be used to produce bioenergy, but are not currently economically viable for commercial-scale production. Considerable research and technology development efforts are underway to produce advanced *drop-in fuels*, such as *butanol* and jet fuel at prices comparable to petroleum oil. Several commercial-scale *cellulosic ethanol* plants have been built in the USA (Iowa and Kansas) and China in recent years that use corn stover/cobs as the major feedstock. Dilute acid and steam explosion are pretreatment technologies adopted in these commercial cellulosic ethanol plants, while enzymes for enzymatic hydrolysis are provided by biotechnology companies—Novozyme or DuPont.

Algal biomass for biofuels, specifically biodiesel and biobased hydrocarbons production, has received significant attention in recent years. Algal technology, however, faces many engineering challenges, including efficient reactor design for light penetration and CO_2 transfer, contamination of open pond systems, algal cell recovery/harvesting, dewatering, and lipids extraction. Among the major bottlenecks for scale-up is the low biological productivity of lipids, the precursor for biodiesel and other hydrocarbons.

Anaerobic digestion is arguably the oldest form of renewable energy production and has been widely applied commercially to produce bioenergy, namely biogas, from food-processing wastes, animal manure, high-strength industrial wastewaters, and even agri-residues and energy crops. The biogas produced is used to generate electricity and heat via combined CHP units. Biogas can also be upgraded to natural gas quality for injection into gas pipelines or implemented as CNG for use as transportation fuel. In addition, anaerobic biotechnology has been employed for ethanol, butanol, and hydrogen production from fermentable sugars and synthesis gas ($CO + H_2$) also known as syngas. The organic acids produced during anaerobic fermentation also have the potential to be converted into gasoline, jet fuel, and diesel through various microbial pathways currently under investigation. In rural areas of developing countries, household anaerobic digesters play an important role and have been providing the energy necessary for cooking and lighting for decades. Anaerobic biotechnology may be an important avenue in providing energy globally.

Somewhat related to anaerobic digestion is the development of microbial fuel cells (MFCs), which exploit the biological electricity-generation capabilities of naturally occurring microbes. However, MFCs face significant challenges as a means for sustainable electricity generation due to the complexity associated with growing and maintaining microbiological cultures, in addition to technological challenges such as low electricity production and maintaining the required pH (potential hydrogen) gradients across membranes.

Combustion, *pyrolysis*, and *gasification* (the major thermal conversion technologies) have been widely studied in recent decades. Biomass combustion and co-firing of biomass with coal for energy (steam and electricity) production have been commercialized. *Pyrolysis* of biomass at a high temperature and pressure under a low oxygen environment produces a liquid fuel called bio-oil and a solid residue called *biochar*. Bio-oil can be refined to produce liquid hydrocarbons and chemicals. Gasification of biomass at high temperature under a controlled environment produces syngas, which can be further converted to electricity or biofuels such as ethanol and liquid hydrocarbons. Coal gasification is commercially available for large-scale production, while most of the biomass gasification units have only been demonstrated at a smaller scale.

1.5.2 Merits of Bioenergy

Bioenergy is considered to be a low carbon footprint and an environmentally friendly energy resource. Its potential merits include providing energy security, benefiting local and national economies by contributing to agricultural sectors, and improving local and global ecology. Some of the merits of bioenergy are highlighted here.

Reducing the Nation's Dependence on Imported Energy Resources

As discussed previously, many countries around the world are heavily reliant on imported petroleum to meet their energy demands. Most often, this non-renewable resource originates from unstable nations, posing a threat to national security and the quality of life. Biorefineries, premised on the fractionation of biomass feedstocks into bioenergy and biochemicals, can reduce our dependence on imported energy resources and mitigate the issue of national security.

Environmental Merits of Bioenergy

The Clean Air Act Amendments of 1990, which mandated the use of oxygenated gasoline in areas with unhealthy levels of air pollution, were the turning point in the boom of biofuel industries in the USA. Since 1992, *methyl tertiary-butyl ether* (MTBE), a fuel oxygenate octane enhancer, has been used at higher concentrations in some gasoline mixes to fulfill the oxygenation requirements. Oxygenated fuel promotes gasoline to burn more completely and cleanly, thereby reducing harmful tailpipe emissions from motor vehicles. In the past, MTBE was chosen over other oxygenates, primarily due to its excellent blending characteristics and low cost. However, groundwater contamination due to the poor biodegradability of MTBE prompted some states, including California, to phase out MTBE use by the end of 2002. This prompted the use of ethanol as an oxygenate to replace MTBE and resulted in a boom in ethanol industries across the USA. Ethanol is an alcohol that contains 35% oxygen and improves fuel combustion when blended with gasoline, resulting in low tailpipe emissions of carbon monoxide, unburned hydrocarbons, and nitrogen oxide (NOx). With favorable properties, such as a higher octane number (108), broader flammability limits, higher flame speeds, and higher heat of vaporization, ethanol has become a valued replacement for MTBE. A lifecycle assessment looking at ethanol produced from renewable resources indicated that compared to gasoline, sugarcane-based ethanol reduces GHG emissions by 84% (on a volume basis), corn ethanol by 30%, sugar beet ethanol by 40%, and cellulosic ethanol by 85%. Consequently, ethanol has been considered to have significant implications as a fuel in addition to being an oxygenate.

In developing countries, biogas has played a critical role in curtailing indoor air pollution by replacing highly polluting firewood and cow dung as a cooking fuel, in addition to providing much-needed energy resources for cooking. Additionally, bioenergy has also provided an opportunity for improving health and sanitation by effective stabilization of biowastes, especially during anaerobic digestion. Anaerobic digestion technology, including methane capture, at landfills also helps reduce GHG emissions from those sites. Methane is an important GHG that has a global warming potential (GWP) of 20 times higher than that of CO_2 on a 100-year timescale (Forster *et al.*, 2007).

Socio-Economic Benefits of Bioenergy

The successful creation of a biobased economy has the potential to create local jobs and improve rural economies. Various entities likely to benefit from the deployment of biorefineries are rural farmers and farm co-operatives; industries involved in agricultural equipment, facility design, and fabrication; construction companies; and biotechnology industries. State and federal governments are also likely to benefit from additional tax revenues. In developing countries, improved access to local energy sources will have significant social merits, such as increased time for women and children to pursue education and other activities rather than collecting fuel biomass.

1.5.3 Demerits of Bioenergy

Although real and/or perceived environmental benefits are important drivers for the greater use of bioenergy, no fuel system is free of environmental ramifications. Multi-dimensional conflicts exist in the use of land, water, energy, and other environmental resources for food, feed, and bioenergy production. Biofuels derived from starch, sugar, and oil-seed crops, and even lignocellulosic biomass, face several challenges. Some of these issues are outlined here.

Potential Impact on Food/Feed Supplies and their Prices

The utilization of first-generation feedstocks for biofuel production has intensified the competition globally for resource utilization (including land) for food, feed, and biofuels. This, among other factors, has contributed to a rise in prices of essential staples in recent years, as starch and sugar-rich cereal grains make up nearly 80% of the world's food supply.

Impact on Biodiversity

The rising demand for bioenergy feedstocks also has led to the increased clearing of forestlands for growing more grains and oilseeds for biofuels, thereby having impacts on biodiversity and ecological balance. This reduction in biodiversity could also be an issue with second-generation biofuel crops such as energy crops. In addition, it is not certain whether the clearing of land for bioenergy feedstocks is advantageous or detrimental to reducing global GHG emissions.

Impact on Water Quantity

Significant amounts of water are needed for both biomass feedstock production and processing. The requirement for fresh, clean, and consistent water supplies is a major issue for emerging biofuel industries.

Impact on Water Quality

The water quality in streams/lakes is likely to deteriorate due to the transport of eroded soils from newly established cropland to various water bodies. In addition, leaching of chemicals from fertilizers, insecticides, and pesticides to surface water and groundwater may also affect water quality. Nutrients in water bodies are responsible for eutrophication (algal growth) and subsequent hypoxia. Water pollution from wastewater disposal from biofuel facilities is another challenge facing the biofuel industries. For example, vinasse, which is produced as waste stream after ethanol recovery from sugarcane-to-ethanol plants, is posing a significant challenge to sugarcane-to-ethanol producers in Brazil.

Degradation of Soil Quality/Soil Erosion

There is potential for the deterioration of topsoil quality with second-generation biofuels, because agricultural residues such as corn stover, wheat, and rice straw provide organic carbon and nutrients when left in the field. Removal also reduces the mulching effect of these biomass feedstocks, which protects the nutrient-rich topsoil against erosion. Adoption of a no-till farming method, although limited in certain areas, may be required in order to reduce the impact of residue removal on soil quality.

References

EIA (2011). *Short-Term Energy Outlook*. Washington, DC: U.S. Energy Information Administration. http://www.eia.gov/forecasts/steo/, accessed April 2016.

EIA (2012). *International Energy Statistics*. Washington, DC: U.S. Energy Information Administration. http://www.eia.gov/cfapps/ipdbproject/iedindex3.cfm?tid=44&pid=44&aid=2&cid=regions,&syid=2005&eyid=2009&unit=QBTU, accessed April 2016.

EIA (2013). *International Energy Outlook 2013*. Washington, DC: U.S. Energy Information Administration. http://www.eia.gov/forecasts/ieo/pdf/0484(2013).pdf, accessed April 2016.

Forster, P., Ramaswamy, V., Artaxo, P., *et al.* (2007). Changes in atmospheric constituents and in radioactive forcing. In: S. Solomon, D. Qin, M. Manning, *et al.* (eds.), *Climate Change 2007: The Physical Science Basis*. Contribution of Working Group I to the Fourth Assessment Report of the Intergovernmental Panel on Climate Change. Cambridge: Cambridge University Press, pp. 114–43.

Gregory, K.B., Vidic, R.D., and Dzombak, D.A. (2011). Water management challenges associated with the production of shale gas by hydraulic fracturing. *Elements* 7: 181–6.

IEA (2010a). *Key World Energy Statistics 2010*. Paris: International Energy Association. doi: 10.1787/9789264095243-en.

IEA (2010b). *World Energy Outlook 2010, Executive Summary*. Paris: IEA. http://www.iea.org/Textbase/npsum/weo2010sum.pdf, accessed April 2016.

IEA (2011a). *Key World Energy Statistics 2011*. Paris: International Energy Agency. doi: 10.1787/key_energ_stat-2011-en.

IEA (2011b). *World Energy Outlook, Executive Summary*. Paris: IEA. http://www.worldenergyoutlook.org/publications/weo-2011/, accessed April 2016.

IEA (2014). *Key World Energy Statistics 2014*. Paris: International Energy Agency. doi: 10.1787/key_energ_stat-2014-en.

OECD NEA (2010). *Uranium 2009: Resources, Production and Demand*. Publication 6891. Paris: OECD Nuclear Energy Agency/International Atomic Energy Agency. https://www.oecd-nea.org/ndd/pubs/2010/6891-uranium-2009.pdf, accessed April 2016.

PEIS Information Center (2012). *About Oil Shale*. Lakewood, CO: Oil Shale & Tar Sands Programmatic EIS. http://ostseis.anl.gov/guide/oilshale/index.cfm, accessed April 2016.

UNFPA (2011). *State of World Population 2011*. New York: United Nations Population Fund. http://foweb.unfpa.org/SWP2011/reports/EN-SWOP2011-FINAL.pdf, accessed April 2016.

Exercise Problems

1.1. Why is energy so important for the socio-economic development of a nation?
1.2. Briefly discuss the current global energy situation.
1.3. Define renewable and non-renewable energy resources, with examples.
1.4. If climate change is associated with energy-derived CO_2, what would be the short- and long-term goals of curtailing CO_2 emissions?
1.5. Briefly discuss shale gas and shale oil and the impacts they are having on the development of renewable energy resources.
1.6. Explain the major motivations for moving toward renewable energy.
1.7. What are the major reasons for the slow development and implementation of renewable energy technologies?
1.8. Differentiate between bioenergy and biofuel, with examples.
1.9. What are some of the potential advantages and disadvantages of bioenergy?
1.10. What are the various feedstocks for producing first- and second-generation bioenergy?

Units and Conversions

Samir Kumar Khanal

What is included in this chapter?

This chapter covers units and conversions relating to bioenergy. Pertinent examples of heating value and unit conversions are also included.

2.1 Introduction

The units and conversions commonly used in science and engineering fields are equally applicable to the bioenergy discipline. It is important, however, to review these units and place an emphasis on their relevance in energy calculations. Although the SI system of units has been adopted virtually worldwide, imperial units are still used in the USA. As many aspiring engineering students know units such as acre, barrel, Btu, bushel, Fahrenheit, foot, gallon, pound, ton, etc. persist in both the literature and everyday life, and unit conversions must be implemented so that comparisons can be made. This chapter aims to lay out a foundation for measurement and to facilitate discussions and calculations in subsequent chapters of this textbook.

2.2 Units of Measurement

There are three systems of units in which most measurements are expressed: the SI system (meter, kilogram, and second); the CGS system (centimeter, gram, and second); and the British (or English) system, also known as the FPS system (foot, pound, and second).

The **SI system** of units (Système international d'unités) was adopted in 1960 as the international standard and is now implemented by all countries with the exception of the USA, Myanmar (formerly Burma), and Liberia. In the SI system, length (L) is measured in meters (m), mass (M) in kilograms (kg), time (T) in seconds (s), and temperature in Kelvin (K). Many other units used in

---

Bioenergy: Principles and Applications, First Edition. Edited by Yebo Li and Samir Kumar Khanal.
© 2017 John Wiley & Sons, Inc. Published 2017 by John Wiley & Sons, Inc.
Companion website: www.wiley.com/go/Li/Bioenergy

energy disciplines are derived from these basic units. For example, the unit of force, Newton (N), is equal to $(kg \cdot m)/s^2$.

Energy and power can also be derived from these fundamental units. Energy is measured in joules (J), with 1 *joule* defined as the amount of work done by a force of 1 Newton moving an object by a distance of 1 meter. In other words, 1 joule = 1 N·m or 1 $kg \cdot m^2/s^2$. The derivation of the unit is shown by:

$$\text{Force (F)} = \text{mass (m)} \times \text{acceleration (a)} = \frac{kg \cdot m}{s^2} = \text{Newton (N)}$$

$$\text{Work (W)} = \text{force (F)} \times \text{distance (d)} = \text{Joule (J)}$$

$$1 \text{ Joule (J)} = 1 \text{ Newton} \times 1 \text{ meter} = 1 \text{ N·m}$$

$$1 \text{ Joule (J)} = 1 \frac{kg \cdot m}{s^2} \times 1 \text{ m} = 1 \frac{kg \cdot m^2}{s^2}$$

Power (P) is defined as the work done per unit time. In SI units, it is written as watt (W).

$$P = \frac{\text{Work done}}{\text{time}} = \frac{\text{Joule}}{s} = W.$$

In application, energy and power often exist in much higher or lower magnitudes than those given here. The higher/lower values can be expressed using prefixes, as shown in Table 2.1.

As mentioned previously, the unit of temperature in the SI system is Kelvin. Degree Celsius (°C) is also used on many occasions, but it is not an SI unit. The increments of temperature are the same for both Kelvin and Celsius, thus (1 °C – 0 °C) = (274.15 K – 273.15 K). Degree Celsius can be converted to Kelvin by the following expression:

$$T(K) = T(°C) + 273.15 \tag{2.1}$$

In the **CGS system** of units, length is measured in centimeters (cm), mass is in grams (g), and time in seconds (s). The CGS system is related to the SI system thus:

$$1 \text{ g} = 10^{-3} \text{ kg}$$
$$1 \text{ cm} = 10^{-2} \text{ m}$$

The unit of force is given as a dyne (dyn), and 1 dyn = 1 $(g \cdot cm)/s^2$ = 1×10^{-5} N.

The unit of work and energy is measured as an erg:

$$1 \text{ erg} = 1 \text{ dyn} \times 1 \text{ cm} = 1 \times 10^{-7} \text{ J}$$

Table 2.1 Higher and lower values

Multiple	Prefix	Symbol	Multiple	Prefix	Symbol
10^{18}	exa	E	10^{-1}	deci	d
10^{15}	peta	P	10^{-2}	centi	c
10^{12}	tera	T	10^{-3}	milli	m
10^{9}	giga	G	10^{-6}	micro	μ
10^{6}	mega	M	10^{-9}	nano	n
10^{3}	kilo	k	10^{-12}	pico	p
10^{2}	hecto	h	10^{-15}	femto	f
10	deka	da	10^{-18}	atto	a

In the **FPS system**, also known as the English system of units, length is measured in feet (ft), mass in pounds (lb_m) and time in seconds (s). Although the SI system is internationally recognized, the USA, Myanmar, and Liberia continue to use their own traditional systems. Energy, fuel, heat, and many other parameters, however, are still commonly expressed in English units of measure. Consequently, the FPS system requires discussion here, as it persists frequently in scientific and engineering literature.

The FPS system is related to the SI system thus:

$$1 \text{ lb mass } (lb_m) = 0.4536 \text{ kg}$$

$$1 \text{ ft} = 0.3048 \text{ m}$$

$$1 \text{ lb force } (lb_f) = \text{mass } (m) \times \text{acceleration due to gravity } (g) = 1 \, lb_m \times 32 \frac{ft}{s^2}$$

$$= 0.4536 \frac{kg}{lb_m} \times 32 \frac{ft}{s^2} \times 0.3048 \frac{m}{ft} = 4.424 \, N$$

Energy or work is the product of force and distance ($lb_f \times$ ft), thus:

$$1 \, lb_f.ft = 4.424 \, N \times 0.3048 \text{ m}$$

$$= 1.3484 \, N \cdot m$$

$$= 1.3484 \, \text{Joule (J)}$$

Temperature in the FPS system is expressed in degrees Fahrenheit (°F). This unit is used heavily in the USA in lieu of degrees Celsius. Conversion between the common temperature units can be achieved as follows:

$$°F = 32 + 1.8 \, (°C) \tag{2.2}$$

$$°C = \frac{1}{1.8} \, (°F - 32) \tag{2.3}$$

$$\text{Degrees Rankine } (°R) = °F + 459.67 \tag{2.4}$$

2.3 Useful Units and Conversions

The units and conversions discussed in this section are commonly used in the bioenergy/biofuel discipline.

Area

$$1 \text{ hectare (ha)} = 10,000 \text{ m}^2$$

$$1 \text{ hectare} = 2.471 \text{ acres}$$

$$1 \text{ acre} = 4,047 \text{ m}^2$$

$$= 43,460 \text{ ft}^2$$

$$1 \text{ m}^2 = 10.76 \text{ ft}^2$$

Volume

1 gallon (gal) = 3.785 liters (L)

$1\,m^3 = 1,000\,L$

$\quad = 35.31\,ft^3$

1 barrel (bbl) = 42 gallons (petroleum fuel)

1 bushel (bu) = $1.244\,ft^3$ (volume of a cylinder 18.5 in diameter and 8 in high)

The bushel is often used for grain measurement and is converted to an equivalent weight. The weight varies depending on the grain being measured. For example, one bushel of corn weighs 56 lb, whereas one bushel of barley weighs 48 lb. This is because the bushel was originally used to measure the dry volume of an agricultural crop. Over time, the unit has evolved to be also a measure of mass.

How much is 1 bushel of grain?

1 bushel corn = 56 lb = 25.4 kg
1 bushel sorghum = 56 lb = 25.4 kg
1 bushel wheat = 60 lb = 27.2 kg
1 bushel soybean = 60 lb = 27.2 kg

Weight

1 kg = 2.205 lb

1 tonne = 1,000 kg = 2,205 lb

A tonne is known as a metric ton to differentiate the unit from US tons and to mitigate confusion in the literature. The measure of a tonne is typically used in countries that implement the SI system.

1 ton (US) = 2,000 lb = 907 kg

The old English ton, known commonly these days as a US ton, is also called a short ton, and is used frequently in the USA and Canada.

1 tonne = 1.1 ton (US)

To add to the confusion, it is worth noting that there is another unit of measure known as the long ton, which was used in the Commonwealth nations that implemented the old imperial system of measurement. Fortunately, the long ton appears less commonly in the contemporary literature. Its relationship to other systems of measurement is shown by:

1 long ton = 1 ton (imperial) = 1.12 ton (US) = 1.02 tonne = 2,240 lb = 1,016 kg

Pressure

Air exerts a pressure on all surfaces with which it is in contact. This is known as atmospheric pressure. The atmospheric pressure varies with altitude: the higher you are, the less air is above you, and consequently the less pressure there is. A device known as a barometer is used to quantify this type of pressure, which can be expressed as follows:

$$p = \frac{F}{A}$$

(2.5)

where F is the force normal to surface area A.

In the SI and metric systems, pressure is measured in Pascals. One Pascal (Pa) = 1 N/m^2 (1 bar = 10^5 Pa). The English system implements the unit pounds (lb_f) per square inch (psi).

One atmospheric pressure (1 atm) = 1.013×10^5 N/m^2; 10.3 m of water; 760 mm of Hg; 1.03 kg (f)/cm^2; 1.013×10^5 bar; or 14.696 lb_f/in^2 (psi).

In practice, all pressure gauges read zero when they are left open to the atmosphere. This is known as gauge pressure. Pressure gauges only read the difference between the pressure of a fluid (either liquid or gas) and the atmospheric pressure. If the pressure of a fluid is below atmospheric, it is known as vacuum pressure or negative gage pressure. When pressure is measured above absolute zero or complete vacuum, it is called absolute pressure.

Absolute pressure = Atmospheric pressure + Gauge pressure

(2.6)

Density

Density is defined as mass per unit volume. It is denoted by the Greek symbol ρ and has a unit of kg/m^3, g/cm^3, or lb_m/ft^3. Density is an important property of material and is often encountered in bioenergy production. This parameter can also be used to convert volume into mass and vice versa. Densities of solids and liquids vary minimally with everyday temperature fluctuations. The density of water at 4 °C is 1.0 g/cm^3 (1,000 kg/m^3) or 62.4 lb_m/ft^3. Ethanol has a density of 789 kg/m^3 at 20 °C. Another important density term in the study of biofuel/bioenergy is known as bulk density. Because bioenergy feedstocks are often granular or grassy (like a bale of corn stover or energy crops), they have varying degrees of empty space between grains/fibers. Bulk density is essentially defined as the overall mass of the feedstock (either wet or dry weight) divided by the total volume that it occupies. The true density of a feedstock varies by section (i.e., leaf, root, stalk), thus it would be both impractical and impossible to refer to the density of a harvest by its true density when talking to farmers or engineers.

Specific Gravity (SG)

This unit of measure is more accurately defined as a ratio of the density of a given solid or liquid substance to the density of a specific reference material at a specific temperature and pressure. Specific gravity is therefore a dimensionless parameter. For most solids and liquids, the reference material is usually water. For gases, air is normally taken as the primary reference material. The specific gravity of water is 1.0. By knowing the specific gravity of a particular substance, its density can easily be calculated. If the specific gravity of ethanol is 0.789 @ 20 °C/4 °C, this means that the specific gravity of ethanol is at 20 °C with respect to the specific gravity of water at 4 °C. Note that the density of water at 4 °C is 1.0 g/cm^3, therefore the density of ethanol at 20 °C is 0.789 g/cm^3 (or 789 kg/m^3).

Viscosity

Viscosity is also a very important parameter to be considered during feedstock pretreatment, hydrolysis, and fermentation, and is associated with slurry flow and mixing. Viscosity is defined as the property of the fluid that tends to resist the movement of one layer of the fluid/slurry over the adjacent layer of the fluid/slurry. Viscosity is often represented by the Greek letter μ and is known as dynamic viscosity or simply viscosity. In SI units, viscosity is expressed in Pa.s, $N \cdot s/m^2$, or $kg/(m \cdot s)$. In the CGS system, viscosity is expressed in $g/(cm \cdot s)$, which is also known as "poise." Sometime centipoise (cp) is also used, which is 0.01 poise. In the FPS system, viscosity is expressed in $lb_m/(ft \cdot s)$ or $lb_f \cdot s/ft^2$.

Sometimes viscosity is given as kinematic viscosity (ν):

$$\nu = \frac{\mu}{\rho} \tag{2.7}$$

The unit of kinematic viscosity is m^2/s, cm^2/s, or ft^2/s. The unit cm^2/s is also known as a stoke.

Moisture Content

Feedstocks for bioenergy production are grown across the world in a number of different climates and environmental conditions. Consequently, the moisture content of the harvested materials changes significantly, and a standard unit of measure must be defined to allow farmers, researchers, and policymakers to compare crops. For example, suppose small farms in Hawaii and Louisiana both grow 100 metric tons of raw sugarcane. The moisture content of the cane is 80% in Hawaii, but only 70% in Louisiana. Which location has the potential to make more biofuel from the sugarcane fibers? The answer is Louisiana, because it produces 30 metric tons of dry fiber per 100 metric tons of raw (wet) sugarcane, compared to 20 metric tons of dry fiber in Hawaii. Mathematically, the moisture content is represented as:

$$\text{Moisture content } (\%) = \frac{g_{water}}{g_{water} + g_{dry\ biomass}} = \left(1 - \frac{g_{dry\ biomass}}{g_{raw\ biomass}}\right) \times 100 \tag{2.8}$$

Note that other units of mass (e.g., lb_m) can be used to calculate the moisture content as well. In many cases, biomass characteristics (like its composition) are reported on an oven-dried basis (at 105 °C). This means that all of the moisture has been removed. In other cases, drying the biomass at an elevated temperature may destroy its native structure (which may be of scientific interest). For these situations, a small amount of the sample is sacrificed to determine the moisture content of the biomass, and the properties of the material are reported as is (with a correlating moisture content). The simplest way to calculate the moisture content is first to determine the total amount of oven-dried solids present in the sample. Once that value is determined, divide it by the wet (raw) weight of the original sample, and subtract this value from 1. Multiply the resulting decimal by 100 to convert it into percent (Equation 2.8).

Mole

A mole (mol) of a pure substance represents a specific number of atoms (defined by Avagadro's number, 6.022×10^{23}), which is coincidentally equal to the combined molecular weights of the elements comprising the molecule of interest. For an individual element, 1 mole is equal to the weight given on a periodic table of elements (e.g., oxygen = 15.9994 g/mol). For a molecule, like methane, we can add up the weights of the individual elements. Units are often reported as g mol or kg mol, which can be confusing. Because the molecular weight is a fixed constant on the periodic table of elements, engineers and scientists must find a way to account for differences in mass when converting between SI and US units. Both g mol and lb mol are used extensively to allow for ease of conversion. By definition, 1 g mol = mass in grams/molecular weight. The same relationship applies for lb mol. Multiplying 1 g mol by the molecular weight (from the periodic table of elements) effectively cancels out the denominator and gives a value for mass. For example:

1 kg mol of CH_4 = 16.04 kg CH_4

1 g mol of CH_4 = 16.04 g CH_4

1 lb mol of CH_4 = 16.04 $lb_m CH_4$

Concentration

A quantity of any substance in an aqueous, gaseous, or solid phase is expressed in units of concentration. Some of the common units encountered in the bioenergy field are discussed here.

Molarity is moles of a solute per liter of solution (mol/L). *Molality* is moles of a component dissolved in a kilogram of solvent, not the whole solution (mol/kg).

The most common method of expressing concentration is mass per unit volume. Units such as mg/L, g/L, kg/m^3, $lb\ mol/ft^3$, and lb_m/gal are used to measure the concentration. Some examples include the amount of sugars in sugarcane juice, the amount of ethanol in fermentation media, the amount of inhibitory compounds in hydrolysate, etc.

Percent by volume or by mass is also common in the bioenergy/biofuel discipline:

$$Percent\ by\ volume\ (\%\ by\ vol) = \left(\frac{\text{Volume of solute}}{\text{Total volume of solution}}\right) \times 100 \tag{2.9}$$

$$Percent\ by\ mass\ (\%\ by\ mass) = \left(\frac{\text{Mass of solute}}{\text{Total mass of solute + solvent}}\right) \times 100 \tag{2.10}$$

Percent (%) is also often used to represent solids in an aqueous phase. 1% is equivalent to 10,000 mg/L. If a fermentation broth contains 10% solids, the concentration would be 100 g/L (10 × 10,000 mg/L). Percent (v/v, volume by volume) is used to represent gas composition. For example, if syngas contains 20% CO, this means that 20% by volume of the total syngas consists of CO.

The mass of a constituent in the solid phase can be expressed as a mass measurement per unit mass, which would be equivalent to a percentage. Units such as mg/g, g/kg, or mg/kg are typically used. Examples include cellulose, hemicellulose, and lignin content in the lignocellulosic biomass. For example, 0.4 g glucose/g biomass can also be written as 40% glucose.

2.4 Energy and Heat

In the SI system, the unit of energy is the joule (J) or kilojoule (kJ). Other common units of energy are the British thermal unit (Btu) and the Calorie (cal).

British Thermal Unit (Btu)

Different fuels and feedstocks generate varying amounts of heat energy when burned. The Btu is commonly used in the USA to assess and compare the heating value (energy content) of different fuels. *1 Btu* is the amount of energy required to raise the temperature of 1 lb of water by 1 °F.

A common mistake when implementing Btu arises when using the prefix "M." 1 *MBtu* is defined as 1,000 Btu, where "M" represents 1,000 from the Roman numerical system. 1 million Btu (10^6), on the other hand, is written as *MMBtu*. Other common terms include *therm* (100,000 or 10^5 Btu) and *quad* (10^{15} Btu). Btu can be converted to SI units using the following relation:

$$1 \text{ Btu} = 1055.6 \text{ J} = 1.0556 \text{ kJ}$$

Calorie

In the context of food energy, the term Calorie actually represents a *kilogram calorie*. (Note that capital C the next time you read the nutritional facts about your food.) In contrast, calorie (with a lowercase c) is 1,000 times smaller, known as a gram calorie, and represents the amount of energy required to raise the temperature of 1 gram of water by 1 °C.

The gram calorie is most often implemented by engineers to measure heat, and can be converted to English and SI units using the following relationship:

$$1 \text{ cal} = 4.184 \text{ J}$$
$$1 \text{ Btu} = 252.16 \text{ cal} = 1055.6 \text{ J or } 1.0556 \text{ kJ}$$

Units such as QBtu and toe (metric ton of oil equivalent) are used extensively in the energy field. For example, the global annual energy consumption in 2010 was reported to be around 524 QBtu, which is equivalent to 13.205 million metric tons of oil equivalent (Mtoe).

1 QBtu is equal to:

- 45 million tons of coal
- 1 trillion cubic feet of natural gas
- 170 million barrels of crude oil

1 toe is equal to:

- 7.4 bbl of crude oil in primary energy
- 1,270 m^3 of natural gas
- 2.3 metric tons of coal

2.4.1 Power

Power is defined as a rate of energy consumption. Implicit in this definition is the unit of time, which appears in the denominator, often in the units of seconds. Some of the common units used to express power are:

$$\text{Watt (W) or kilowatt (kW)} = \text{J/s or kJ/s}$$

$$1 \text{ Megawatt (MW)} = 3.415 \text{ MMBtu/hr}$$

$$1 \text{ horsepower (hp)} = 745.7 \text{ W} = 2,546 \text{ Btu/hr}$$

$$1 \text{ kilowatt-hour (kWh)} = 3.6 \times 10^6 \text{ J}$$

The power rating may be expressed as MW_{el} or MW_{th}, where the subscripts "el" and "th" refer to electrical and thermal energy, respectively.

Most of the appliances we use today have a power rating. This will help us to calculate the amount of energy consumed if the appliance is run/used for a given period of time. For example, if the power rating of a water heater is 1 kW and it runs for 1 hour, the energy consumption by the heater would be 1 kilowatt-hour (kWh). One kW is 1,000 W or 1,000 J/s and the unit is running for 1 hour or 3,600 seconds, so we have:

$$1\,kWh = 1000\,(J/s) \times 3600\,(s) = 3.6 \times 10^6\,J = 3.6\,MJ$$

2.4.2 Heating Value

The heating value is the net enthalpy released upon reaction of a particular fuel with oxygen under isothermal (constant temperature) conditions. The heating values of some common materials are given in Table 2.2.

The **higher heating value (HHV)** is defined as the gross heat (enthalpy) measured if all of the water vapor produced is recondensed into water. Thus, the HHV considers the recovery of all of the heat energy generated. Heating values are obtained experimentally for various fuels from bomb calorimetric experiments.

The **lower heating value (LHV)** is the total heat released when water vapor is not recondensed into water and is vented off as steam. The lower heating value can be estimated as follows:

$$LHV = [HHV - (\Delta H_v)(9H)] \tag{2.11}$$

where ΔH_v is the heat of the vaporization of water (2.42 kJ/kg), and H is the hydrogen content.

Table 2.2 Heating value of common fuels

Fuel types	Heating value	
	SI system	British system
Natural gas	38.37 MJ/m^3	1030 Btu/ft^3
Propane	93.14 MJ/m^3	2500 Btu/ft^3
Methane	37.26 MJ/m^3	1000 Btu/ft^3
Ethanol	21.18–23.53 MJ/m^3	76,000–84,400 Btu/gal
Gasoline	34.84 MJ/m^3	125,000 Btu/gal
# 2 Diesel	38.61 MJ/m^3	138,500 Btu/gal
Biodiesel (B100)	34.45 MJ/m^3	120,000 Btu/gal
Charcoal	26.3 MJ/kg	11,330 Btu/lb
Hard coal (anthracite)	30.1 MJ/kg	13,000 Btu/lb
Soft coal (bituminous)	27.85 MJ/m^3	12,000 Btu/lb
Municipal solid wastes	10.5 MJ/kg	4,524 Btu/lb
Oak wood	13.3–19.3 MJ/kg	5,730–8,315 Btu/lb
Pine wood	14.9–22.3 MJ/kg	6,419–9,607 Btu/lb
Corn (shelled) at 15% moisture	16.25–18.57 MJ/kg	7,000–8,000 Btu/lb
Plastic	41.7–46.4 MJ/kg	18,000–20,000 Btu/lb

Source: Data from http://www.hrt.msu.edu/Energy/pdf/Heating%20Value%20of%20Common%20Fuels.pdf; http://www.uwsp.edu/CNR/wcee/keep/Mod1/Whatis/energyresourcetables.htm

Example 2.1: Calculate the heating value of 1 bushel of corn and compare it with the heating value of ethanol derived from 1 bushel of corn. Corn has a moisture content of 15%.

Solution:

1. **Heating value of 1 bushel of corn:**

   1 bushel of corn = 56 lb_m
   At 15% moisture content, the heating value of corn (shelled) = 7,000 Btu/lb (Table 2.2)

   Total Btu produced from a bushel of corn = 7,000 Btu/lb × 56 lbs/bushel

   = **392,000 Btu/bushel**

2. **Heating value of ethanol from 1 bushel of corn:**

   1 bushel of corn produces ~2.7 gallons of ethanol (see Chapter 24)
   1 gallon of ethanol has a heating value of 76,000 Btu (see Table 2.2)
   Total Btu from ethanol = 76,000 Btu/gal × 2.7 gal/bushel = **205,200 Btu/bushel**

The direct burning of shelled corn produces a higher heating value than ethanol derived from corn. This calculation, however, does not consider the co-products such as distillers' dried grains, which offset some of the energy used to produce animal feed. Also, it is vital to note that liquid fuel is needed to run our transportation sector.

2.4.3 Heat Capacity

The heat capacity or specific heat capacity (c_p) of a material is the amount of heat necessary to increase the unit mass of a substance by a temperature of 1 degree. In SI units, heat capacity is expressed in J/(kg·K). Other units include cal/(g·°C) or Btu/(lb_m·°F).

It is important to point out that the numerical value of heat capacity is the same in the different units. That is, 1 cal/(g·°C) = 1 Btu/(lb_m·°F). A substance with a heat capacity of 0.8 Btu/(lb_m·°F) = 0.8 cal/(g·°C).

Given that 1.8 °F = 1 °C or 1 K; 251.16 cal = 1 Btu; and 453.6 g = 1 lb_m, we have the following calculations:

Heat capacity (0.8 Btu/(lb_m·°F))

$$= \left(\frac{0.8\,\text{Btu}}{lb_m \cdot °F}\right)\left(\frac{252.16\,\text{cal}}{\text{Btu}}\right)\left(\frac{1}{453.6\,\dfrac{\text{g}}{lb_m}\cdot\dfrac{1°C}{1.8°F}}\right)$$

$$= 0.8\,\text{cal}/(g\cdot °C)$$

The heat capacity of a material is important to calculate the energy required to heat or cool the substance. The heat needed can be calculated using the following expression:

$$\text{H} = m \times c_p \times \Delta T \tag{2.12}$$

where H is the heat required or to be removed (J, cal, or Btu) and m is the mass of the substance (lb_m or kg).

> **Example 2.2:** In a fermentation plant, one million gallons per day of medium is cooled from 85° to 35 °C. Calculate the heat removed from the medium.
>
> Assume that the specific gravity of medium is 1.1, and heat capacity is 3.9 kJ/(kg·K).

Solution:

$$\text{Mass of substrate} = \left(1 \times 10^6 \text{gal/day} \times 3.785 \text{L/gal} \times 10^{-3} \text{m}^3/\text{L}\right) \times \left(1{,}000 \text{ kg/m}^3 \times 1.1\right)$$

$$= 4{,}163{,}500 \text{ kg/day} = 2891 \text{ kg/min}$$

$$\text{Heat removed (from eq. (2.6))} = m \times c_p \times \Delta T$$

$$= 2891 \text{ kg/min} \times 3.9 \text{ kJ/(kg·K)} \times (85 - 35)°\text{C}$$

$$\mathbf{= 563{,}745 \text{ kJ/min}}$$

2.5 Volume–Mass Relationship

The volume of slurry (semi-solid) can be converted into mass and vice versa if the percentage solids and specific gravity are known using the following expression:

$$\text{Mass} (M_s) = V \left(\rho_w \cdot SG \cdot P_s\right) \tag{2.13}$$

where M_s is the mass of dry solids (kg), V is the volume (m^3), ρ_w is the density of water (kg/m^3), SG is the specific gravity of slurry, and P_s is the percentage solids in fraction.

> **Example 2.3:** A pilot-scale fermenter is used to ferment 1,000 m^3/day of biomass-derived slurry. The slurry has a specific gravity of 1.2 and total solids (TS) of 25%. Calculate the mass of the slurry.

Solution:

$$\text{Mass of slurry (from eq. (2.13))} = V \left(\rho_w \cdot SG \cdot P_s\right)$$

$$= 1{,}000 \text{ m}^3/\text{day} \left(1{,}000 \text{ kg/m}^3 \times 1.2 \times 0.25\right)$$

$$\mathbf{= 300{,}000 \text{ kg/day}}$$

2.6 Ideal Gas Law

The relationship among the volume, temperature, and pressure of gases is governed by the ideal gas law. According to this law, the volume of a gas is directly proportional to the absolute temperature and inversely proportional to the absolute pressure:

$$pV = nRT \tag{2.14}$$

where p is the absolute pressure, V is the volume of the gas, n is the number of moles of the gas, T is the absolute temperature, and R is the universal gas constant.

The numerical value of R depends on the unit chosen for p, V, and T. When absolute pressure p is in N/m^2, the volume of the gas V is in m^3, n is in kg mole, and absolute temperature T is in K, the gas law constant, R is 8,314.3 kg·m²/(kg mol·s²·K). Depending on the unit chosen, R can be 0.082057 L·atm/(g mol·K), 0.7302 ft³·atm/(lb mol·°R), or 82.057 cm³·atm/(g mol·K).

Using the ideal gas law, it is possible to convert the volume of any gas into its mass and vice versa. This is quite important, as several bioenergy products are in gaseous form. For better comparison of the volume, standard conditions of temperature (0 °C or 273.15 K) and pressure (1 atmosphere or 101.325 kPa) are used in the calculations.

Thus, under standard temperature and pressure (STP):

1 g mol of an ideal gas occupies 22.414 L

1 kg mol of an ideal gas occupies a volume of 22.414 m^3

1 lb mol of an ideal gas occupies a volume of 359.05 ft^3

> **Example 2.4:** Show that 1 g mol of any gas occupies volume of 22.4 L under standard conditions.
>
> **Solution:** Given temperature, T is 273.15 K, pressure, p is 1.0 atm, n is 1 g mole and gas constant, R is 0.082057 L·atm/(g mol·K). From eq. (2.14), we have:
>
> $$V = \frac{nRT}{p}$$
>
> $$= \frac{1.0 \, \text{g mol} \times 0.082057 \, \text{L·atm/g mol·K} \times 273.15 \, \text{K}}{1 \, \text{atm}}$$
>
> $$= \mathbf{22.414 \, L \sim 22.4 \, L}$$

2.7 Henry's Law

Some feedstocks such as syngas (H_2 + CO), (H_2 + CO_2), and CO_2 are in gaseous form, which need to be dissolved in the aqueous phase before they can be converted into bioenergy products such as ethanol, butanol, biodiesel, and methane. The equilibrium or saturation concentration of gas dissolved in a liquid phase is governed by the partial pressure of the gas adjacent to the liquid. The relationship between the partial pressure of the gas in the atmosphere above liquid and its concentration in the liquid phase is given by Henry's Law:

$$p_g = H C_L \tag{2.15}$$

where p_g is the partial pressure of the gas (atm), C_L is the concentration of the gas dissolved in mole fraction, and H is Henry's law constant (atm/mole fraction).

If C_L is expressed as mol/m^3, H has the units of atm·m^3/mol.

In Equation 2.15, the equilibrium mole fraction C_L can be given by:

$$\frac{mol_{gas}\left(n_g\right)}{mol_{gas}\left(n_g\right) + mol_{water}\left(n_w\right)} \tag{2.16}$$

Henry's constant can also be presented as the ratio of the concentration in the gas to the concentration in the liquid:

$$H = \frac{C_g}{C_L} \tag{2.17}$$

Henry's constant can be expressed in different units. To convert Henry's constant in atm·m^3/mol into a dimensionless unit, we use:

$$H_c = \frac{H}{RT} \tag{2.18}$$

where Hc is Henry's constant (dimensionless), H is Henry's constant (atm·m^3/mol), R is the universal gas constant (0.000082057 m^3·atm/(mol·K)), and T is the temperature (K or (273 + °C)).

Example 2.5: Determine the concentration (mg/L) of carbon dioxide in the upper liquid layer in a closed algal bioreactor operating at 30 °C and 1 atmosphere pressure, if the headspace CO_2 concentration is 44%.

Solution: From eq. (2.15), $p_g = H\,C_L \rightarrow; C_L = \frac{P_g}{H}$

Henry constant H for CO_2 at 30 °C is 0.186×10^4 atm/mol fraction.

(Note: H can also be calculated using Van't Hoff equation, $H_T = H_{298.15} \times e^{-C\left(\frac{1}{T} - \frac{1}{298.15}\right)}$)

Thus, $C_L = \dfrac{0.44\,\text{atm}}{0.186 \times 10^4\,\text{atm/mol fraction}} = 2.366 \times 10^{-4}$ mol fraction.

In eq. (2.16), n_w is mol water $= \dfrac{1000\,\text{g/L}}{18\,\text{g/mol}} = 55.6\,\text{mol/L}$. Substituting all values, we have:

$$2.366 \times 10^{-4} = \frac{\text{mol gas}\,(n_g)}{\text{mol gas}\,(n_g) + 55.6} \rightarrow n_g = \left(n_g + 55.6\right) \times 2.366 \times 10^{-4}$$

$n_g \times 2.366 \times 10^{-4}$ is very small, we can ignore the term, we have

$$ng = 55.6 \times 2.336 \times 10^{-4} = 0.01299\,\text{mol/L}$$

Saturation concentration of CO_2 in aqueous phase is

$$= (0.01299\,\text{mol/L}) \times (44\,\text{g/mol}) = 0.57\,\text{g/L} = \mathbf{570\,mg/L}$$

References and Further Reading

Doran, P.M. (1995). *Bioprocess Engineering Principles*. London: Academic Press.
Geankoplis, C.J. (2003). *Transport Processes and Separation Process Principles*. Upple Saddle River, NJ: Prentice Hall.
IEA (2008). *From 1st- to 2nd-Generation Biofuel Technologies: An Overview of Current Technologies and RD&D Activities*. Paris: International Energy Agency. https://www.iea.org/publications/freepublications/publication/2nd_Biofuel_Gen.pdf, accessed April 2016.

Exercise Problems

2.1. Calculate the amount of ethanol produced from 1 metric ton of corn in L, gal, m^3, cm^3, and ft^3, if 1 bushel of corn produces 2.7 gal of ethanol.

2.2. If a biogas plant generates 1 million gallons of methane gas daily, calculate the amount of energy produced in J, kJ, MBtu, kcal, and kWh.

2.3. Biomass-derived hydrolysate contains 150 g/L of glucose. Calculate the molar concentration of glucose. If the fermentation of sugar produces 76.6 g/L of ethanol, what is the % ethanol (v/v) in the beer?

2.4. What should be the glucose concentration to obtain 15% (v/v) ethanol in fermentation media if 1 g glucose generates 0.51 g ethanol?

2.5. Palm oil (0.5 million gallon/day) at 25 °C is being heated in a heat exchanger to 70 °C by hot water. How much heat is needed? Assume that the specific gravity of palm oil is 0.92 and the average heat capacity is 1.9 kJ/(kg·°C).

2.6. An energy provider uses 200 dry metric tons of coal/day. It wants to replace a part of coal by 50 dry metric tons/day of solid waste to meet Renewable Portfolio Standards (RPS). How much energy does the solid waste contribute?

2.7. A pilot-scale fermenter is being proposed to ferment 400 m^3/day of sugarcane-derived sugar at 35 °C. What is the dry weight (total solids, TS) fed daily to the fermenter? Assume that the specific gravity is 1.01 and the TS content is 2.0%.

2.8. Calculate the heating value of oil and biodiesel produced from 1 bushel of soybean. Assume that soybean seeds contain 20% oil on a dry weight basis, and that oil recovery and transesterification efficiency are 100%.

2.9. If 1,000 kg of hydrogen gas is to be stored in a container at a pressure of 20 atm and a temperature of 25 °C, how large should the container be in m^3?

2.10. Syngas consisting of 30% CO and 10% H_2 is injected continuously into a bioreactor for biofuel production via syngas fermentation. If the bioreactor is operated at 35 °C and 1 atm pressure, calculate the aqueous phase CO and H_2 concentrations.

Mass and Energy Balances

Devin Takara and Samir Kumar Khanal

What is included in this chapter?

The concepts of mass and energy balances as applied to bioenergy are discussed. The relevant topics in stoichiometry are also covered.

3.1 Introduction

As with any unit operation and process, mass and energy balances are extremely important to quantify the efficacy and efficiency of systems and to enable insightful economic analyses. In practice, however, it is often difficult to conduct direct mass and energy balances due to the (often hazardous) chemical reactions taking place during the production of materials. The goal of this chapter is to introduce the basic concepts of mass, energy, and stoichiometry to equip you with strategic approaches that will help you in later in-depth investigations of bioenergy.

3.2 Mass Balances

Mass balances are based soundly on the law of conservation of mass, which states that matter can neither be created nor destroyed (except during nuclear reactions). This foundation allows engineers and scientists to design, optimize, and analyze various systems and their unit operations.

Control volumes are often applied to mass balance problems to simplify the calculations surrounding enormous systems into manageable sizes. Although abstract and arbitrary in concept, control volumes are extremely powerful if implemented properly. To establish a control volume, an imaginary boundary is drawn around the perimeter of a system of interest to isolate it from other operations that may be occurring simultaneously. (Note that the term *system* is also somewhat arbitrary and can be defined as an entire process or a particular unit operation within a larger group of

Bioenergy: Principles and Applications, First Edition. Edited by Yebo Li and Samir Kumar Khanal.
© 2017 John Wiley & Sons, Inc. Published 2017 by John Wiley & Sons, Inc.
Companion website: www.wiley.com/go/Li/Bioenergy

concurrent processes.) Matter both entering and exiting through the defined boundaries is accounted for using the general mass balance equation:

$$\dot{m}_{in} - \dot{m}_{out} + \dot{m}_g - \dot{m}_c = \dot{m}_a \tag{3.1}$$

where $\dot{m}_{in}$ is the rate of mass into the system (kg/hr or lb$_m$/hr), $\dot{m}_{out}$ is the rate of mass out of the system, $\dot{m}_g$ is the rate of mass generated within the system, $\dot{m}_c$ is the rate of mass consumed within the system, and $\dot{m}_a$ is the rate of mass accumulated within the system. In many cases, an engineer is only interested in the system under steady-state conditions, during which the mass accumulation term becomes zero.

Note that although we stated previously that mass cannot be generated or destroyed following the law of conservation of mass, there is a mass generation and consumption term in Equation 3.1. This is because mass balances can be performed on unit operations such as a reactor, which may include molecules that are created (or depleted) by chemical reactions. An elemental mass balance at steady state, however, would not have a generation, consumption, or accumulation term, in contrast to a molecular mass balance. For example, if we were conducting a mass balance on the sugar fed to a fermenter cultivating microbes, we would expect to see less sugar leaving the system compared to the amount entering the system as a result of sugar consumption. Correlating to this decrease in sugar concentration would be the increase in microbial biomass. If we were to perform a mass balance on elemental carbon (instead of sugar) at steady-state conditions, we would see that the carbon from the sugar is being incorporated into the cell biomass, and that no carbon is being created or lost in the process.

Example 3.1: The mass flow rate into and out of a bioreactor is known along with the mass fraction of the effluent. Based on this information, calculate the amount of components A and B entering the system in streams 1 and 2 under steady-state conditions. Assume that no reactions occur within the vessel.

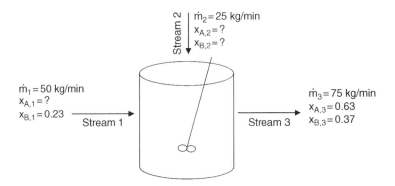

Solution: First, a control volume needs to be established around the system of interest. Next, valid assumptions must be made regarding the nature of the bioreactor. The problem statement asks for values at steady-state conditions. Thus, immediately we know that the rate of accumulation ($\dot{m}_a$) from the general equation is equal to zero. Also, because there are no reactions taking place within the system, we can assume that the generation ($\dot{m}_g$) and consumption terms ($\dot{m}_c$) are equal to zero. The general equation is then simplified to

$$\dot{m}_{in} = \dot{m}_{out}$$

By close analysis, we see that there are four unknowns. We can write two independent mass balance equations based on the influent and effluent of components A and B, but two more equations are still required. Because x_A and x_B represent the only mass fraction of the streams, we can conclude that the sum of x_A and x_B equals 1.

$$\dot{m}_1 x_{A,1} + \dot{m}_2 x_{A,2} = \dot{m}_3 x_{A,3}$$

$$\dot{m}_1 x_{B,1} + \dot{m}_2 x_{B,2} = \dot{m}_3 x_{B,3}$$

$$x_{A,2} + x_{B,2} = 1$$

Plugging in the known values and moving all the variables to the left-hand side, we have:

$$50 x_{A,1} + 25 x_{A,2} = 47.25$$

$$50(0.23) + 25 x_{B,2} = 27.75$$

$$x_{A,2} + x_{B,2} = 1$$

Since these are independent equations, recall from linear algebra that we can set up a matrix to solve for the unknowns. (Note that this problem can easily be solved by manual calculation, but in later problems using a matrix will simplify and expedite the process.)

$$\begin{pmatrix} 50 & 25 & 0 \\ 0 & 0 & 25 \\ 0 & 1 & 1 \end{pmatrix}^{-1} \bullet \begin{pmatrix} 47.25 \\ 16.25 \\ 1 \end{pmatrix} = \begin{pmatrix} 0.77 \\ 0.35 \\ 0.65 \end{pmatrix}$$

From the matrix, we calculate the following:

$$x_{A,1} = 0.77, \ x_{A,2} = 0.35, \ x_{B,2} = 0.65$$

The example given is relatively simple compared to most real-world applications, but the concepts are still applicable. Solving these types of problems can be approached by:

1. Defining a system.
2. Sketching boundaries around the system.
3. Identifying all species that cross the system boundary.
4. Considering whether there is an accumulation of mass within the system (not steady-state).
5. Applying the conservation of mass concept (Equation 3.1).

3.3 Enthalpy

A clear understanding of *enthalpy* and its properties is crucial for conducting an energy balance. There are two types of enthalpy for non-reactive systems: sensible heat, and latent heat. (Note that the word *heat* is often used interchangeably with enthalpy, since enthalpy, by definition, means *heat*.) Sensible heat exists when a temperature change occurs for a given species. For example, the change in temperature of water from 20 °C to 25 °C has a specific (measurable) value of sensible heat associated with it. In contrast, latent heat exists when there is a phase change of a particular species. The change from water to steam, for example, would have a specific latent heat value associated with it. Sensible and latent heat must be determined when examining the change in temperature of water from 20 °C to 100 °C at atmospheric conditions.

Reactive systems (i.e., those in which chemical reactions occur) can become more complicated because two additional enthalpy terms are added: heat of combustion, and heat of reaction. The heat of combustion refers to the enthalpy required or generated by completely combusting a reactant into carbon dioxide and water (and heat). The heat of reaction, on the other hand, is the heat required/released when two reactants are combined to form a new compound. The heat of combustion and the heat of reaction are especially important in bioenergy production processes (e.g., gasifiers).

3.4 Energy Balances

The law of conservation of energy, similar to the conservation of mass, states clearly that energy also can neither be created nor destroyed. If you recall from thermodynamics, *energy balances* are slightly trickier to perform on systems compared to mass balances because there are many different forms in which energy can exist. Three of the most common forms of energy include potential, kinetic, and internal energy. A detailed study of energy transformations can be quite extensive. Here, only the basics are covered for examining bioenergy-related systems and topics.

Because many of the energy values (like enthalpy) are reported on a mass basis, energy and mass balances are often coupled in the analyses of most systems within a defined control volume. The general equation for energy balances is often given by:

$$\Delta E = \dot{Q} - \dot{W} + \dot{m}_{in}\left(h_{in} + \frac{1}{2}v_{in}^2 + gz_{in}\right) - \dot{m}_{out}\left(h_{out} + \frac{1}{2}v_{out}^2 + gz_{out}\right) \tag{3.2}$$

where ΔE is the net change in energy within the control volume with respect to time (J/hr or Btu/hr), $\dot{Q}$ is the change in heat added/removed with respect to time (J/hr or Btu/hr), $\dot{W}$ is the work added to the system with respect to time (J/hr or Btu/hr), $\dot{m}$ is the mass flow rate (kg/hr or lb_m/hr), g is the acceleration due to gravity (m/sec² or ft/sec²), h is the specific enthalpy (J/kg or Btu/lb_m), v is the velocity per mass (m/(sec·kg) or ft/(sec·lb_m)), and, finally, z is the height with respect to mass (m/kg or ft/lb_m).

In many bioprocesses, steady-state conditions are of primary interest in determining the efficacy of a system. Examples of this can be seen in the analysis of bioreactors, gasifiers, and/or specific unit operations found within a biorefinery. Under steady-state conditions, the rate of change in energy within a control volume becomes zero (by definition) and the general energy balance equation introduced previously simplifies to:

$$\dot{Q} - \dot{W} = \dot{m}_{out}\left(h_{out} + \frac{1}{2}v_{out}^2 + gz_{out}\right) - \dot{m}_{in}\left(h_{in} + \frac{1}{2}v_{in}^2 + gz_{in}\right) \tag{3.3}$$

Further simplification can occur if we consider that in most cases, potential and kinetic energies are negligible, at least in the case of bioenergy production. The energy balance equation reduces to:

$$\dot{Q} - \dot{W} = \dot{m}_{out}h_{out} - \dot{m}_{in}h_{in} \tag{3.4}$$

Although it might not seem obvious, Equation 3.4 is often used to account for energy required and/ or released in chemical reactions. Before continuing further, however, we must first review the simple practice of balancing chemical equations, namely stoichiometry. In many cases, the chemical relationship, which illustrates the generation and consumption of particular compounds, is known or characterized to enable accurate analyses of the system. In basic chemistry, we are taught to check whether chemical equations are balanced before conducting any other calculations. The same is true for mass and energy balance problems. Consider the equation for the combustion of biomass:

$$C_xH_yO_z + p(O_2 + 3.762N_2) \rightarrow jCO_2 + kH_2O + mO_2 + nN_2 \tag{3.5}$$

where x, y, and z represent the different ratios of carbon, hydrogen, and oxygen found in the biomass, respectively, and the coefficients p, j, k, m, and n represent values that balance the overall equation. Simple algebra can be implemented by conducting a mass balance on the elemental species present in this reaction. For example, if we were combusting glucose ($C_6H_{12}O_6$) at standard ambient conditions (25 °C [298.15 K] and 1 atm), we might start with a carbon balance, keeping in mind that carbon is neither created nor destroyed on both sides of the equation. (In other words, the amount of carbon on the left-hand side is equal to the amount of carbon on the right-hand side.) This holds true for the other elements as well.

C : $x = j$

H : $y = 2k$

O : $z + 2p = 2j + k + 2m$

N : $(2)(3.762)p = 2n$

We briefly discussed earlier how energy is required to form and combust chemical species. In the case being discussed, –2,805,000 kJ/kmol is the amount of energy released when glucose is completely combusted into carbon dioxide and water. Note that a negative sign indicates a release of energy (as is customary in the study of thermodynamics as well). Since the reaction occurs at standard ambient conditions, we know that the heat of formation of oxygen and nitrogen on the right-hand side of Equation 3.5 are equal to zero. (Oxygen and nitrogen are not generated on the left-hand side of the equation and exist at room temperature, thus their enthalpies equal zero. Energy is required, however, to form water and carbon dioxide from combusted glucose.) We calculate the heat of reaction by Equation 3.6. Notice that Equation 3.6 is similar to Equations 3.3 and 3.4, except that the work ($\dot{W}$) and heat ($\dot{Q}$) terms are excluded as we are only interested in the reaction.

$$\Delta H_{rxn} = \sum n_r h_r - \sum n_p h_p \tag{3.6}$$

where n represents the number of moles of a particular species, h represents the heat of combustion, and the subscripts p and r represent products and reactants, respectively. If we substitute values in the general equation, we obtain the following:

$$\Delta H_{rxn} = -2,805,000 \, kJ/kmol - [(0)(6 \, kmol) + (0)(6 \, kmol)]$$

Because the heat of combustion values were given on a molar basis, we were able to directly use the coefficients found from balancing the chemical equation. However, if the enthalpy were reported on a mass basis, a simple conversion from moles to mass would be required. Since the heat of reaction is negative in value, the overall process is exothermic. This is in agreement with the fact that glucose is used by the microbes as a source of (food) energy for metabolic processes.

Notice that if Equation 3.6 is written in terms of mass (instead of moles), it can be substituted into Equation 3.4 as shown in Equation 3.7. This assumes steady state ($\Delta E = 0$)and that kinetic and potential energy are negligible.

$$-\Delta H_{rxn} = \dot{Q} - \dot{W} \tag{3.7}$$

Equation 3.7 and its various forms have important applications in the field of biofuel and bioenergy, and may help in estimating the total energy produced by combusting feedstocks versus producing liquid biofuel. In situations where biochemical reactions are present, Equation 3.7 is best implemented. For other conditions, where reactions are not present, Equation 3.3 is best used. If the change of a reactor with respect to a specific time interval must be considered, Equation 3.2 must be integrated.

> **Example 3.2:** Bioenergy feedstocks are known as lignocellulosic biomass because they are made up of three primary components: lignin, cellulose, and hemicellulose. Both cellulose and hemicellulose are polysaccharides of great interest due to their importance in bioenergy production. Concentrated sulfuric acid serves as a catalyst to promote the following hydrolysis reaction and releases glucose from the homogeneous polysaccharide, cellulose:
>
> $$(C_6H_{10}O_5)_3 + H_2O \rightarrow C_6H_{12}O_6$$
>
> This process is critical for quantifying how much glucose is available in a particular feedstock for biofuel production. Is this reaction endothermic or exothermic at 25 °C? If you were to supply 2.0 kJ/mol of work to agitate the solution, would you also need to supply heat? If so, how much?
>
> **Solution:** The first step in approaching this type of problem is to balance the chemical equation. Fortunately, this equation is rather straightforward:
>
> $$(C_6H_{10}O_5)_3 + 3H_2O \rightarrow 3C_6H_{12}O_6$$
>
> Next, we need to calculate the heat of reaction (ΔH_{rxn}), using the respective heats of combustion at 25 °C, and determine whether the reaction is endothermic or exothermic.

$$\Delta H_{rxn} = \sum n_r h_r - \sum n_p h_p$$

$$\Delta H_{rxn} = [3(0) + 3(-2,808.0)] - [3(-2805.0)]$$

$$\quad\quad\quad h_c \text{ cellulose} \quad h_c \text{ water} \quad h_c \text{ glucose}$$

$$\Delta H_{rxn} = -9 \frac{kJ}{mol}$$

The reaction is exothermic because the heat of reaction is negative. This means that even with the help of the sulfuric acid catalyst, this reaction will not proceed at a temperature of 25 °C.

Because we now know the heat of reaction, we can calculate the heat required. We assume steady-state conditions; that is, "ΔE is equal to zero. (Recall from thermodynamics that by convention, heat added into a system is positive, while work added to a system is negative.)

$$-\Delta H_{rxn} = Q - W$$

$$-\Delta H_{rxn} = -(-9)\frac{kJ}{mol} = Q - \left(-2\frac{kJ}{mol}\right)$$

$$Q = 9.0\frac{kJ}{mol} - 2\frac{kJ}{mol} = 7\frac{\mathbf{kJ}}{\mathbf{mol}}$$

You need to remove 7 kJ/mol of heat to promote the hydrolysis reaction in the forward direction.

References and Further Reading

1. Doran, P.M. (1995). *Bioprocess Engineering Principles*. London: Academic Press.
2. Moran, M.J., and Shapiro, H.N. (2008). *Fundamentals of Engineering Thermodynamics*. Hoboken, NJ: John Wiley & Sons, Inc.

Exercise Problems

3.1. Consider the system shown in Figure 3.1. Write a mass balance for the system assuming steady state.

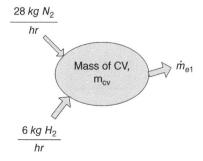

FIG. 3.1

3.2. In Problem 3.1, assume that N_2 and H_2 are completely converted to NH_3 via the following reaction:

$$N_2 + 3H_2 \rightarrow 2NH_3$$

What would be the mass flow rate of NH_3 exiting the system at steady state?

3.3. Freshly harvested feedstock must be completely dried before being stored for later use in bioenergy production. You have been assigned to estimate the amount of energy required to dry this biomass. Which property do you need to consider, latent or sensible heat? Why?

3.4. In Figure 3.2, an aqueous solution of glucose enters the reactor via Stream 1 at a mass fraction of 40%. Additional water is supplied continuously through Stream 2. Analyses of the output stream found that it contained 5% glucose. Assuming that all of the glucose was converted into ethanol by microorganisms inside of the reactor, what is the mass fraction of ethanol in the output stream?

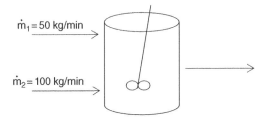

FIG. 3.2

3.5. There appears to be a problem with the fermentation unit shown in Figure 3.3. What control volume could be drawn to best conduct a mass/energy balance for this unit operation? Based

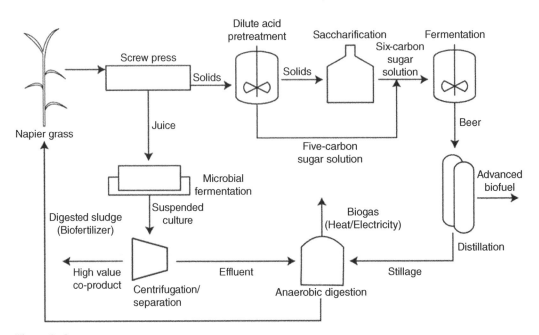

FIG. 3.3

on what you know about fermentation, is this diagram complete? How many output streams should cross the system boundary for the fermenter? How many inputs should there be?

3.6. Using Equation 3.5, balance the chemical reaction used to combust lignin if its elemental constituents were found to be $C_{11}H_{14}O_4$.

3.7. A gasifier is used to produce heat using biogas. 10 m^3 per day of biogas – 60% methane (CH_4) (v/v) and 40% CO_2 (v/v) – is burned completely using air. The density of CH_4 gas is 0.668 kg/m^3 at 20 °C and 1 atm. Show the complete mass balance (inputs and outputs).

3.8. Calculate the heat of reaction (ΔH_{rxn}) for the combustion of lignin at standard ambient conditions. The higher heating value (HHV) of lignin can be assumed to be 21 MJ/kg.

3.9. Acetic acid (20% by weight) is added with water through the input stream in Figure 3.4. A special mixed culture inside the reactor is able to convert the acetic acid into a hydrocarbon (C_4H_6) under anaerobic conditions. Interestingly, the output stream is 5% C_4H_6, 7% methane, and water. If work, in the form of agitation, is supplied to the system at 9 kW, determine the amount of heat that must be supplied/removed from the system.

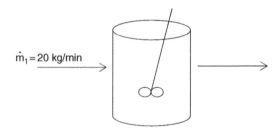

FIG. 3.4

3.10. The complete combustion of feedstocks releases more energy than if the biomass were to be converted into liquid biofuel. Moreover, biogas, evolved from the anaerobic digestion (decomposition) of biomass, is also known to produce more usable energy than liquid biofuels. Such being the case, why has there been so much emphasis on the generation of liquid biofuel? What evidence and/or justification is required to silence opponents of commercial biofuel production? (Relate your response to the topic of energy.)

Thermodynamics and Kinetics of Basic Chemical Reactions

Devin Takara and Samir Kumar Khanal

What is included in this chapter?

This chapter discusses the fundamental principles underlying homogeneous chemical reactions as relevant to bioenergy production. In particular, the chapter covers reaction thermodynamics and basic reaction kinetics.

4.1 Introduction

Chemical reactions play a critical role in a number of commercial bioenergy production facilities, ranging from the processing of raw materials to product manufacturing. Under many circumstances, engineers are tasked with the arduous challenge of mathematically representing these important reactions to enable process optimization and/or cost analyses. Although this kind of work typically belongs to chemical engineers, it is important to note that the fundamental principles of reaction chemistry are of particular importance for the production of bioenergy as well.

In general, chemical reactions are represented by two different, but significant components. The first part is reaction thermodynamics; put simply, whether or not a chemical reaction will take place spontaneously and continue to completion. Similar to the topic of Chapter 3, endothermic reactions require a net input of energy while exothermic reactions result in a net release of energy (to the system). The second important component of reaction chemistry is kinetics. For industrial/commercial applications, reaction kinetics play a significant role in determining the rate at which substrates are consumed (e.g., biomass-derived sugars) and products are generated (e.g., bioenergy). As you will soon see, both reaction thermodynamics and kinetics are topics for high-level and in-depth discussion. In this chapter, however, we aim to provide and review the basics of chemical reactions within

---

Bioenergy: Principles and Applications, First Edition. Edited by Yebo Li and Samir Kumar Khanal.
Companion website: www.wiley.com/go/Li/Bioenergy

the context of bioenergy. We encourage the reader to consult some of the classic references included at the end of this chapter for further reading.

4.2 Reaction Thermodynamics

Reaction thermodynamics define systems in terms of the net energy required or released to the surrounding environment (i.e., beyond the system boundaries). Most continuous operations like the production of biofuel deal with the steady-state conditions of a system and the characteristics observed during chemical equilibrium. When the system attains chemical equilibrium, useful thermodynamic properties can be calculated. Recall from Chapter 3 that quantities often used to characterize reaction thermodynamics (e.g., enthalpy, entropy, and internal energy) are known as state functions and are independent of process pathways.

A useful thermodynamic entity often used to describe the tendencies of chemical reactions is known as *Gibbs free energy*, named after the nineteenth-century scientist Willard Gibbs. Gibbs free energy can be calculated in several different ways, but it is most commonly implemented in this form:

$$\Delta G = \Delta H - T\Delta S \tag{4.1}$$

where G represents Gibbs free energy (kJ/mol), H represents enthalpy (kJ/mol), T represents temperature (K), and S represents entropy (kJ/(mol·K)). By convention, when the change in Gibbs free energy is negative, the reaction is said to be spontaneous and proceeds in the forward direction. When the change in Gibbs free energy is positive, the reaction proceeds in the reverse reaction or energy is needed for the reaction to proceed. A net zero change in Gibbs free energy is indicative of a reaction or system at equilibrium. Thermodynamic state functions are often derived empirically from experimentation. It is common to see a superscript next to enthalpy and Gibbs free energy symbols, H° and G°, respectively, to represent that the values were empirically determined under standard conditions (25 °C and 1 atm). Deviations from standard conditions require that different values (determined at corresponding temperature and pressure) be implemented instead of standard values. Example 4.1 illustrates how Gibbs free energy is used to determine the spontaneity of a biochemical reaction.

> **Example 4.1:** Bioethanol is one type of liquid biofuel that is generated by this simplified chemical reaction:
>
> $$C_6H_{12}O_6 \rightarrow 2CH_3CH_2OH + 2CO_2$$
>
> Calculate the Gibbs free energy of formation of ethanol at 25 °C and 1 atm. Is this reaction spontaneous? How can the spontaneity of the reaction be increased?
>
> **Solution:** Recall that one can calculate the change in Gibbs free energy by using Equation 4.1:
>
> $$\Delta G = \Delta H - T\Delta S$$
>
> It was stated that the reaction will occur at the standard reference temperature of 25 °C, thus we can ignore sensible heat (namely, the heat required to raise the temperature of

the reaction above a reference temperature). First, let us calculate the change in enthalpy (i.e., the heat of reaction):

$$\Delta H_{rxn} = \Delta H_{f,\,products} - \Delta H_{f,\,reactants}$$

There are two products from this reaction: ethanol and carbon dioxide. From enthalpy tables (found in the appendix section of most engineering textbooks), we find that the heat of formation of ethanol is –278 kJ/mol, and the heat of formation of carbon dioxide is –394 kJ/mol. The reactant, glucose, has a heat of formation of –1,271 kJ/mol.

Substituting these values into our enthalpy equation yields the following:

$$\Delta H = 2\left(-278\,kJ/mol - 394\,kJ/mol\right) - \left(-1,271\,kJ/mol\right) = -73\,kJ/mol$$

To calculate entropy, we follow a slight variation of our earlier enthalpy equation. Looking up the entropy values from engineering tables, we are able to calculate ΔS by subtracting the entropy of the reactant from the combined entropies of the products:

$$\Delta S = \Delta S_{products} - \Delta S_{reactants}$$
$$\Delta S = 2\left[160\,J/\left(mol \cdot K\right) + 214\,J/\left(mol \cdot K\right)\right] - 209\,J/\left(mol \cdot K\right) = 539\,J/\left(mol \cdot K\right)$$

Next, we have to convert 25 °C to Kelvin (25 °C + 273 = 298 K). Substituting all of these values into Equation 4.1, and adjusting our units to the same order of magnitude, we obtain the following:

$$\Delta G = -73\,kJ/mol - \left(298K\right)\left(0.539\,kJ/\left(mol \cdot K\right)\right) = \textbf{–234 kJ/mol}$$

Because the change in Gibbs free energy is negative, the reaction is indeed spontaneous. The easiest way to increase the spontaneity of the reaction is to increase the temperature.

Gibbs Free Energy and the Equilibrium Constant

Many commercial chemical processes often rely on reversible reactions, expressed by this generic equation:

$$\alpha A + \beta B \leftrightarrows \gamma C + \delta D \tag{4.2}$$

where A and B are the substrates, C and D are the products, and α, β, γ and δ are stoichiometric coefficients. If the reaction is allowed to proceed for an indefinite amount of time, without interruption, the reaction will eventually achieve *chemical equilibrium* such that the forward and reverse reactions are both likely to occur. Mathematically, chemical equilibrium can be represented by this expression:

$$K_{eq} = \frac{C^{\gamma} D^{\delta}}{A^{\alpha} B^{\beta}} \tag{4.3}$$

Because the chemical equilibrium constant K_{eq} is a ratio of the concentration of products over the concentration of substrates, it has no units. This property inherently suggests that units of A, B, C, and D (whether in SI or English units) must completely cancel each other out.

The chemical equilibrium constant, K_{eq} can also be related to Gibbs free energy. A general representation of this relationship is given as:

$$\Delta G^\circ = -RT \ \ln\left(K_{eq}\right) = -RT \ \ln\left(\frac{C^\gamma D^\delta}{A^\alpha B^\beta}\right) \tag{4.4}$$

where R is the universal gas constant (kJ/(mol·K)) and T is temperature (K). Interestingly, irrespective of how one decides to calculate K_{eq}, the same value is obtained. For example, if a system were to contain solid substrates and products, one could still apply Equation 4.4 (and the universal gas constant) to arrive at K_{eq}. Be aware that Equation 4.4 is a special case of Gibbs free energy, where equilibrium has been achieved. Under non-equilibrium conditions, the non-standard change in Gibbs free energy (ΔG) must be subtracted by the standard change in Gibbs free energy. The term $C^\gamma D^\delta/A^\alpha B^\beta$ shown in Equation 4.5 would represent the real-time concentrations of products and reactants, and the following relationship can be made about the reaction.

$$\Delta G - \Delta G^\circ = RT \ \ln\left(\frac{C^\gamma D^\delta}{A^\alpha B^\beta}\right) \tag{4.5}$$

Example 4.2: Assume that the reaction in Example 4.1 reaches equilibrium at 37 °C. Calculate the equilibrium constant, K_{eq}.

Solution: Equation 4.5 can be used to solve this problem:

$$\Delta G - \Delta G^\circ = RT \ \ln\left(\frac{C^\gamma D^\delta}{A^\alpha B^\beta}\right)$$

Assuming that the reaction is in equilibrium, we can set ΔG equal to zero. This is because the change in Gibbs free energy of a reaction is equal to zero once equilibrium is established. Note, however, that the change in the standard Gibbs free energy (ΔG°) is not zero. Again, we know that $\Delta G = 0$ since the reaction has reached a state of equilibrium. Thus:

$$\Delta G^\circ = -RT \ \ln\left(K_{eq}\right)$$

Solving for the equilibrium constant, K_{eq}:

$$K_{eq} = e^{-\frac{\Delta G^\circ}{RT}}$$

To solve this problem, we can reuse some of the thermodynamic properties given in Example 4.1, since the conditions given previously were standard conditions (25 °C and 1 atm). Recall this:

$$\Delta G^\circ = -73\,\text{kJ/mol} - (298\,\text{K})(0.539\,\text{kJ/(mol} \cdot \text{K})) = -234\,\text{kJ/mol}$$

R is the universal gas constant, $0.008314\,\text{kJ/(mol·K)}$, and the temperature is $37\,°\text{C} + 273 = 310\,\text{K}$.

$$-\frac{\Delta G^\circ}{RT} = -\left(\frac{-234\,\text{kJ/mol}}{(0.008314\,\text{kJ/(mol} \cdot \text{K})(310\,\text{K}))}\right) = 90.8\,\text{kJ/mol}$$

$$\mathbf{K_{eq} = e^{90.8} = 2.7 \times 10^{39}}$$

In the study of bioenergy and bioproducts, it is often non-trivial to apply concepts like Gibbs free energy due to the complexity of cellular metabolism and the dynamic (and sometimes spontaneous) nature of the microbes responsible for generating products. Additionally, in practice, every effort is made to optimize biochemical reactions such that they are virtually irreversible and proceed to completion. For example, in the production of ethanol, strains of *Saccharomyces cerevisiae* have been selected to achieve near-theoretical yields of ethanol. System relationship data, such as Gibbs free energy and enthalpy values, are often collected and examined under steady-state conditions.

4.3 Reaction Kinetics

While reaction thermodynamics help determine whether or not a reaction will proceed to completion (or near-completion), reaction kinetics are useful in determining how fast the reaction will occur (see Chapter 13 for details). This is particularly significant in commercial facilities and is perhaps best summarized by the famous adage "Time is money." Industries typically do not favor chemical reactions that require long processing times. Labor and operational costs (e.g., electricity, water, and heat) both increase significantly with time, which ultimately drives up production costs. For biofuels, high production costs translate to higher consumer prices at the pump, making it difficult for biofuels to be cost competitive with conventional petroleum-based fuels.

Reaction rates can be measured in several different units, but they are most commonly reported in the form of concentration per unit time (mol/(L·sec)). Equation 4.6 mathematically describes the reaction rate:

$$r_\alpha = \frac{-dC_\alpha}{dt} = -\frac{1}{V}\frac{dn_\alpha}{dt} \qquad (4.6)$$

where r_α is the reaction rate of species α (mol/(L·sec)), C_α is the concentration of α (mol/L), and t is time (sec). More often than not, these reactions take place in a reactor of fixed volume. Thus, the volume component (V) can be treated as a constant, and the change in molar concentration (n_α) can be considered with respect to time (t). A negative sign exists when the molecular species is a substrate and is consumed. This occurrence can be confirmed by applying the law of conservation of mass to each molecule participating in a given reaction. Products of the reaction will have a positive value since they are generated. Equation 4.6 portrays a somewhat intuitive example of how to define the rate of a particular reaction, but there are other components that also affect the speed at which a reaction proceeds to completion, or near-completion. For example, the addition of heat, catalysts, and inhibitors can positively or negatively affect the reaction rate. The reaction rate constant, k, is implemented to account for some of these other factors and is independent of substrate (or product) concentrations. The reaction rate constant is, however, affected by temperature. Fortunately, most biochemical reactions occur in isothermal environments to preserve the biological activity of microbial communities. The reaction rate constant, under isothermal conditions, can be calculated by the Arrhenius equation:

$$k = Ae^{-E/RT} \tag{4.7}$$

where k is the first-order reaction rate constant (1/sec), A is the Arrhenius constant independent of temperature (1/sec), E is the activation energy of the reaction (J/mol), R is the ideal gas constant (J/(mol · K)), and T is temperature (K). The Arrhenius constant (also known as the *frequency factor*) is an empirically determined value that represents the likelihood that two substrates will come into contact and react with each other. Reaction rates incorporating the constant, k, take the form shown in Equation 4.8:

$$r_A = -kC_A^x C_B^y C_C^z \ldots \tag{4.8}$$

The exponents do not necessarily correlate with the stoichiometric relationships of the reacting molecules, and are represented in Equation 4.8 by x, y, and z to emphasize the fact that they are (often) unknown. In some cases, the exponents *do* correspond to stoichiometric relationships. These reactions are said to be elementary reactions. When a given reaction is not elementary, it is referred to as a non-elementary reaction. When describing a reaction like that in Equation 4.8, one can say that it is of order x with respect to species A or of order y with respect to species B (and so on). The overall order of the reaction is given by:

$$n = x + y + z \tag{4.9}$$

The units of the reaction rate constant vary with the order of the reaction, and must cancel out to give the final units of a reaction rate (i.e., concentration/time). If a reaction rate constant for a particular reaction is known, it can be used to determine the overall order of the reaction. As a self-check, one can determine the units of a rate constant by the following:

$$(\text{units of } k) = \frac{C_x^{1-n}}{t} \tag{4.10}$$

where t is time, C_x represents the units of concentration, and n is the overall reaction order.

More on the topic of reaction orders, as they pertain to bioenergy production, is provided in greater detail in Chapter 13.

Catalysts in Chemical Reactions

The origins of the word *catalyst* come from the Greek word *kataleuin*, which means "dissolve." By definition, a catalyst is a compound or chemical that actively promotes a reaction in the desired direction without being consumed in the process. A catalyst may be dissolved in solution or present in virtually any state of matter (i.e., solid, liquid, or gas). Catalysts function by lowering the free energy (energy of activation) required to overcome the transition state (indicated by the peak of the parabola

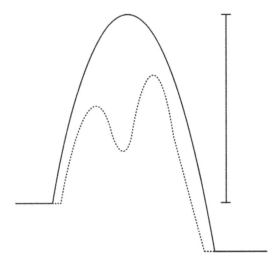

FIG. 4.1   The function of catalysts

in Figure 4.1), but do not change the total amount of free energy required to go from substrate to product. Recall that Gibbs free energy is a state function. The total change in Gibbs free energy is independent of the process or number of steps required to transform a substrate into product. A catalyst simply provides an alternative path for substrate-to-product formation. Potassium hydroxide (KOH), among other strong bases, is an important catalyst for promoting the transesterification of oil and animal fat into biodiesel.

Catalysts also exist in biology and are a critical part of everyday life. These biological catalysts are known as enzymes, and coincidentally also play a crucial role in bioenergy production. As you will see in Chapter 13, special enzymes produced by a select group of fungi are generated commercially to dissolve the structural carbohydrates (namely, cellulose and hemicellulose) of plant biomass and starch into fermentable sugars for subsequent bioethanol and advanced biofuel production. Without these enzymes, many of the technologies currently used in research and development for renewable fuels would not exist.

References and Further Reading

Fogler, H.S. (2005). *Elements of Chemical Reaction Engineering*. Upper Saddle River, NJ: Prentice Hall Professional Technical Reference.

Hill, C.G., Jr. (1977). *An Introduction to Chemical Engineering Kinetics and Reactor Design*. New York: John Wiley & Sons, Inc.

Levenspiel, O. (1999). *Chemical Reaction Engineering*. New York: John Wiley & Sons, Inc.

Exercise Problems

4.1. The reaction A + B → C occurs at 30 °C, but is also known to occur at 10 °C. Under which condition would ΔH_{rxn} be the greater? Explain your answer.

4.2. Calculate ΔH_{rxn} for ethanol formation from glucose. Is this an exothermic process? How about the formation of ethanol from xylose?

4.3. In Problem 4.2, what would be ΔH_{rxn} for butanol formation from glucose? How about the formation of butanol from xylose?

4.4. Biogas produced through anaerobic digestion is being upgraded to 100% methane (v/v). If the methane gas is combusted, determine both the net and gross heat of combustion.

4.5. Nitrogen dioxide (NO_2) is an air pollutant that is created from the reaction of nitric oxide (NO) with oxygen. Is this an exothermic reaction?

4.6. Calculate the Gibbs free energy of the formation of butanol at 25 °C and 1 atm. Is this reaction spontaneous?

4.7. Hydrogen is considered to be an energy carrier. Calculate the net heat value when it is burned.

4.8. Calculate the equilibrium constant of the reaction in Problem 4.5 if NO_2 is produced at 25 °C.

4.9. Give two detailed examples of how one might increase the rate constant k of a given reaction.

4.10. Suppose that the rate of a particular reaction takes this form:

$$r = -kC_A^2 C_B^1 C_C^1$$

What is the overall order of the reaction? Give at least two examples of the units of the rate constant k. (Hint: Concentration can be measured in more than one unit.)

Organic and Carbohydrate Chemistry

Xiaolan Luo and Yebo Li

What is included in this chapter?

This chapter covers basic organic and carbohydrate chemistry as applied to bioenergy. The history of organic chemistry and the sources, nomenclature, structures, properties, and reactions of organic compounds, including carbohydrates, are discussed. Some typical organic compounds important to bioenergy production are also introduced.

5.1 Introduction

Organic chemistry deals with the study of the structures, physical and chemical properties, functional groups, and reactions of organic compounds. Originally, organic compounds were considered to be derived from living organisms; that is, plants and animals. However, organic compounds were also found to be formed from inorganic material when Friedrich Wöhler discovered the chemical conversion of ammonium cyanate into urea in 1828. The conventional understanding of organic compounds was further clarified with the syntheses of many organic compounds using petroleum resources as the raw material.

Organic compounds comprise three major elements: carbon, hydrogen, and oxygen. Some organic compounds also contain nitrogen, phosphorus, sulfur, halogens, and/or metals. Compared to inorganic compounds, organic compounds have different physical and chemical properties, and different reactivity. In general, organic compounds are more combustible, with lower melting and boiling points, and have lower reactivity with other chemicals; this is in contrast to the rapid ionic reactions of inorganic compounds. Some organic compounds also form isomers, which have a different chemical structure but have the same molecular formula.

Currently, there are over 13 million known organic compounds, most of which are produced synthetically. Some organic compounds, such as animal fats, vegetable oils, carbohydrates, and proteins, are obtained from natural sources. Some organic compounds, including alcohols, organic acids, and antibiotics, are derived from fermentation processes.

Bioenergy: Principles and Applications, First Edition. Edited by Yebo Li and Samir Kumar Khanal.
© 2017 John Wiley & Sons, Inc. Published 2017 by John Wiley & Sons, Inc.
Companion website: www.wiley.com/go/Li/Bioenergy

Carbohydrate chemistry, an important subdiscipline of organic chemistry, deals with the study of carbohydrates; which are the major constituent of foods, and the building units of lignocellulosic feedstocks and microbial biomass. In this chapter, the fundamentals of organic and carbohydrate chemistry involved in the nomenclature, structures, properties, functional groups, and reactions of organic compounds are introduced. The basic properties and reactions of proteins and lipids are also discussed. An understanding of organic and carbohydrate chemistry helps readers recognize differences in the chemical composition and structure of *bioenergy feedstocks*, and the basic chemical reactions that occur during biochemical and thermochemical conversions. It also provides students with a guide for selecting feedstocks and optimizing processes in the production of bioenergy.

5.2 Structural Formulas and Classification of Organic Compounds

The structural formula of an organic compound is a drawing used to depict its molecular structure, showing the arrangement of atoms and bonds. Compared to molecular formulas of organic compounds, which indicate the atomic composition, structural formulas provide chemists with a visual representation of the molecular structure used for the identification of functional groups and potential reactions.

In a structural formula, all the chemical bonds between atoms of a compound are shown. Each straight line represents a single bond. The number of lines between atoms represents the number of bonds: for example, two or three parallel lines between atoms represent a double or triple bond, respectively. A simplified structural formula, which only shows bonds without linkages to hydrogen atoms, is called a *condensed structural formula*. Condensed structural formulas can be further simplified to a single line of text without showing graphic covalent bonds. For example, the condensed structural formula for methacrylic acid (2-methylpropenoic acid) can also be depicted as CH_2CCH_3COOH. In some cases, there may be two or more instances of the same group in a structural formula. In such a case, a parenthesis and subscript are used to indicate the identical groups, and their quantity. For example, the condensed structural formula for 2-methyl-1-propanol (isobutanol) can be depicted as $(CH_3)_2CHCH_2OH$. *Skeletal formulas* are another type of simplified structural formula used to depict organic molecules. In skeletal formulas, a zigzag line is often used to illustrate the carbon–carbon bonds and/or carbon–oxygen single bonds of a main chain. The bonds linking carbon to atoms other than hydrogen need to be clearly drawn. Moreover, all atoms other than hydrogen and carbon, and all hydrogen attached to atoms other than carbon, need to be shown in the skeletal formula. As it is easy to identify the functional groups of an organic compound in its skeletal formula, skeletal formulas of organic compounds are widely used, especially in the study of reaction mechanisms. Figure 5.1 shows molecular, structural, condensed structural, and skeletal formulas of methacrylic acid, to exemplify their differences.

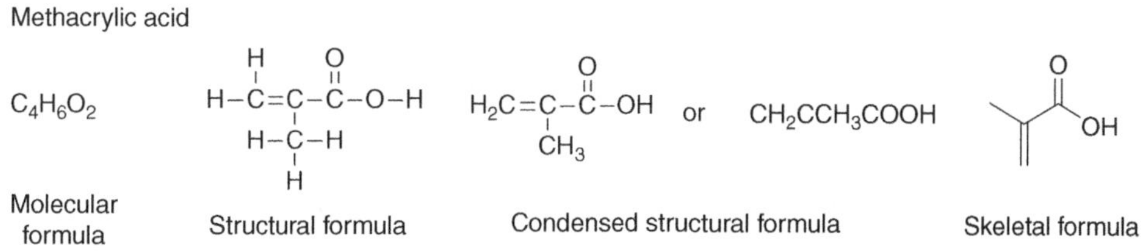

FIG. 5.1   Molecular, structural, condensed structural, and skeletal formulas of methacrylic acid.

Generally, organic compounds are classified into three major categories: aliphatic, aromatic, and heterocyclic. These categories are discussed in the following sections. Natural compounds, including carbohydrates, fats, oils, and proteins, are introduced in Sections 5.6 and 5.7.

5.3 Aliphatic Compounds

Aliphatic compounds are those that are based on linear or branched carbon chains. In terms of their characteristic functional groups, aliphatic compounds can be subdivided into alkanes, cycloalkanes, alkenes, alkynes, alcohols, ethers, aldehydes, ketones, carboxylic acids and their derivatives, and other compounds. The simplest aliphatic compounds are known as hydrocarbons, including saturated hydrocarbons without specific functional groups as well as unsaturated hydrocarbons with carbon–carbon double or triple bonds.

5.3.1 Alkanes and Cycloalkanes

Alkanes

Alkanes are saturated hydrocarbons that have the general molecular formula C_nH_{2n+2} and do not have characteristic functional groups. Normal alkanes have a straight carbon chain and are named by using the term associated with the number of carbon atoms with *-ane* added to the end. The prefix *n-* may be used (e.g., *n*-butane) to indicate a normal alkane, but it is usually omitted. An alkyl is the residual group that exists after the removal of one hydrogen atom from the alkane. The name of an alkyl group is derived from its parent alkane, with *-yl* replacing the *-ane* ending. R- is generally used to represent a non-specific alkyl group. Therefore, the general formula of alkanes is R–H.

Alkanes with simple branched carbon chains are called isoalkanes or neoalkanes depending on their structure and are usually named by replacing the *n-* with iso- or neo- as the prefix. For example, *n*-butane is a normal alkane with a straight carbon chain, whereas isobutane, which has the same molecular formula as *n*-butane, has branched carbon chains. Figure 5.2 shows the condensed structural formulas of some simple alkanes and their alkyl groups.

Alkanes with complex branched carbon chains are named according to the International Union of Pure and Applied Chemistry (IUPAC) system. In the IUPAC system, the compound is named based on the longest carbon chain (i.e., the main chain) in the molecule. The names of alkyl groups attached to the main chain are added alphabetically as a prefix to the IUPAC name of the compound. Positions of these alkyl groups are also added as the prefix to the names of alkyl groups via numbering carbon atoms in the main chain and making the alkyl group position number as small as possible. In addition, if two or more identical alkyl groups are attached to the main chain, di-, tri-, and so on are used as the prefix of the alkyl group's name. Figure 5.3 shows several examples of this IUPAC nomenclature.

Alkanes are colorless, odorless, and insoluble in water, but are soluble in most organic solvents. Alkanes containing fewer than five carbon atoms and neopentane are gases at room temperature (Table 5.1). the boiling point of *n*-alkanes increases as their molecular weight increases. Compared to *n*-alkanes, branched alkanes with the same carbon atoms have a lower boiling point. In other words, the boiling point of branched alkanes decreases as the degree of branching increases. In the petroleum industry, alkanes with high molecular weights are usually pyrolyzed to produce low molecular-weight compounds, such as gasoline or raw materials for the synthesis of industrial

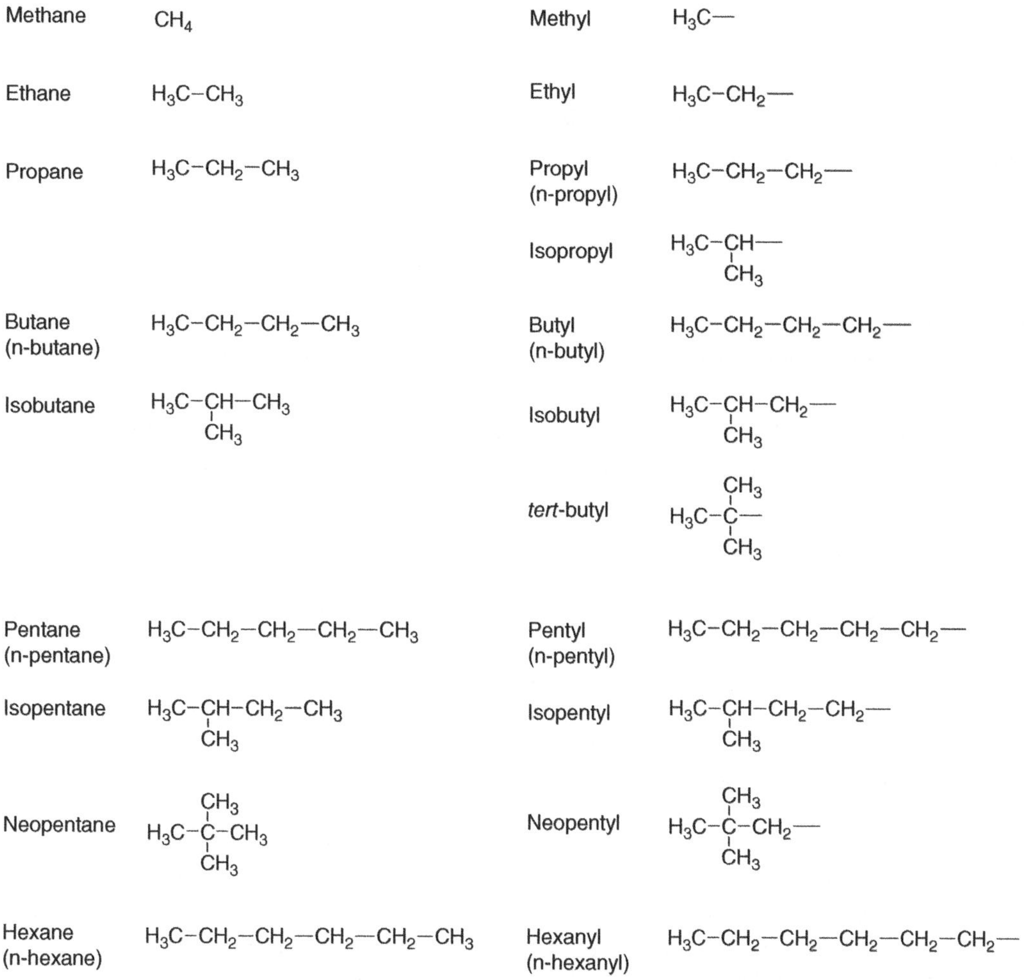

Methane	CH_4	Methyl	H_3C-
Ethane	H_3C-CH_3	Ethyl	H_3C-CH_2-
Propane	$H_3C-CH_2-CH_3$	Propyl (n-propyl)	$H_3C-CH_2-CH_2-$

Isopropyl

H_3C-CH-
         $|$
         CH_3

| Butane (n-butane) | $H_3C-CH_2-CH_2-CH_3$ | Butyl (n-butyl) | $H_3C-CH_2-CH_2-CH_2-$ |

Isobutane

$H_3C-CH-CH_3$
       $|$
       CH_3

Isobutyl

$H_3C-CH-CH_2-$
       $|$
       CH_3

tert-butyl

$\quad\quad CH_3$
$\quad\quad |$
H_3C-C-
$\quad\quad |$
$\quad\quad CH_3$

| Pentane (n-pentane) | $H_3C-CH_2-CH_2-CH_2-CH_3$ | Pentyl (n-pentyl) | $H_3C-CH_2-CH_2-CH_2-CH_2-$ |

Isopentane

$H_3C-CH-CH_2-CH_3$
       $|$
       CH_3

Isopentyl

$H_3C-CH-CH_2-CH_2-$
       $|$
       CH_3

Neopentane

$\quad\quad CH_3$
$\quad\quad |$
$H_3C-C-CH_3$
$\quad\quad |$
$\quad\quad CH_3$

Neopentyl

$\quad\quad CH_3$
$\quad\quad |$
$H_3C-C-CH_2-$
$\quad\quad |$
$\quad\quad CH_3$

| Hexane (n-hexane) | $H_3C-CH_2-CH_2-CH_2-CH_2-CH_3$ | Hexanyl (n-hexanyl) | $H_3C-CH_2-CH_2-CH_2-CH_2-CH_2-$ |

FIG. 5.2   Condensed structural formulas of alkanes and their alkyl groups.

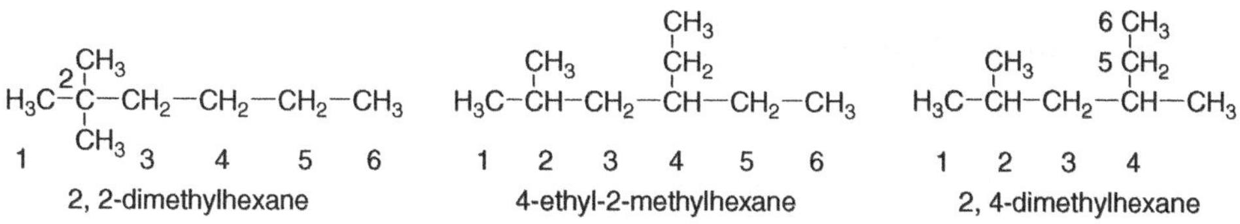

$$\underset{1}{H_3C}\overset{2}{-}\underset{|}{\overset{CH_3}{\underset{CH_3}{C}}}\overset{}{-}\underset{3}{CH_2}-\underset{4}{CH_2}-\underset{5}{CH_2}-\underset{6}{CH_3}$$

2, 2-dimethylhexane

$$\underset{1}{H_3C}-\underset{2}{\overset{CH_3}{CH}}-\underset{3}{CH_2}-\underset{4}{\overset{CH_2}{\underset{}{CH}}}\overset{CH_3}{-}\underset{5}{CH_2}-\underset{6}{CH_3}$$

4-ethyl-2-methylhexane

$$\underset{1}{H_3C}-\underset{2}{\overset{CH_3}{CH}}-\underset{3}{CH_2}-\underset{4}{\overset{6\,CH_3}{\underset{5\,CH_2}{CH}}}-\underset{}{CH_3}$$

2, 4-dimethylhexane

FIG. 5.3   Examples of the IUPAC nomenclature of alkanes.

Table 5.1 Boiling points (BP) of some simple alkanes

Name	Methane	Ethane	Propane	Butane	Isobutane	Pentane	Isopentane	Neopentane	Hexnae
n^a	1	2	3	4	4	5	5	5	6
$BP(^\circ C)^b$	−161.5	−88.6	−42.1	−0.5	−11.7	36.0	27.7	9.5	68.7

^{a} Number of carbon atoms
^{b} Boiling point

Cyclopropane   Cyclobutane   Cyclohexane

FIG. 5.4   Skeletal formulas of cyclopropane, cyclobutane, and cyclohexane.

FIG. 5.5   Chair conformation of cyclohexane.

chemicals. Methane is a dominant component of natural gas. It can also be produced through anaerobic digestion of lignocellulosic feedstocks, manure, organic waste, and other organic feedstocks.

Cycloalkanes

Cycloalkanes are saturated cyclic hydrocarbons with the general molecular formula C_nH_{2n}. They are named using the prefix cyclo- followed by the name of the alkane with the same number of carbon atoms. For example, a cycloalkane containing three carbon atoms is called cyclopropane, that containing four carbon atoms is called cyclobutane, and that containing six carbon atoms is called cyclohexane. Figure 5.4 shows the structures of cyclopropane, cyclobutane, and cyclohexane.

The six-carbon ring of cyclohexane is an important structural unit for many natural compounds. Its structural formula is commonly drawn as a chair conformation, as shown in Figure 5.5. Cycloalkanes have physical properties similar to alkanes: for example, they are insoluble in water, while soluble in most organic solvents. However, cycloalkanes have higher boiling points and densities than the corresponding alkanes. Due to the lack of functional groups, both cycloalkanes and alkanes have low reactivity, except for some cycloalkanes with a very low number of carbon atoms, particularly cyclopropane. Cycloalkanes and alkanes with a relatively low boiling point, such as cyclohexane, hexane, or heptane, are often used as extracting solvents.

5.3.2 Alkenes and Alkynes

Alkenes

Alkenes are unsaturated hydrocarbons with the general molecular formula C_nH_{2n} and have one characteristic carbon–carbon double bond. The general formula can be represented by $R–CH=CH_2$ or $R–CH=CH–R'$ ($R = R'$ or $R \neq R'$). Compared to alkanes, alkenes with the same carbon atoms have two fewer hydrogen atoms. Alkenes with a straight carbon chain are named following a nomenclature similar to the alkanes, in which the term associated with the number of carbon atoms is used with an -*ene* ending. Due to the presence of a characteristic double bond, the position of the double bond is numbered to make the number as small as possible, which needs to be added as the prefix to an alkene, except for ethene and propene. For the nomenclature of an alkene containing a complex structure, the longest carbon chain with the carbon–carbon double bond in the molecule is used as the main chain. The positions and names of alkyl groups attached to the main chain are successively added as the prefix to the alkene, with the names of different alkyl groups added alphabetically.

Likewise, a prefix of di-, tri-, etc. is used in cases in which two or more identical alkyl groups are attached to the main chain of the compound. Figure 5.6 shows some examples of this nomenclature.

Like alkanes, lower-carbon atom alkenes with 2–4 carbon atoms and 3-methyl-1-butene are gas, while high-carbon atom alkenes with 5–15 carbon atoms, except for 3-methyl-1-butene, are liquid at room temperature. In contrast, alkenes with a higher number of carbon atoms are solid at room temperature. Compared to alkanes, alkenes have higher reactivity. The typical reactions of alkenes are addition and polymerization. For example, ethylene, the simplest alkene, can be polymerized to produce polyethylene (PE), which is extensively used in the plastics industry.

In addition to alkenes with one carbon–carbon double bond, there are some unsaturated hydrocarbons with two or more carbon–carbon double bonds. In terms of the distribution of double bonds in the main chain, hydrocarbons with two double bonds can fall into three types: allenes, which have adjacent double bonds, e.g., $CH_2{=}C{=}CH{-}CH_3$; conjugated dienes, which have alternating single and double bonds, e.g., $CH_2{=}CH{-}CH{=}CH{-}CH_3$; and dienes, which have isolated double bonds, e.g., $CH_2{=}CH{-}CH_2{-}CH{=}CH_2$.

Alkynes

Alkynes are another type of unsaturated hydrocarbon with the general molecular formula C_nH_{2n-2}, and with one characteristic carbon–carbon triple bond. Their general formula can be shown as $R{-}C{\equiv}CH$ or $R{-}C{\equiv}C{-}R'$ ($R = R'$ or $R \neq R'$). The nomenclature of alkynes follows the same IUPAC rules as alkenes, except with an -yne ending. Figure 5.7 contains several examples of the nomenclature of alkynes. Alkynes are similar to alkenes in their general physical and chemical properties. The typical reaction of alkynes is addition reaction. Ethyne, the simplest alkyne, can produce extremely high energy when it combusts with oxygen. Due to this property, ethyne is widely used in the welding industry.

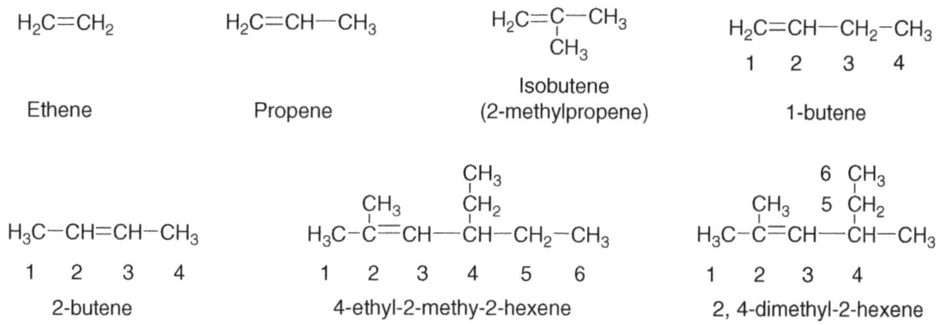

FIG. 5.6   Nomenclature of alkenes.

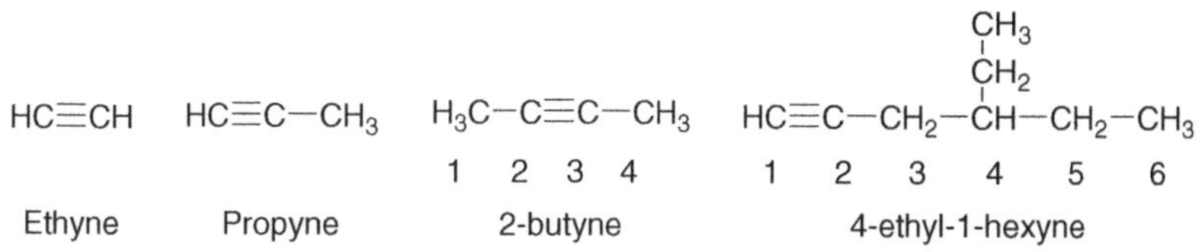

FIG. 5.7   Nomenclature of alkynes.

5.3.3 Alcohols and Ethers

Alcohols

Alcohols are compounds with hydroxyl functional groups. The general formula of alcohol is R–OH. Depending on the number of hydroxyl groups in a molecule, alcohols can be classified into monohydric or polyhydric alcohols. Monohydric alcohols are compounds with one hydroxyl group. There are two types of monohydric alcohols: saturated and unsaturated. A saturated monohydric alcohol is considered to be derived from an alkane via the substitution of one hydrogen atom of an alkane by a hydroxyl group. It is named by using the *-ol* ending to replace the *-e* ending of the alkane. Like alkenes and alkynes, the position of the hydroxyl functional group needs to be shown in the name of the alcohol. The rule for numbering the position of the hydroxyl group in the main chain of an alcohol is to obtain the smallest number. For the nomenclature of unsaturated monohydric alcohols, the *-ol* endings are used to replace the *-e* endings of the corresponding alkenes, and all the positions of the hydroxyl group and unsaturated bonds should be shown so that the number representing the position of the hydroxyl group is as small as possible. Figure 5.8 shows some examples of the nomenclature of alcohols.

Monohydric alcohols can be classified as primary, secondary, or tertiary alcohols (Figure 5.9) in terms of the position of the hydroxyl group in the molecule; that is, which carbon atom is linked to the hydroxyl group. Figure 5.9 shows the general formulas of these three types of alcohols, in which R, R', and R'' can be identical or different.

Monohydric alcohols with lower carbon atoms are miscible with water, and compared to alkanes have increased polarity due to the existence of the hydroxyl group. With the increase in carbon atoms, the polarity of monohydric alcohols and their miscibility with water decrease. For example, methanol, ethanol, 1-propanol, and 2-propanol are all soluble in water. However, 1-butanol has limited solubility in water. Monohydric alcohols have higher boiling points than alkanes because of the intermolecular hydrogen bonding of alcohols. Alcohols can undergo three major types of reaction: dehydration, for the formation of ethers and alkenes; oxidation, for the formation of aldehydes, ketones, or acids; and esterification, with acids and transesterification for the formation of esters. Alcohols are commonly produced by the hydration of alkenes, an addition reaction of alkenes with water. Currently, bioethanol and 1-butanol are produced by fermentation processes, and the former is extensively used as a biofuel.

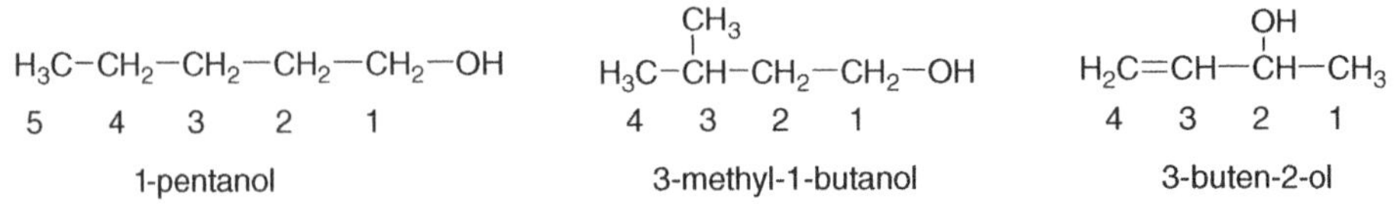

$H_3C-CH_2-CH_2-CH_2-CH_2-OH$

5   4   3   2   1

1-pentanol

$\underset{\underset{4}{}}{H_3C}-\underset{\underset{3}{}}{\overset{\overset{CH_3}{|}}{CH}}-\underset{\underset{2}{}}{CH_2}-\underset{\underset{1}{}}{CH_2}-OH$

3-methyl-1-butanol

$\underset{\underset{4}{}}{H_2C}=\underset{\underset{3}{}}{CH}-\underset{\underset{2}{}}{\overset{\overset{OH}{|}}{CH}}-\underset{\underset{1}{}}{CH_3}$

3-buten-2-ol

FIG. 5.8   Nomenclature of alcohols.

$R-CH_2-OH$

Primary alcohol

$R'-\overset{\overset{R}{|}}{CH}-OH$

Secondary alcohol

$R'-\overset{\overset{R}{|}}{\underset{\underset{R''}{|}}{C}}-OH$

Tertiary alcohol

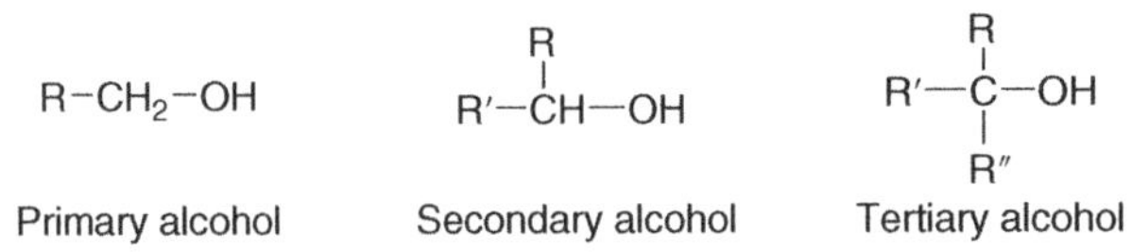

FIG. 5.9   General structures of monohydric alcohols.

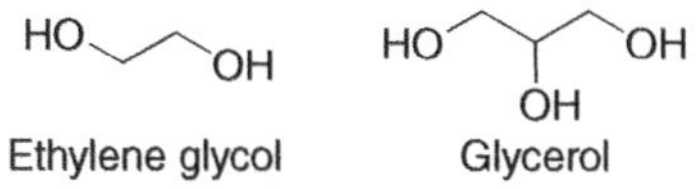

Ethylene glycol   Glycerol

FIG. 5.10   Chemical structures of ethylene glycol and glycerol.

Polyhydric alcohols are known as polyols, which have two or more hydroxyl groups. Depending on the number of hydroxyl groups, polyhydric alcohols are called diols, triols, and so on. Ethylene glycol and glycerol are the simplest diol and triol, respectively, as shown in Figure 5.10. Ethylene glycol is typically used as an antifreeze to cool automotive engines and as a de-icing agent. It is also a useful precursor for the production of polyesters, such as polyester polyols and polyester resins. Glycerol, a valuable feedstock derived from either petroleum or natural sources, is extensively used in food, pharmaceuticals, cosmetics, tobacco, polyurethane, and other industries. Crude glycerol is currently being generated as a byproduct in large quantities by the biodiesel industry.

Ethers

Ethers are formed by the dehydration of two alcohol molecules, and are represented by the general formula R–O–R'. If the two alkyl groups are identical (R = R'), ethers are defined as simple or symmetrical ethers. If the two alkyl groups are different (R ≠ R'), ethers are defined as mixed or asymmetrical ethers. An ether is commonly named by alphabetically showing the name of the alkyl groups as the prefix of the ether, such as $CH_3CH_2OCH_2CH_3$, diethyl ether, or $CH_3OCH_2CH_3$, ethyl methyl ether. In the IUPAC nomenclature system, the group of R–O– or R'–O– in the general formula of ethers is called the alkoxy group. Therefore, ethers may alternatively be named alkoxyalkanes, in which the alkyl group with the higher number of carbon atoms is defined as the main carbon chain. For example, ethyl methyl ether is also known as methoxyethane.

Ethers have lower polarity than alcohols, and higher polarity than alkanes. Due to the absence of the intermolecular hydrogen bonds from hydroxyl groups, ethers have lower boiling points than alcohols with the same molecular formulas. Ethers are of low chemical reactivity and used extensively as solvents.

5.3.4 Aldehydes and Ketones

Both aldehydes and ketones are organic compounds with carbonyl functional groups, but they have different molecular structures. A carbonyl group is formed by the linkage of an oxygen atom and a carbon atom via a double bond. In the structure of an aldehyde, at least one hydrogen atom is linked to the carbonyl carbon atom. When two hydrogen atoms are linked to the carbonyl carbon atom of an aldehyde, the compound is known as formaldehyde, the simplest aldehyde. All the other aldehydes have the general formula R–CHO, in which a hydrogen atom and an alkyl group are linked to the carbonyl group.

In the IUPAC nomenclature system, an aldehyde is considered to be a derivative of an alkane with the same number of carbon atoms in the main chain as an aldehyde. An aldehyde is named by using the -al ending as a substitute for the final -e of the alkane. For example, formaldehyde is also known as methanal. The nomenclature of complex aldehydes is similar to those of the compounds previously introduced in this chapter. The position and type of the alkyl groups attached to the main chain are successively added in alphabetical order as prefixes. For example, a hexanal chain with a methyl group

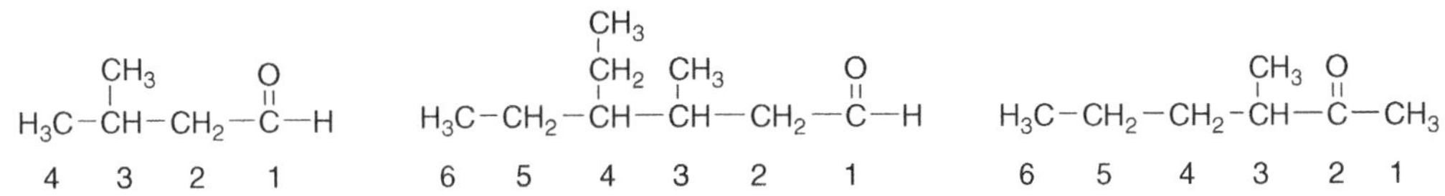

FIG. 5.11   Nomenclatures of aldehydes and ketones.

at its third carbon atom and an ethyl group at its fourth carbon atom would be named 4-ethyl-3-methyl-hexanal (Figure 5.11). The carbon atoms in the main chain are always numbered beginning with the carbonyl carbon atom at the lowest numbered position by default. Thus, it is not necessary to indicate the position of the carbonyl group in the name of the aldehyde.

Ketones have the general formula R–C(=O)–R', in which two of the same or different alkyl groups are linked to the carbonyl group. The nomenclature of ketones resembles that of aldehydes except with the *-one* ending. The position of the carbonyl group of a ketone needs to be listed in its name. Acetone, a common name of the simplest ketone, is also called propanone. Figure 5.11 shows some examples of the structural formulas and names of aldehydes and ketones.

Aldehydes and ketones with a lower number of carbon atoms are soluble in water due to the hydrogen-bonding interaction of the carbonyl group with water. The carbonyl group can serve as a hydrogen acceptor, but not as a hydrogen donor. Compared to alcohols, there are no intermolecular hydrogen bonds among aldehydes and ketones. Therefore, aldehydes and ketones are more volatile than alcohols. Ketones with a low number of carbon atoms are less toxic than aldehydes, and have been widely used as solvents for polar organic compounds. For example, acetone is a common solvent as a paint thinner and for cleaning laboratories.

Both aldehydes and ketones can be obtained by either oxidation of primary and secondary alcohols, respectively, or partial reduction of carboxylic acids. They can undergo four types of reactions, including oxidation reaction into carboxylic acids, reduction reaction into alcohols, addition reactions with alcohols to form hemiacetals/hemiketals or acetals/ketals, and condensation reactions with amines to form imines with carbon–nitrogen double bonds. In general, the reactions for the formation of hemiacetals or hemiketals are reversible, except in some cases such as the intramolecular hemiacetal or hemiketal formations. These stable cyclic structures of hemiacetals or hemiketals are very important in carbohydrate chemistry, and will be introduced in Section 5.6. Aldehydes and ketones show different chemical reactivity during oxidation. Compared to aldehydes, ketones are more difficult to oxidize. A strong oxidant such as $KMnO_4$ is required to break the carbon–carbon bonds of ketones for the formation of carboxylic acids.

5.3.5 Carboxylic Acids and Derivatives

Carboxylic Acids

Carboxylic acids are compounds with carboxyl functional groups. A carboxyl group is formed by attaching a hydroxyl group to a carbonyl carbon atom. The general formula for carboxylic acids is R–C(=O)OH. Depending on the number of the carboxyl group in a compound, carboxylic acids can be classified as monocarboxylic or polycarboxylic acids. Most monocarboxyl acids are called by their common names. However, these common names cannot provide structural information. In the IUPAC system, a monocarboxylic acid is named by using the *-oic acid* ending as a substitute for the final -*e* of an alkane that has the same number of carbon atoms in the main chain. For example, acetic acid (a common name), CH_3COOH, is named ethanoic acid based on the IUPAC system.

Some carboxylic acids have other functional groups, such as hydroxyl groups, in addition to the carboxyl groups. They are named by showing the position and the name of the additional functional group in the prefix of the carboxylic acid. The carboxyl carbon atom is used as a starter for numbering the main chain of a carboxylic acid. Lactic acid is a propanoic acid with a hydroxyl group that is attached to the carbon atom adjacent to the carboxyl carbon atom. Based on the IUPAC nomenclature, it is named 2-hydroxylpropanoic acid. Some carboxylic acids have carbon–carbon double bonds in their main chains. They are named by replacing the final -*e* of a corresponding alkene with the -*oic acid* ending and showing the position of the double bond in the prefix. The exceptions are propenoic acid and 2-methylpropenoic acid.

Some carboxylic acids have long saturated and unsaturated carbon chains and are commonly called fatty acids. Palmitic and stearic acids are typical saturated fatty acids; while oleic, linoleic, and linolenic acids are typical unsaturated fatty acids. Generally, Cm:n is used to represent fatty acids in the technical literature. Cm indicates the number of carbon atoms and n indicates the number of double bonds. Thus, C16:0 represents palmitic acid, while C18:1 represents oleic acid. Most fatty acids can be produced from the hydrolysis of vegetable oils and animal fats. Figure 5.12 shows the structures of several typical carboxylic acids and fatty acids.

Polycarboxylic acids are compounds with two or more carboxyl groups. They are often referred to by the number of carboxyl groups in a simplified term: for example, diacids or triacids. Succinic acid,

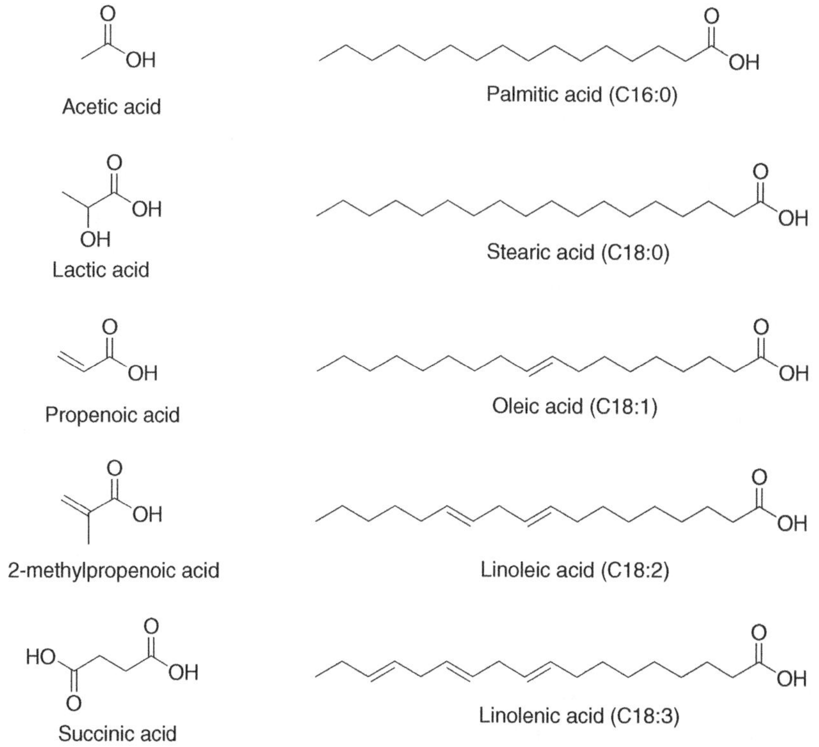

Acetic acid

Palmitic acid (C16:0)

Lactic acid

Stearic acid (C18:0)

Propenoic acid

Oleic acid (C18:1)

2-methylpropenoic acid

Linoleic acid (C18:2)

Succinic acid

Linolenic acid (C18:3)

FIG. 5.12   Structures of typical carboxylic acids and fatty acids.

a well-known diacid (Figure 5.12), has been widely used in the production of polyester polymers for plastic applications. Recently, succinic acid has been produced commercially using microbial fermentation.

A carboxyl group consists of a carbonyl group and a hydroxyl group, and is able to act as both hydrogen acceptor and donor. Like alcohols, carboxylic acids can also form intermolecular hydrogen bonds with each other or with water, which contributes to a high boiling point and high solubility in water. Due to the specific structures of carboxylic acids, a dimer with two hydrogen bonds is formed between the two carboxylic acid molecules. Thus, carboxylic acids have higher boiling points than alcohols with comparable molecular weights. Generally, liquid carboxylic acids with six or fewer carbon atoms have low boiling points and unfavorable odors, and are commonly known as volatile fatty acids (VFAs). Carboxylic acids are weak acids that react with alkaline compounds such as sodium hydroxide to form salts of carboxylic acids. The typical reactions of carboxylic acids include esterification with alcohols, reduction into alcohols, and dehydration into carboxylic anhydrides.

Derivatives of Carboxylic Acids

Carboxylic acid derivatives are a series of different classes of compounds with modified carboxyl groups, such as acid anhydrides, esters, acyl halides, and amides, and are illustrated in Figure 5.13.

Acid anhydrides are compounds formed by the dehydration of two carboxylic acid molecules. The general formula of acid anhydrides is R–C(=O)–O–C(=O)–R′, in which R and R′ can be the same or different alkyl groups. Acid anhydrides are named by using the term anhydride as the ending to replace acid: for example, acetic anhydride, $(CH_3CO)_2O$. If R and R′ are different, both groups need to be alphabetically listed as prefixes to the name of the acid anhydride. Like aldehydes and ketones, acid anhydrides cannot self-associate, and therefore have lower boiling points than acids with comparable molecular weights. Acid anhydrides containing short carbon chains are colorless liquids with strong smells and can react with water to produce acids.

Esters are compounds formed by the esterification of carboxylic acids (RCOOH) and alcohols (R′OH). The general formula of esters is R–C(=O)–O–R′ (R = R′ or R ≠ R′). An ester is named an alkyl alkanoate, in which the alkyl is from the alcohol and the alkanoate is from the carboxylic acid, and the -ate ending is used to replace the -ic acid ending of the acid: for example, ethyl ethanoate or ethyl acetate ($CH_3COOCH_2CH_3$). Esters containing a low number of carbon atoms are colorless liquids with pleasant fruit smells. They have lower boiling points and solubility in water compared to the corresponding carboxylic acids. Esters, such as ethyl acetate, are usually used as solvents for the extraction of organic compounds. Esters can undergo *hydrolysis* to form carboxylic acids, *transesterification* to form other esters, or *reduction* to alcohols.

Acyl halides are compounds derived by replacing the hydroxyl groups of carboxylic acids with halogen atoms, such as chlorine and bromine. The general formula of acid halides is R–C(=O)–X, in which R–C(=O)– is called the acyl group and X represents a halogen atom. An acyl halide is named by suffixing the corresponding halide to the acyl group, which is named by replacing the -ic acid ending of the parent acid with an -yl ending. Acid halides have extremely high chemical

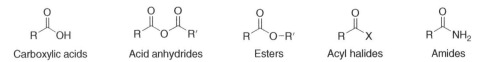

FIG. 5.13 General structures of carboxylic acids, acid anhydrides, esters, acyl halides, and amides.

reactivity and have often been used as acylating agents. Amides are compounds with the general formula R–C(=O)–NH$_2$. When hydrogen atoms attached to the nitrogen atom are substituted with alkyl groups, they are known as N-substituted amides.

5.3.6 Other Aliphatic Compounds

Besides the compounds already mentioned, there are some aliphatic compounds with other functional groups, such as alkyl halides, amines, alkanethiols, and thioethers. Alkyl halides are compounds with the general structural formula R–X, which can be derived by replacing a hydrogen atom of the alkane with a halogen atom. Aliphatic amines are compounds with the general structural formula of R–NH$_2$. When the hydrogen atoms of the amino group (–NH$_2$) are replaced with one or two alkyl groups, they are called secondary and tertiary amines. Alkanethiols and thioethers are compounds with the general formulas R–S–H and R–S–R′ (R = R′ or R ≠ R′), respectively. They have structures similar to alcohols and ethers, but sulfur atoms are attached to carbon atoms to form C–S bonds rather than C–O bonds.

5.4 Aromatic Compounds

Aromatic compounds are compounds with one or more benzene rings, or with chemical behaviors similar to benzene. This section mainly focuses on benzene and its derivatives. The other aromatic compounds, such as aromatic heterocyclic compounds, will be discussed along with aliphatic heterocyclic compounds in Section 5.5.

Benzene is a six-membered cyclic hydrocarbon with the molecular formula C$_6$H$_6$. The structure of benzene is generally described as a ring consisting of three alternating C–C and C = C bonds. The six carbon atoms of the benzene ring have completely identical functions. Benzene is a relatively stable molecule, and has distinctly different chemical reactivity than an alkene. For example, benzene tends to undergo substitution reactions rather than addition reactions.

Benzene and its derivatives are important aromatic compounds. There are three major types of benzene derivatives. One type is formed by directly attaching alkyl or other functional groups to the benzene ring, including aromatic hydrocarbons, phenols, aryl ethers, aromatic aldehydes and ketones, aromatic carboxylic acids, halobenzenes, arylamines, and so on. Another type of benzene derivative is formed by attaching functional groups to the alkyl carbon chain of alkylbenzenes, such as aryl alcohols, benzyl ethers, aryl aliphatic acids, halogenated alkylbenzenes, and benzylamines. These benzene derivatives have chemical reactivity similar to aliphatic compounds, except for the additional reaction sites in the benzene ring. Like an alkyl group derived from an alkane, an aryl group is derived from an aromatic compound; *Ar-* is generally used to represent an aryl group. In aromatic compounds, there are two important aryl groups: the phenyl group, a functional group formed by removing one hydrogen atom from the benzene ring; and the benzyl group, a functional group formed by removing one hydrogen atom from the methyl group of toluene (methylbenzene). The third type of benzene derivatives is a polycyclic aromatic compound, which consists of two or more fused benzene rings. Figure 5.14 shows examples for each type of benzene derivative.

Lignin is a highly branched aromatic polymer with various phenylpropane-derived units that are formed from three major monomers: *p*-coumaryl alcohol, coniferyl alcohol, and sinapyl alcohol (Figure 5.15). Lignin has multiple functional groups, which primarily include hydroxyl groups in the phenol or aryl alcohol structures, methoxy groups attached to a benzene ring, and aldehyde

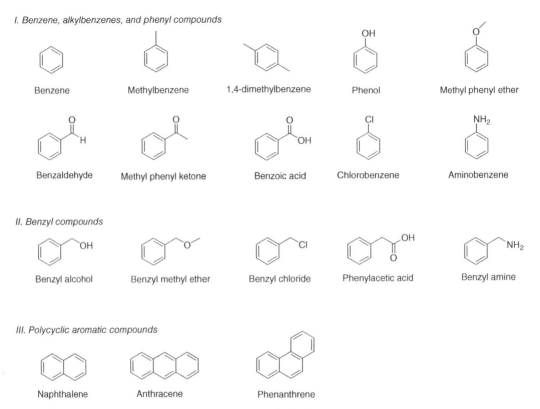

FIG. 5.14 Structures of three types of benzene derivatives.

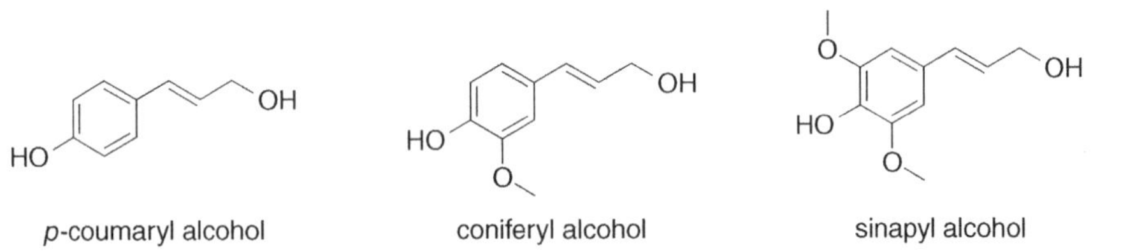

FIG. 5.15 Chemical structures of *p*-coumaryl alcohol, coniferyl alcohol, and sinapyl alcohol.

groups. Lignin is a very important component of ligocellulosic feedstocks, and is extensively studied for bioenergy production through both biochemical and thermochemical platforms.

5.5 Heterocyclic Compounds

Heterocyclic compounds have heteroatoms in their cyclic structures. Heteroatoms are atoms other than carbon and hydrogen atoms, and primarily include nitrogen, oxygen, and sulfur atoms. Depending on their structures and properties, heterocyclic compounds can be classified as aliphatic or aromatic heterocyclic.

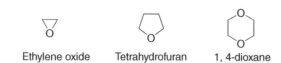

Ethylene oxide   Tetrahydrofuran   1, 4-dioxane

FIG. 5.16   Chemical structures of ethylene oxide, tetrahydrofuran, and 1,4-dioxnae.

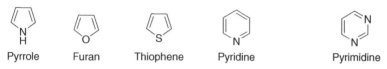

Pyrrole   Furan   Thiophene   Pyridine   Pyrimidine

FIG. 5.17   Chemical structures of pyrrole, furan, thiophene, pyridine, and pyrimidine.

Aliphatic heterocyclic compounds can be considered cycloalkane derivatives with heteroatoms. For example, ethylene oxide is a three-membered ring with one oxygen and two carbon atoms that has been extensively used for the production of polyether polyols (e.g., polyethylene oxide, PEO) in the polyurethane and pharmaceutical industries. Tetrahydrofuran and 1,4-dioxane are a five-membered ring with one oxygen atom and a six-membered ring with two oxygen atoms, respectively. Figure 5.16 shows their chemical structures.

Aromatic heterocyclic compounds are cyclic compounds with one or more heteroatoms, which also have chemical behaviors similar to benzene. For example, pyrrole, furan, and thiophene are five-membered aromatic heterocyclic compounds with one nitrogen, oxygen, and sulfur atom, respectively, as shown in Figure 5.17. Pyridine is a six-membered aromatic heterocyclic compound with one nitrogen atom. Pyrimidine is a six-membered one with two nitrogen atoms. Like benzene, aromatic heterocyclic compounds tend to undergo substitution reactions.

5.6 Carbohydrates

As the major components of food and building units for *lignocellulosic biomass*, *carbohydrates* are the most abundant bioresource on earth. They consist of carbon, hydrogen, and oxygen atoms with the empirical molecular formula $C_n(H_2O)_n$. Depending on the number of monosaccharide units in a molecule, carbohydrates can be classified into monosaccharides, oligosaccharides, and polysaccharides. A brief general overview of carbohydrates is given here and a more detailed discussion, especially on the carbohydrate chemistry of plant biomass, is presented in Chapter 6.

5.6.1 Monosaccharides

Monosaccharides, also known as simple sugars, are polyhydroxyl aldehydes or ketones. In general, sugars with aldehyde carbonyl groups are called aldoses, while those with ketone carbonyl groups are called ketoses. As a result of the number of carbons in a molecule, sugars containing three carbon atoms are called aldotriose or ketotriose; those containing four carbon atoms are called aldotetrose or ketotetrose; those containing five carbon atoms are called aldopentose or ketopentose; and those containing six carbon atoms are called aldohexose or ketohexose. For example, 2,3-dihydroxylpropanal (also called glyceraldehyde) is an aldotriose.

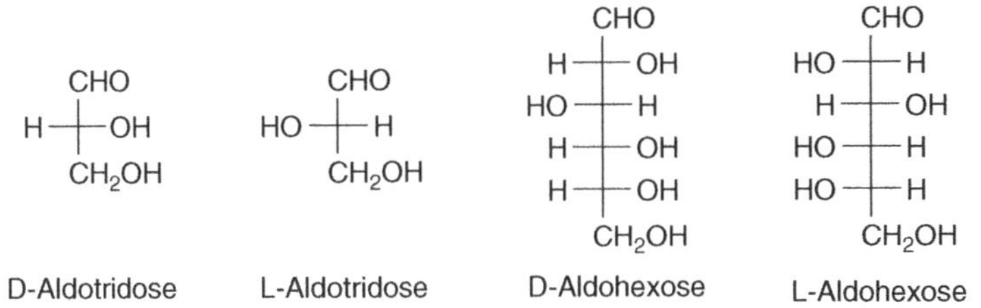

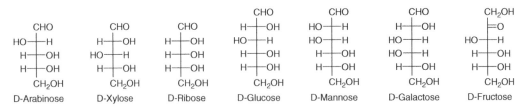

FIG. 5.18   Configurations of D- and L-sugars in Fischer projections.

FIG. 5.19   Configurations of different D-sugars with five or six carbon atoms in Fischer projections.

Due to the existence of one asymmetrical carbon, aldotriose has two enantiomers, D-aldotriose or L-aldotriose. One enantiomer is a mirror image of the other. In carbohydrate chemistry, a monosaccharide is defined as a D-sugar if the asymmetrical carbon atom farthest from the carbonyl group has the same configuration as D-aldotriose. Similarly, a monosaccharide is defined as an L-sugar if the asymmetrical carbon atom farthest from the carbonyl group has the same configuration as L-aldotriose. Fischer projections are commonly used to describe the structures of carbohydrates. Figure 5.18 shows several examples of the configurations of D- and L-sugars in Fischer projections. Most naturally occurring monosaccharides are D-sugars. The most biologically relevant sugars among them are pentoses and hexoses, including arabinose, xylose, ribose, glucose, mannose, galactose, and fructose, which are depicted in Figure 5.19.

As mentioned in Section 5.3.4, the hydroxyl and carbonyl groups in most carbohydrates tend to form intramolecular hemiacetals or hemiketals with five- or six-membered rings. These five- or six-membered cyclic hemiacetal or hemiketal isomers of monosaccharides are called furanoses and pyranoses, respectively. Their steric structures can be presented as Haworth projections. In the formation of a cyclic structure, the carbonyl carbon of a carbohydrate becomes a new asymmetrical carbon atom, which is called an anomeric carbon. This results in each cyclic hemiacetal or hemiketal isomer having two conformations, an α and β form. Both forms coexist with the intermediate open-chain monosaccharide in aqueous solution. In a Haworth projection, the hydroxyl group attached to the anomeric carbon is projected down for the α form, whereas the hydroxyl group is projected up for the β form. Like cyclohexane in a chair conformation, the six-membered rings of monosaccharides can also be presented as chair conformations. In this case, the hydroxyl group attached to the anomeric carbon is vertical and the other substituents are horizontal for the α form, whereas all the substituents are horizontal for the β form. Figure 5.20 shows the Haworth projection and chair conformation of D-glucopyranose.

Monosaccharides undergo a series of reactions, including oxidation reactions into carboxylic acids or dicarbonyl compounds, reduction reactions into polyols, condensation reactions with amine

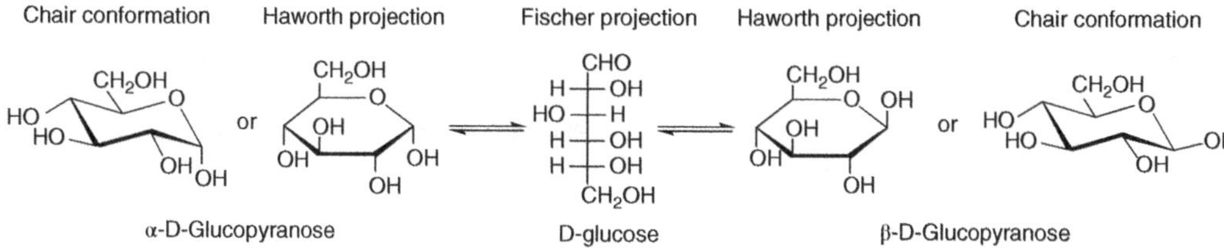

FIG. 5.20   The Haworth projection and chair conformation of D-glucopyranose.

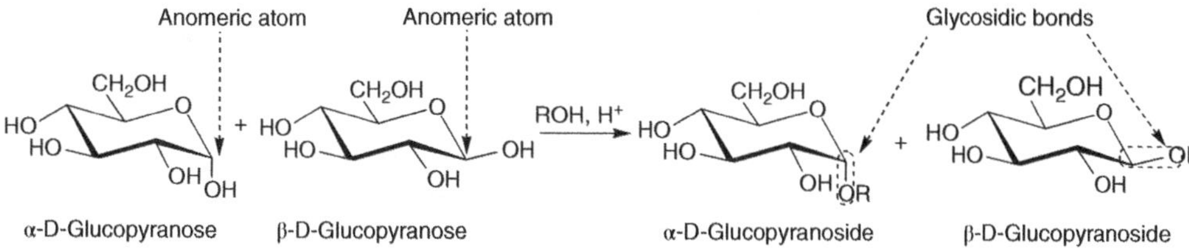

FIG. 5.21   Selective reactions for the formation of glycosides.

derivatives to form osazones, esterification, etherification, and selective reactions with alcohols under an acidic condition to form acetals, which are called glycosides in carbohydrate chemistry. A glycosidic bond is a chemical bond linking an anomeric carbon atom of a carbohydrate and a hydroxyl oxygen atom of an alcohol via the loss of one molecule of water (i.e., a dehydration reaction) during the synthesis of a glycoside. The glycosidic bond of a glycoside is an important chemical bond by which two or more monosaccharides can be connected to form oligosaccharides, such as disaccharides and polysaccharides. Figure 5.21 illustrates the selective reactions for the formation of glycosides.

5.6.2 Oligosaccharides

Oligosaccharides are polymeric carbohydrates consisting of two to ten monosaccharide units. Disaccharides are the most common oligosaccharides. A disaccharide is formed by the connection of two monosaccharides via a glycosidic bond with the loss of one molecule of water (i.e., a dehydration reaction). Depending on the type of monosaccharide units and the position of the glycosidic bond in *disaccharides*, they have different structures and physical properties. For example, sucrose obtained from sugarcane or sugar beet is formed by the connection of α-glucose and β-fructose in an α-C1–O–C2′ bond; lactose produced from animal milk is formed by the connection of β-galactose and α-glucose in a β-C1-O-C4′ bond; maltose from malted barley is formed by the connection of two α-glucoses in an α-C1–O–C4′ bond; and *cellobiose* derived from cellulose is formed by the connection of two β-glucoses in a β-C1–O–C4′ bond. Figure 5.22 shows the chemical structures of these four disaccharides.

5.6.3 Polysaccharides

Polysaccharides are polymeric carbohydrates consisting of dozens to thousands of monosaccharide units. In terms of their function in living organisms, polysaccharides can be classified into storage polysaccharides (e.g., starch) and structural polysaccharides (e.g., cellulose and hemicellulose).

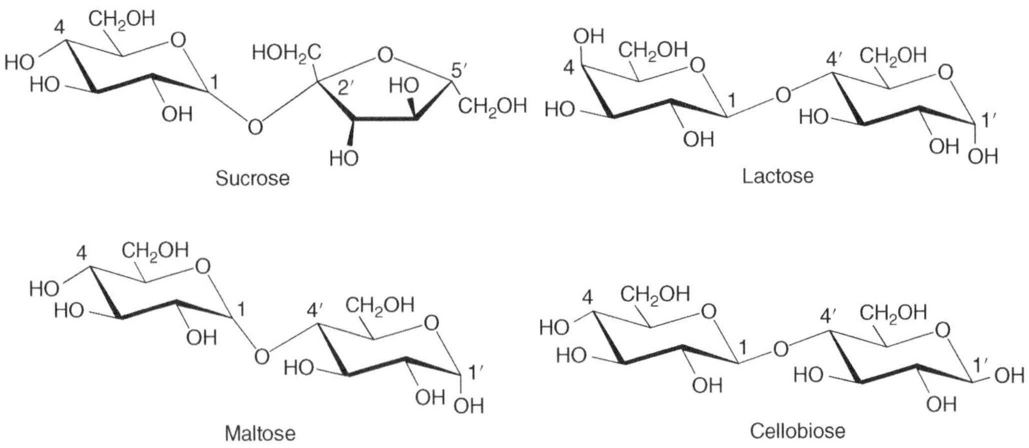

FIG. 5.22   Chemical structures of sucrose, lactose, maltose, and cellobiose.

Commercially, starch is mainly produced from crops such as potatoes, cassava, and grains, and is composed of approximately 20% amylose and 80% amylopectin. Amylose is a linear polymer formed by the linkage of multiple α-glucose units via α-C1–O–C4′ bonds. Amylopectin is a branched polymer in which α-glucose side chains are attached to a linear α-glucose main chain via α-C1–O–C6′ bonds. The linear α-glucose chains of amylopectin are formed by the linkage of multiple α-glucose units via α-C1–O–C4′ bonds. Figure 5.23 shows the chemical structures of amylose and amylopectin. Cellulose, one of the major components of lignocellulosic biomass, is a linear polymer of β-glucoses connected via multiple β-C1–O–C4′ bonds. Hemicellulose, another major component in lignocellulosic biomass, is a branched heteropolymer with pentose and hexose units, such as xylose, arabinose, glucose, mannose, and galactose. Xylan is a major component of hemicellulose. Its general structure is a linear backbone consisting of D-xylopyranose units via β-C1–O–C4′ linkages (Figure 5.23).

5.7 Proteins and Lipids

5.7.1 Proteins

As essential components of organisms, *proteins* are *biomacromolecules* formed by linking many α-amino acids via amide bonds, which are also referred to as peptide bonds in this context. The α-amino acids are carboxylic acids in which the amino group is attached to the α-carbon atom adjacent to the carboxyl group. Their general formula is R–CH(NH$_2$)–COOH, in which R can be hydrogen or other groups with carbon atoms. The simplest α-amino acid is glycine, NH$_2$–CH$_2$–COOH. The α-amino acids have amphoteric properties and zwitterionic structures (Figure 5.24). They can act as either acids or bases. An α-amino acid shows different structures in an aqueous solution, depending on the pH. At its isoelectric point (pI, also called isoelectric pH), the α-amino acid is charge-neutralized and has equal amounts of positive and negative charges.

The amide bond formation is a typical reaction that occurs among amino acids. In general, a compound formed by linking two α-amino acids is called a dipeptide, while that formed by linking three α-amino acids is called a tripeptide. Compounds formed by linking more α-amino acids are called polypeptides (Figure 5.24). Compared to polypeptides, proteins have higher molecular weights,

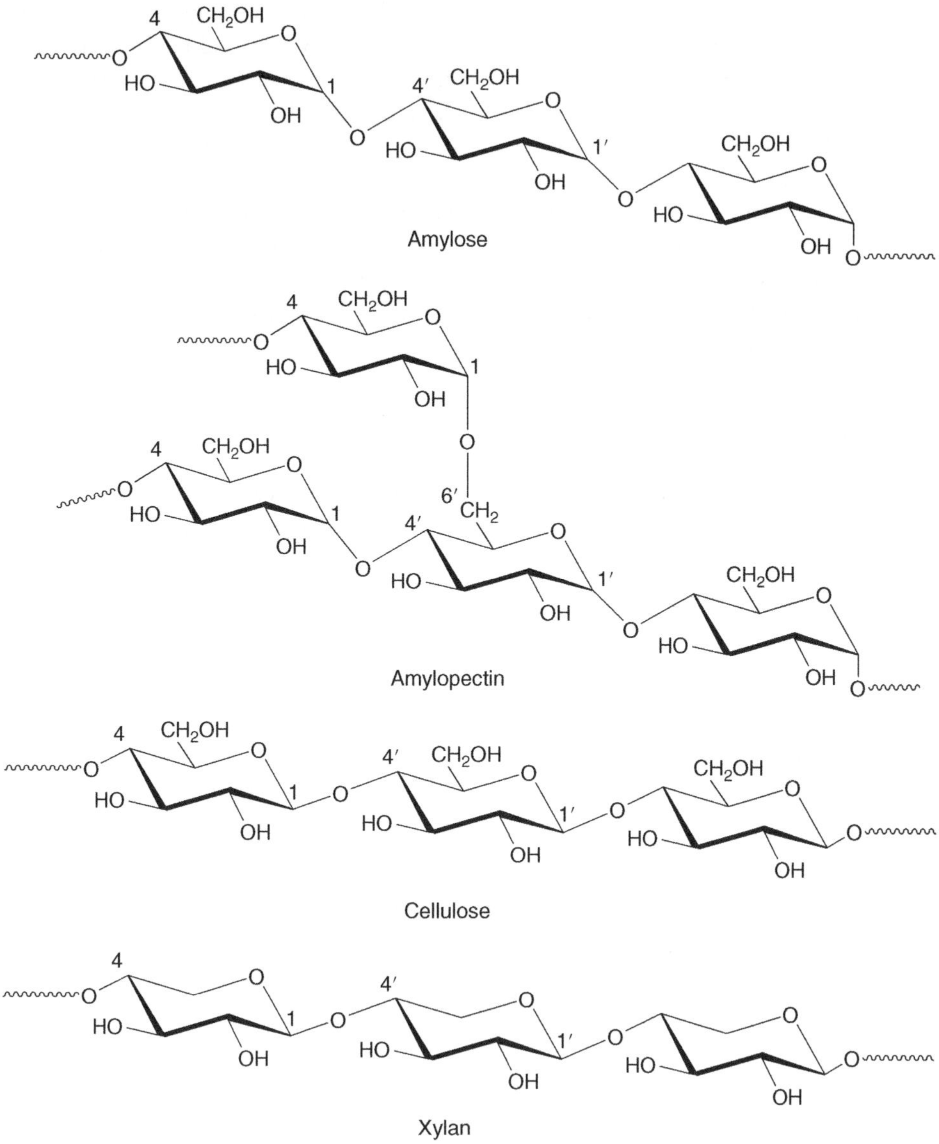

FIG. 5.23   Chemical structures of amylose, amylopectin, cellulose, and xylan.

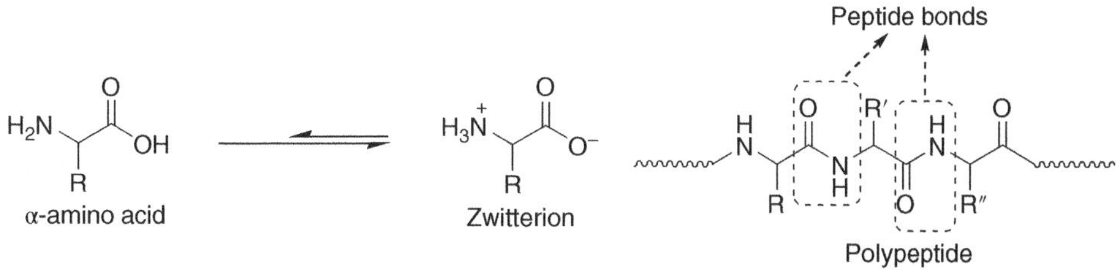

FIG. 5.24   General structures of α-amino acid, its zwitterion and polypeptide.

commonly more than 10,000 Da (1 dalton $\approx$ 1 g/mol), and are considered longer and more complex polypeptide chains. Most proteins consist of carbon, hydrogen, oxygen, and nitrogen atoms. Some proteins contain phosphorus and sulfur atoms in addition to those already mentioned.

Proteins have four levels of structure: a primary structure, which shows the amino acid sequence of a polypeptide chain; a secondary structure, which shows the folding pattern of a polypeptide chain and is formed mainly by hydrogen bonding, such as α-helix and β-pleated sheet; a tertiary structure, which is the 3D structure of a polypeptide chain that includes the spatial arrangement of its side chains; and a quaternary structure, which shows the spatial arrangement of two or more polypeptide chains in a protein. Due to the diversity of structures, proteins have various functions and biological activities, such as catalysis, regulation, storage, movement, and defense. Proteins are abundantly available in animals and plants: for example, fish, milk, eggs, algal biomass soybeans, and so on.

5.7.2 Lipids

Lipids are a class of hydrophobic compounds with different structures and functions, including animal fats, vegetable oils, waxes, phospholipids, and so on. Both animal fats and vegetable oils are triglycerides with various fatty acid compositions. Most *fatty acids* in animal fats and vegetable oils are C16 and C18 chains without double bonds, or with one or two double bonds. Compared to vegetable oils, animal fats have more saturated fatty acid chains in their *triglyceride* structures. This causes animal fats to be solid at room temperature, while vegetable oils are liquid. Table 5.2 shows the compositions of the fatty acids of some animal fats and vegetable oils.

Currently, animal fats and vegetable oils are extensively used for the production of *biodiesel* (*fatty acid esters*) via transesterification with alcohols. Methanol is the most commonly used alcohol during the production of biodiesel; that is, *fatty acid methyl esters*. Figure 5.25 shows an example of the production of biodiesel. As the production of biodiesel from vegetable oils competes with food supply,

Table 5.2 Fatty acid compositions of some animal fats and vegetable oils

Fats or oils	C14:0 (wt %)	C16:0 (wt %)	C18:0 (wt %)	C18:1 (wt %)	C18:2 (wt %)	C18:3 (wt %)
Butter	7 ~ 10	24 ~ 26	10 ~ 13	28 ~ 31	1 ~ 2.5	0.2 ~ 0.5
Lard	1 ~ 2	28 ~ 30	12 ~ 18	40 ~ 50	7 ~ 13	0 ~ 1
Soybean oil	N/A	6 ~ 10	2 ~ 5	20 ~ 30	50 ~ 60	5 ~ 11
Corn oil	1 ~ 2	8 ~ 12	2 ~ 5	19 ~ 49	34 ~ 62	trace
Rapeseed oil	N/A	3.49	0.85	64.4	22.30	8.23
Sunflower oil	N/A	6.08	3.26	16.93	73.73	N/A

Source: Data from Linstromberg, 1970; Canakci and Gerpen, 2001.
N/A: Not available

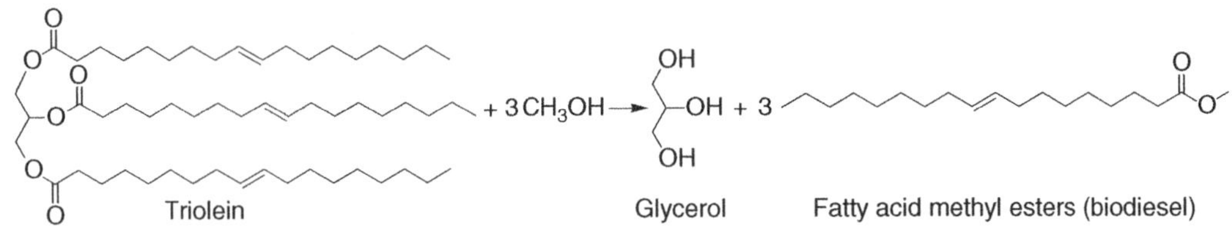

FIG. 5.25   The production of biodiesel from triolein.

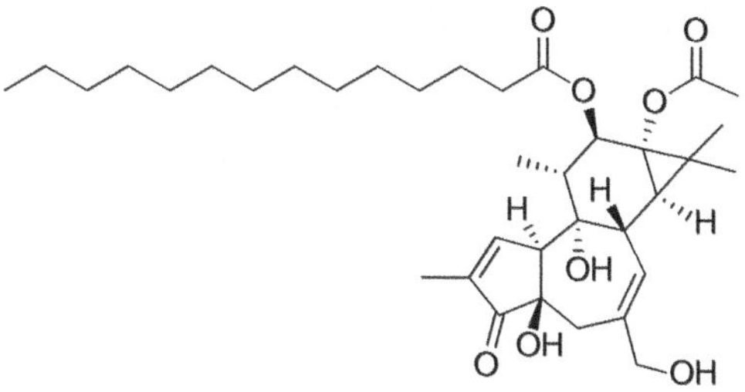

FIG. 5.26   Chemical structure of phorbol-12-myristate-13-acetate.

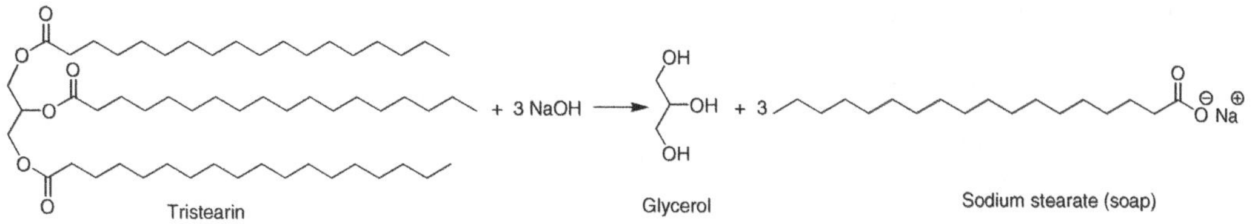

FIG. 5.27   The production of soap from tristearin.

increased attention has been paid to the use of non-edible plant oils as feedstocks for biodiesel production, such as *jatropha* and *algal oils*. Phorbol esters are toxic lipids that are responsible for the non-edible property of jatropha oil. They are extracted from jatropha oil for various applications in medicinal, pharmaceutical, and agricultural areas. Phorbol-12-myristate-13-acetate is the most common phorbol ester in jatropha oil. Its chemical structure is shown in Figure 5.26.

Animal fats have also been used to produce soaps, which are sodium or potassium salts of corresponding fatty acids, via *saponification* reactions with sodium hydroxide or potassium hydroxide. Figure 5.27 shows an example of the production of soap.

References and Further Reading

Amen-Chen, C., Pakdel, H., and Roy, C. (2001). Production of monomeric phenols by thermochemical conversion of biomass: A review. *Bioresource Technology*, 79: 277–99.

Berg, J.M., Tymoczko, J.L., and Stryer, L. (2002). *Biochemistry*, 5th edn. New York: W.H. Freeman.

Brown, R.C. (2003). *Biorenewable Resources: Engineering New Products from Agriculture*. Ames, IA: Blackwell Professional.

Canakci, M., and Gerpen, J.V. (2001). Biodiesel production from oils and fats with high free fatty acids. *Transactions of the ASAE*, 44(6): 1429–36.

Devappa, R.K., Maes, J., Makkar, H.P.S., Greyt, W.D., and Becker, K. (2010). Quality of biodiesel prepared from phorbol ester extracted Jatropha curcas oil. *Journal of the American Oil Chemists' Society*, 87: 697–704.

France, J., and Dijkstra, J. (2005). Volatile fatty acid production. In: J. Dijkstra, J.M. Forbes, and J. France (eds.), *Quantitative Aspects of Ruminant Digestion and Metabolism*, 2nd edn., Cambridge, MA: CABI Publishing, pp. 157–76.

Linstromberg, W.W. (1970). *Organic Chemistry: A Brief Course*, 2nd edn. Lexington, MA: D.C. Heath.

Pagliaro, M., and Rossi, M. (2008). *The Future of Glycerol: New Uses of a Versatile Raw Material*. Cambridge: RSC Green Chemistry Book Series.

Sawyer, C.N., McCarty, P.L., and Parkin, G.F. (2003). *Chemistry for Environmental Engineering and Science.* New York: McGraw-Hill.

Spiridon, I., and Popa, V.I. (2008). *Hemicellulose: Major Sources, Properties and Applications.* Oxford: Elsevier.

Vollhardt, K.P.C., and Schore, N.E. (2002). *Organic Chemistry: Structure and Function,* 4th edn. New York: W.H. Freeman.

Walter, W. (1997). *Organic Chemistry.* Chichester: Albion Publishing.

Exercise Problems

5.1. What are the sources of organic compounds?

5.2. How are the properties of organic compounds different from those of inorganic compounds?

5.3. List all alkanes that are gases at room temperature.

5.4. Ethanol is an important compound, which can be produced from either petroleum or natural resources. List the major reactions that alcohols undergo using ethanol as an example.

5.5. Lignin is a complex aromatic polymer and exists abundantly in biomass. List the primary lignin monomers and functional groups in the structure of lignin.

5.6. Draw the chemical structures of four common C18 fatty acids.

5.7. Define carbohydrates and describe their classification.

5.8. Draw the Fischer projection, Haworth projection, and chair conformation of D-glucose.

5.9. The glycosidic bonds play key roles in the formation of oligosaccharides and polysaccharides. Draw the structures of four disaccharides, such as sucrose, lactose, maltose, and cellobiose, and indicate the glycosidic linkages between their monosaccharide units.

5.10. Explain the similarities and differences between proteins and peptides.

5.11. Animal fats and vegetable oils have been extensively used in the biodiesel industry. Write the reaction for the production of biodiesel using linoleic acid to represent the fatty acid chain in the triglyceride and methanol as the solvent.

Plant Structural Chemistry

Samir Kumar Khanal, Saoharit Nitayavardhana, and Rakshit Devappa

---

What is included in this chapter?

This chapter focuses on the cell wall of plants, its composition and structure. The architecture, carbohydrate classification, carbohydrate chemistry, and critical role of carbohydrates in the cell wall are also discussed.

6.1 Introduction

Plant growth primarily takes place through a complex process known as photosynthesis, wherein energy from the sun (light energy) is harvested and converted into chemical energy – ATP (adenosine triphosphate) and NADH (nicotinamide adenine dinucleotide). Plants use this chemical energy as a resource to produce carbohydrates and other organic compounds from carbon dioxide (inorganic carbon) and water. Photosynthesis has been responsible for producing food for millions of years, and continues to sustain an ever-growing world population in addition to countless other animals and insects. In fact, life on Earth is infinitely fortunate, since plant biomass is the most abundant (renewable) resource on the planet. Globally, over 200 billion metric tons of terrestrial plant biomass are available each year, and about 70–80% of the biomass consists of cell wall components (cellulose, hemicellulose, and lignin), which are considered to be renewable bioresources for producing bioenergy and biobased products. Terrestrial plants also fix nearly 56 billion metric tons of CO_2 annually, a major greenhouse gas that has been linked to climate change (Kalluri *et al.*, 2010). Presently, humans only utilize 2% of all plant cell wall biomass. Therefore, there is significant potential to exploit these abundant bioresources as raw material for the generation of sustainable bioenergy and other biobased products.

The plant cell wall consists of a rigid and tough layer located outside of the plasma membrane, which surrounds the organelles. The wall provides structural support, protects against toxic substances/invaders, and controls permeability and intracellular pressure. Virtually all plant cell walls are made up of complex (water-insoluble) polysaccharides, namely cellulose and hemicellulose,

---

Bioenergy: Principles and Applications, First Edition. Edited by Yebo Li and Samir Kumar Khanal.
© 2017 John Wiley & Sons, Inc. Published 2017 by John Wiley & Sons, Inc.
Companion website: www.wiley.com/go/Li/Bioenergy

covalently linked and reinforced with polyphenols such as lignin (Yarbrough *et al.*, 2009; Pauly and Keegstra, 2008; Higuchi, 1997). The efficient conversion of plant biomass into bioenergy and bio-based products requires an in-depth understanding of plant structures and their chemical composition, which is the primary focus of this chapter. It is equally important to note that although many similarities exist among a vast array of plant structures, cell wall compositions can and will differ in response to climatic conditions (e.g., soil, nutrient availability, water, sunlight, and stress, among others), species, and age. These variabilities, however, are comprehensive and cannot be fully covered in this chapter.

6.2 Carbohydrates and Their Classification

Carbohydrates (i.e., sugars) are the primary building blocks of plant biomass, consisting mostly of carbon, hydrogen, and oxygen as basic elements with the empirical formula $(CH_2O)_n$. Chemically, they are known as polyhydroxy aldehydes or ketones. Carbohydrates can be broadly classified into three major classes, namely monosaccharides, oligosaccharides, and polysaccharides. The word "saccharide" is derived from the Greek word *sakcharon*, meaning sugar. Carbohydrates have different functions in nature, ranging from energy storage and production to mechanical support and physical protection, among others (Izydorczyk, 2005). In particular, the insoluble structural carbohydrates, which are found in cell walls, are of significant importance for bioenergy and biobased products generation. A brief discussion of carbohydrate chemistry with a focus on lignocellulosic biomass is presented here, together with the general overview of carbohydrate chemistry presented in Chapter 5.

6.2.1 Monosaccharides

The term *carbohydrate* can be used to refer to sugars in their simplest fundamental form; that is, a monosaccharide. Monosaccharides can be further classified based on their number of carbon atoms, as shown in Table 6.1, or by their aldehyde (aldose) and ketone (ketose) groups. If the carbonyl group is present at the terminal end of the carbon chain (an aldehyde group), the monosaccharide is known as an aldose. If the carbonyl group is at other positions (in a ketone group), the monosaccharide is termed a ketose. All common monosaccharides have names ending with the suffix "*-ose*". The simplest monosaccharides are the two three-carbon sugars: glyceraldehydes (an aldose) and dihydroxyacetone (a ketose).

Six-carbon sugars (hexoses) such as glucose, fructose, galactose, and mannose, and five-carbon sugars (pentoses) such as xylose and arabinose, are the most common monosaccharide building

Table 6.1 Classification of monosaccharides and their derivatives based on number of carbon atoms

No. of carbon	Category name	Example
3	Triose	Glyceraldehyde, dihydroxyacetone
4	Tetrose	Erythrose, threose, erythrulose
5	Pentose	Ribose, ribulose, xylose, xylulose, arabinose
6	Hexose	Glucose, galactose, mannose, fructose
7	Heptose	Sedoheptulose
8	Octose	Octulose
9	Nonose	Neuraminic acid, also known as sialic acid (derivative)

blocks present in plants, and are important monosaccharides for both bioenergy and biobased products generation.

All *hexoses* have the same chemical formula, $C_6H_{12}O_6$, but their atoms can be arranged differently. Monosaccharides of more than four carbons tend to have a cyclic structure, but can exist either in a ring form or in an open chain form. In an open chain form, glucose, galactose, and mannose have aldehyde functional groups, whereas fructose, in comparison, has a ketone functional group (Figure 6.1). Glucose, galactose, and mannose differ only in their placement of the H and OH groups in one of the carbon atoms (Figure 6.1). This difference, however, can be significant in determining what role the sugar plays in the plant structure. Similarly, all *pentoses* also have the same chemical formula, $C_5H_{10}O_5$, with different arrangements of atoms. Pentose sugars, including xylose and arabinose, are all aldoses (Figure 6.1).

6.2.2 Oligosaccharides

Carbohydrate molecules consisting of 2–10 monosaccharides are commonly known as *oligosaccharides*, but to avoid confusion it should be noted here that the exact number of carbons in oligosaccharides can be somewhat subjective from textbook to textbook. Technically speaking, all disaccharides are oligosaccharides, but the opposite is not necessarily true. The most common disaccharides of interest in commercial processes include sucrose, cellobiose, and maltose (Figure 6.1). To form a disaccharide, two monosaccharides are joined together by a condensation reaction, also known as a dehydration reaction. Examples of oligosaccharides and their covalent linkages are presented in Table 6.2.

6.2.3 Polysaccharides

Most carbohydrates found in nature occur as *polysaccharides*. As mentioned previously, they are the main constituents of plant biomass (making up the cell wall) and are of significant importance for bioenergy and biobased products generation. Polysaccharides are long polymers of more than 20 monosaccharide units joined together by glycosidic bonds. Polysaccharides, also called *glycans*, are further classified into *homopolysaccharides* (containing only a single type of monomer) or *heteropolysaccharides* (containing two or more different monomers). Both homopolysaccharides (cellulose) and heteropolysaccharides (hemicellulose and pectin) in plant cells serve as a structural component of the cell wall. The hydrolysis (namely, the addition of a water molecule to lyse glycosidic bonds) of cellulose produces exclusively glucose, whereas galactose, mannose, xylose, and arabinose are released from the hydrolysis of hemicellulose (Figure 6.1) (Vanholme *et al.*, 2013; Yarbrough *et al.*, 2009; Izydorczyk *et al.*, 2005; Minorsky, 2002). A detailed discussion of these structural carbohydrates is provided in the following sections.

6.3 Main Constituents of Plant Biomass

The main architecture of most plants comprises of four major chemical compounds: *cellulose, hemicellulose, pectin*, and *lignin*. These components are arranged in a complex cell wall structure providing rigidity to the plant cell. For effective utilization of *lignocellulosic biomass* for bioenergy and bioproduct generation, the non-carbohydrate compound (lignin) of the plant either needs to be removed from the structural polysaccharides (cellulose, hemicellulose, and pectin) or the interaction of cellulose, hemicellulose, and lignin needs to be disrupted. Cellulose and hemicellulose are the primary polysaccharides of interest since they serve as precursors for a number of end-user products, which include an array of renewable transportation fuels. Although present in small quantities, pectin also has uses in a number of food applications.

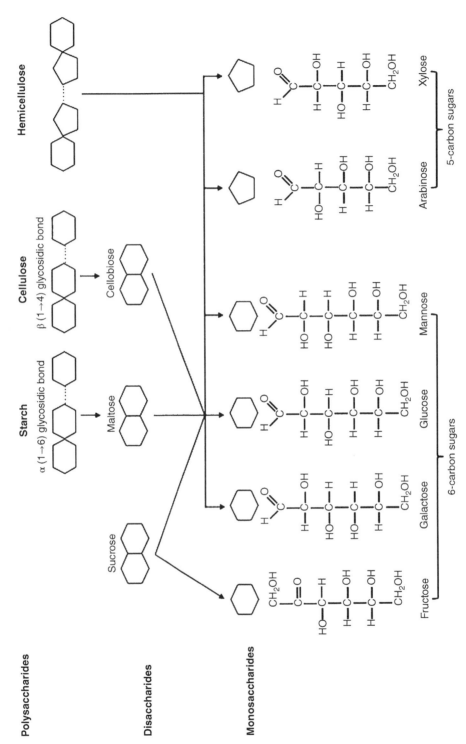

FIG. 6.1 Storage and structural carbohydrates in plants.

Table 6.2 Examples of oligosaccharides and their linkages

Oligosaccharide	Molecule	Linkage
Sucrose	α-glucose and β-fructose	$\alpha(1\rightarrow2)$
Lactose	galactose and glucose	$\beta(1\rightarrow4)$
Maltose	α-D-glucose and α-D-glucose	$\alpha(1\rightarrow4)$
Trehalose	α-D-glucose and α-D-glucose	$\alpha(1\rightarrow1)$
Cellobiose	β-D-glucose and β-D-glucose	$\beta(1\rightarrow4)$

6.3.1 Structural Carbohydrates

You may recall from the earlier discussion that carbohydrates are among the most abundant polymers on Earth. However, carbohydrates themselves exist in several different forms and serve a range of functions: some carbohydrates are incorporated into a plant's structure, while others exist as a means for energy storage. For example, during active *photosynthesis*, triose phosphates are synthesized for producing both energy and metabolic precursors. Any excess generated is converted to sucrose (a soluble non-structural sugar) and various other monosaccharides. Some plants are able to store their energy in the form of a water-insoluble carbohydrate, namely starch. This type of carbohydrate, however, does not contribute to the plant cell structure and will not be discussed in this chapter.

Chemical composition of wood

Like many plant-derived resources, wood is also composed of cellulose, hemicellulose, and lignin, among other compounds. The distribution and quantity of these constituents, however, vary depending on the wood species. **Cellulose** is the major component of all woody biomass and represents 40–45% (dry wt.). Cellulose consists of linear chains of D-glucose connected by $\beta(1\rightarrow4)$ glycosidic bonds, with a degree of polymerization of 10,000 units in native wood. The strong tendency of cellulose to form intra- and intermolecular hydrogen bonds via hydroxyl groups along the cellulose chains imparts crystallinity to the structure. In contrast, **hemicellulose** has a much lower degree of polymerization (50–300 units) with side groups on the chain molecule and is relatively amorphous. The thermal and chemical stability of hemicellulose is thus much lower than cellulose due to its lack of crystallinity. The hemicelluloses of softwoods are primarily galactoglucomannans and arabinoglucuronoxylan, while hardwoods contain mainly glucuronoxylan. **Lignin** is another important constituent of wood, which contributes to water resistance and structural integrity. Softwoods tend to have a higher lignin content of 25–30% compared to hardwoods (20–25%).

How does an engineer visualize lignocellulose?

As depicted in Figure 6.2, the plant structure can be beautifully compared with reinforced man-made structures. The cellulosic microfibrils resemble the vertical rebars, and the hemicellulose, which cross-links the cellulose fibrils, resembles the cross-sectional tie bars. Finally lignin, which acts as a binding agent in plants, can be thought of as the cement.

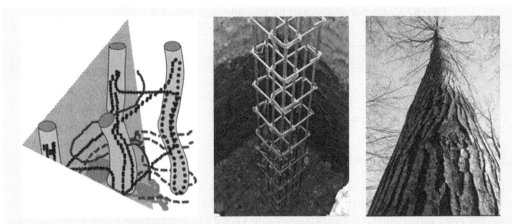

FIG. 6.2   Plant and man-made structures compared. Photos courtesy of Prachand Shrestha, Iowa State University.

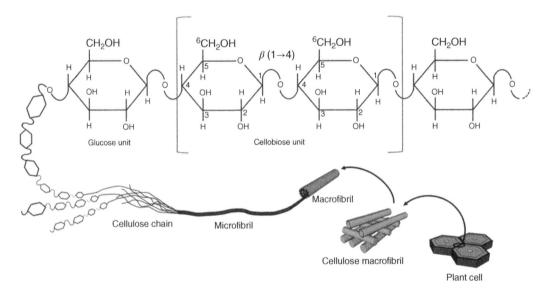

FIG. 6.3   Cellulose structure.

Cellulose

Cellulose is a linear (unbranched) homopolysaccharide consisting of 10,000–15,000 D-glucose units linked by $\beta(1 \rightarrow 4)$ bonds, and is represented by the chemical formula $(C_6H_{10}O_5)_n$. Although cellulose is exclusively made up of glucose monomers, its fundamental building block is considered to be repeating cellobiose dimers (i.e., two glucose molecules linked by a $\beta(1 \rightarrow 4)$ bond; Figure 6.3). The β configuration of the glucose residues in cellulose and cellobiose results in very different structures and physical properties when compared to starch, another homopolysaccharide of glucose. The orientation of the chemical bonds also makes cellulose indigestible for most animals (except ruminants), as they do not have the proper enzymes (cellulases) to hydrolyze $\beta(1 \rightarrow 4)$ linkages. Similar to other biochemical molecules, cellulose subunits are organized into three-dimensional formations for biological functions. As shown in Figure 6.3, because polysaccharides have several hydroxyl groups, hydrogen bonding has an especially important influence on the three-dimensional structure. Cellulose chains (about 36 chains) associate through H-bonds, resulting in the formation of microfibrils, and the microfibrils can further interact with each other to form macrofibrils (visible to the naked eye). It is the H-bonding that gives high tensile strength to the cellulosic structures (i.e., fibers). This is not to be confused with the β configuration of the glucose residues, which provides rigidity to the cellulose chain at the molecular level. The position of H-bonds within and between the strands results in different types of crystalline structures, which will be discussed later. The decomposition of cellulose to produce glucose molecules requires two specific enzymes, cellulase and cellobiase. Cellulase first breaks down cellulose chains into the disaccharide cellobiose. Cellobiose is then hydrolyzed by the cellobiase, resulting in two molecules of glucose.

Hemicellulose

Hemicellulose is a heteropolysaccharide that can be removed from the plant cell wall by alkaline or acid pretreatment. Unlike cellulose, the hemicellulose structure can be relatively complex. It is a highly branched polymer containing a number of five- and six-carbon sugars. Because hemicellulose is a highly branched molecule, it forms a rather amorphous structure in the plant cell wall, and interacts with cellulose and lignin to reinforce the cell wall. The sugar composition of hemicellulose varies considerably depending on the plant species, but, in general, the monosaccharides obtained from the hydrolysis of hemicellulose include glucose, galactose, mannose, arabinose, xylose, and rhamnose.

Pectin

Pectin is a heteropolysaccharide usually composed of α-D-galacturonic acid (Gal*p*A) residues linked together by α(1 → 4) glycosidic linkages. Some pectin polysaccharides contain a number of L-rhamnose residues, resulting in a much more highly branched structure. Unlike cellulose, pectin polysaccharides, similar to hemicellulose, have no defined structure and the composition of pectin varies significantly between plant species. Because pectin has the ability to gel in the presence of high sugar concentrations, it is found in the primary cell wall and middle lamella to help bind adjacent cells together. In fact, the word pectin is derived from the Greek word *pēktikós*, which means to congeal or thicken. There is a detailed discussion of pectin in Section 6.4.

6.3.2 Lignin

Lignin is a complex phenolic compound containing a large number of aromatic rings without defined repetitive units (Figure 6.4). The word lignin is derived from the Latin word *lignum*, meaning wood. It is a phenylpropane-based polymer and is the largest non-carbohydrate fraction of most plant biomass. In fact, lignin is the second most abundant polymer present in the cell wall of vascular plants on Earth (just behind cellulose). It provides mechanical support and water impermeability to the secondary cell walls (Figure 6.5). Lignified cells can vary in size, lignin content, and location in the plant. The lignin content can vary from 10–30% (dry wt.) for most lignocellulosic biomass feedstocks. The precursors of lignin are three p-hydroxycinnamyl alcohols or monolignols (p-coumaryl, coniferyl, and sinapyl) and their acylated forms. These lignols are integrated into the lignin polymer as phenylpropanoids – p-hydroxyphenyl (H), guaiacyl (G), and syringyl (S), respectively. Lignin in gymnosperms contains entirely G with small quantities of H, whereas dicot angiosperms more often contain a mixture of G and S, and low quantities of H. In monocot angiosperms, lignin contains a mixture of H, G, and S. Most of the grass species commonly used for bioenergy production are lignin rich in G. In addition, lignin also contains incomplete/modified monolignols in smaller proportions.

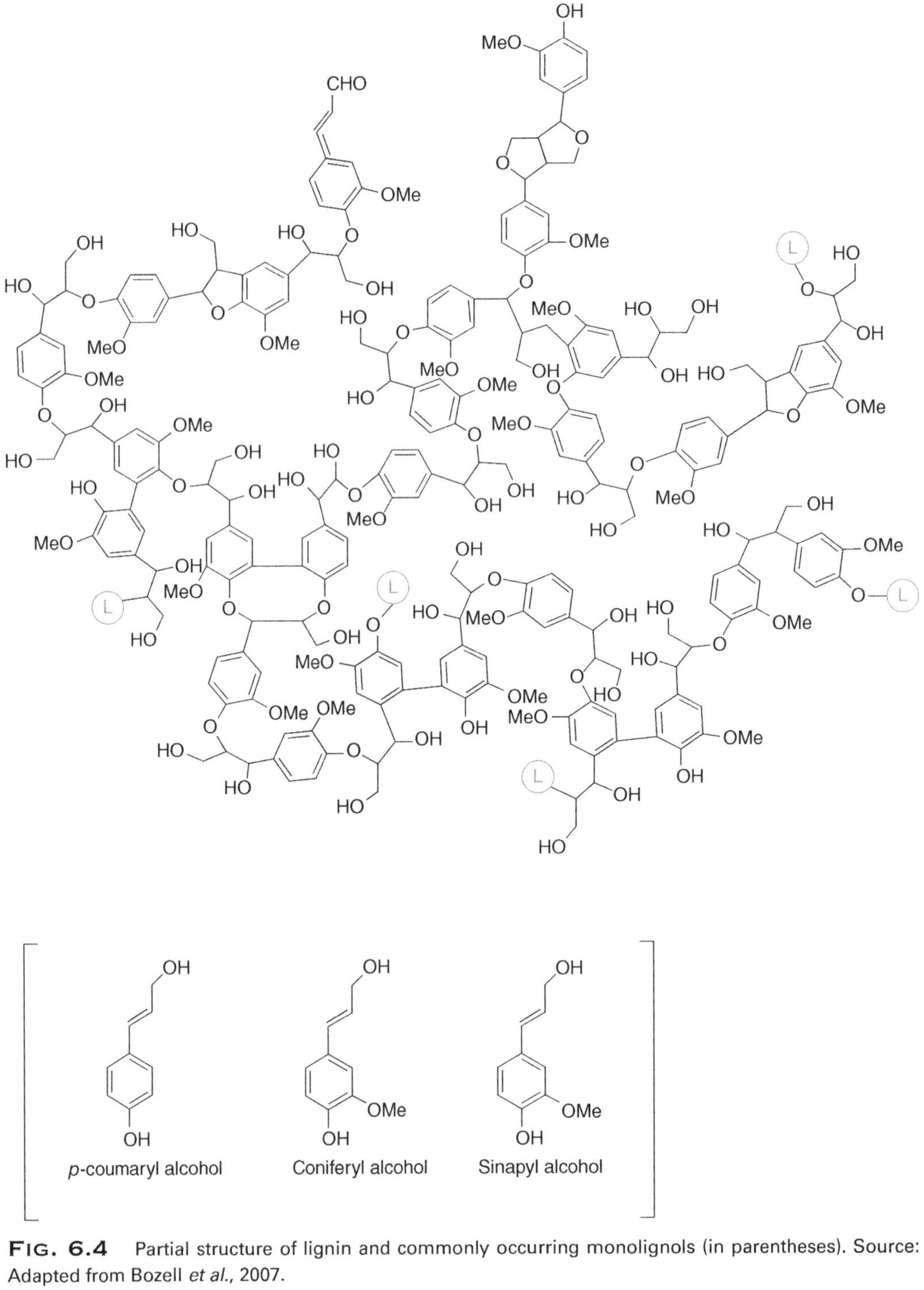

FIG. 6.4 Partial structure of lignin and commonly occurring monolignols (in parentheses). Source: Adapted from Bozell *et al.*, 2007.

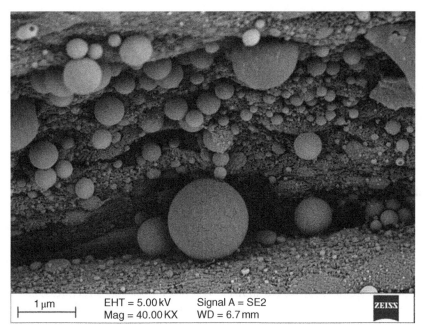

FIG. 6.5   Lignin droplets on liquid hot water–treated sugarcane bagasse. Photo courtesy of Prashant Reddy, Sugar Milling Research Institute NPC, Durban, South Africa.

Lignin polymerization is formed due to many ether and carbon–carbon interunit linkages, resulting in complex substructures. The lignin further binds to hemicelluloses by both covalent and non-covalent bonds. This forms a compact network surrounding the cellulose molecule by extensive chemical and hydrogen bonding. The interactions of cellulose, hemicellulose, and lignin impart high recalcitrance to the biomass, and pose an obstacle for the commercial bioconversion processes of biofuel and biobased product generation, as they hinder the enzyme-driven hydrolysis of structural carbohydrates (Ragauskas *et al.*, 2014; Martínez *et al.*, 2008; Ralph *et al.*, 2004; Mooney *et al.*, 1998).

Plant-stored carbohydrates for ethanol production

Plants store energy in the form of carbohydrates such as sugars and starch, which have been commercially exploited for bioethanol production. The most well-known sugar found in plants is a disaccharide called sucrose. This molecule consists of one glucose monomer and one fructose monomer linked by a glycosidic bond, and can be extracted from sugarcane, sugar beet, and sweet sorghum by pressing for ethanol production. Like most organisms, traditional brewing yeast, namely *Saccharomyces cerevisiae*, must first break down sucrose into monomers before metabolizing the sugar to ethanol. The yeast releases a special enzyme called invertase, located in its periplasm, to hydrolyze sucrose into glucose and fructose. The monosaccharides are then transported into the cell and undergo glycolysis and fermentation.

In some plants, such as corn and cassava, energy is stored in the form of starch. Starch consists of thousands of covalently bonded glucose monomers, thus making it a homopolysaccharide. Unlike sucrose fermentation into ethanol, ethanol production from starch requires an additional step prior to fermentation, called hydrolysis, to break down the starch polymer into fermentable sugars. Traditional yeast does not contain enzymes for starch hydrolysis, thus extracellular enzymes known as amylases (alpha- and gluco-amylases) are needed.

6.4 Plant Cell Wall Architecture

Lignocellulosic feedstocks are likely to play an important role in the emerging bioeconomy as the primary substrate for bioenergy, and in the production of a plethora of biochemicals and biobased products. Because the overall plant comprises microscopic cells working synergistically, it is crucial to have a deeper understanding of these fundamental "building blocks" in order to appreciate what we can observe macroscopically. Figure 6.6 shows a typical plant cell wall structure. Unlike in animal cells, the outermost boundary of each individual plant cell is the cell wall, which primarily consists of cellulose, hemicellulose, lignin, and pectin. The wall contains up to three layers: primary cell wall, secondary cell wall, and middle lamella. The primary and secondary cell walls surround the cytoplasm of each individual plant cell, while the pectin-rich middle lamella forms the outermost layer and acts as a glue to join adjacent plant cells (Ek *et al.*, 2009). The cell wall covers the plasma membrane, and provides strength and protection against mechanical and osmotic stress (Cooper, 2000). Given its chemical (carbohydrate-rich) composition, the precursors essential for biofuel production are derived mostly from the cell wall.

6.4.1 Primary Cell Wall

The *primary cell wall* is present in most growing (younger) cells and is extensible. It gives structural and mechanical support, resistance to internal cellular turgor pressure, and protection against pathogens; controls the rate and direction of growth; maintains and determines the cell shape; and facilitates cell-to-cell interactions, among others. The primary cell wall consists of a rigid skeleton of

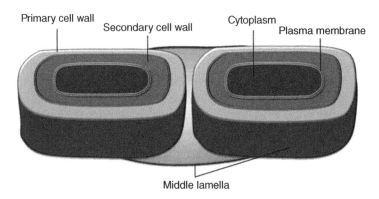

FIG. 6.6   Schematics of a typical plant cell wall structure.

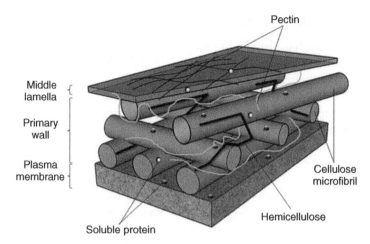

FIG. 6.7   Schematics of the primary cell wall structure. Source: Sticklen, 2008.

cellulose microfibrils connected via hemicellulosic chains to form the cellulose–hemicellulose network, which is embedded in the pectin matrix (Brett and Waldron, 1996), as shown in Figure 6.7. In regard to biofuel and biobased products applications, the following section will cover only the main carbohydrates contributing to the primary cell wall structure. The major polysaccharides in the primary cell wall are cellulose, hemicellulose, and pectin (see subsequent sections). Xyloglucan is the most common hemicellulose found in the primary cell wall. In grass cell walls, however, both xyloglucan and pectin are found in low quantities, and are often partially replaced by glucuronarabinoxylan, another type of hemicellulose.

Cellulose in the Primary Cell Wall

Cellulose is the main fraction in both the primary and secondary cell walls. Usually primary and secondary cell walls contain 20–30% and 45–50% cellulose, respectively, except in some specialized plants in which the cell wall is mostly the secondary cell wall (e.g., cotton, >95%; wood, ~40–50%; and hemp, 75%). As mentioned previously, the H-bonding results in the formation of cellulose macrofibrils, and the position of H-bonds within and between the strands results in different types of cellulosic structures. Natural cellulose is called cellulose-I with structures I_α and I_β, whereas cellulose present in regenerated cellulose fibers is called cellulose-II. Cellulose-II is more stable than cellulose-I, resulting from H-bonding variations. A number of chemical pretreatments can also produce structures like cellulose-III and cellulose-IV. In addition, the chain length and degree of polymerization influence the properties of cellulose. Cellodextrin (the breakdown product of cellulose) is generally soluble in water and organic solvents, whereas cellobiose (a disaccharide) is soluble in water, but not in organic solvents such as ether and chloroform. In wood, cellulose fibers are bound to a complex polymer called lignin (Section 6.3.2), adding a degree of hydrophobicity. To remove the lignin, wood pulp is treated with alkali or bisulfites in pulp and paper industries. The pulp is then pressed to matte the cellulose fibers together. Interest in this and similar processes is not only limited to the use of cellulose fiber for paper. Cellulose and hemicellulose are currently being examined for biofuel and biobased products generation (Mittal *et al.*, 2011; Caffall and Mohnen, 2009; Gajera *et al.*, 2008).

Hemicellulose in the Primary Cell Wall

Hemicellulose is a heteropolysaccharide consisting of hexoses and pentoses that are cross-linked to cellulose fibrils, as shown in Figures 6.7 and 6.8. It is abundantly present in both the primary and secondary cell walls, and is often highly branched, hydrophilic, and amorphous. Hemicellulose can be easily hydrolyzed by dilute acid, alkaline, or hot water pretreatments to release its monomeric sugars. The polysaccharide contains glucose, galactose, rhamnose, arabinose, and mannose, but xylose is often the primary sugar. Hemicellulose, which has a high quantity of xylose or arabinose, is referred to as xyloglucans or arabinoglucans, respectively. Softwood hemicelluloses typically have more mannose and galactose, and fewer xylose units than hard woods (Caffall and Mohnen, 2009; Gajera *et al.*, 2008).

Xyloglucans (XG; molecular weights up to 200,000 kD) account for nearly 30% of the primary cell walls in dicotyledons and <2% in *Poaceae* (grass family). They strongly bind to cellulose through H-bonds and attach to the surface of cellulose microfibrils in the growing plant cell wall. This permits the formation of xyloglucan bridges between the cellulose microfibrils and helps to reduce the ability of cellulose microfibrils to aggregate laterally into very large fibrils. The resulting cellulose–hemicellulose network forms a framework around which other cell wall macromolecules become organized. The hemicellulose–cellulose network also serves to limit the flexibility of the cell wall by linking adjacent microfibrils and preventing them from sliding against each other. Xyloglucans consist of a β(1 → 4)-linked glucose chain as a backbone, in which 75% of glucose molecules are attached (at C-6 position) to a side chain comprising α-D-xylose (Figure 6.8). Furthermore, this α-D-xylose side chain residue has a tendency to attach to β-D-galactosyl or α-L-fucosyl-(1 → 2)-β-D-galactosyl units at the C-2 position. The side chain composition and its acetylation vary with different plant species. For example, xyloglycans from the *Solanaceae* family do not contain a fucosyl residue, but have galactosyl residue (small amounts) in their side chains. In addition, some of the xylosyl side chain residues contain α-L-arabinosyl residues at the C-2 position. In *Poaceae*, xyloglucans lack fucosyl and arabinosyl residues, but contain a few galactosyl residues. Also in *Poaceae*, the majority of xyloglucan side chains are single-terminal α-D-xylosyl residues (Taiz and Zeiger, 2010; Caffal and Mohnen, 2009; Hon and Shiraishi, 2000).

Arabinoxylans (AG) are mainly found in the primary cell walls of the grass family (*Poaceae*). They have a backbone of β(1 → 4)-linked xylosyl residues. The side chains are attached to the xylosyl backbone at positions C-2 and/or C-3 (Figure 6.9) and may include arabinosyls, galactosyls,

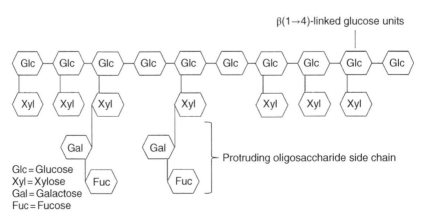

FIG. 6.8   Structure of hemicellulose (xyloglucan).

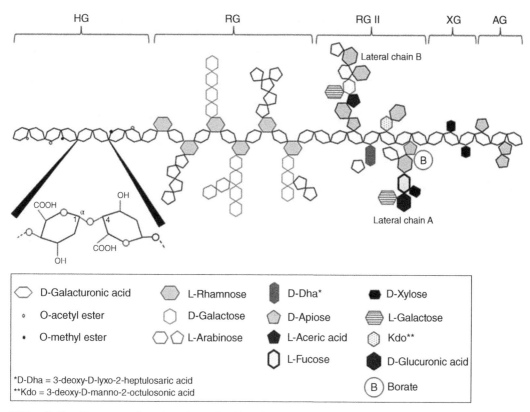

FIG. 6.9   Structure of pectin polysaccharides. Source: Adapted from Leclere *et al.*, 2013.

glucosyluronic acids, and 4-*O*-methulgluosyluronic acid residues. A majority of the arabinoxylans contain glucosyluronic acid residues or 4-*O*-methylglucosyluronic acid residues, thus imparting acidic functionality to the arabinoxylan, which can then influence its interaction with other cell wall molecules. In addition, many arabinoxyans contain phenolic acids (e.g., ferulic acid and p-coumaric acid) cross-linking the cell wall polysaccharides (Taiz and Zeiger, 2010; Caffall and Mohnen, 2009).

Pectin in the Primary Cell Wall

Pectin is a mixture of polysaccharides mostly present in the primary cell walls of lignocellulosic biomass and middle lamella, as shown in Figure 6.7. This group of polysaccharides contains $(1 \rightarrow 4)$-linked α-D-galactopyranosyluronic acid (Gal*p*A) residues. Primary cell walls generally contain three major pectic polysaccharides – homogalacturonan (HG), rhamnogalacturonan-I, and rhamnogalacturonon-II. Homogalacturonan (HG) is a linear chain consisting of $α(1 \rightarrow 4)$-linked Gal*p*A residues in which some of the carboxyl groups are methylated (Figure 6.9). HGs may also be partially *O*-acetylated at positions C-2 or C-3 (Ridley *et al.*, 2001).

Rhamnogalacturonan-I (RG-I) type polysaccharides contain alternating rhamnosyl and galactosyluronic acid residues [→4)-α-D-Gal*p*A-(1 → 2)-α-L-Rha*p*-(1→] attached to side chains. The C-2 position of rhamnosyl residues is linked to galactosyluronic acid residues and the C-4 position of rhamnosyl residues is attached to side chains. The side chain may be galactosyl, arabinosyl, and less frequently, fucosyl and glucosyluronic acid residues. The backbone Gal*p*A residues may be *O*-acetylated on C-2 and/or C-3 (Eskin and Hoehn, 2012; Komalavilas and Mort, 1989).

Rhamnogalacturonan II (RG-II) contains only $(1 \rightarrow 4)$-linked α-D-Gal$p$A residues. Some of the galactosyluronic acid residues present in the backbone of RGII are found esterified with methyl groups. The disaccharide units 3-deoxy-D-manno-octulosonic acid (KDO) and 3-deoxy-D-lyxo-2 heptulosaric acid (DHA) are attached to C-3 of the backbone, and two oligosaccharides (Figure 6.9) are attached to C-2 of the backbone. The exact sequence of the side chains attached to the backbone is not known. Two RG-II exist in the primary wall as a dimer that is cross-linked by a borate–diol diester formed between the apiosyl residues in side chain A of each monomer sub-unit, as shown in Figure 6.9 (Ridley *et al.*, 2001).

6.4.2 Secondary Cell Wall

The *secondary cell wall* is deposited when the cell elongation process ceases. Thus, the secondary cell wall is mostly found in mature cells. The secondary cell wall constitutes the major structural component of wood that holds trees upright against the force of gravity. It is generally thicker than the primary cell wall and stacked in multiple layers, as shown in Figure 6.10 (Sarkar and Bosneaga, 2009; CCRC, 2007). The most dominant polysaccharides in the secondary cell walls of common biomass feedstocks include cellulose (40.6–51.2%), hemicelluloses (28.5–37.2%), and lignin (13.6–28.1%). The detailed structural chemistry of cellulose (Sections 6.3.1 and 6.4.1) and lignin (Section 6.3.2) were previously described in this chapter and only hemicellulose in the secondary cell wall is discussed here.

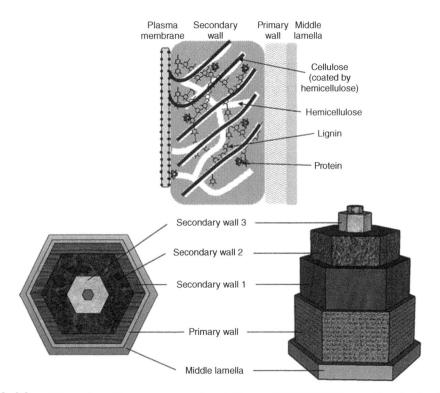

FIG. 6.10 Schematics of the secondary cell wall. Source: CCRC, 2007. Plant cell walls. Available at: http://www.ccrc.uga.edu/~mao/intro/ouline.htm#Secondary cell walls (Accessed on 14th June 2014).

The primary role of hemicellulose is to crosslink cellulose microfibrils through hydrogen bonding within the cell wall matrix. It also limits the extensibility of cell walls, thereby regulating the extent of enlargement after cell division. Xylans and glucomannans are the dominant hemicelluloses present in the secondary cell wall. In hardwood secondary cell walls, up to 15–30% glucuronoxylans and 2–5% glucomannans can be found. In softwood secondary walls, up to ~20% galactoglucomannans and 5–10% arabinoglucuronoxylans can be found.

Chemically, glucuronoxylan is similar to the xylan found in the primary cell wall, consisting of β-1,4-linked xylopyranosyl residues. However, a single-terminal 4-O-methyl α–D-glucosyluronic acid residue is frequently found attached at the C-2 position of the xylosyl residue; and infrequently, arabinosyl residues have also been identified. Approximately 70% of the backbone xylosyl residues contain one O-acetyl group at the C-2 or C-3 position; whereas glucomannan hemicellulose consists of linear β(1 → 4)-linked D-mannose and D-glucose residues.

References

Bourbonnais, Gilles (2009). Les molécules de la vie: Les glucides. http://www.cegep-ste-foy.qc.ca/profs/gbourbonnais/pascal/fya/chimcell/notesmolecules/glucides_3.htm, accessed June 2014.

Bozell, J.J., Holladay, J.E., Johnson, D., and White, J.F. (2007). *Top Value Added Chemicals from Biomass, Volume II.* PNNL-16983. Richland, WA: Pacific Northwest National Laboratory.

Brett, C.T., and Waldron, K.W. (1996). *Physiology and Biochemistry of Plant Cell Walls.* London: Chapman & Hall.

Caffall, K.H., and Mohnen, D. (2009). The structure, function, and biosynthesis of plant cell wall pectic polysaccharides. *Carbohydrate Research* 344(14): 1879–900.

CCRC (2007). *Plant Cell Walls.* Athens, GA: Complex Carbohydrate Research Center, University of Georgia. Available at: http://www.ccrc.uga.edu/~mao/intro/ouline.htm#Secondary cell walls, accessed June 2014.

Cooper, G.M. (2000). *The Cell: A Molecular Approach,* 2nd edn. Sunderland, MA: Sinauer Associates.

de Vries, R.P., and Visser, J. (2001). *Aspergillus* enzymes involved in degradation of plant cell wall polysaccharides. *Microbiology and Molecular Biology Reviews* 65(4): 497–522.

Ek, M., Gellerstedt, G., and Henriksson, G. (2009). *Pulp and Paper Chemistry and Technology, Volume 1: Wood Chemistry and Wood Biotechnology.* Boston, MA: De Gruyter.

Eskin, N.A.M., and Hoehn, E. (2012). Fruits and vegetables. In: N.A.M. Eskin and F. Shahidi (eds.), *Biochemistry of Foods,* 3rd edn, Boca Raton, FL: Taylor & Francis, pp. 49–126.

Gajera, H.P., Patel, S.V., and Golakiya, B.A. (2008). *Fundamentals of Biochemistry: A Textbook, Student Edition.* Lucknow: International Book Distributing Company.

Higuchi, T. (1997). *Biochemistry and Molecular Biology of Wood.* Berlin: Springer.

Hon, D.N.S., and Shiraishi, N. (2000). *Wood and Cellulosic Chemistry.* Boca Raton, FL: CRC Press.

Izydorczyk, M. (2005). Understanding the chemistry of food carbohydrates. In: S.W. Cui (ed.), *Food Carbohydrates: Chemistry, Physical Properties and Applications.* Boca Raton, FL: Taylor & Francis, pp. 1–65.

Izydorczyk, M., Cui, S.W., and Wang, Q. (2005). Polysaccharide gums: Structures, functional properties, and applications. In: S.W. Cui (ed.), *Food Carbohydrates: Chemistry, Physical Properties and Applications.* Boca Raton, FL: Taylor & Francis, pp. 263–307.

Kalluri, U.C., and Keller, M. (2010). Bioenergy research: A new paradigm in multidisciplinary research. *Journal of the Royal Society Interface* 7(51): 1391–401.

Komalavilas, P., and Mort, A.J. (1989). The acetylation of O-3 of galacturonic acid in the rhamnose-rich portion of pectins. *Carbohydrate Research* 189: 261–72.

Krempels, D.M. (n.d.). The plant cell is special. http://www.bio.miami.edu/dana/226/226F07_3.html, accessed June 2014.

Leclere, L., Cutsem, P.V., and Michiels, C. (2013). Anti-cancer activities of pH- or heat-modified pectin. *Frontiers in Pharmacology* 4: 128.

Martínez, A.T., Rencoret, J., Marques, G., *et al.* 2008. Monolignol acylation and lignin structure in some nonwoody plants: A 2D NMR study. *Phytochemistry* 69: 2831–43.

Minorsky, P.V. (2002). The wall becomes surmountable. *Plant Physiology* 128: 345–53.

Mittal, A., Katahira, R., Himmel, M.E., and Johnson, D.K. (2011). Effects of alkaline or liquid-ammonia treatment on crystalline cellulose: Changes in crystalline structure and effects on enzymatic digestibility. *Biotechnology for Biofuels* 4: 41.

Mooney, C.A., Mansfield, S.D., Touhy, M.G., and Saddler, J.N. (1998). The effect of initial pore volume and lignin content on the enzymatic hydrolysis of softwoods. *Bioresource Technology* 64: 113–19.

O'Neill, M.A., and York, W.S. (2003). The composition and structure of plant primary cell walls. In: J.K.C. Rose (ed.), *The Plant Cell Wall*. Boca Raton, FL CRC Press, pp. 1–54.

Pauly, M., and Keegstra, K. (2008). Cell-wall carbohydrates and their modification as a resource for biofuels. *The Plant Journal* 54: 559–68.

Ragauskas, A.J., Beckham, G.T., Biddy, M.J., *et al.* (2014). Lignin valorization: Improving lignin processing in the biorefinery. *Science* 344(6185): 1246843.

Ralph, J., Lundquist, K., Brunow, G., *et al.* (2004). Lignins: Natural polymers from oxidative coupling of 4-hydro-xyphenylpropanoids. *Phytochemistry Reviews* 3, 29–60.

Ridley, B.L., O'Neill, M.A., and Mohnen, D. (2001). Pectins: Structure, biosynthesis, and oligogalacturonide-related signaling. *Phytochemistry* 57(6): 929–67.

Rolin, L, Nielsen, B.U., and Glahn, P.E. (1998). Pectin. In: S. Dumitriu (ed.), *Polysaccharides: Structural Diversity and Functional Versatility*. Boca Raton, FL: CRC Press, pp. 377–431.

Sarkar, P., Bosneaga, E., and Auer, M. (2009). Plant cell walls throughout evolution: Towards a molecular understanding of their design principles. *Journal of Experimental Botany* 60(13): 3615–35.

Sjostrom, E. (1993). *Wood Chemistry: Fundamentals and Applications*, 2nd edn. San Diego, CA: Academic Press.

Sticklen, M.B. (2008). Plant genetic engineering for biofuel production: Towards affordable cellulosic ethanol. *Nature Reviews Genetics* 9: 433–43.

Taiz, L., and Zeiger, E. (2010). Cell walls: Structure, biogenesis, and expansion. In: L. Taiz and E. Zeiger (eds.), *Plant Physiology*. Sunderland, MA: Sinauer Associates, pp. 313–38. Available at: http://www.psu.edu/dept/cellwall/cell-wall.pdf, accessed June 2014.

Vanholme, B., Desmet, T., Ronsse, F., *et al.* (2013). Towards a carbon-negative sustainable bio-based economy. *Frontiers in Plant Science* 4: 174.

Yarbrough, J.M., Himmel, M.E., and Ding, S.Y. (2009). Plant cell wall characterization using scanning probe microscopy techniques. *Biotechnology for Biofuels* 2: 17.

Exercise Problems

6.1. Plant biomass plays an important role in the emerging bioeconomy. Understanding the plant cell structure is critically important. Elaborate on this statement.

6.2. Draw a schematic diagram of a typical plant cell and show its various components.

6.3. Show schematically how cellulose, hemicellulose, and lignin are arranged in the plant cell wall. What kind(s) of bonds hold these molecules together?

6.4. What are primary and secondary cell walls? When does the formation of these cell walls take place?

6.5. What makes a plant so recalcitrant for biochemical conversion into biofuel?

6.6. Give the typical composition of some common lignocellulosic biomass and state the main differences between cellulose and hemicellulose. Discuss the structure of starch as well. Is it possible for a plant to have cellulose and starch? If yes, how? Give examples.

6.7. Discuss the various monosaccharides that could be obtained from plant biomass and their origins.

6.8. What is the difference between hardwood and softwood in terms of composition? Provide insight into their suitability for biochemical and/or thermochemical conversion into biofuels.

6.9. As shown in Figure 6.A, the biomass composition of energy crops (grasses) changes with age. As a bioprocess engineer, how do you use the information presented in the figure in select the conversion technology for biofuel/biobased products generation?

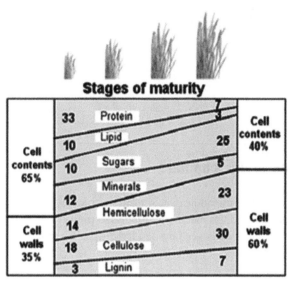

FIG. 6.A   Source: Adapted from Holmes, 1980.

6.10. Chemical analyses of samples from some wood species are presented in Table 6.A. Based on the given composition, classify the types of wood and also discuss how this information could be helpful in deciding the use of the wood species for bioenergy and biochemical products generation.

Table 6.A The chemical composition of some wood species

Constituents, %	Wood A	Wood B	Wood C
Cellulose	40.0	39.5	41.0
Hemicellulose			
– Glucomannan	16.0	17.2	2.3
– Glucuronoxylan	8.9	10.4	27.5
– Other polysaccharides	3.6	2.1	3.0
Lignin	27.7	27.5	22
Total extractives	3.5	2.1	3.0

Source: Adapted from Sjostrom, 1993.

Microbial Metabolisms

Arul M. Varman, Lian He, and Yinjie J. Tang

What is included in this chapter?

This chapter covers microbial metabolism as it relates to biofuel production. There is discussion of microbial (both kinetic and metabolic) models to predict cell growth, the dynamics of product secretion, metabolite turnover rates, and yield coefficients. Pertinent examples and calculations are also provided.

7.1 Introduction

Microbial metabolism involves complex sets of biochemical reactions catalyzed by different enzymes. These enzyme-mediated metabolic reactions are grouped into two categories: catabolism – reactions responsible for the generation of cellular energy by the breakdown of complex molecules into simpler ones; and anabolism – energy-consuming reactions that synthesize complex molecules (Figure 7.1). For supporting cell growth and biofuel synthesis, a microbial culture requires carbon substrates and other essential nutrients (nitrogen, phosphorus, potassium, and trace elements). The cell biochemical reactions for biofuel synthesis can be summarized as the production of microbial biomass and other products. For biofuel production, anaerobic microbial metabolism is preferred, since more carbons from the substrate can be converted into products (instead of CO_2 and microbial biomass) and no aeration is required during oxygen-limited fermentations (see Equation 7.1, $y_1 > y_2$, $n_2 < n_1$).

$$Substrate + O_2 + Nutrients \xrightarrow{Enzymes} y_1 \, Microbial \, biomass + n_1 CO_2 + Products \, (aerobic \, condition)$$

$$(7.1a)$$

$$Substrate + Nutrients \xrightarrow{Enzymes} y_2 \, Microbial \, biomass + n_2 CO_2 + Products \, (anaerobic \, condition)$$

$$(7.1b)$$

Bioenergy: Principles and Applications, First Edition. Edited by Yebo Li and Samir Kumar Khanal.
Companion website: www.wiley.com/go/Li/Bioenergy

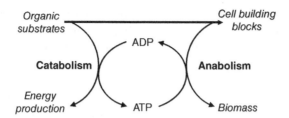

FIG. 7.1   The two groups of metabolic reactions anabolism and catabolism, coupled through the generation and consumption of ATP.

Microbial metabolism operates like a very complex cell factory in which thousands of reactions take place. Microbial fermentations enable the production of diverse bioenergy resources from energy crops, waste streams (such as municipal, industrial, agricultural, or forestry residues), syngas ($H_2 + CO$), $H_2 + CO_2$, or CO_2. In recent years, molecular biology technologies have been extensively employed to improve cell metabolic capabilities for converting renewable carbon compounds into advanced biofuels (such as biodiesel and higher alcohols). Therefore, a good understanding of common metabolic pathways related to biofuel conversion is important for the development of the bioenergy industry. Moreover, metabolic models play a key role in designing and controlling microbial-based bioenergy production processes. These models can quantify and predict microbial cell/biofuel synthesis under various cultivation conditions. As an example, Monod-based kinetic models can describe the rate of cell growth as a function of substrate concentration. Other models, such as stoichiometry flux analysis, have been used to measure intracellular reaction rates and the carbon flows from substrates to the microbial cell, as well as final products.

This chapter introduces several useful microbial models. In-depth discussion of these models is beyond the scope of this textbook, and students can refer to several excellent books in the biochemical engineering field for further reading (Blanch and Clark, 1997; Shuler and Kargi, 2002; Stephanopoulos *et al.*, 1998).

7.2 Carbon Metabolisms

Microbial metabolisms can utilize inorganic and organic substrates for biomass and biofuel synthesis. To support cellular functions, microbial metabolisms also generate energy molecules, such as ATP (adenosine triphosphate), NADH (nicotinamide adenine dinucleotide), and NADPH (nicotinamide adenine dinucleotide phosphate).

Autotrophs grow on inorganic carbon sources (such as CO_2, $H_2 + CO$, and $H_2 + CO_2$). Photoautotrophs (e.g., plants, algae, and cyanobacteria) can absorb inorganic carbon (CO_2) to produce alcohol, lipids, and H_2. They generate O_2 (from water) and energy molecules (ATP and NADPH) by photosynthesis. Typical photoautotrophic CO_2 fixation is via the Calvin cycle, which involves the enzyme ribulose bisphosphate carboxylase/oxygenase (RuBisCO) that converts ribulose-1,5-bisphosphate and CO_2 into 3-phosphoglycerate (3PG). 3PG is the starting building block for microbial biomass and biofuel synthesis. Besides the Calvin cycle, microbes may employ other autotrophic pathways for microbial biomass growth or biofuel synthesis (e.g., the reductive citric acid cycle, the 3-hydroxypropionate cycle, etc.). For example, some *Clostridium* species can carry out syngas (CO_2, CO, and H_2) fermentation to produce alcohols (such as ethanol or butanol). They can oxidize CO and H_2 to obtain NADPH or NADH, which powers the Wood-Ljungdahl pathway to convert CO and CO_2 into AcCoA (acetyl-coA; see Chapter 16). In addition, methanogens, especially hydrogenotrophic methanogens, can generate CH_4 from H_2 and CO_2. Methanogenesis also employs enzymes in the Wood-Ljungdahl pathway.

Heterotrophs grow on organic carbon sources. The catabolism of organic substrates can efficiently produce both energy molecules – ATP and NAD(P)H – and building blocks for microbial biomass and product biosynthesis. Glucose is one of the most common carbon sources used by heterotrophs for biofuel production. Glucose degrades into pyruvate via three primary pathways: the Embden-Meyerhof-Parnas (EMP) pathway; the pentose phosphate (PP) pathway; and the Entner–Doudoroff (ED) pathway. Pyruvate is the key metabolic node to synthesize many microbial biomass building blocks (e.g., amino acids) and biofuel molecules (e.g., ethanol). Pyruvate can also be completely oxidized into CO_2 via the TCA cycle under aerobic conditions, which generates ATP and NAD(P)H. Microbial species can also use other carbon substrates, such as acetate, glycerol, xylose, and cellulose. The most common industrial microbes, *Escherichia coli* and *Saccharomyces cerevisiae*, are heterotrophs.

Mixotrophs are microbes that grow on both inorganic and organic carbon sources. Many microalgal species can employ *mixotrophic metabolisms* to utilize CO_2 and organic substrates (e.g., sugar or acetate) simultaneously. Such a metabolic process alleviates the burdens for CO_2 fixation and resolves light limitations in large-scale algal biorefineries. Thereby, mixotrophic cultures (also called photofermentation) often show higher microbial biomass and biofuel yields than autotrophic or heterotrophic cultures alone.

The primary pathways of microbial metabolism are listed in Table 7.1. Microbes use these pathways to generate energy in the form of ATP and reducing equivalents (NADH, NADPH, and $FADH_2$). These pathways also provide chemical precursors that are essential for the synthesis of biofuel products and microbial biomass components.

Sugars, such as glucose, sucrose, and starch, have been widely used for biofuel production. These carbohydrates can be obtained either from food crops (e.g., corn, sugarcane, and sugar beet) or from lignocellulosic biomass polymers (e.g., cellulose and hemicelluloses). Non-sugar-based substrates, such as glycerol, lactate, acetate, CO_2, protein, and syngas (CO, CO_2, and H_2), can also be converted to biofuels, as shown in Figure 7.2. As for biofuel products, ethanol (via yeast fermentation) and butanol (via clostridium ABE fermentation; see Chapter 15) are produced at the largest scale among

Table 7.1 Primary pathways for substrate metabolism

Pathway	Starting metabolite	Ending metabolite	ATP and reducing equivalents	Microbial biomass precursors
EMP pathway	Glucose	2 Pyruvate	2 ATP, 2 NADH	3PG, Pyruvate, PEP
PP pathway	G6P	C5P and CO_2	2 NADPH	C5P, E4P
Glucose degradation via ED pathway: 6PG → Pyruvate + GAP	Glucose	2 Pyruvate	1ATP, 1NADH, 1 NADPH	3PG, Pyruvate, PEP
TCA pathways	Acetyl-CoA	2 CO_2	3 NADH, 1 $FADH_2$, 1 GTP	α-ketoglutarate, Oxaloacetate
Calvin cycle	3 CO_2	3PG	–6 NADPH, –9 ATP	3PG

Note: A negative sign (-) means the consumption of energy molecules.

Abbreviations: C5P, Ribose-5-phosphate; E4P, Erythrose-4-phosphate; $FADH_2$, Flavin adenine dinucleotide; G6P, Glucose 6-phosphate; GTP, Guanosine-5'-triphosphate; PEP, Phosphoenolpyruvate

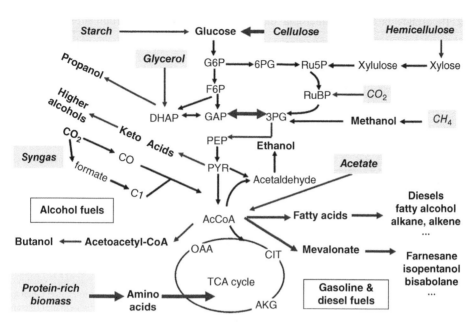

FIG. 7.2   Substrates, products and microbial pathways for biofuel synthesis. Abbreviations: 6PG, 6-phosphogluconolactone; AKG, α-ketoglutarate; C1, C1 pathway; CIT, Citrate; DHAP, Dihydroxyacetone phosphate; F6P, Fructose 6-phosphate; GAP, Glyceraldehyde 3-phosphate; OAA, Oxaloacetate; PYR, Pyruvate; Ru5P, Ribulose-5-phosphate; RuBP, Ribulose-1,5-bisphosphate.

all biofuel compounds. In addition, advanced biofuels with properties similar to those of gasoline and diesel can be synthesized using genetically modified microbial strains. Several biofuel pathways are demonstrated in Figure 7.2. First, the engineered bacteria or yeast can use the keto acid pathway to produce higher alcohols, such as butanol and isobutanol. Second, the fatty acid pathway can be modified to produce biodiesels and hydrocarbons. Third, the mevalonate pathway can be extended to branched and cyclic hydrocarbons, which are promising jet fuels with a low freezing point and high energy content.

Yield coefficients are defined as the ratio of the mass of product or microbial biomass formed to the mass of substrate consumed. The theoretical yield dictates the maximum amount of product, such as biofuel, that can be produced from a given carbon source. Examples of theoretical yields are presented in Table 7.2. Among all liquid biofuels, ethanol has the highest yield (0.51 g ethanol/g glucose) from yeast fermentation. Advanced biofuels (hydrocarbons and biodiesels) have been synthesized using genetically modified microbes. However, these advanced biofuel molecules have lower theoretical yields (0.3–0.4 g/g glucose) and poor production titers (below 10 g/L). Under anaerobic conditions, microbial biomass yield from yeast fermentation is less than 0.1 g microbial biomass/g glucose. Under aerobic conditions, microbial biomass yield can be as high as 0.5 g/g glucose. It is important to note that there is always a trade-off between biofuel production and microbial biomass growth using the same amount of carbon substrates. To ensure enough microbial cells for fuel synthesis, fermentation processes often grow microbial biomass in the early stage, and then switch to product synthesis during the late growth phase.

Microbial physiologies vary as a function of species. *Saccharomyces cerevisiae* and *Escherichia coli* are widely used in biofuel industries because they efficiently ferment sugars, are amenable to genetic modifications, and pose minimal problems during scale-up. *S. cerevisiae* has excellent tolerance to

Table 7.2 Theoretical biofuel and microbial biomass yields for selected carbon substrates

Substrate	Product	Product yield (g/g)	Microbial biomass yield* (g microbial biomass/g substrate)
Glucose	Ethanol	0.51	0.51
Glucose	Isobutanol	0.41	0.51
Glucose	Lipid	0.33	0.51
Acetate	Fatty acids	0.29	0.36
Acetate	Methane	0.27	0.36
Glycerol	Ethanol	0.50	0.50
CO_2 (light)	Ethanol	0.52 (photoautotrophic)	∼0.60 (photoautotrophic)

* Aerobic growth condition

Table 7.3 Examples of microbes for biofuel production

Species	Substrates	Biofuel products	Features
Saccharomyces cerevisiae	Glucose, fructose, galactose, etc.	Ethanol	Easy genetic manipulations and high ethanol tolerance; Crabtree effect
Zymomonas mobilis	Glucose, fructose, and sucrose	Ethanol	High ethanol tolerance and yield
Clostridium thermocellum	Glucose, cellulose, and cellobiose	Ethanol	Growth at high temperature and mixed fermentation pathways
Clostridium acetobutylicum	Glucose and xylose	Ethanol and butanol	Acetone, ethanol, and butanol (ABE) fermentation
Escherichia coli	Glucose, xylose, glycerol, and others	Alcohols, diesels, and other biofuels	Easy genetic manipulations and fast growth
Cyanobacteria	CO_2	Alcohols, H_2, and fatty acids	CO_2 fixation and mixotrophic growth
Yarrowia lipolytica	Glucose, acetate, and fatty acids	Lipids	Oleaginous yeast that accumulates lipids
Some acetogens (e.g., Clostridium carboxidivorans)	Syngas (CO, CO_2, H_2)	Ethanol	Use of gaseous substrates (constrained by mass transfer limitations)

ethanol (>100 g/L) and can perform alcohol fermentation under aerobic conditions (*Crabtree effect*). *E.coli* has fast carbon and energy metabolism, and is often used for aerobic biodiesel fermentation. Microalgal systems show unique advantages due to their strong photosynthesis capabilities. For example, cyanobacteria are promising hosts for biofuel production from CO_2. Table 7.3 shows common microbial species that produce biofuels, either via the native biofuel pathway or via a metabolically engineered pathway.

In addition, metabolic engineering becomes an important way to create new microbial platforms to synthesize biofuel. Via computer modeling and recombinant DNA technology, high-performance microbial species can be engineered. However, there are still some bottlenecks for metabolic engineering applications in biofuel production. First, genetic instability in the engineered strain is often poor for high-performance strains, as self-mutations may deactivate biofuel production pathways. Second, as heterologous pathways become more complicated, optimizations of biofuel yield are increasingly difficult due to metabolic burdens from enzyme overexpression and metabolite imbalances. Third, the production of biofuel with low profit margins, such as ethanol, is constrained by economic viability. Besides, advanced biofuel (e.g., biodiesels) using genetically modified microbes is also difficult to make profitable because of low production yield and titer, as well as expensive biosynthesis processes (using a rich medium and aerobic fermentation conditions). Last, poor mixing in large bioreactors often causes suboptimal growth conditions (e.g., unfavorable pH and oxygen levels), and the resulting metabolic stresses may greatly reduce microbial productivities during bioprocess scale-up.

Crabtree effect

This is a phenomenon wherein yeasts are able to produce ethanol under aerobic conditions when provided with high concentrations of sugars. For example, fast glucose utilization allows *Saccharomyces cerevisiae* to enhance its glycolysis pathway for ATP/NADH generation, thus minimizing the TCA cycle flux and aerobic respiration rate.

7.3 Metabolic Models

7.3.1 Microbial Growth in Batch Culture

In batch systems, microbial growth takes place in a closed system without inflow of liquid streams. Gases, on the other hand, are exchanged. Microbial growth can be estimated by monitoring the dry weight, the turbidity (or optical density) of the culture medium, and/or the number of colony-forming units (CFUs). For slow-growing cultures at low cell density or cultures containing suspended solids, cell density can be indirectly determined by measuring the concentrations of DNA, RNA, or total proteins. Microbial growth usually undergoes five growth phases: lag phase, exponential growth phase, declining growth phase, stationary phase, and death phase, as shown in Figure 7.3. However, depending on the cultivation conditions, microbial growth curves may not strictly follow the five growth phases.

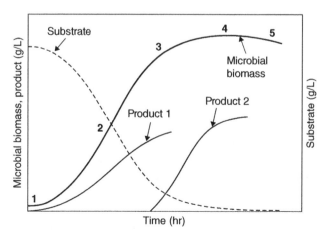

FIG. 7.3   Microbial biomass growth and product secretion curves. Product 1 is growth associated, while product 2 is likely non-growth associated.

In the **lag phase**, cell numbers remain unchanged for a certain time after the initial inoculation. The cells are metabolically active during this period and the cell size may increase. The length of the lag phase can be reduced by using an inoculum that is actively growing, usually stemming from the exponential phase; increasing the inoculation ratio; or improving the culture medium (e.g., by adding extra nutrients).

The **exponential growth phase**, also known as the logarithmic growth phase, starts when cells divide at a constant rate. In this stage, the metabolic activity (e.g., metabolic fluxes) and chemical composition of all the cells can be assumed to be in a pseudo-steady state. The exponential growth rate can be described as:

$$\frac{dX}{dt} = \mu X \tag{7.2}$$

where X is the cell concentration (g/L), t is the time (hr), and μ is the specific growth rate (1/hr). Integration of Equation 7.2 from the initial time and cell concentration (t_0, X_0) to any time point and cell concentration (t, X) in the exponential growth phase yields:

$$\ln\left(\frac{X}{X_0}\right) = \mu(t - t_0) \tag{7.3}$$

Plotting Equation 7.3 with ln(X) and time (t) as "X" and "Y" coordinates results in a straight line with slope μ. The time required for cells to double their population (t_d, doubling time) is given by:

$$t_d = \frac{\ln(2)}{\mu} = \frac{0.693}{\mu} \tag{7.4}$$

The cell doubling time is dependent on growth conditions. For example, *E.coli* has a doubling time of approximately 20 mins in the LB medium at a cultivation temperatures of 37 °C. On the other hand, microalgae have much longer doubling time, which can be greater than 10 hours for cyanobacterial species. A modified form of Equation 7.2 includes the lag phase of cell growth, with lag time expressed as t_L (hr):

$$\frac{dX}{dt} = \mu X \left(1 - e^{-t/t_L}\right) \tag{7.5}$$

In the **declining growth phase**, at the end of the exponential phase, the limitation of certain nutrient(s) reduces microbial biomass growth rates. For example, nitrogen-limiting conditions cause yeast cells to decrease protein synthesis and to accumulate lipids during the late growth stage. In addition, accumulations of biofuels and waste byproducts (such as acetate) in the culture medium may also stress cell metabolism and interfere with microbial biomass growth.

During the **stationary phase**, the cell population remains relatively constant, because certain key nutrients are used up and toxic products become inhibitory to microbial biomass growth. At this stage, the microbial host can still actively produce non-growth-associated biofuel products, such as lipids and higher alcohols. Cells may continue to grow slowly, which is counterbalanced by cell death.

During the **death phase**, cell lysis occurs and the cell population starts to decline. In the death phase, the oxygen utilization rate in aerobic fermentations drops to near zero. The cell population in the death phase can be described by a first-order rate equation:

$$\frac{dX}{dt} = -k_d X \tag{7.6}$$

where k_d is the death rate constant or endogenous decay rate (1/hr).

> **Example 7.1:** Optical density (OD) is normally used for monitoring the growth of organisms. Growth of a blue-green algal strain was monitored by measuring the absorbance of the culture at 730 nm and the data recorded are as follows (cells enter the exponential growth phase after t >10 hr):
>
> | Time (hr) | 0 | 21 | 96 | 128 |
> |-----------|-----|-----|-----|-----|
> | OD_{730} | 0.4 | 0.6 | 3.2 | ? |
>
> 1. Calculate the specific growth rate and doubling time of this algal strain.
> 2. Estimate the OD at time point 128 hr.
>
> *Solution:*
> 1. Specific growth rate $(\mu) = \dfrac{\ln(OD_2) - \ln(OD_1)}{t_2 - t_1} = \dfrac{\ln(3.2) - \ln(0.6)}{96 - 21} = \textbf{0.022/hr}$ Doubling time $= \dfrac{\ln(2)}{\mu} = \dfrac{0.693}{0.022} = \textbf{31.5 hr}$
> 2. $OD[128hr] = OD[96hr]\ e^{\ [\mu(128-96)]} = 3.2\ e^{\ [0.022(128-96)]} = \textbf{6.5}$
>
> Note: The first data point is not used for rate calculation due to the presence of the lag phase.

7.3.2 Monod Equation for Microbial Growth

The *Monod equation* describes cell growth as a function of substrate concentrations. The kinetics are similar to those of the Michaelis–Menten equation developed for enzymatic reactions. The basic Monod equation for cell growth limited by a single substrate is:

$$\mu = \frac{\mu_{max} S}{K_S + S} \tag{7.7}$$

where μ_{max} is the maximum specific growth rate (1/hr), and K_S is defined as the saturation constant or the Monod constant (g/L), and is equal to the substrate concentration at which the specific growth rate is half of the maximum. The values of μ_{max} and K_S are dependent on the organism selected, types of substrate used, and cultivation conditions. The model has two simpler forms, depending on the substrate concentration. At high substrate concentration $(S >> K_S)$:

$$\mu = \mu_{max} (\text{zero order}) \tag{7.8}$$

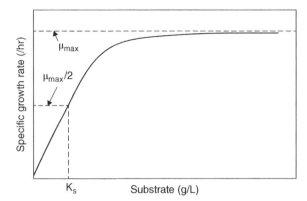

FIG. 7.4   Microbial specific growth rates as a function of substrate concentrations.

At low substrate concentration ($S << K_S$):

$$\mu = \frac{\mu_{max}}{K_s} S \text{ (first order)} \tag{7.9}$$

At low substrate concentrations, the simplified Monod equation is a first-order kinetic reaction, or, in other words, the specific growth rate is proportional to the substrate concentration. At high substrate concentrations, the Monod equation is a zero-order equation. The Monod equation is graphically represented in Figure 7.4. The Monod model can also be modified to describe different cell cultures. For example, it may include the term of microbial biomass decay rate constant, k_d (1/hr):

$$\mu = \frac{\mu_{max} S}{[K_S + S]} - k_d \tag{7.10}$$

The Monod-based models assume a cell population as a single homogeneous system. There are other, more complex growth models. For example, segregated models describe each cell individually and view cell populations as a heterogeneous mix of individual cells. Structured models consider microbial biomass components within a cell, such as the concentration of intracellular metabolites, DNA and RNA. These complex models have been described elsewhere (Blanch and Clark, 1997; Shuler and Kargi, 2002; Stephanopoulos *et al.*, 1998).

7.3.3 Inhibition and Multiple Substrate Models

Chemical compounds in the culture medium may inhibit cell growth. The inhibitors can be substrates, products, or other chemicals in the culture medium. For example, a high concentration of glucose (>40 g/L) decreases *E.coli* growth due to osmotic stress. Biofuel, such as alcohols and biodiesel products, may damage cell membrane integrity and interfere with cell physiology. Common approaches to model inhibition account for the inhibitors, I (g/L), as competitive, uncompetitive, or non-competitive inhibitions – inhibition coefficient K_I (g/L).

A model describing competitive inhibition is given by:

$$\mu = \frac{\mu_{max}S}{K_S\left(1 + \frac{I}{K_I}\right) + S} \tag{7.11}$$

A model describing uncompetitive inhibition is given by:

$$\mu = \frac{\mu_{max}S}{K_S + S\left(1 + \frac{I}{K_I}\right)} \tag{7.12}$$

A model describing non-competitive inhibition is given by:

$$\mu = \frac{\mu_{max}S}{K_S\left(1 + \frac{S}{K_S}\right)\left(1 + \frac{I}{K_I}\right)} \tag{7.13}$$

Another common inhibitory model can be expressed by using the inhibitory constant in an exponential form:

$$\mu = \frac{\mu_{max}S}{[K_S + S]}e^{-K_I I} \tag{7.14}$$

By using the maximum inhibitor concentration (I_m), Equation 7.15 can be derived:

$$\mu = \frac{\mu_{max}S}{[K_S + S]}\left(1 - \frac{I}{I_m}\right)^n \tag{7.15}$$

where n is a parameter that indicates the toxicity of the inhibitor (unitless). As the value of I approaches I_m, μ becomes zero and cell growth ceases.

The modified Monod model for growth with multiple substrates (S_1, S_2, S_3, ... S_n) can be expressed either in an additive or a multiplicative form. Equation 7.16 describes cell culture under additive carbon substrate consumption. The cultivation of *E.coli* provides a good example of this concept: when glucose (S_1) and yeast extract (S_2) are available in the medium, *E.coli* can consume both substrates for its growth. Equation 7.17 describes multiple-nutrient-controlled cell growth. For example, aerobic microbial growth is dependent on two rate-limiting substrates in a multiplicative manner (S_1: glucose and S_2: oxygen).

$$\mu = \frac{\mu_{max1}S_1}{[K_1 + S_1]} + \frac{\mu_{max2}S_2}{[K_2 + S_2]} + \cdots + \frac{\mu_{maxn}S_n}{[K_n + S_n]} \tag{7.16}$$

$$\mu = \mu_{max}\left[\frac{S_1}{[K_1 + S_1]}\frac{S_2}{[K_2 + S_2]}\cdots\frac{S_n}{[K_n + S_n]}\right] \tag{7.17}$$

7.3.4 Monod-Based Kinetic Model in Batch Bioreactors

Kinetic models can describe microbial biomass growth, substrate utilization, and product formation. Yield coefficients link microbial biomass growth with substrate consumption and product formation in kinetic models. Theoretical product and microbial biomass yields are determined by the stoichiometry of metabolic pathways. In reality, the actual yields of products or microbial biomass are always much smaller than the theoretical yields, depending on growth conditions and microbial metabolic performance. For example, when certain microbial hosts are under oxygen limitations, they will produce acetate and other waste compounds, and thus reduce product yields.

This section introduces a kinetic model to predict isobutanol (IB) fermentation using an engineered *E.coli*. In an aerobic batch reactor, *E.coli* uses glucose (G) as the sole carbon source to produce microbial biomass and acetate (A). The IB synthesis by *E.coli* is via a keto acid pathway. This pathway also involves amino acid metabolism. During cell growth, glucose is the main carbon source for IB synthesis. Microbial cells also have the capability to synthesize IB using microbial biomass components (e.g., amino acids). In this model, isobutanol formation is described by both growth-associated and non-growth-associated kinetics. The model consists of four time-dependent variables, X, A, IB, and G, which represent the concentrations of microbial biomass, acetate, isobutanol, and glucose, respectively.

$$\text{Microbial biomass:} \quad \frac{dX}{dt} = R_X - k_d X \tag{7.18}$$

$$\text{Acetate:} \quad \frac{dA}{dt} = R_A \tag{7.19}$$

$$\text{Isobutanol accumulation:} \quad \frac{dIB}{dt} = R_{IB} + \beta X - k_{IB} IB \tag{7.20}$$

$$\text{Glucose consumption:} \quad \frac{dG}{dt} = -\frac{R_X}{Y_{XG}} - \frac{R_A}{Y_{AG}} - \frac{R_{IB}}{Y_{IBG}} \tag{7.21}$$

$$\text{Microbial biomass growth:} \quad R_X = \frac{\mu_{max} G}{(K_S + G)\left(1 + \frac{A}{k_{iA}}\right)} X \tag{7.22}$$

$$\text{Acetate production:} \quad R_A = \alpha_{AX} R_X \tag{7.23}$$

$$\text{Isobutanol production:} \quad R_{IB} = \alpha_{IBX} R_X \tag{7.24}$$

where R_x, R_A, and R_{IB} are the reaction rates as shown earlier (g/(L·hr)); k_d is the cell death rate constant (1/hr); Y_{XG}, Y_{AG}, and Y_{IBG} are the actual yield coefficients (g/g) for the production of microbial biomass, acetate, and isobutanol from glucose, respectively (note: actual yield coefficients are smaller than the theoretical yields); and k_{IB} (1/hr) represents the rate of IB loss from the fermentor (due to product vaporization). The presence of acetate inhibits cell growth, and hence a non-competitive inhibition constant k_{iA} (g/L) is included in the model. α_{AX} and α_{IBX} (g/g) are the growth-associated yield coefficients of acetate and IB, respectively. β is the coefficient of non-growth-associated IB produced by the degradation of amino acids. Thereby, Equation 7.21 has ignored the glucose consumption by non-growth-associated IB production. Example 7.2 illustrates the use of these model equations to predict isobutanol fermentation processes.

Example 7.2: For the parameter values given in Table 7.4, we predict the isobutanol concentration as a function of time, for a 50-hour production cycle. Assume the initial concentrations of biomass (X), acetate (A), isobutanol (IB), and glucose (G) to be 0.12 g/L, 0.08 g/L, 0 g/L, and 20 g/L, respectively.

Table 7.4 Parameters used in Example 7.2

Parameters	Units	Values
K_S, Monod constant	g/L	0.305
μ_{max}, maximum specific growth rate	1/hr	0.073
K_{iA}, acetate inhibition constant	g/L	47.69
k_d, death rate constant	1/hr	0.007
Y_{XG}, yield coefficient for microbial biomass	g biomass/g glucose	0.161
Y_{IBG}, yield coefficient for IB	g IB/g glucose	0.163
Y_{AG}, yield coefficient for acetate	g acetate/g glucose	0.080
α_{IBX}, growth-associated yield of IB	g IB/g biomass	0.528
α_{AX}, growth-associated yield of acetate	g acetate/g biomass	0.614
k_{IB}, IB loss rate	1/hr	0.123
β, non-growth-associated IB production rate	g IB/(g biomass·hr)	0.006

Solution: The model equations 7.18–7.24 are solved using these parameters by the MATLAB function "ode23." The simulation results are presented in Figure 7.5. MATLAB codes related to this problem can be found in Appendix 7.1.

Discussion: Monod-based equations (with an inhibition term) are often used for kinetic modeling of fermentations. This study employed four ODE equations, which predict microbial biomass, substrate, and product curves. Figure 7.5 shows that IB production is mainly at cell growth stage, while acetate is a major byproduct during fermentation.

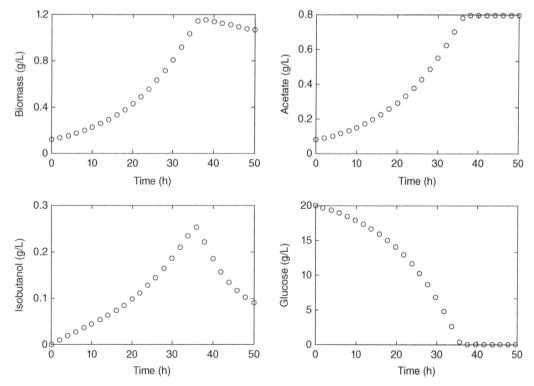

FIG. 7.5   Simulated results from the isobutanol kinetic model, showing cell growth, acetate production, isobutanol production, and glucose consumption over time.

When glucose is used up, cell growth approaches its stationary phase and IB production ceases. In the late fermentation stage, IB concentration starts declining because significant IB is lost through the bioreactor outlet due to vaporization. This example shows that a kinetic model is useful to predict time-dependent biofuel and microbial biomass accumulations, product loss, and growth inhibitions. By adjusting the model parameters, researchers can analyze and predict the key factors for optimizing biofuel fermentation.

7.3.5 Monod Model Coupled with Mass Transfer

Cellular growth is also affected by cell mixing conditions and substrate transport processes. For example, the presence of a liquid film around a cell alters the substrate concentration to which the microbe is exposed. Hence, the substrate concentration at the cell surface (S_s) will be lower than the substrate concentration in the bulk medium (S_o), and microbial biomass (X) growth is dependent on the diffusion of the substrate from the bulk solution to the cell surface. The rate of substrate consumption for cell growth is:

$$\text{Rate of substrate consumption} = \frac{\mu_{max} S_S}{[K_S + S_S]} \frac{X}{Y_{X/S}} \tag{7.25}$$

where $Y_{x/s}$ is the microbial biomass yield. The rate of substrate transfer from the bulk solution to the cell surface is given by:

$$\text{Rate of substrate transport} = k_L a \left(\frac{X}{\rho}\right)(S_0 - S_s) \tag{7.26}$$

where k_L is the mass transfer coefficient (m/hr), a is the interfacial area per unit volume of the cell (m^2/m^3 or $1/m$), and (X/ρ) is the volume ratio of microbial cells to solution (unitless). At steady state, there is no accumulation of substrate within the liquid film, and the rate of substrate transfer to the cell surface will be equal to the rate at which substrate is consumed. Hence, by equating Equation 7.25 to Equation 7.26, we have:

$$\frac{\mu_{max} S_S}{[K_S + S_S]} \frac{X}{Y_{X/S}} = k_L a \left(\frac{X}{\rho}\right)(S_0 - S_S) \tag{7.27}$$

After rearranging and simplifying Equation 7.27, we can obtain a Monod-based equation:

$$\mu = \frac{\mu_{max} S_0}{\left[S_0 + K_S + \frac{\mu_{max}\rho}{k_L a Y_{X/S}}\right]} \tag{7.28}$$

The detailed explanations and examples involved in Equation 7.28 can be found in Blanch and Clark (1997). Equation 7.28 reveals that the actual rate of growth is dependent on both mass transfer coefficients and intrinsic growth rate constants (μ_{max} and K_S).

7.3.6 Mass Balances and Reactions in Fed-Batch and Continuous-Stirred Tank Bioreactors

In a fed-batch bioreactor, fresh media is added to the bioreactor without product removal. A fed-batch configuration reduces substrate inhibitions. In a continuous-stirred bioreactor, fresh media is continuously added to the bioreactor, and at the same time culture medium–containing products, wastes, and cells are removed from the vessel. The continuous bioreactor operation can increase the biofuel production rate.

Figure 7.6 illustrates a continuous-flow bioreactor, with S_{in} and S_{out} representing the concentration of substrate in inlet and outlet, respectively. The variables Q_{in} and Q_{out} represent the substrate flow rate (L/hr) at inlet and outlet, respectively. The variables V, r_s, and X represent the volume of the reactor (L), the substrate consumption rate by cells (1/hr), and the microbial biomass concentration (g/L), respectively (Stephanopoulos *et al.*, 1998).

A general mass balance on substrate is given by:

$$\frac{dS}{dt} = -r_s X + \frac{Q_{in}}{V} S_{in} - \frac{Q_{out}}{V} S_{out} - S \frac{1}{V} \frac{dV}{dt} \tag{7.29}$$

In a continuous-stirred tank bioreactor under chemostat conditions, it is reasonable to assume that:

$$\frac{dS}{dt} = 0, \frac{dV}{dt} = 0, Q_{in} = Q_{out} = Q, S_{out} = S, \text{ and } D = \frac{Q}{V}$$

where D is the dilution rate (1/hr). Substituting these values in Equation 7.29, we obtain:

$$r_s X = D(S_{in} - S_{out}) \tag{7.30}$$

In a fed-batch bioreactor, $Q_{out} = 0$, and by substituting this in Equation 7.29, we obtain:

$$\frac{dS}{dt} = -r_s X + \frac{Q_{in}}{V} S_{in} - S \frac{1}{V} \frac{dV}{dt} \tag{7.31}$$

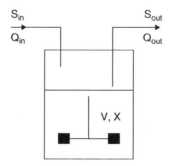

FIG. 7.6 Continuous-flow bioreactor operations.

By substituting $dV/dt = Q_{in}$ along with the definition of the dilution rate in Equation 7.31, we have:

$$\frac{dS}{dt} + r_s X = D(S_{in} - S) \tag{7.32}$$

Mass balance expressions for microbial biomass and product formations in batch, fed-batch, and chemostat bioreactors can be derived using similar approaches based on Equation 7.29, as described elsewhere (Stephanopoulos *et al.*, 1998).

> **Example 7.3:** In a continuous bioreactor (i.e., chemostat), "washout" is a condition in which the dilution rate D (inverse of hydraulic retention time) is too high and the cells cannot grow sufficiently fast, leading the microbial biomass concentration in the reactor to be continuously decreased during cultivation ($X \rightarrow 0$). Assuming the parameters in a chemostat culture as $\mu_{max} = 0.20$ 1/hr, $K_S = 1$ g/L, and $S_{in} = 9$ g/L, discuss the flow rate within which a one-liter bioreactor must be operated without washout.
>
> **Solution:** A general mass balance on microbial biomass concentration is:
>
> $$\frac{dX}{dt} = \mu X - \frac{Q}{V} X$$
>
> At steady state, V=1 L, dX/dt =0, and thus μ = D. Under a washout condition, the substrate concentration inside the bioreactor is $S \rightarrow S_{in}$.
>   At the maximum dilution rate:
>
> $$D_{max} = \frac{\mu_{max} S_{in}}{[K_S + S_{in}]} = \frac{(0.2)(9)}{(9+1)} = 0.18/hr$$
>
> Maximum flow rate $Q_{max} = (V)(D_{max}) = \textbf{0.18 L/hr}$
>   Since cells consume substrate, the substrate concentration in the bioreactor has to be maintained below 9 g/L for microbial biomass accumulation and product formation, and therefore the flow rate into the bioreactor should be controlled below 0.18 L/hr to avoid any washout.

7.3.7 Elemental Balance and Stoichiometric Models

Elementary balance model

Elementary balance models belong to the so-called black box approach, where all cellular reactions are lumped into elemental balances. It is rather easy to apply such a model to calculate the overall conversion of substrates to microbial biomass and product. For example, the approximate elemental formula of microbial biomass is $CH_{1.8}O_{0.6}N_{0.2}S_{0.004}P_{0.01}$. Carbon constitutes nearly 50% of the dry cell weight of microbial biomass. Based on the mass conservation of each element, the following reaction represents the general aerobic growth of a microorganism ($CH_cO_dN_e$) on an organic carbon substrate (CH_aO_b) and a nitrogen source (NH_3):

$$CH_aO_b + \alpha O_2 + \beta NH_3 = \gamma CH_cO_dN_e + \delta H_2O + \varepsilon CO_2 \tag{7.33}$$

The elemental balances include C, N, H, and O. To determine the five unknown coefficients (α, β, γ, δ, and ε), the respiratory quotient (RQ) has to be measured. The respiratory quotient is defined as the ratio of the CO_2 production rate to the O_2 consumption rate:

$$RQ = \frac{\varepsilon}{\alpha} \tag{7.34}$$

The elementary balance equations and the RQ can be written in a matrix form and the unknowns calculated via a matrix algorithm.

$$
\begin{matrix}
\text{Carbon} \\
\text{Oxygen} \\
\text{Hydrogen} \\
\text{Nitrogen} \\
\text{RQ}
\end{matrix}
\begin{pmatrix}
0 & 0 & 1 & 0 & 1 \\
-2 & 0 & d & 1 & 2 \\
0 & -3 & c & 2 & 0 \\
0 & -1 & e & 0 & 0 \\
-RQ & 0 & 0 & 0 & 1
\end{pmatrix}
\begin{pmatrix}
\alpha \\
\beta \\
\gamma \\
\delta \\
\varepsilon
\end{pmatrix}
=
\begin{pmatrix}
1 \\
b \\
a \\
0 \\
0
\end{pmatrix}
\tag{7.35}
$$

In the biofuel field, the elementary balance model can predict biofuel or microbial biomass yields from a microbial host when the compositions of substrates, microbial biomass, and products are known. Such a model may also be useful for designing the optimal growth medium for biofuel fermentation.

Stoichiometric Model

Metabolic fluxes are rates of intracellular enzyme reactions, g metabolite/g microbial biomass.hr. Flux analysis can determine the carbon flow from substrate to final production through the metabolic network, and offer insights into intracellular reactions related to biosynthesis. The estimation of metabolic fluxes in a simplified metabolic network is based on stoichiometric equations:

$$\frac{dc}{dt} = S \cdot v - b \tag{7.36}$$

where dc/dt is the net accumulation rate of all metabolites inside of cells (g metabolite/g microbial biomass·hr), S is the stoichiometry matrix for a specific metabolic network, v is the vector for the flux variables, and b is the vector for the substrate uptake, microbial biomass synthesis, and product formation rates. In the matrix S, each column contains the stoichiometry of specific reactions in the metabolic network, and the row corresponds to the mass balance of metabolites, substrates, or products. At steady-state metabolic conditions, the inflow rates of all the metabolites are equal to their outflow rates. Hence, the equation can be simplified as:

$$S \cdot v = b \tag{7.37}$$

The stoichiometric model is unable to find a unique flux solution if there are fewer stoichiometric equations than unknown flux variables. To resolve this underdetermined model, constraint-based flux balance analysis (FBA) can be employed, in which the unknown fluxes are predicted by an objective function (i.e., the maximum microbial biomass yield). Such FBA models are described in detail elsewhere (Stephanopoulos *et al.*, 1998).

Example 7.4: Consider the simple network shown here. We can calculate the internal fluxes v_1, v_2, v_3, and v_4 using the stoichiometric model, assuming that v_a and v_b are determined experimentally and the metabolism is under steady state.

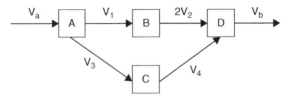

The fluxes for v_1, v_2, v_3 and v_4 can be calculated with a stoichiometric model, as follows:

$$\begin{pmatrix} 1 & 0 & 1 & 0 \\ 1 & -2 & 0 & 0 \\ 0 & 0 & 1 & -1 \\ 0 & 1 & 0 & 1 \end{pmatrix} \begin{pmatrix} v_1 \\ v_2 \\ v_3 \\ v_4 \end{pmatrix} = \begin{pmatrix} v_a \\ 0 \\ 0 \\ v_b \end{pmatrix}$$

In summary, metabolic models can be developed based on both cell metabolisms and bioreactor operations. In the biofuel industry, good metabolism analysis and bioreaction modeling can provide important knowledge and guidelines for creating effective microbial cell factories. Because of the complexity of biological systems, all models need assumptions to simplify the mathematical descriptions of biological or cellular processes. Remember, it is always challenging to build accurate and applicable models for bioprocesses. As statistician Dr. George Box said, "Essentially, all models are wrong, but some are useful."

References

Bailey, J.E. (1998). Mathematical modeling and analysis in biochemical engineering: Past accomplishments and future opportunities. *Biotechnology Progress* 14: 8–20.

Blanch, H.W., and Clark, D.S. (1997). *Biochemical Engineering*, 2nd edn. London: Taylor & Francis.

Dellomonaco, C., Fava, F., and Gonzalez, R. (2010). The path to next generation biofuels: Successes and challenges in the era of synthetic biology. *Microbial Cell Factories* 9: 3.

Kauffman, K.J., Prakash, P., and Edwards, J.S. (2003). Advances in flux balance analysis. *Current Opinion in Biotechnology* 14: 491–6.

Lee, J.M., Gianchandani, E.P., and Papin, J.A. (2006). Flux balance analysis in the era of metabolomics. *Briefings in Bioinformatics* 7: 140–50.

Lin, J., Lee, S.-M., Lee, H.-J., and Koo, Y.-M. (2000). Modeling of typical microbial cell growth in batch culture. *Biotechnology and Bioprocess Engineering* 5: 382–5.

Orth, J.D., Thiele, I., and Palsson, B.Ø. (2010). What is flux balance analysis? *Nature Biotechnology* 28: 245–8.

Pelczar, M.J., Jr., Chan, E.C.S, and Krieg, N.R. (1993). *Microbiology*. Delhi: Tata McGraw-Hill.

Peralta-Yahya, P.P., Zhang, F., del Cardayre, S.B., and Keasling, J.D. (2012). Microbial engineering for the production of advanced biofuels. *Nature* 488: 320–8.

Pramanik, J., and Keasling, J.D. (1997). Stoichiometric model of *Escherichia coli* metabolism: Incorporation of growth-rate dependent biomass composition and mechanistic energy requirements. *Biotechnology and Bioengineering* 56: 398–421.

Pratt, C.W., Voet, D., and Voet, J.G. (2008). *Fundamentals of Biochemistry*. Chichester: John Wiley & Sons Ltd.

Shuler, M.L., and Kargi, F. (2002). *Bioprocess Engineering: Basic Concepts*. Upper Saddle River, NJ: Prentice Hall.

Stephanopoulos, N., Aristidou, A., and Nielsen, J. (1998). *Metabolic Engineering Principles and Methodologies*. San Diego, CA: Academic Press.

Xiao, Y., Feng, X., Varman, A.M., He, L., Yu, H., and Tang, Y.J. (2012). Kinetic modeling and isotopic investigation of isobutanol fermentation by two engineered *Escherichia coli* strains. *Industrial and Engineering Chemistry Research* 51: 15855–63.

Exercise Problems

7.1. For inhibition models, which type of inhibition reduces the model's apparent μ_{max}?

7.2. *Zymomonas mobilis* uses the ED pathway for the fermentation of glucose. How many NADHs are produced when the cells convert glucose to pyruvate via the ED pathway?

7.3. Describe two different butanol-producing pathways.

7.4. How many ATPs/NADHs are generated by glycolysis? Which pathway produces the majority of NADPHs?

7.5. One microbial species can use either acetate or glycerol as the sole carbon source for growth. Can you predict which carbon substrate yields the higher microbial biomass?

7.6. What is the advantage of using *Clostridium thermocellum* to produce ethanol? Can yeast ferment ethanol under aerobic conditions?

7.7. What is the model equation for uncompetitive substrate inhibition for microbial growth?

7.8. The microbial biomass formation equation for a microorganism using glucose and NH_3 as carbon and nitrogen sources is:

a $C_6H_{12}O_6$ + b NH_3+ c $O_2 \rightarrow CH_{1.7}N_{0.18}O_{0.47}$ (microbial biomass) + d CO_2 + e H_2O

Balance this equation and calculate the glucose yield coefficient for microbial biomass (c/d = 1.9).

7.9. An engineered *E.coli* has a mass doubling time of 3 hr when grown on 40 g/L acetate. The Monod constant using acetate is 1 g/L, and cell yield on acetate is 0.4 g cell/g acetate. If we operate a chemostat on a feed stream containing 40 g/L acetate, find the substrate concentration and cell productivity (g cell/hr) when the chemostat dilution rate is 60% of the theoretical maximum dilution rate.

7.10. Bioreactions are often carried out in batch reactors. Use the available information to determine how much time is required to achieve a 95% conversion of the substrate. Assume that the volume V of the reactor content is constant, and that the reaction rate follows the Monod model.

Initial conditions: Microbial biomass $X(0) = 0.05$ g/L, substrate $S(0) = 15$ g/L, product $P(0) = 0$ g/L

Parameter values: $V = 1$ L, $\mu_{max} = 0.2$ 1/hr, Ks = 1 g/L

Yield coefficients are assumed to be constant: $Y_{x/s} = 0.5$ g microbial biomass/g substrate, $Y_{P/s} = 0.1$ g product/g substrate.

Appendix 7.1 Code Useful for Example 7.2

MAIN FILE

```
clear;
% initial conditions
y0 = [0.12 0.08 0 20]; %x A IB G
tspan = [0:2:50];
% solving the differential equations "ff"
```

```
[t1,y] = ode23(@ff,tspan,y0);
y1 = y(:,1); %X
y2 = y(:,2); %A
y3 = y(:,3); %Ib
y4 = y(:,4); %G
%Plot
figure(1)
plot(t1,y(:,1),'ko')
set(gca,'linewidth',1.5,'fontsize',12,'YTick',0:0.4:1.2)
xlabel('Time (h)','fontsize',16,'fontweight','b')
ylabel('Biomass (g/L)','fontsize',16,'fontweight','b')
figure(2)
plot(t1,y(:,2),'ko')
set(gca,'linewidth',1.5,'fontsize',12,'YTick',0:0.2:1)
xlabel('Time (h)','fontsize',16,'fontweight','b')
ylabel('Acetate (g/L)','fontsize',16,'fontweight','b')
figure(3)
plot(t1,y(:,3),'ko')
set(gca,'linewidth',1.5,'fontsize',12,'YTick',0:0.1:0.3)
ylim([0 0.3])
xlabel('Time (h)','fontsize',16,'fontweight','b')
ylabel('Isobutanol (g/L)','fontsize',16,'fontweight','b')
figure(4)
plot(t1,y(:,4),'ko')
set(gca,'linewidth',1.5,'fontsize',12,'YTick',0:5:20)
ylim([0 20])
xlabel('Time (h)','fontsize',16,'fontweight','b')
ylabel('Glucose (g/L)','fontsize',16,'fontweight','b')
```

FUNCTION FILE

```
  function dy = ff(t,y)
   param = [0.073 0.305 47.69 0.007 0.614 0.528 0.123 0.161 0.080
   0.163 0.006];
    Ug = (param(1)*y(4)/(param(2) + y(4)))*(1/(1 + y(2)/param(3)));
    Rx = y(1)*Ug;
    dy(1) = y(1)*(Ug-param(4)); %microbial biomass
    dy(2) = param(5)*Rx; % acetate
    dy(3) = param(6)*Rx + param(11)*y(1) -param(7)*y(3); % isobutanol
    dy(4) = -Rx/param(8)-dy(2)/param(9)-param(6)*Rx/param(10); %
    glucose
    dy = dy';
   end
```

SECTION II
Bioenergy Feedstocks

Starch-Based Feedstocks

Xumeng Ge and Yebo Li

What is included in this chapter?

This chapter covers important starch-based feedstocks, especially their growing conditions and yields for biofuel production. Feedstocks plantation, harvesting, and storage are also discussed.

8.1 Introduction

Starch crops are usually grown for food and feed supply, and include cereals (such as corn, sorghum, millet, wheat, rice, barley, oats, teff, and quinoa) and starchy roots (such as cassava, sweet potato, yam, taro, potato, and arrowroot) (Latham, 1997). The current global production of starch feedstocks is around 3.07 billion metric tons (2.35 billion metric tons of cereals and 0.72 billion metric tons of starchy roots). Approximately 79% is used for food/feed applications, about 6% for biofuel production, and the remaining 15% for other uses, such as seeds and stocks (FAOSTAT, 2009). About 40% of corn is used for bioethanol production in the USA (FAOSTAT, 2012), while cassava is commonly grown in tropical countries like Thailand, Nigeria, Indonesia, and Congo for bioethanol production (Kuiper *et al.*, 2007). In Germany, corn is planted as an energy crop where the whole plant is harvested before maturation, and is used as a feedstock for biogas production through anaerobic digestion (see Chapters 17 and 18). Currently, about 98% of starch-based bioethanol is globally produced from corn starch (Vivekanandhan *et al.*, 2013). However, as corn is one of the major cereal crops after wheat and rice, expanding bioethanol production using corn starch may contribute to rising food/feed prices and may lead to global hunger. There could also be environmental concerns associated with increased nutrient and herbicide inputs. Tuber-based starch crops, such as sweet potato and cassava, which require lower inputs, could be a better choice for bioethanol production (Ziska *et al.*, 2009). This chapter focuses on the agronomic aspects of growing major *starch-based feedstocks* (e.g., corn, sweet potato, and cassava), input requirements (water and fertilizer), yields, harvesting, and storage.

Bioenergy: Principles and Applications, First Edition. Edited by Yebo Li and Samir Kumar Khanal.
© 2017 John Wiley & Sons, Inc. Published 2017 by John Wiley & Sons, Inc.
Companion website: www.wiley.com/go/Li/Bioenergy

8.2 Corn

Although corn (*Zea mays* subsp. *mays*) is the most widely cultivated cereal crop in the world, it is actually a tropical plant. It has taken a century of breeding to make corn well adapted to, and productive in, temperate climates (Troyer, 2006). Currently, the USA is the leading corn producer, followed by China and Brazil, contributing about 35%, 21%, and 8% of global production (1,017 million metric tons), respectively, with production yields of 9,970 kilograms per hectare (kg/ha) or 149 bushels per acre (bu/ac), 6,175 kg/ha (92 bu/ac), and 5,258 kg/ha (78 bu/ac), respectively, based on 2013 data (FAOSTAT, 2013).

The corn kernel consists of four major parts: pericarp (bran coat or hull), endosperm, germ (embryo), and tip cap (Figure 8.1). The pericarp is the outer part of the seed, which acts as a barrier against water loss and protects the seed from disease and insects. The endosperm is inside the pericarp, and primarily contains starch. The germ is embedded in the endosperm and mainly consists of oil. The tip cap connects the kernel to the cob, and allows the passage of water and minerals to the kernel (Roozeboom *et al.*, 2007). Yellow dent corn (*Zea mays* var. *indentata*), which is also known as commodity corn, is most commonly used for ethanol production.

8.2.1 Growth and Development of Corn

Corn is generally grown from selected cultivars. After planting, the seed absorbs water, causing the nutrients in the endosperm to dissolve and be absorbed by the embryo, which grows into the new plant. Typically, the seedling emerges within 5–21 days depending on the depth of planting, soil moisture, and soil temperature. Corn growth is classified into emergence (E), vegetative (VEG), and reproductive (REP) stages. The VEG stage is further designated numerically as VEG1, VEG2, and so on, with the number representing the highest leaf that has a visible collar (a light-

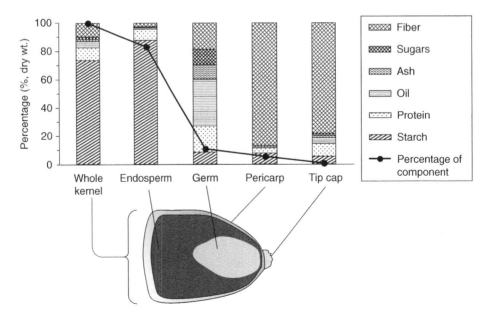

FIG. 8.1   Composition of components of dent corn kernel. Source: Data from Sprague and Dudley, 1988.

Table 8.1 Growth stages of a corn plant

Symbol	Stage	Days after planting	Description	Risks
E	Emergence	0	–	–
VEG1	1st leaf	–	–	–
VEG2	2nd leaf	–	–	–
VEG3	3rd leaf	8–10	Leaves and ear shoots are visible	Flood
VEG6–8	6th–8th leaves	21–36	Growing point above ground	Susceptible to wind and hail damage. Moisture and nutrient stress should be prevented
VEG12–17	12th–17th leaf	36–60	The number of rows per ear and ear size is visible	Moisture and nutrient stress can cause unfilled kernels
VT–REP1	Tasseling–silking	54–62	Tassel and ear shoots visible	Water stress may cause loss in yield
REP2	Blister	66–74	Kernel moisture ~85%	Not applicable
REP3	Milk	76–86	Kernel moisture ~80%	Stress can reduce corn yield
REP4	Dough	84–88	Kernel moisture ~70%	Not applicable
REP5	Dent	90–100	Kernel moisture ~55%	Stress may reduce kernel mass
REP6	Physiological maturity	105–120	Formation of black layer with maximum dry wt. of kernels	Not applicable

Source: Data from Hanway, 1971.

colored, narrow band at the base of the leaf). The last VEG stage, VT, denotes tasseling. At the VT stage, the last branch of the tassel is visible, while the silks are not. The REP stage is related to kernel development, and is designated REP1, REP2, and so on, as defined in Table 8.1. Identifying the correct growth stage is essential for scheduling fertilization, irrigation, pest management, and harvesting operations.

8.2.2 Growing Degree Days for Corn Growth

Plant growth is heavily influenced by daily ambient temperatures. Plants do not grow at temperatures below their base temperature. When the air temperature is above the base temperature, plant growth increases with increasing temperature as long as the levels of light, moisture, and nutrients are sufficient. However, when the air temperature rises above the ceiling temperature, plant growth decreases significantly. The base and ceiling temperatures for corn are 10 °C and 30 °C, respectively (Espinoza and Ross, 2004).

The growing degree day (GDD) calculation reflects the accumulated heat for a given period of time and is used to predict stages of plant growth. It is defined as the average daily temperature minus the base temperature. The GDDs determined for each day are added together to calculate the total GDD. The equation for calculating total GDD is as follows:

$$Total\ GDD = \sum \left(\frac{Daily\ max\ temp. + Daily\ min\ temp.}{2} - Base\ temp. \right) \tag{8.1}$$

Note: If the daily maximum temperature is higher than the ceiling temperature, the ceiling temperature is used as the daily maximum. GDDs can be calculated using either Celsius or Fahrenheit, and can be converted as $5\ GDD^C = 9\ GDD^F$ (GDD^C is used in this chapter).

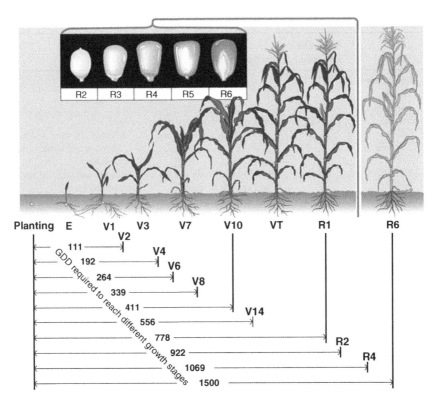

FIG. 8.2 GDD requirements for different growth stages of a typical hybrid of corn. Source: Adapted from https://extension.cropsci.illinois.edu; http://agron-www.agron.iastate.edu.

One of the most important applications of GDD is tracking the development of crops. Figure 8.2 shows the GDDs required to reach different growth stages for a typical hybrid of corn.

Example 8.1: If the daily maximum and minimum temperatures are 27 °C and 17 °C, respectively, what is the GDD for corn growth for this day?

Solution: Since the maximum and minimum temperatures are between the base and ceiling temperatures for corn, the GDD of this day can be calculated as: $(27 + 17)/2 - 10 =$ **12 GDD**.

8.2.3 Cultivation Practices in Corn Production

Selection of Seed

Seed selection is essential for achieving high yields of corn. Seeds from well-adapted hybrids have demonstrated increased yields of 0.3 kg or more per square meter (50 bushels or more per acre) (Roozeboom *et al.*, 2007). Seeds may also be selected for improved tolerance to abiotic conditions.

For example, an early-maturing hybrid may be suitable for non-irrigated soil with low fertilization. Conversely, a full-season hybrid seed may do well under full irrigation, heavy fertilization, and a long growing season. Also, some seeds are stable under varying environmental conditions, while others may be more sensitive to changes. In all cases, it is important to use multi-year data for growing conditions before selecting seeds.

Seedbed Preparation

Corn has been successfully cultured on a wide range of soil types, but grows best on well-drained, medium- to coarse-textured soils. Corn needs warm, moist soil that is well aerated and fine enough to provide contact between the soil and seed to insure rapid germination. To minimize soil erosion caused by wind and water, conservation tillage methods are commonly used. In conservation tillage, crop residues are left on the field after harvest, increasing surface roughness and/or soil permeability to reduce the erosive effects of wind, rain, and flowing water. Several factors, including climatic conditions, soil erodibility, surface roughness, field length, and length and steepness of slope, need to be considered to determine the amount of residue that should be left on the field. When water erosion is the main concern, 30% residue cover is recommended (Al-Kaisi *et al.*, 2009). In contrast, relatively flat fields need only 12–20% residue cover (Al-Kaisi *et al.*, 2009). For fields with steeper or longer slopes, 50–60% residue cover may be required (Al-Kaisi *et al.*, 2009).

Drainage

Ideally, corn fields should not have standing water for 24 hours after rainfall or irrigation. The effect of cold and wet conditions on crop development can be reduced by planting in raised rows or beds, especially for relatively flat fields. Grading to slope fields toward drainage outlets may be possible. However, the soil should be evaluated before grading, since some areas with poor soil quality may be exposed by deep cuts. As discussed earlier, crop residues used in conservation tillage can improve infiltration, especially for silty soils that may seal or crust after rainfall or irrigation, which restricts water infiltration into the root zone. In areas with high clay content and minimal slopes, drainage tiles are installed to help remove excess water.

Planting

Soil temperature is crucial for optimal corn production. Early planting utilizes the entire growing season, which can improve yields (Espinoza and Ross, 2004). Planting can be started when the soil temperature reaches 12.8 °C at a 5 cm (~2 in) depth by 9:00 AM for three consecutive days (Espinoza and Ross, 2004). Therefore, the calendar date may vary depending on geographical location. The seeding rate and planting depth also affect yields. Since not all seeds will germinate or mature, the seeding rate should be higher than the desired harvest; however, too high a seeding rate can also reduce yields. Typically, the harvest is about 85% of the planting rate (Roozeboom *et al.*, 2007). For example, in order to obtain 24,000 plants per acre, 28,200 seeds per acre should be planted. The normal planting depth is from 3.8–5 cm (~1.5–2 in), with depths greater than 8.9 cm (~3.5 in) negatively affecting emergence (Roozeboom *et al.*, 2007). However, if seeds are planted too shallowly, the crown may develop at the soil surface, which inhibits the development of roots and limits nutrient uptake.

Water Requirement

During the growing season, corn crops require 51–76 cm (~20–30 in) of water, depending on weather conditions, plant density, fertility, soil type, and days to maturity (Roozeboom *et al.*, 2007).

Table 8.2 Rates of nitrogen (N) required before crop emergence with or without in-season fertilization

	N application rate in kg N/ha (lbs N/ac)	
Crop category	No in-season fertilization	In-season fertilization
Corn on recently manured soils	0–100 (0–90)	0–34 (0–30)
Corn after established alfalfa	0–34 (0–30)	0–34 (0–30)
2nd-year corn after alfalfa	0–67 (0–60)	0–34 (0–30)
Corn after corn	168–224 (150–200)	56–140 (50–125)
Corn after soybean (no manure)	112–168 (100–150)	0–84 (0–75)

Source: Data from Blackmer *et al.*, 1997.

The agronomic water requirement depends on the geographical location. In the USA, over 95% of corn farms are not irrigated. In the first 3–4 weeks of vegetative growth (VEG1–VEG6), the water demand can usually be met by rainfall. However, at the stage when the corn has approximately 8 fully developed leaves (VEG8), the growth rate of corn greatly increases and irrigation is usually required to avoid moisture stress if rain is sporadic. The most critical period to insure adequate moisture is from approximately 2 weeks before silking (VT) until 2–3 weeks after silking (REP2–REP5).

Fertilizer Inputs

Nutrients, such as nitrogen (N), phosphorus (P), and potassium (K), must be supplied in the form of fertilizers, manures, and/or legume rotations to sustain profitable corn production. Table 8.2 shows recommended N rates before crop emergence with or without in-season fertilization (Blackmer *et al.*, 1997).

P and K sufficiency recommendations for corn production can be calculated using the following equations (Roozeboom *et al.*, 2007):

$$P = 56 + 3.57 \times 10^{-3} \, Y - 2.8 P_0 - 1.78 \times 10^{-4} \, Y \cdot P_0 \tag{8.2}$$

$$K = 81.76 + 3.75 \times 10^{-3} \, Y - 0.63 \, K_0 - 2.86 \times 10^{-5} \, Y \cdot K_0 \tag{8.3}$$

where P is the phosphate application rate (kg/ha), Y is the expected yield (kg/ha), P_0 is the phosphate content in soil (ppm), K is the potassium application rate (kg/ha), and K_0 is the potassium content in soil (ppm).

8.2.4 Harvesting and Storage of Corn

Harvesting dates vary in different locations depending on the regional climates. In the USA, harvesting begins in October and finishes by the end of November. In China, corn harvesting dates are from August through October, whereas in Brazil, corn is harvested from February through May. Corn grain harvesting has been fully mechanized in developed countries, where the goal is to separate the desired product from other plant residues (i.e., corn kernels from cob, husk, leaves, and stalk). This separation process is known as threshing. A combine harvester, also known simply as a combine, is employed to integrate both harvesting and threshing of corn grain. Such equipment mainly recovers the grain, with the major parts of the plants left in the field (Figure 8.3). The total grain loss from a combine can be as low as 3% (Proctor, 1994). However, the average corn harvest losses may vary from 130–630 kg/ha (2–10 bu/ac), depending on the operator's skills (Roozeboom *et al.*,

1 Cab	6 Feed conveyor	10 Cleaning sieve
2 Engine	7 Threshing cylinder speed variator	11 Grain auger trough
3 Guide wheels	8 Threshing cylinder	12 Grain elevator
4 Hydraulic speed variator	9 Fan	13 Hopper
5 Maize header		14 Unloading auger

FIG. 8.3   Combine harvester used for corn harvesting. Source: http://www.fao.org/docrep/t1838e/ T1838E13.GIF. Reproduced with permission of Cirad from doc. Rivierre-Casalis.

Table 8.3 Corn harvest loss in kg/ha (bu/ac) for different operators

Parameters	Average	Expert
Ear loss	251 (4.0)	63 (1.0)
Loose kernel loss	88 (1.4)	31 (0.5)
Cylinder loss	44 (0.7)	19 (0.3)
Total	383 (6.1)	113 (1.8)

Source: Data from Roozeboom *et al.*, 2007.

2007). There are three types of loss: ear losses where ears are left on the stalks or dropped from the header after being snapped; loose kernel losses where kernels are left on the ground, either by threshing or by being discharged from the rear of the harvester; and cylinder losses where kernels are left on the cob due to incomplete shelling. Table 8.3 shows the losses expected from the average combine operator compared to an expert. Achieving the "expert" level of yield requires the determination of losses and making the appropriate adjustments.

Harvested corn should be dried to a moisture level of about 12% for long-term storage (several months) (Roozeboom *et al.*, 2007). Corn drying is necessary to reduce the damage and loss of corn during storage. Drying is performed by passing large quantities of dry air over the corn as it is placed in bins. Different corn drying and storage facilities are shown in Figure 8.4. For the continuous and automatic batch dryer, corn near the inside of the column is at a lower moisture content and a higher temperature than grain near the outside of the column. For the batch-in-bin dryer, the moisture content of corn at the floor of the bin is much lower than that of the top layer. Batch-stirring dryers produce a more uniform moisture content and corn temperature due to the adoption of a stirring auger. The sweep auger in the continuous-flow bin dryer removes the corn on the bottom of the bin to prevent overdrying of the corn. A high drying temperature may affect the yields of products during the corn wet milling process.

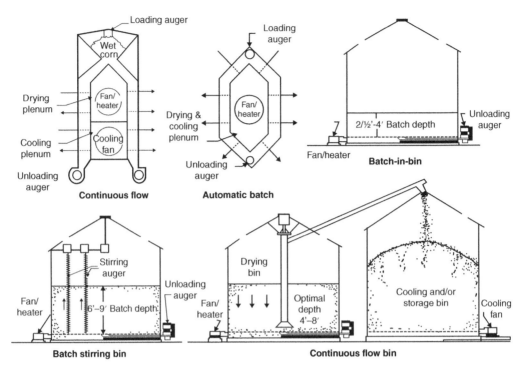

FIG. 8.4 Schematic diagrams for corn drying and storage facilities. Source: http://www.extension. umn.edu/agriculture/corn/harvest/dryeration-and-cooling-for-corn-drying/. Reproduced with permission of the University of Minnesota.

Key considerations that can improve long-term storage

- Cool grain to about 16 °C as soon as possible after harvesting (Espinoza and Ross, 2004).
- Continue to cool grain uniformly down to -1–4 °C if possible, which will prevent internal moisture migration and insect activity (Espinoza and Ross, 2004).
- Maintain uniform temperature and moisture levels by monitoring the grain and aerating monthly during storage.
- Keep the grain cool for as long as possible.
- Limit air, moisture, and insect movement by covering fans and openings that are not in use.
- Inspect the grain surface at least every week.

8.3 Sweet Potato

Sweet potato (Ipomoea batatas) is a starch-rich storage root originating from South America. Global sweet potato production was 102.7 million metric tons in 2013, mainly from China (~69%) (FAOSTAT, 2013). Storage roots of sweet potato (unprepared raw material) contain (on a dry wt. basis) 56% starch, 18% sugar, 13% fiber, 7% protein, and 4% ash, as well as small amounts of lipids and vitamins (USDA, 2013).

8.3.1 Growth and Development of Sweet Potato

Sweet potatoes can be propagated vegetatively either by producing transplants from roots (Figure 8.5) or by planting vine cuttings. The period from transplanting to harvesting may vary from 90–190 days (Martin *et al.*, 2013), depending on cultivars and environmental conditions. Sweet potato growth can be divided into three stages (Van de Fliert and Braun, 1999), as shown in Figure 8.5.

8.3.2 Cultivation Practices in Sweet Potato Production

Selection of Cultivars

Several cultivars of sweet potato are listed in Table 8.4 with characteristics that may affect bioethanol potential (Garrett, 1987).

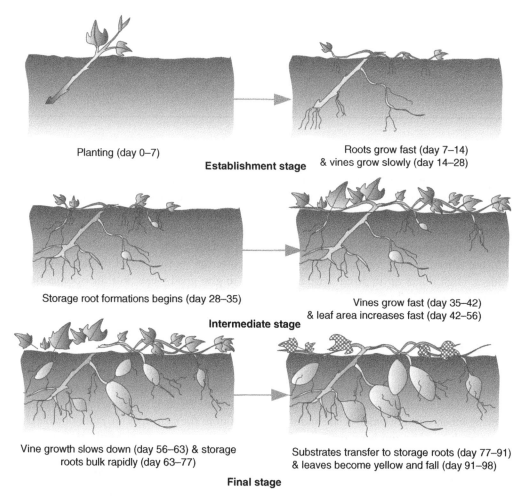

Planting (day 0–7)

Establishment stage

Roots grow fast (day 7–14)
& vines grow slowly (day 14–28)

Storage root formations begins (day 28–35)

Intermediate stage

Vines grow fast (day 35–42)
& leaf area increases fast (day 42–56)

Vine growth slows down (day 56–63) & storage
roots bulk rapidly (day 63–77)

Substrates transfer to storage roots (day 77–91)
& leaves become yellow and fall (day 91–98)

Final stage

FIG. 8.5   Growth stages of a sweet potato plant (Van de Fliert and Braun, 1999).

Table 8.4 Characteristics of different cultivars of sweet potato (Garrett, 1987)

Cultivars	Advantages	Disadvantages
Beauregard	Resistant to soil pox; stores well	Slow sprouting
Hernandez	Resistant to root knot, soil pox, and Fusarium wilt	Late sporadic sprouting
Jewel	Resistant to root knot nematode; stores well	Crack with variable soil moisture
Centennial	Resistant to Fusarium wilt, internal cork, and wireworms	Poor sprout potential
Goldstar	Resistant to root knot and Fusarium wilt; early, prolific sprouting; stores well	Not applicable
White Hayman	Not applicable	Probably susceptible to nematodes, disease, and insects

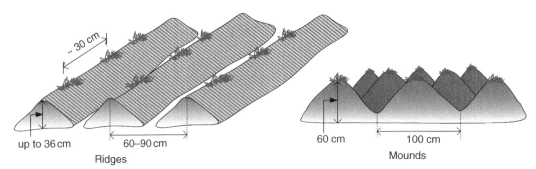

FIG. 8.6 Preparation of field for sweet potato cultivation. Source: Adapted from Onwueme, 1978; http://www.fao.org.

Field Preparation

Sweet potatoes can grow in a variety of soil types, but the most desirable is a well-drained, moderately deep, and sandy loam (Garrett, 1987). Heavier soils (clay or fine textured) may cause sweet potato cracking, which increases the chance of rotting during storage. In tropical climates, sweet potatoes may be grown on ridges, heaped piles of earth (mounds), or flat ground (Figure 8.6). However, cultivation on flat ground may result in low yields. Sweet potatoes grown in mounds generate good yields, but labor costs for mound-making are high. Thus, the recommended method is planting on ridges. High ridges up to 36 cm have been shown to give good yields, but the optimal height is dependent on the soil type and cultivar of sweet potato (Onwueme, 1978).

Planting Material

Sweet potato tubers should be well shaped; free of diseases, insects, and root mutations; and have high yields. Generally, vine cuttings are better suited to the commercial propagation of sweet potato. This method has several advantages, such as reducing soil-borne diseases, achieving higher yields, and producing tubers of more uniform size and shape. Pieces from the stem apex are more suitable than those from the middle and basal portions of the stem. If there is a shortage of planting materials, middle and basal vine cuttings may also be acceptable; however, there will be a slight reduction in yield. Longer vine cuttings usually give higher tuber yields, while shorter cuttings establish more slowly and result in lower yields. However, cuttings longer than the recommended length of 30 cm tend to waste planting material.

Field Planting

The planting date is usually 1–2 weeks after the frost-free date, or when soil temperatures are at least 18 °C. The optimum temperatures for sweet potato planting are 24–29 °C (Martin *et al.*, 2013). The vine cuttings can be planted using different methods, as shown in Figure 8.7.

Water Requirement

Although sweet potato is tolerant to drought conditions, additional water may be needed to optimize yields. If the soil is dry before planting, it may be irrigated. About 225 m^3/ha of water can be applied every 7–10 days after planting when the roots are developing. When the plant becomes mature, only moderate amounts of water are needed (Ramirez, 1991). Irrigation may be stopped about half a month before harvest (Ramirez, 1991).

Fertilizer Inputs

There are different ways of applying fertilizer for sweet potato, as shown in Table 8.5.

8.3.3 Harvesting and Storage of Sweet Potato

Frequent checking of root development is necessary, since there are no above-ground indicators of ripening or maturity. Commonly, a potato plow or mechanical harvester (Figure 8.8) is used to lift the roots to the soil surface.

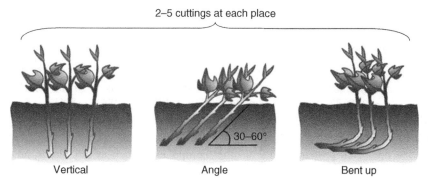

2–5 cuttings at each place

Vertical   Angle   Bent up

30–60°

FIG. 8.7   Methods of sweet potato planting. Source: Adapted from http://www.uq.edu.au.

Table 8.5 Method and timing of fertilization for sweet potato

Time	Location	Methods
20–30 days before planting	In sandy soils	Distribute using a harrow after spreading
At planting	In the furrow	Place in the furrow and cover with soil from the hill
At planting	In the hill	Place at points 30 cm apart within the furrow
30 days after planting	On the hill	(a) Spread along one side of the hill
		(b) Cover with soil using a plow

Source: Adapted from Ramirez, 1991.

(a)   (c)

(b)

FIG. 8.8   Machines for sweet potato harvesting (a) Chopper for removing vines; (b) Digger for digging sweet potatoes; and (c) Combine that can harvest sweet potatoes without vine removal. Source: http://www.caes.uga.edu/commodities/fruits/veg/pubs/documents/sweetpotato.pdf. Reproduced with permission of Paul Sumner.

Care must be taken during harvesting and grading, because injury to the tender outer skin increases the risk of storage diseases. When sweet potatoes are cured, they can be stored in baskets or bulk bins for long periods. After harvest, curing should be done immediately at 27–29 °C with a relative humidity of 90% for 4–8 days (Martin *et al.*, 2013). The sweet potato skin develops a waxy-like layer during the curing process, which is resistant to moisture loss and microorganism invasion. The storage room or building should provide a controlled temperature of 13 °C, humidity of 85%, and ventilation throughout the fall and winter months. Exposure to temperatures below 13 °C can cause chilling injury (Martin *et al.*, 2013).

8.4 Cassava

Cassava (Manihot esculenta) originated in Brazil and Paraguay, and has enlarged, starch-filled roots (Cock, 1985; Figure 8.9). The global cassava production in 2012 was 256.5 million metric tons, with 21% produced in Nigeria, followed by significant production in Indonesia, Brazil, and Thailand, which each contributed about 9%. Cassava roots (raw) contain (on a dry wt. basis) 86% starch, 4% sugar, 5% fiber, 3% protein, and 2% other components, such as ash, lipids, and vitamins (USDA, 2013).

8.4.1 Growth and Development of Cassava

Cassava seed, which is difficult to germinate, is used only for breeding work. However, cassava can also grow vegetatively from the stem segments. The time required from cassava planting to harvesting ranges from 9–24 months, depending on climate and soil conditions. Typically, at least 9–12 months are required to obtain good yields, although some cultivars can grow faster and be harvested in 6–7 months. In general, the stems can be cut after 6 months and planted into other fields, while the roots remain underground. Within several weeks, new vegetative growth will develop from the transplanted stems.

There are two distinct growth phases of the cassava plant. During the first phase, from planting to 8 weeks, stems, leaves, and root systems begin to develop. During this phase, the cassava tuber begins to form. The second growth phase is from week 8 to week 72. During this phase, stems and leaves

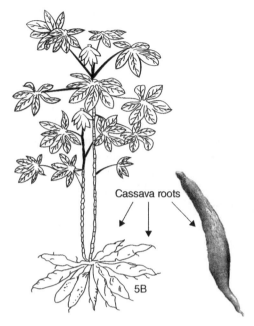

FIG. 8.9   Cassava roots. Source: International Starch Institute, Science Park Aarhus, Denmark.

grow rapidly, and the cassava tuber swells and bulks (Titus *et al.*, 2011). Cassava has no designated mature stage, and is ready for harvest when the tuber reaches the required size. The tubers will continue to grow if left in the ground, but the central portion becomes woody and inedible (Moore and Lawrence, 2003).

8.4.2 Cultivation Practices in Cassava Production

Cassava cultivars vary in yield, root size, disease and pest resistance, time to harvest, and temperature adaptation. There are common criteria for selecting cassava cultivars as a starch-based feedstock for bioethanol production (FiBL, 2011):

- Adaptation to local weather conditions
- High dry matter content (>30%, w/w)
- Ability to bulk early
- Ability to be stored underground for a long time
- Tolerance to local weeds, pests, and disease

Cultivation Conditions

Cassava grows best at 25–27 °C, and can grow with rainfall ranging from 50–500 cm (Moore and Lawrence, 2003). Cassava is a good crop in low or uncertain rainfall areas due to its resistance to prolonged periods of drought. Light sandy loams of medium fertility are the best for cassava production. Clay soils are more suitable for stem and leaf growth; however, this growth is at the expense of roots, which can lead to a reduction in total tuber yield. Saline and swampy soils are unsuitable for cassava cultivation.

Seed Bed Preparation

In traditional agriculture, unplowed land and mounds are the most common methods for planting cassava. Typically, planting on unplowed land occurs after brush is removed via burning. In West Africa and several other cassava-producing areas, it is common to plant on mounds that are built to heights of 30–60 cm. Cassava plants grown on mounds may give greater yields than those grown on unplowed land. Also, mounding can reduce the risk of waterlogging and ease harvesting. However, labor is required to prepare mounds due to the difficulty in mechanizing the mound-making process. In industrial agriculture, land is first plowed and then harrowed. The cassava can be planted on the flat ground before ridging or on ridges after ridging (Figure 8.10). Although planting on flat ground can improve tuber yield compared to planting on ridges, it is not suitable for heavy soils since tubers may rot. Thus, ridge planting is recommended on heavy soils, while planting on flat ground can be practiced on sandy soils.

Spacing

Generally, cassava is planted in rows, with spacing of 80–100 cm within and between the rows. The exact spacing depends on the cultivar and the growing conditions. Cultivars that tend to spread should be planted farther apart than those that tend to grow upright. Cassava should be spaced farther apart in areas with high soil fertility and high rainfall. In order to suppress weeds and prevent soil wash or desiccation, it is important that the canopies of adjacent plants merge within the first 2–3 months after planting.

FIG. 8.10   Seed bed preparation for cassava cultivation. Source: International Starch Institute, Science Park Aarhus, Denmark.

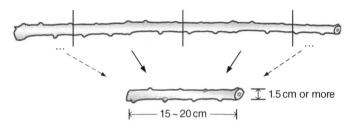

FIG. 8.11   Stem cuttings of cassava planting materials.

Stem Preparation

Healthy stem cuttings 15–20 cm long are taken from mature mid-plant sections (Figure 8.11). It is important that disease-free stems be selected for planting, as disease has the greatest impact on profitability. However, selecting stems cannot be mechanized, thus it is a labor-intensive operation.

Planting

It is important to plant cassava when soils have sufficient moisture, and additional rainwater can be expected. For regions that have distinct wet/dry seasons, planting is recommended as early as possible in the wet season. Early planting in the warm season is also important in cooler, sub-tropical or high-altitude regions. In various parts of the world, particularly Brazil and Mexico, mechanization of cassava planting has been successfully implemented.

Fertilization and Irrigation

Cassava responds primarily to phosphorus and potassium fertilization. Excessive foliage may be produced, at the expense of storage root development, when nitrogen fertilization is higher than 100 kg/ha (89 lb/ac) (Moore and Lawrence, 2003). Fertilization is only required in the first few months of growth. Irrigation may be used when rainfall is limited (Moore and Lawrence, 2003).

8.4.3 Harvesting and Storage of Cassava

Harvesting of cassava usually begins 7–10 months after planting, although it can be left longer in drier soils. Manual harvesting is commonly used, and an example technique is shown in Figure 8.12. Mold-board plows have also been used for hand harvesting.

Mechanical harvesting of cassava is difficult because the roots can spread horizontally over 1 m and can penetrate to depths of 50–60 cm. A feasible harvesting method is to cut stalks by a mid-mounted mower or a topping machine, and then lift the roots mechanically with a mid-mounted disk terrace.

Fresh cassava roots cannot be stored for long periods because they begin to spoil once they are out of the ground. Physiological deterioration begins within 3 days. Broken roots during harvesting allow the entry of microorganisms that increase the rate of root rot. One strategy to minimize physiological deterioration in the tubers is to preprune aerial portions of the plant 2–3 weeks before harvesting. Packing in moisture-absorbent material, such as sawdust, can minimize deterioration as well as invasion by pathogens, and can extend the storage life of fresh roots to about 4 weeks. For long-term storage (up to 1 year), cassava roots can be sliced and dried until the moisture content is about 14% (Cock, 1985). Sun-drying is generally practiced due to its low cost.

FIG. 8.12   Method of manually harvesting cassava roots practiced in Thailand.

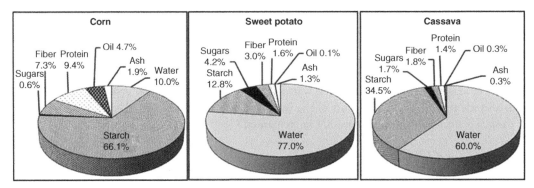

FIG. 8.13   Compositions of corn, sweet potato, and cassava (wet wt. basis). Source: USDA National Nutrient Database, http://ndb.nal.usda.gov/ndb.

8.5 Comparison of Composition, Yield, and Energy Potential of Corn, Sweet Potato, and Cassava

The relative compositions of corn (yellow), sweet potato root (raw, unprepared), and cassava root (raw, unprepared) are shown in Figure 8.13. The most common biofuel product from starch-based feedstocks is bioethanol. Normally, starch is enzymatically hydrolyzed into glucose, followed by ethanol fermentation with yeast. In addition, the soluble sugars in corn, sweet potato, and cassava are mainly sucrose, which can also be readily fermented into ethanol by yeast, *Saccharomyces cerevisiae*.

An important parameter that is typically considered for biofuel production from crops is energy potential (e.g., how much energy can potentially be produced). Energy potential can be calculated based on crop yield, composition, and conversion efficiency. Table 8.6 shows the ethanol yield and energy potential of corn, sweet potato, and cassava based on global crop yields. It should be noted that the energy potential for different regions may vary. For example, it has been reported that sweet potato grown in Maryland may have a greater ethanol potential than corn due to the high yield and starch content of sweet potato in that location (Ziska *et al.*, 2009).

Table 8.6 Comparison of yield and energy potential between corn, sweet potato, and cassava

Feedstocks	Fresh yield[a] (wet kg/ha)	Ethanol yield[b] (kg/ha)	Energy potential[c] (GGE/ha)
Corn	4,944	1,676	374
Sweet potato	12,891	1,156	258
Cassava	12,833	2,337	522

[a] Global crop yield (http://faostat3.fao.org).

[b] Approximately 1 kg of starch (or sucrose) can produce about 0.5 kg of ethanol with a total conversion efficiency of 90%.

[c] GGE is gasoline gallon equivalent, where 1 GGE = 1.5 gallons of ethanol (or 4.48 kg of ethanol).

References

Al-Kaisi, M.M., Hanna, M., Peterson, T.S. (2009). *Residue Management and Cultural Practices*. Pm.1901a. Ames, IA: Iowa State University, University Extension.

Blackmer, A.M., Voss, R.D., and Mallarino, A.P. (1997). *Nitrogen Fertilizer Recommendations for Corn in Iowa*. Pm.1714. Ames, IA: Iowa State University, University Extension.

Cock, J.H. (1985). *Cassava: New Potential for a Neglected Crop*. Boulder, CO: Westview Press.

Espinoza, L., and Ross, J. (2004). *Corn Production Handbook*. MP437-2M-2-03N. Little Rock, AR: University of Arkansas, Division of Agriculture Cooperative Extension Service.

FAOSTAT (2009). *Food Balance Sheets*. Rome: Food and Agriculture Organization of the United Nations, Statistics Division. http://faostat3.fao.org/faostat-gateway/go/to/download/FB/FB/E, accessed January 2014.

FAOSTAT (2012). *Food Outlook: Global Market Analysis*. Rome: Food and Agriculture Organization of the United Nations, Statistics Division. http://www.fao.org/docrep/016/al993e/al993e00.pdf, accessed January 2014.

FAOSTAT (2013). *Production*. Rome: Food and Agriculture Organization of the United Nations, Statistics Division. http://faostat3.fao.org/browse/Q/QC/E, accessed January 2015.

FiBL (2011). *African Organic Agriculture Training Manual*, Version 1.0, G. Weidmann and L. Kilcher (eds.). Frick: Research Institute of Organic Agriculture. http://www.organic-africa.net/fileadmin/documents-africamanual/training-manual/chapter-09/Africa_Manual_M09-06-low-res.pdf, accessed January 2014.

Garrett, W.T. (1987). *Growing Sweet Potatoes*. Fact sheet. College Park, MD: University of Maryland, Cooperative Extension Service.

Hanway, J.J. (1971). *How a Corn Plant Develops*. Ames, IA: Iowa State University of Science and Technology, Co-operative Extension Services.

Kuiper, L., Ekmekci, B., Hamelinck, C., Hettinga, W., Meyer, S., and Koop, K. (2007). *Bio-ethanol from Cassava*. Project no. PBIONL062937. Utrecht: Ecofys Netherlands.

Latham, M.C. (1997). *Human Nutrition in the Developing World*. Food and Nutrition Series 29. Rome: Food and Agriculture Organization of the United Nations.

Martin, D.A., Myers, R.D., and McClurg, C.A. (2013). *Growing Sweet Potatoes*. Fact Sheet 464. College Park, MD: University of Maryland, Cooperative Extension Service.

Moore, L.M., and Lawrence, J.H. (2003). *Plant Guide*. Baton Rouge, LA: USDA NRCS National Plant Data Center.

Onwueme, I.C. (1978). *The Tropical Tuber Crops: Yams, Cassava, Sweet Potato, and Cocoyams*. Chichester: John Wiley & Sons Ltd.

Proctor, D.L. (1994). *Grain Storage Techniques: Evolution and Trends in Developing Countries. No. 109*. Rome: Food and Agriculture Organization of the United Nations.

Ramirez, G.P. (1991). Cultivation harvesting and storage of sweet potato products. *Seed* 6: 6.

Roozeboom, K., Duncan, S., Fjell, D.L., *et al.* (2007). *Corn Production Handbook*. C-560. Manhattan, KS: Kansas State University.

Sprague, G.F., and Dudley, J.W. (1988). Corn marketing, processing, and utilization. In: G.F. Sprague and J.W. Dudley (eds.), *Corn and Corn Improvement*. Madison, WI: ASA-CSSA-SSSA, pp. 881–940.

Titus, P., Lawrence, J., and Seesahai, A. (2011). *Commercial Cassava Production*. Technical bulletin. St Augustine: CARDI.

Troyer, A.F. (2006). Adaptedness and heterosis in corn and mule hybrids. *Crop Science* 46: 528–43.

USDA (2013). *USDA National Nutrient Database for Standard Reference, Release 26.* Beltsville, MD: United States Department of Agriculture, Agricultural Research Service. http://www.ars.usda.gov/services/docs.htm?docid=8964, accessed January 2014.

Van de Fliert, E., and Braun, A.R. (1999). *Farmer Field School for Integrated Crop Management of Sweetpotato: Field Guides and Technical Manual.* Lima: International Potato Center.

Vivekanandhan, S., Zarrinbakhsh, M., Misra, M., and Mohanty A.K. (2013). Coproducts of biofuel industries in value-added biomaterials uses: A move towards a sustainable bioeconomy. In: F. Zhen (ed.), *Liquid, Gaseous and Solid Biofuels-Conversion Techniques.* Croatia: InTech. doi: 10.5772/55382.

Ziska, L.H., Runion, G.B., Tomecek, M., Prior, S.A., Torbet, H.A., and Sicher, R. (2009). An evaluation of cassava, sweet potato and field corn as potential carbohydrate sources for bioethanol production in Alabama and Maryland. *Biomass and Bioenergy* 33: 1503–8.

Exercise Problems

8.1. Define starch-based energy crops. Besides the three major categories of starch-based energy crops discussed in this chapter, list 2–3 other starch-based crops and compare them with the crops discussed in this chapter in terms of water and nutrient requirements and yields.

8.2. If daily maximum and minimum temperatures are 22 °C and 8 °C, respectively, what is the GDD for corn growth for this day?

8.3. If the base and ceiling temperatures for sweet potato growth are 15 °C and 27 °C, respectively, and daily maximum and minimum temperatures are 24 °C and 35 °C, respectively, what is the GDD for sweet potato growth for this day?

8.4. According to Table 8.6, how many hectares (ha) are required to meet the feedstock requirement for a 50,000 ton ethanol per year ethanol plant using (a) corn, (b) sweet potato, and (c) cassava?

8.5. Discuss the composition, values, and uses of byproducts after starch is converted to ethanol for corn, sweet potato, and cassava, respectively.

8.6. Select a country that produces corn, sweet potato, and cassava. Calculate and compare the energy potentials of these feedstocks based on their yields in this country (yield values can be found at http://faostat3.fao.org).

8.7. Compare the ethanol potential of rice and corn based on their yields in the USA (rice is composed of approximately 80% starch; yield values can be found at http://faostat3.fao.org).

8.8. Can we harvest cassava roots using the same machinery that is used for harvesting sweet potato? Explain your answer. Develop an idea for mechanical harvesting of cassava roots.

8.9. List three factors that affect the energy potential of a particular starch-based feedstock, and propose strategies to improve its energy potential.

Oilseed-Based Feedstocks

Chengci Chen and Marisol Berti

What is included in this chapter?

This chapter covers the basics of oilseed-based feedstocks production. Major considerations are the agronomic aspects of oilseed-based feedstocks, e.g., planting, inputs (water, fertilizer, and pesticides), and managing the crops, harvesting, and storage.

9.1 Introduction

Plant oils have long been used for food and fuel by human civilizations. As an example, the ancient Egyptians used plant oil, such as castor (*Ricinus communis* L.) oil, as fuel in oil lamps. At the beginning of the petroleum era in the 1900s, the earliest compression-ignition (diesel) engines were tested using vegetable oils. More recent developments of biodiesel derived from plant seed oils and animal fats are based on the transesterification process developed in the 1940s (Walton, 1938). Using esterification, the large and branched triglyceride molecules of plant oils and animal fats are transformed into smaller, straight-chain molecules, the fatty acid methyl esters (FAMEs), or simply biodiesel (Quick, 1989; Knothe, 2001).

Plant oils may be extracted from seeds, fruits, nuts, wood sap, or algae. However, fewer than 20 plant species are recognized as the principal sources of the world's plant oil production (Pryde and Rothfus, 1989). Some of the important oil-bearing plant species are soybean (*Glycine max* L. Merr.), oil palm (*Elaeis guineensis* Jacq.), rapeseed (*Brassica napus* L.), sunflower (*Helianthus annuus* L.), peanut (*Arachis hypogaea* L.), cottonseed (*Gossypium hirsutum* L.), coconut (*Cocos nucifera* L.), olive (*Olea europaea* L.), linseed (*Linum usitatissimum*), corn (*Zea mays* L.), mustard (*Brassica juncea*, *B. carinata*), sesame (*Sesamum indicum* L.), safflower (*Carthamus tinctorius* L.), castor, and tung (*Aleurites fordii* L.).

Different plant species adapt to different climate and soil environments, therefore climate, geography, and economics determine which type of oilseed is most appropriate for biodiesel production.

Bioenergy: Principles and Applications, First Edition. Edited by Yebo Li and Samir Kumar Khanal.
© 2017 John Wiley & Sons, Inc. Published 2017 by John Wiley & Sons, Inc.
Companion website: www.wiley.com/go/Li/Bioenergy

In the USA, soybean is the major feedstock for biodiesel production, while in Europe rapeseed is grown for this purpose. In tropical zones, such as Malaysia, Indonesia, and Colombia, oil palm is the major oil crop. Many other oilseeds, such as castor, grape seed (*Vitis vinifera* L.), camelina (*Camelina sativa* L.), jatropha (*Jatropha curcas* L.), hemp seed (*Cannabis sativa* L.), linseed, and cottonseed, also have potential for biodiesel production.

World vegetable oil production reached 170 million metric tons in 2013 (USDA-FAS, 2014); the oil production from major oilseed crops was estimated at 45 million metric tons (26.5%) for soybean oil, 26 million metric tons (15.3%) for rapeseed oil, 59 million metric tons (34.7%) for palm oil, and 7 million metric tons (4.1%) for palm kernel oil.

The major soybean producers are the USA (31.5%), Brazil (30.8%), and Argentina (19.0%), whereas the major rapeseed-producing countries are those of the European Union (29.6%), Canada (25.3%), China (20.3%), and India (10.3%). Indonesia and Malaysia together produced an estimated 86% (52.3% and 33.6%, respectively) of the world's palm oil (USDA-FAS, 2014). Although corn is not considered an oilseed crop due to the relatively low oil content of the seed (3–5%), the total oil production of this crop is notable due to the volume of corn harvested. For example, the USA produced 35.4 million metric tons of corn oil in 2013 (USDA-NASS, 2013).

The following sections will discuss the agronomic aspects of growing major oilseed-based feedstocks (soybean, rapeseed, oil palm, jatropha, and camelina), such as input requirements (water, fertilizer, and herbicides/insecticides), yield, harvesting, and storage.

9.2 Soybean

9.2.1 Feedstock Production and Handling

Originating in China, *soybean* is a warm-season, annual leguminous crop (Figure 9.1). It can be grown in tropical, subtropical, and temperate climate zones. It is critical to choose a cultivar with an appropriate maturity date, because in northern regions the fall frost will likely kill late-maturing cultivars before they reach physiological maturity. Soybean plant development can be divided into vegetative stages (VE, VC, V1–V5) and reproductive stages (R1–R8; Figure 9.1). The flowering of a soybean cultivar is controlled by both day length and temperature (growing degree days). The late-maturing cultivars are more sensitive to day length than to growing degree days. Potential for high yields, and resistance to lodging and diseases, are additional factors that must be considered in cultivar selection. Both conventional and herbicide-tolerant soybean cultivars are available.

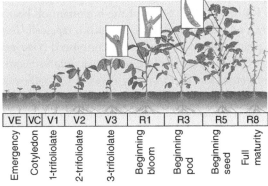

FIG. 9.1   Closer view of soybean plants and flowers (left) and soybean growth stages (right). Photo reproduced with permission of Nathan Mueller, Nebraska Extension; soybean growth stages adapted from University of Illinois Extension.

FIG. 9.2   Gravity-drop (left) and air-drill (right) row planters. Photos courtesy of Chengci Chen, Montana State University.

Soybean is propagated by seed. It can be grown in a wide range of well-drained soils with an optimum pH range of 6.0–6.5. A gravity-drop or air-drill row planter (Figure 9.2) may be used to plant soybean into tilled, reduced tillage, or even no-till soil. Soybean should be planted into a weed-free, moist, and uniform seedbed after the soil temperature reaches 10 °C in the spring (in southern hemisphere countries, such as Brazil, soybean is planted in September) at a seeding depth of 2.5–4.0 cm. Planting soybean into cool and wet soil may delay germination and result in disease infestation. Soybean seeds are generally inoculated with *Bradyrhizobium* and treated with fungicide and insecticide. Soil compaction and surface crust reduce soybean emergence.

The seeding rate of soybeans represents the germination rate along with some plant stand loss during emergence and establishment (Kandel, 2010). In North Dakota, the optimum seeding rate for soybeans is 370,500 seeds/ha (150,000 seeds/ac), with a row spacing ranging from 30–60 cm (11.8–23.6 in). In Brazil, farmers plant soybeans at 395,000 seeds/ha (159,919 seeds/ac), with a row spacing of 30–42 cm (11.8–16.5 in). Narrow row spacing is used for short growing season varieties. Narrow row spacing also helps to control soil erosion and improve crop competitiveness over weeds.

Weed control, as well as insect and disease management, is critical for optimum crop productivity. In the early growth stages, soybean plants cannot compete with weeds. Therefore, weeds need to be controlled until the crop develops a canopy. Weed control can be achieved either by cultivation, by spraying herbicides (Figure 9.3), or both. Many pre- and postemergence herbicides can be used on soybean to control weeds. Glyphosate-tolerant cultivars have been developed, in which glyphosate can be applied directly to the crop. However, different groups of herbicides should be used in rotation to avoid developing herbicide-resistant weeds. Soybean fields must be scouted on a regular basis to minimize insect and disease pest damage (Kandel, 2010). In Brazil, farmers spray insecticides to control caterpillars and stink bugs, and spray fungicides to control diseases like rust and powdery mildew.

Example 9.1: The desired plant density of soybean is 370,500 plants/ha. If the seed weight is 6,600 seeds/kg with a 95% germination rate and 10% stand loss during stand establishment, what seeding rate (kg/ha) should be used to achieve the desired plant population?

Solution:

$$\text{Seeding rate} = \frac{\left(\dfrac{370{,}500\,\dfrac{plants}{ha}}{0.95 \times (1.0 - 0.10)} \right)}{\left(6{,}600\,\dfrac{seeds}{kg} \right)} = \mathbf{66\,kg/ha}$$

FIG. 9.3   Self-propelled herbicide sprayer. Photo courtesy of Chengci Chen, Montana State University.

FIG. 9.4   Combine harvester for soybean. Photo courtesy of Chengci Chen, Montana State University.

Soybean is directly harvested using a combine harvester (Figure 9.4) when seeds are mature, with moisture content of about 13%. In Brazil, some farmers harvest soybean at a moisture content of 18%. Although soybean may be harvested when the moisture content is below 20%, it must be stored with a moisture content of less than 13%. When harvesting soybean at a moisture content below 12%, harvest losses and seed damage may occur. The airflow and ground speed of the combine harvester must be set appropriately to reduce harvest losses.

With a moisture content below 13%, soybean may be directly loaded into a bin and stored at a temperature below 16 °C (Figure 9.5). Soybean seeds may become moldy if the moisture content

FIG. 9.5   Storage bins for soybean. Photo courtesy of Chengci Chen, Montana State University.

---

is high and the bin temperature is warm (Hellevang, 2010). An aeration fan may be installed to lower the bin temperature and dry the seeds to prevent mold and discoloration. Soybean should be handled with care, because it may split and damage during handling, especially transport through augers.

9.2.2 Nutrient and Water Use

Soybean is a leguminous crop that can biologically fix nitrogen from the atmosphere for plant growth. This process involves the symbiotic relationship between soybean and the bacterium *Bradyrhizobium japonicum*. Soybean plants provide carbohydrates, energy in the form of adenosine triphosphate (ATP), and protection to the bacteria, while the bacteria fix N_2 from the atmosphere into ammonia (NH_3), which is quickly incorporated into amides or other N-containing compounds. The N-containing compounds are then transported via xylem to the shoots, being incorporated into amino acids and then into plant proteins.

$$N_2 + 8\,e^- + 8\,H^+ + 16\,ATP \rightarrow 2NH_3 + H_2 + 16\,ADP + 16\,Pi \tag{9.1}$$

Like most other crops, soybeans require other macro- and micronutrients, such as phosphorus (P), potassium (K), sulfur (S), calcium (Ca), magnesium (Mg), zinc (Zn), manganese (Mn), copper (Cu), iron (Fe), boron (B), chloride (Cl), nickel (Ni), and molybdenum (Mo). Most soils can provide sufficient micronutrients, while soils should be tested for macronutrients to determine nutrient deficiency. If the soil tested is low in P and K, fertilizer should be applied. The fertilizer requirements are 15–30 kg P_2O_5/ha and 25–60 kg K_2O/ha. A starter fertilizer of 10–20 kg N/ha at planting is beneficial for good early growth. Water requirements for producing 2.4–3.0 Mg/ha (2.4–3.0 metric ton/ha) of soybean seed vary between 500 and 650 mm/season depending on climate and length of growing period (Tacker and Vories, 2000).

To reduce the incidence of disease, soybean should not be planted in the same field every year. Soybean planted after a previous soybean season usually shows an increased incidence of root rot fungi, such as *Phytophthora* and *Rhizoctonia*. In the USA, soybean is commonly grown in rotation with corn, cotton, sorghum (*Sorghum bicolor* L.), or wheat (*Triticum aestivum* L.).

9.3 Rapeseed and Canola

9.3.1 Feedstock Production and Handling

Rapeseed and canola (Figure 9.6) are same species (*Brassica napus Brassica rapa, or Brassica juncea*) but rather cultivars containing different fatty acid profiles in the seeds. Rapeseed is commonly referred to as industrial rapeseed with high erucic acid and glucosinolates, whereas canola is primarily used for human consumption with very low erucic acid and glucosinolates concentrations in the seeds. Rapeseed is an annual oilseed crop, which can be grown in various soils in the temperate zone due to its ability to germinate and grow in cooler environments. Rapeseed has little tolerance to heat and drought, therefore water must be supplied through irrigation for a better yield. Rapeseed has winter- and spring-type cultivars. Winter-type cultivars, planted in the fall, have a longer growing season and may also flower before the weather becomes too hot the following summer. This gives winter-type cultivars a greater yield potential than spring-type cultivars. Herbicide-tolerant cultivars with resistance to a specific herbicide have been developed, including glyphosate- and glufosinate-tolerant cultivars (genetically modified) and Clearfield cultivars (non-genetically modified Imidazolinone-tolerant). The cultivar chosen should be high yielding, able to mature in the selected climate, and resistant to lodging and diseases (Kandel and Knodel, 2011).

Rapeseed plant development can be divided into vegetative stages (seedling, rosette, bud) and reproductive stages (flower and ripening; Figure 9.6). Rapeseed grows well in a clay-loam soil with good drainage and a pH from 6.0–7.0. It is propagated by seed and can be planted in a no-till or conventionally tilled soil using a gravity-drop or air-drill planter (Figure 9.2). The optimum seeding depth is 1.5–2.5 cm at a row spacing of 15–40 cm. The seed bed must be weed free, uniform, and moist. Seeding too deep or soil crust formation can reduce canola emergence. In the northern Great Plains of the USA, such as Montana and North Dakota, early seeding in the spring (April) usually leads to a higher seed yield.

Rapeseed seedlings can tolerate temperatures as low as –4 °C with a minimum 3 °C of soil temperature for germination. Rapeseed is very susceptible to heat and drought during flowering. The ideal planting density is 80–108 plants/m^2 (7.4–10.0 plants/ft^2). Rapeseed has a seed weight ranging from 165,000–441,000 seeds/kg depending on the cultivar (Kandel and Berglund, 2011). The seeding rate can be calculated according to seed weight while accounting for germination rate, and for stand losses during emergence and establishment. Wider row spacing is used in drier areas, but narrower row spacing can shorten the time for canopy closing and help control weeds.

Like soybean, rapeseed is not very competitive with weeds in the early growth stages, but it becomes more competitive after the late-rosette and bolting stages. Uniform and vigorous seedlings

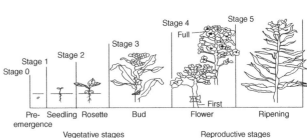

FIG. 9.6   Closer view of rapeseed flowers (left) and canola plant growth stages (right). Photo courtesy of Marisol Berti, North Dakota State University. Diagram adapted from Canola Council of Canada.

FIG. 9.7   Swather used to swath rapeseed. Photo courtesy of Chengci Chen, Montana State University.

are helpful to compete with weeds. Several pre- and postemergence herbicides may be used for weed control in rapeseed. For herbicide-tolerant cultivars, specific herbicides, such as glyphosate, can be applied with a sprayer (Figure 9.3) to control weeds. However, different herbicides with different modes of action should be used alternately to delay the development of herbicide-resistant weeds. Precautions should be taken to avoid plant injury by residual herbicide used in previous crops.

In the early seedling stage, leaf flea beetles (*Phyllotreta cruciferae* Goeze) can severely damage rapeseed. Therefore, insecticides need to be applied as seed treatment prior to planting or as a foliar spray during the seedling stage. Once the rapeseed plant reaches the four- to six-leaf stage, it can usually withstand flea beetle feeding. In the northern Great Plains, rapeseed should also be scouted for diseases, such as blackleg (*Phoma lingam* (Tode) Desmaz) and Sclerotinia (*Sclerotinia sclerotiorum* (Lib.) Debary) stem rot (Markell *et al.*, 2011).

Rapeseed must be harvested at the appropriate time, either by direct combining or by swathing and combining. Early harvest may result in excessive green seed, low oil content, and high moisture, because the pods at the bottom of the stem mature before those at the top. Late harvest can cause harvest losses due to shattering of the pods. Commonly, rapeseed is first swathed with a swather (Figure 9.7), then the windrows are left in the field for a few days to allow the seed to mature and dry, which is followed by threshing the swathed windrows using a combine harvester. Plants should be swathed when seeds have turned brown on the lower third of the main stem. Some producers use the herbicide paraquat to desiccate the crop before direct combining. Producers must follow the herbicide label and apply paraquat at the proper time. The cylinder speed, air flow, and ground speed of the combine harvester should be set at an appropriate level to prevent cracking the seed or blowing seed out.

Rapeseed is usually stored at a moisture content of less than 8% in bins like those shown in Figure 9.5. For the first week of storage, rapeseed goes through a high respiration period that produces heat and moisture, and that heat can reduce rapeseed protein content and increase the free fatty acid content in the seed (Hellevang, 2011). Therefore, aeration should be used in storage to lower the temperature, and the temperature should be monitored. The container and storage bin must be sealed tightly to prevent spilling or leakage due to the small size and round shape of the seeds.

9.3.2 Nutrient and Water Use

Like most other crops, rapeseed requires macro- and micronutrients to obtain a high yield. Fortunately, most soils contain sufficient micronutrients for growing rapeseed. Macronutrients such as N, P, K, and S must be supplied from fertilizers. Unlike legume crops, rapeseed requires N fertilizer. Depending on the soil type and rainfall, it generally requires 5.0–7.5 kg N to produce 100 kg of seed. The N rate is set based on precipitation amount and expected yield. Soil needs to be tested to estimate the available N in the soil profile to adjust the fertilizer application rate. If the soil test results show less than 15 mg P/kg and 75 mg K/kg, fertilizer should be applied to supply both of these nutrients. Sulfur is important to rapeseed, with about 1 kg S needed to produce 100 kg seed; therefore, 20–30 kg S/ha fertilizer is often required. If the soil test shows less than 0.5 mg/kg of boron, 1–2 kg/ha of B-containing fertilizer is recommended; however, overapplication of B is toxic to the crop (Brown *et al.*, 2009).

Rapeseed requires 125–152 mm of baseline water use. After reaching the baseline water use, it produces 7.6 kg of seed with each additional millimeter of water used (Nielsen, 1997; Hergert *et al.*, 2011). Therefore, using the average baseline water use (139 mm), rapeseed seed yield and water use may be predicted by the following equation:

$$\text{Yield} (\text{kg/ha}) = 7.6 \, \text{kg}/(\text{ha} \cdot \text{mm}) \times (W - 139) \, \text{mm} \tag{9.2}$$

where W is the total water used by the crop (mm).

> **Example 9.2:** Based on rainfall, the expected rapeseed yield is 2,500 kg/ha. About 7.5 kg nitrogen is needed to produce 100 kg canola seed yield. Soil test results show 30 kg/ha nitrogen remaining from the previous crop. How much nitrogen should be applied to achieve the expected yield?
>
> *Solution:*
>
> $$\text{Amount of nitrogen fertilizer} = (2,500 \, \text{kg seed/ha} \times 7.5 \, \text{kg N}/100 \, \text{kg seed})$$
> $$-30 \, \text{kg N/ha} = \textbf{157.5 kg/ha}$$

To reduce soil-borne diseases such as Sclerotinia stem rot and blackleg, rapeseed should not be grown after a previous rapeseed crop. In fact, the recommendation is to plant rapeseed in the same field once every four years. It is often rotated with other crops such as wheat, barley (*Hordeum sativum* L.), grass seed, sorghum, and potato (*Solanum tuberosum* L.).

> **Example 9.3:** Rapeseed is grown in a field with an expected growing season rainfall of 584 mm. If rapeseed yield increases at a rate of 7.6 kg per mm of water used above the threshold rainfall of 139 mm, what is the expected rapeseed seed yield in this location?
>
> *Solution:*
>
> $$\text{Yield} = 7.6 \, \text{kg}/(\text{ha} \cdot \text{mm}) \times (584 \, \text{mm} - 139 \, \text{mm}) = \textbf{3,382 kg/ha}$$

9.4 Oil Palm

9.4.1 Feedstock Production and Handling

Oil palm trees (Figure 9.8) are believed to have originated from West Africa, although currently Indonesia and Malaysia are the main producers worldwide. South American countries, especially Colombia, produce the majority of their biodiesel from oil palm. These perennial trees are best adapted to tropical areas, with evenly distributed rainfall of between 1,800 and 2,000 mm throughout the year and no distinct dry season. If oil palm is grown in drier areas, the soil must have a high water-holding capacity. Supplemental irrigation may be needed. The ideal soil is moist, deep, and well drained, with a medium texture and high humus content. Oil palm does not grow well in sandy soils. For the best yield, the oil palm tree needs ample sunshine, ideally with 1,800–2,000 sunlight hours annually, and constant sunlight of more than 5 hours per day, as well as a maximum average temperature of 29–33 °C and a minimum average temperature of 22–24 °C throughout the year (TNAU, 2013).

Oil palm seedlings are usually raised in nurseries for about 12–14 months before they are transplanted into palm plantations. The nursery is prepared by planting germinated seeds either in a seedbed or in polybags. After 12–14 months of growth, the well-developed seedlings are transplanted. A well-developed seedling will have a height of 1.0–1.3 m from its base, and more than 13 functional leaves.

Before planting oil palm trees, the land must be prepared by removing past vegetation. The recommendation is to prepare the land at least 3 months before transplanting the seedlings. Transplanting to the main field must be done during the onset of the rainy season. Drainage ditches may be dug to prevent water lodging if the soil has low permeability. Before planting, the plant positions are marked and the holes are dug to a volume of 60 cm^3. The seedlings are then planted with basic fertilizers at the recommended rates. Only healthy seedlings are selected, and any deformed and elongated seedlings or seedlings with signs of disease must be discarded. Oil palm should be planted in a triangular pattern at a spacing of 9 m × 9 m × 9 m to maximize sunlight use, which results in about 143 plants/ha (Basiron, 2007; TNAU, 2013).

FIG. 9.8   Oil palm tree plantation (left) and oil palm fruit and kernels (right). Photos courtesy of Mickey Vincent, Universiti Malaysia Sarawak, Malaysia.

Weed control by both cultivation and herbicide may be adopted. The basin area should be kept weed free, especially for young palm trees. Herbicides should be chosen and applied carefully. Contact burn-down herbicide is preferred over systemic herbicides (Basiron, 2007; TNAU, 2013). Cover crops are often planted to establish ground cover for preventing soil erosion and the growth of weeds. Oil palm is a hardy plant and is not affected by many pests and diseases, but integrated pest management practice, such as the use of barn owls, may be employed to reduce the rodent population in the plantation.

Oil palm trees need appropriate pruning to remove excessive leaves (2 leaves are produced each month). Severe pruning may result in the abortion of female flowers, and reduce tree growth and yield. It is recommended to keep 6–7 leaves per spiral for 4–7-year-old palms, 5–6 leaves per spiral for 8–14-year-old palms, and 4–5 leaves per spiral for those older than 15 years (TNAU, 2013).

Oil palm trees start bearing fruit bunches at around 30 months after planting and continue to produce fruit for more than 30 years. Palm trees bear fruit bunches year round. The oil palm produces two types of oil from the fruit: crude palm oil from the flesh, and mesocarp and palm kernel oil from the kernel (Figure 9.8). The fruit bunches typically ripen 6 months after flowering, and ripening occurs sooner with bunches at the top of the tree. After ripening, palm fruit bunches are harvested using a chisel on a short pole, or a sickle on a longer pole for taller palms. After harvest, the fruit bunches are collected and sent to a mill for oil extraction. Harvest timing is important, because the oil content is considerably higher in ripe fruits (Basiron, 2007; TNAU, 2013).

9.4.2 Nutrient and Water Use

Nutrient management is important in order to achieve the maximum growth and yield potential of oil palm trees. It is recommended that rock phosphate should be applied at 200 g per planting pit, and N is applied after 4–6 weeks of planting. For first-, second-, and third-year palm trees, an N:P:K combination of 400:200:400, 800:400:800, and 1,200:600:2,700 g per palm, respectively, is required. In Mg-deficient soil, 500 g of $MgSO_4$ per palm per year needs to be applied. The fertilizers are usually applied in two equal split doses during the May–June and September–October periods within a 2 m circle around the base of the palm tree (TNAU, 2013). About 25% of its harvested biomass may be returned to the field to provide nutrients and ground mulch. In order to meet the monthly evapotranspiration loss of the oil palm trees, about 120–150 mm water is needed (TNAU, 2013). Irrigation may be applied through flood, sprinkler, and drip irrigation in areas with dry seasons.

9.5 Jatropha

9.5.1 Feedstock Production and Handling

Jatropha also known as physic nut or purging nut, is a drought-resistant, non-food, oil-bearing, perennial hardy shrub (Figure 9.9) belonging to the *Euphorbiaceae* family, which grows well in tropical and subtropical climates (Makkar *et al.*, 1998). Jatropha has two genotypes: toxic and non-toxic. The toxic genotype is most studied because of its oil (e.g., as a source of biofuel) and associated co-product utilization (Makkar *et al.*, 1998; Makker and Becker, 1997), and is found in many tropical and subtropical regions of South America, Africa, and Asia. A non-toxic genotype of jatropha has also been reported, which is found only in Mexico (Makkar *et al.*, 1998). Jatropha grows under a wide range of

FIG. 9.9   Jatropha trees (left), seeds (middle), and fruits (right). Photos courtesy of Samir Khanal, University of Hawaii.

rainfall regimes, from a low of 250 mm/year to over 1200 mm/year (Kumar and Sharma, 2008). One major trait associated with the plant is its hardiness and sustainability in warm and arid climates. It is found typically as a small perennial tree or a large shrub, normally reaching a height of 3–5 m, but it can attain 8–10 m under favorable conditions (Gaur, 2009).

Jatropha can be propagated from seeds or cuttings. The seeds should be selected from mature yellow fruits. Before planting, the seeds are soaked in water for 24 hours and any floating seeds are removed. Jatropha seedlings may be raised in a nursery for 2–3 months and then transplanted into the main field. Seedlings are often raised in containers with at least 2 L of volume. Tissue culture technology may be used to propagate seedlings from buds, leaves, or stems. This technology can produce a large number of genetically identical seedlings under controlled conditions (FACT, 2011; NL Agency, 2010).

Direct seeding may be done in moist soil with a warm temperature (>20 °C). Two to four seeds are placed in one hole and covered with 4–7 cm of soil. After they emerge, the plants are thinned to one plant per hole. The recommended plant density is 3,030 plants per hectare, with plant spacings of 2 m × 2 m, or 3 m × 3 m. Mycorrhizal fungi are sometimes inoculated into the seeds in order to enhance young jatropha plant growth and nutrient uptake (FACT, 2011; NL Agency, 2010).

Pruning is needed for jatropha to correctly shape the tree (formative pruning) and remove excess biomass (maintenance pruning). When the transplanted seedlings reach 30–60 cm in height, formative pruning is performed to reduce apical growth and induce branching. Additional pruning is performed after a year of growth and later to control the tree to 35–40 branches at 1.2 m in height. Maintenance pruning is carried out periodically after the plant reaches 1.2 m in height (FACT, 2011; NL Agency, 2010).

Although jatropha trees start bearing fruits after 8 months of planting, full production is only achieve in their fifth year of plantation (FACT Foundation, 2011; NL Agency, 2010). Jatropha trees are expected to have a life span of 50 years. The fruits are harvested by hand when ripe. After harvest, the fruits must be processed within 24 hours.

9.5.2 Nutrient and Water Use

Macro- and micronutrients are required for optimal growth and fruit yield of jatropha. Most soils provide sufficient micronutrients, but macronutrients, such as N, P, and K, must be provided by fertilization.

Jatropha may be planted as individual trees, hedges, or living fences, monoculture plantations, or intercropping systems. Shade-tolerant crops may be planted between jatropha trees and legumes may be used to intercrop with jatropha. Legumes can fix nitrogen from the atmosphere and, after harvest, the leguminous plant residues are left in the field to provide nitrogen for jatropha.

Toxins in jatropha seeds

Jatropha seeds contain toxic compounds such as phorbolesters, curcin (lectin), trypsin inhibitors, and phytates. About 70% of the toxic compounds end up in the extracted oil and the remaining 30% are in the seedcake matrix. The seedcake is poisonous to animals even at very small percentages. As a result, the direct use of seedcake as an animal feed is currently not feasible. Therefore, there is a critical need for an environmentally friendly method for the detoxification of jatropha-derived seedcake without losing the integrity and functionality of its protein content.

9.6 Camelina

9.6.1 Feedstock Production and Handling

Camelina is an ancient oilseed crop, native to the Mediterranean regions of Europe and Asia. It is a small-seeded oilseed crop belonging to the *Brassicaceae* family, and has the potential to be used as a non-food oil feedstock. Camelina is a cool-season oilseed crop, which adapts better to the cooler and drier areas of the temperate zone than rapeseed, and has a short growing season (85–100 days). Like rapeseed, camelina has spring and winter annual types. The plants grow to 1.5 m tall and produce small yellow-brown seeds that have a high oil content (Figure 9.10).

The camelina plant development stages are similar to rapeseed, shown in Figure 9.6. Camelina grows well in various soil textures under tilled or no-till conditions, with a soil pH from 6.0–7.5. A weed-free field is required to grow this crop, although burn-down herbicides may be used to control weeds prior to planting. Camelina planting can be done using the same planter as for rapeseed (Figure 9.2). Camelina must be planted early (mid-March to early April) in the Pacific Northwest and the northern Great Plains, with a seeding rate of 3–6 kg/ha to obtain a high yield. Since camelina has small seeds (about 880,000 seeds/kg), it should be seeded at 1–2 cm in depth (McVay and Lamb, 2008).

Although no severe insect and disease infestations and damage have been found since this crop was introduced to Montana and other regions in the Pacific Northwest of the USA, a field scout is

  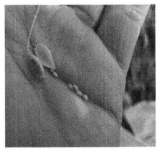

FIG. 9.10 Camelina plants (left), seed pods (middle), and seeds (right). Photos courtesy of Marisol Berti, North Dakota State University.

recommended for any disease or insect infestation. A few incidences of camelina infected by downy mildew (*Peronospora camelinae*) and Sclerotinia stem rot have been observed (Ehrensing and Guy, 2008).

Camelina is normally harvested after ripening and with a seed moisture below 8%, using the same combine harvester as for rapeseed (Figure 9.4). Combine settings are similar to those used for rapeseed, except the fan speed must be reduced to prevent seed losses, and small-opening screens must be used for effective separation of camelina seeds and hulls. Camelina seeds must be stored at a moisture content less than 8%. Storing seed at a moisture content higher than 10% may result in seed binding and spoilage. Because of the small size of camelina seeds, the equipment and storage bins must be tightly sealed to reduce seed losses during harvest, transport, and storage (Ehrensing and Guy, 2008).

9.6.2 Nutrient and Water Use

Like most of the other crops, camelina responds to N, P, and S fertilizers. Fertilization with 90 kg N/ha and 67 kg P_2O_5/ha, respectively, is needed to obtain maximum yield. Application of S tends to increase oil content, but it does not increase seed yield. Higher N rates are recommended in areas with more available water (Wysocki *et al.*, 2013; Solis et al., 2013).

Field experiments have shown that camelina has a minimal threshold water use of about 122 mm; after this threshold is met, camelina produces 7.0 kg seed/ha with each additional millimeter of water used (Hergert *et al.*, 2011). The water use and seed yield can be expressed as:

$$\text{Yield}\,(\text{kg/ha}) = 7.0\,\text{kg}/(\text{ha}\cdot\text{mm})\,(W - 122)\,\text{mm} \tag{9.3}$$

where W is the total water use (mm).

Camelina is commonly rotated with wheat or barley in the northern Great Plains (Chen and Bekkerman, 2012). It is not recommended to grow camelina in succession or after another broadleaf crop, because there is no herbicide registered for use with camelina. A successful camelina crop relies on good broadleaf weed control in the previous cereal crop. Rotation of camelina with cereal crops will also reduce disease infestation.

9.7 Yield and Oil Content of Major Oilseed Feedstocks

Soybean seeds typically contain about 21% oil and 40% protein (on a dry wt. basis; Table 9.1). After the oil is extracted, the remaining soybean meal can be used as a valuable protein supplement in animal feed or as a source of soy lecithin. In fact, the meal accounts for over 50% of the revenue for the soybean oil crushing facility, which has allowed soybean to become dominant in the oilseed market. In the USA, nationwide average soybean seed yield is about 2.6 metric ton/ha. With 21% oil content, it takes 1.83 hectares of land to produce 1 metric ton of soybean oil.

Rapeseed and canola seed contain 40% oil and 23% protein (dry wt. basis). The average seed yield for rapeseed is 1.98 metric tons/ha in the USA. At 40% oil content, 1.26 hectares of land are required to produce 1 metric ton of oil.

Oil palm produces about 4 metric tons of crude palm oil/ha annually, and the kernels hold about 10% of the total oil produced. Therefore, oil palm is a very efficient oil producer. Only 0.27 hectares of land are required to produce 1 metric ton of oil, and palm trees do not require annual planting. After the extraction of oil, the palm kernel meal is mainly used for animal feed.

In the USA, the average corn yield was about 9,950 kg/ha in 2013 (~158 bu/ac). However, corn is mainly grown for starch, and oil is simply a byproduct extracted from corn germ. With 4% oil

Table 9.1 Oil content and oil yield of major oilseed feedstocks

Feedstocks	Oil content (%)	Oil yield in kg/ha (lb/ac)
Soybean	21	546 (486)
Rapeseed	40	793 (706)
Palm flesh	42	3,780 (3,367)
Palm kernel	50	430 (383)
Corn	4	398 (354)
Jatropha	30	1,350 (1,202)
Camelina	40	400 (356)

Source: Data from USDA-NASS, 2013; USDA-FAS, 2014.

content, the corn oil yield is about 398 kg/ha. Thus, 2.51 hectares of land are required to produce 1 metric ton of oil.

Jatropha seed yield varies greatly depending on soil type, nutrient, and water input, with a range of 2–7 Mg/ha (2–7 metric ton/ha). Jatropha seeds and kernels (seed without shells) contain 25–35% and 55–60% oil (dry wt. basis), respectively.

Camelina seeds contain about 30–40% oil. The plant has the potential to produce 0.8 metric ton of oil/ha. The average oil yield in the northern Great Plains of the USA is 0.4 metric ton/ha. Camelina meal or cake, the remaining material after the oil has been removed, has 10–12% oil and 40% protein. It has the potential to be used as feed or supplement for fish, chicken, and dairy cows. However, camelina meal contains anti-nutrient glucosinolates, which can reduce livestock performance and health at high concentrations (McVay and Lamb, 2008; Ehrensing and Guy, 2008).

Example 9.4: Oil palm fruit bunch, soybean seed, and camelina seed have yields of 17.0, 2.6, and 1.0 metric ton/ha, respectively. The oil contents of oil palm fruit bunch, soybean seed, and camelina seed are 22%, 21%, and 40%, respectively. How much land is needed for each crop to produce 1 metric ton of oil?

Solution:

Oil yield of palm = 17.0 metric ton/ha × 22% = 3.74 metric ton/ha
Oil yield of soybean = 2.6 metric ton/ha × 21% = 0.55 metric ton/ha
Oil yield of camelina = 1.0 metric ton/ha × 40% = 0.40 metric ton/ha

Land required for palm = 1 metric ton/3.74 (metric ton · ha) = **0.27 ha**
Land required for soybean = 1 metric ton/0.55 (metric ton · ha) = **1.82 ha**
Land required for camelina = 1 metric ton/0.40 (metric ton · ha) = **2.50 ha**

References

Akbar, E., Yaakob, Z., Kamarudin, S.K., Ismail, M., and Salimon, J. (2009). Characteristic and composition of *Jatropha curcas* oil seed from Malaysia and its potential as biodiesel feedstock feedstock. *European Journal of Scientific Research* 29(3): 396–403.

Basiron, Y. (2007). Palm oil production through sustainable plantations. *European Journal of Lipid Science and Technology* 109: 289–95. doi:10.1002/ejlt.200600223.

Brown, J., Davis, J.B., Lauver, M., and Wysocki, D. (2009). *United States Canola Association Canola Growers' Manual*. Moscow, ID: University of Idaho.

Chen, C., and Bekkerman, A. (2012). Wheat-based crop rotations for camelina sativa oilseed bioenergy feedstock production in central Montana. ASA-CSSA-SSSA 2012 International Annual Meeting, October 21–24, Cincinnati, OH.

Edem, D.O. (2002). Palm oil: Biochemical, physiological, nutritional, hematological, and toxicological aspects: A review. *Plant Foods for Human Nutrition* 57(3–4): 319–41.

Ehrensing, D.T., and Guy, S.O. (2008). *Camelina.* EM 8953–E. Corvallis, OR: Oregon State University Extension Service.

FACT (2011). *The Jatroph Handbook: From Cultivation to Application.* Eindhoven: FACT Foundation.

Gaur, S. (2009). Development and evaluation of an effective process for the recovery of oil and detoxification of meal from *Jatropha curcas.* Master's thesis, Missouri University of Science and Technology.

Hellevang, K. (2010). Soybean drying, handling, and storage. In: H. Kandel (ed.), *Soybean Production Field Guide for North Dakota and Northwestern Minnesota.* A1172. Fargo, ND: North Dakota State University Extension Service.

Hellevang, K. (2011). Canola drying and storage management. In: H. Kandel and J.J. Knodel (eds.), *Canola Production Field Guide.* A1280. Fargo, ND: North Dakota State University, Extension Service. http://www.ag. ndsu.edu/pubs/plantsci/crops/a1280.pdf, accessed April 2013.

Hergert, G.W., Margheim, J., Pavlista, A., *et al.* (2011). Yields and ET of deficit to fully irrigated canola and camelina. 23rd Annual Central Plains Irrigation Conference, Burlington, CO, February 22–23.

Kandel, H. (ed.) (2010). *Soybean Production Field Guide for North Dakota and Northwestern Minnesota.* A1172. Fargo, ND: North Dakota State University Extension Service.

Kandel, H., and Berglund, D.R. (2011). Rates and establishment. In H. Kandel and J.J. Knodel (eds.), *Canola Production Field Guide.* A1280. Fargo, ND: North Dakota State University, Extension Service. http://www.ag.ndsu. edu/pubs/plantsci/crops/a1280.pdf, accessed April 2013.

Kandel, H., and Knodel, J.J. (eds.) (2011). *Canola Production Field Guide.* A1280. Fargo, ND: North Dakota State University, Extension Service. http://www.ag.ndsu.edu/pubs/plantsci/crops/a1280.pdf, accessed April 2013.

Knothe, G. (2001). Historical perspectives on vegetable oil-based diesel fuels. *Inform* 12: 1103–7.

Kumar, A., and Sharma, S. (2008). An evaluation of multipurpose oil seed crop for industrial uses (*Jatropha curcas* L.): A review. *Industrial Crops and Products* 28: 1–10.

Makkar, H.P.S., and Becker, K. (1997). Potential of *Jatropha curcas* seed meal as a protein supplement to livestock feed: Constraints to its utilisation and possible strategies to overcome constraints. In: G.M. Gubitz, M. Mittelbach, and M. Trabi (eds.), *Biofuel and Industrial Products from Jatropha Curcas*, Graz: Dbv-Verlag for the Technical University of Graz Uhland Street, pp. 190–205.

Makkar, H.P.S., Aderibigbe, A.O., and Becker, K. (1998). Comparative evaluation of non-toxic and toxic varieties of *Jatropha curcas* for chemical composition, digestibility, protein degradability and toxic factors. *Food Chemistry* 62: 207–15.

Markell, S., Mendoza, L.D.R., and Wunsch, M. (2011). Disease management. In: H. Kandel and J.J. Knodel (eds.), *Canola Production Field Guide.* A1280. Fargo, ND: North Dakota State University, Extension Service. http:// www.ag.ndsu.edu/pubs/plantsci/crops/a1280.pdf, accessed April 2013.

McVay, K.A., and Lamb, P.F. (2008). *Camelina Production in Montana.* MontGuide MT200701AG. Bozeman, MT: Montana State University Extension Service.

Moser, B.R. (2010). Camelina (*Camelina sativa* L.) oil as a biofuels feedstock: Golden opportunity or false hope? *Lipid Technology* 22: 12.

Nielsen, D.C. (1997). Water use and yield of canola under dryland conditions in the central Great Plains. *Journal of Production Agriculture* 10: 307–13.

NL Agency (2010). *Jatropha Assessment: Agronomy, Socio-economic Issues, and Ecology.* Ministry of Economic Affairs, Agriculture and Innovation. Utrecht: NL Agency.

Pryde, E.H., and Rothfus, J.A. (1989). Industrial and nonfood uses of vegetable oils. In: G. Robbelen *et al.* (eds.), *Oil Crops of the World: Their Breeding and Utilization.* New York: McGraw-Hill.

Quick, G.R. (1989). Oilseeds as energy crops. In: G. Robbelen, R.K. Downey, and A. Ashri (eds.), *Oil Crops of the World: Their Breeding and Utilization.* New York: McGraw-Hill, pp. 118–31.

Solis, A., Vidal, I., Paulino, L., Johnson, B.L., and Berti, M.T. (2013). Camelina seed yield response to nitrogen, sulfur, and phosphorus fertilizer in South Central Chile. *Industrial Crops and Products* 44: 132–8.

Tacker, P., and Vories, E. (2000). Irrigation. In: *Arkansas Soybean Handbook, MP197.* Division of Agriculture and Cooperative Extension Service, University of Arkansas.

TNAU (2013). *Crop Production: Oil Seeds: Oil Palm.* TNAU Agritech Portal. Coimbatore: TamilNadu Agricultural University. http://agritech.tnau.ac.in/agriculture/oilseeds_oilpalm.html, accessed April 2013.

USDA-FAS (2014). *Oilseeds: World Markets and Trade*. Washington, DC: United States Department of Agriculture, Foreign Agricultural Service. http://apps.fas.usda.gov/psdonline/circulars/oilseeds.pdf, accessed September 2014.

USDA-NASS (2013). United States Department of Agriculture, National Agricultural Statistics Service. http://www.nass.usda.gov, accessed September 2014.

Velasco, L., and Fernandez-Martinez, J.M. (2002). Breeding oilseed crops for improved oil quality. *Journal of Crop Production* 5(1–2): 309–44.

Vollmann, J., Moritz, T., Kargl, C., Baumgartner, S., and Wagentristl, H. (2007). Agronomic evaluation of camelina genotypes selected for seed quality characteristics. *Industrial Crops and Products* 26: 270–77.

Walton, J. (1938). The fuel possibilities of vegetable oils. *Gas Oil Power* 33: 167–8.

Wysocki, D.J., Chastain, T.G., Schillinger, W.F., Guy, S.O., and Karow, R.S. (2013). Camelina: Seed yield response to applied nitrogen and sulfur. *Field Crops Research* 145: 60–66.

Exercise Problems

9.1. A farmer plants soybean at 40 cm row spacing. After soybean emergence, the farmer finds a plant density of 37 plants/m². What is the soybean population expressed in plants/ha? Assuming that soybean has a seed weight of 6,600 seeds/kg and the seed has a 95% germination rate, what seeding rate did the farmer use (expressed in kg/ha)?

9.2. Where did the name canola come from? How is canola different from rapeseed?

9.3. Rapeseed is planted in a field that is expected to receive 500 mm of rainfall during the growing season. It is expected that the rapeseed will yield 7.6 kg seed/ha after reaching the water use threshold of 139 mm. What seed yield can the farmer expect for this crop? If the rapeseed requires 0.07 kg of nitrogen to produce 1 kg of seed, and the soil test shows 40 kg/ha nitrogen in the soil profile, how much nitrogen should the farmer apply to achieve the expected yield?

9.4. Describe the differences between soybean and rapeseed in fertilizer inputs, and explain why the requirements are different.

9.5. If a palm tree is planted at 9 m × 9 m × 9 m triangle spacing, each palm tree produces 150 kg of fruit bunches with 20% oil content. What is the oil yield of this palm plantation (expressed in kg/ha)?

9.6. Describe the differences in jatropha genotypes. Why must caution be used when feeding jatropha oil and seedcake to livestock?

9.7. Camelina has a seed weight of 1.1 g per 1,000 seeds. If a farmer plants camelina at 4 kg/ha, assuming that camelina has a 90% germination rate and 20% stand loss during establishment, what is the plant population expressed in plants/ha? Camelina has a minimal threshold water use of about 122 mm, and produces 7.0 kg seed/ha with each additional mm of water used. If camelina is planted in a field with 450 mm of rainfall, how much seed yield (kg/ha) can be produced and how much land is needed to produce 1 metric ton of camelina (assuming that camelina seed has a 40% oil content)?

9.8. What are the climatic zones to which soybean, rapeseed, camelina, oil palm, and jatropha are adapted? Also compare the land area required for each of these oilseed crops to produce 1 million gallons of biodiesel per year (MGY), assuming the oil contents as given in this textbook and a conversion efficiency of 100%.

9.9. What is meant by annual and perennial oilseed crops? Name two annual and two perennial oilseed crops, and discuss the advantages and disadvantages of growing these crops as oilseed feedstocks.

9.10. Why should annual oilseed crops, such as soybean, rapeseed, and camelina, be rotated with cereal crops, such as corn, wheat, and sorghum?

Lignocellulose-Based Feedstocks

Sudhagar Mani

What is included in this chapter?

This chapter covers lignocellulosic feedstock availability, and the supply logistics system including operations such as harvesting, baling, transportation, and storage of crop residues, energy crops, and forest biomass. The principles, operating procedures, energy consumption, cost of operations, and applications are discussed. Pertinent examples and calculations are also included.

10.1 Introduction

Lignocellulosic feedstock is described as organic matter derived from plant origins. Examples include crop residues, dedicated energy crops, forest biomass, and short rotational woody crops. Biomass is the oldest renewable source of energy used by humankind for cooking applications. It is still used as a cooking fuel in several developing countries in Africa and Asia. Biomass is one of the most abundant resources in the world, with an annual global energy potential ranging from 100–270 EJ (exajoules; 5.5–15 billion ton/yr). Woody biomass from forestlands alone contributes more than 40% of the available global biomass (Skytte *et al.*, 2006). The US Department of Energy (DOE) has estimated that more than 1.3 billion tons of lignocellulosic feedstock can potentially be available for bioenergy production in the USA. If dedicated energy crops can be grown globally on land not suitable for agricultural food production (marginal, mined, and fallow land), biomass could supply more than half of the global annual energy demand of 500 EJ (Ladanai and Vinterback, 2009).

Lignocellulosic feedstock is primarily composed of cellulose (30–50%), hemicellulose (15–35%), lignin (5–30%), inorganic minerals or ash (0.5–8%), and other minor fractions. Cellulose is the most abundant natural polymer and has a straight chain molecule of glucose (C-6 sugar) with more than 100–1,000 units. Hemicellulose is a branched chain molecule of xylose (predominantly C-5 sugar). The combination of cellulose and hemicellulose, often called holocellulose, accounts for more than

---

Bioenergy: Principles and Applications, First Edition. Edited by Yebo Li and Samir Kumar Khanal.
© 2017 John Wiley & Sons, Inc. Published 2017 by John Wiley & Sons, Inc.
Companion website: www.wiley.com/go/Li/Bioenergy

Table 10.1 Chemical composition (% dry wt.) of common lignocellulosic feedstocks

Biomass	Cellulose	Hemicellulose	Lignin
Crop residues			
Corn stover	35–40	17–35	7–18
Wheat straw	33–50	24–36	9–17
Rice straw	36–47	19–25	10–24
Sugarcane bagasse	35–45	23–27	19–32
Dedicated energy crops			
Switchgrass	32–36	24–27	15–18
Miscanthus	38–44	18–27	22–25
Energy cane	43	24	22
Napier grass	36–39	20–22	18–20
Forest biomass			
Softwood	40–50	11–20	27–30
Hardwood	40–50	15–20	20–25
Hybrid popular	42–56	18–25	21–25
Eucalyptus	49–50	13–14	27–28

two-thirds of biomass composition. Lignin is the second most abundant biopolymer, with a building block of phenol molecules glued with holocellulose.

Table 10.1 lists the chemical compositions of common lignocellulosic feedstock types. Although lignocellulosic feedstock is abundantly available in many parts of the world, efficient and economical collection, transport, and storage operations are challenging due to its sparse distribution in croplands, low bulk density, and high susceptibility to spoilage. Systematic and continuous supply of biomass to a biorefinery/bioenergy conversion facility requires fundamental knowledge and assessment of availability, harvesting, transport, and storage options for major biomass types. This chapter discusses the production and availability of lignocellulosic feedstock, supply logistics options with machinery operations, and cost estimations with examples.

10.2 Feedstock Availability and Production

10.2.1 Crop Residues

Crop residues include plant stems, leaves, and other plant parts left in the field after grain or seed harvesting from agricultural crops (e.g., corn, wheat, barley, rice, soybean, sunflower). Examples of crop residues include corn stover, wheat straw, soybean stalk, and rice straw. The availability of crop residues left in the field is highly dependent on grain or seed yield. Corn stover (Figure 10.1), the most abundant biomass available in the USA, is composed of approximately 50% stalk, 22% leaves, 15% cobs, and 13% husks. Wheat straw is the most abundant biomass available in Canada, followed by barley and oat straw. The availability of crop residues is highly dependent on crop type, grain yield, and geographical region.

Crop Residue Availability

Availability of crop residue is usually calculated based on the grain/seed yields reported in bu/ac (bushels per acre) for cereal and oil seed crops or using a harvest index (Z). The exception is cotton plant, where the cotton stalk yield depends on the cotton fiber yield. The harvest index for agricultural crops ranges from 0.4 to 0.55. A harvest index of 0.5 represents a yield of grain equal to that of the residue, or a residue to grain ratio (R) of 1.0.

FIG. 10.1   Corn stover. Photo courtesy of Ryan Billman, Ohio State University.

Relationship between harvest index and residue to grain ratio

The harvest index (Z) is defined as the ratio of the weight of grain or seed (X) over the weight of the above-ground total biomass at maturity (Y). It is also often referred to as the ratio of economic productivity over biological productivity. The residue to grain ratio (R) is defined as the ratio of the weight of straw (S) over the weight of grain or seed (X).

$$Z = \frac{X}{Y} = \frac{X}{X+S} \tag{10.1}$$

$$R = \frac{S}{X} \tag{10.2}$$

$$Z = \frac{1}{1+R} \tag{10.3}$$

$$R = \frac{1-Z}{Z} \tag{10.4}$$

The availability of residues after grain harvesting can be calculated either from the crop harvest index or from the residue to grain ratio. Table 10.2 lists the average grain yield and residue to grain ratio for common agricultural crops, which can be used to estimate total available residues in crop fields.

Complete removal of above-ground residues is not often practiced, as exposure of topsoil to water and wind will deteriorate soil quality. Crop fields are often covered with residues as mulch to prevent soil erosion, to retain soil moisture, and to improve overall soil quality and crop yield. The Natural Resource Conservation Services (NRCS) in the USA recommends leaving at least 1.57 Mg/ha (0.7 ton/ac) of residues in the field to prevent soil erosion. The amount of residues required for soil

Table 10.2 Grain yield and residue to grain ratio for various agricultural crops

Crops	Average grain/seed yield in bu/ac (metric ton/ha)	Grain/seed weight per bushel (lb/bu)	Moisture content (%, wet basis)	Residue to grain ratio (R)
Corn	155.3 (9.8)	56	15.5	1.0
Wheat – winter	47.8 (3.2)	60	13.5	1.7
Wheat – spring	42.8 (2.9)	60	13.5	1.3
Soybean	41.8 (2.9)	60	11	1.5
Sorghum	65.1 (4.1)	56	14	1.0
Barley	63.1 (3.4)	48	14.5	1.5
Oat	60.9 (2.2)	32	14	2.1
Rice	75.1 (8.4)	100	12	1.5
Cotton	24.9 (0.9)	32	7.5	4.5

Source: Data from NASS, 2013; US DOE, 2011.

conservation varies with a number of factors, including land slope, climate conditions, soil type, and quality. It is always recommended that about 30–60% of available above ground residues are left in the field to prevent soil erosion, build soil organic matter, maintain soil health, and preserve soil moisture. Assessment of the available crop residues from any cropland is required for the sustainable production of biofuels, bioenergy, and bioproducts. Example 10.1 demonstrates the estimation of potentially available crop residues for harvesting and collection from a corn field.

Example 10.1: Calculate the amount of corn stover available from a 2,000-acre corn field in Iowa. Also calculate the amount of sustainably harvestable corn stover for bioenergy production. Assume that 40% of the available residue is left in the field for soil conservation.

Solution:

$$
\text{Amount of crop residue available} (dry\ tons/ac) = Grain\ yield \left(\frac{bu}{ac} \right)
$$

$$
* Grain\ test\ weight \left(\frac{lb}{bu} \right) * R * \frac{[100 - \text{moisture content (wb)}]}{100 \times 2,000 \left(\frac{lb}{ton} \right)}
\tag{10.5}
$$

From Table 10.1, average grain yield for corn (bu/ac) = 155; grain test weight (lb/bu) = 56; moisture content (wb)= 15.5; residue to grain ratio (R) = 1.0

$$
\text{Amount of available corn stover in the field} (ton/ac) = 155 \times 56 \times 1.0 \times \frac{(100 - 15.5)}{100 \times 2,000}
$$

$$
= 3.68
$$

If 40% of available corn stover is required to be left in the field for soil conservation:

$$
\text{Sustainably harvestable amount of corn stover} (ton/ac) = 0.6 \times 3.68 = 2.2
$$

$$
\text{Total amount of corn stover from a 2,000 – acre corn field is} (dry\ tons) = 2,000 \times 2.2
$$

$$
= \mathbf{4,400}
$$

FIG. 10.2   Switchgrass. Photo courtesy of Sudhagar Mani, University of Georgia.

10.2.2 Dedicated Energy Crops

Dedicated energy crops are purposely grown annual and perennial plant species for bioenergy applications. Examples of annual energy crops include sweet sorghum, forage sorghum, and energy cane. Perennial grasses such as switchgrass, napier grass, and *Miscanthus* can regrow after harvesting, with a 10–15-year growth cycle from a single planting. Other perennial grasses such as alfalfa, Bermuda grass, and other hay grasses that are grown for animal feed can also be used for bioenergy applications. Unlike agricultural crops, energy crops can be grown in marginal, grass, and reclaimed lands to improve soil organic matter and water quality. The US Department of Energy (DOE) has established a research program to extensively study the cultivation of switchgrass (Figure 10.2) as a model energy crop in marginal lands for bioenergy application. Similarly, the UK and several other European countries have been conducting research on the cultivation of giant reed, *Miscanthus*, and reed canary grass as potential bioenergy feedstocks.

Switchgrass (*Panicum virgatum L.*) is a perennial, warm-season grass native to North America. It is recognized as a major bioenergy feedstock because of its high yield, adaptation to marginal lands, tolerance to water deficit, and low soil nutrient concentrations, and many environmental benefits including soil erosion reduction, soil carbon sequestration, wildlife habitat, and water quality improvement. Cultural practices to produce switchgrass vary widely depending on land conditions, climate conditions, farming practices, and many other factors. Switchgrass is broadly classified into upland and lowland varieties. Upland varieties grow usually 1.5–1.8 m tall and are better adapted to well-drained soils in northern latitudes, whereas lowland varieties grow up to 3.7 m tall and are typically adapted to lower latitudes. The switchgrass varieties Alamo, Kenlow, and Cave-In-Rock are the most successful cultivars in North America.

Switchgrass is established by seed propagation, with a typical recommended seed rate of 4–10 kg/ha (3.5–9.0 lb/ac) to achieve a stand density of 278 plants/m^2 (Teel *et al.*, 2003). Both conventional and no-tillage operations are practiced for initial establishment. Complete establishment of switchgrass requires 1–2 years depending on proper weed management. Switchgrass is usually thrifty in its use of P and K fertilizers, but P, K, and lime fertilizers are often applied to maintain soil nutrient balance, both during establishment and in production years. After establishment, the plant may last

FIG. 10.3   Miscanthus. Photo courtesy of Jose G. Guzman, Ohio State University.

for 10 years or more. Regular fertilizer application and proper harvesting strategies are required to achieve maximum switchgrass yield. Nitrogen and other nutrients that contribute to plant ash content generally decline to their lowest levels after first frost and senescence. The nitrogen in the whole plant tissue is translocated to the crown/rhizome and root tissues during senescence, which reduces overall nutrient use and thus reduces the ash content of biomass feedstock to make it suitable for bioenergy production. The loss of biomass during this period is mainly due to the translocation of N-rich materials and non-carbohydrates from senescing shoots to below-ground storage. Switchgrass yield potential across the USA ranges between 7.6 and 15.1 dry Mg/ha-yr (McLaughlin *et al.*, 2002).

Miscanthus (*Miscanthus Spp.*) is a warm-season, cold-tolerant grass native to Asia (Figure 10.3). Cultivation of miscanthus has been extensively studied in Europe for bioenergy production, and biomass yield varies in the ranges of 13–34 dry Mg/ha (6–15 ton/ac) depending on the region (Lewandowski *et al.*, 2003). Current research in the USA is mainly focused on the cultivation of giant miscanthus (*Miscanthus x Giganteus*), a sterile natural hybrid, in marginal and strip-mined lands in northeast Ohio, Illinois, and several other regions. Stem cuttings, rhizome cuttings, and micropropagation techniques are mainly used for establishing the crop. It can grow over 2 m in the establishment year and as high as 4 m in subsequent years, with root penetration over 1 m into the ground and a plant density of 2 plants/m^2 (El Bassam, 1998). After establishment, miscanthus can grow for 15–20 years. With a single harvesting in early spring or just after the frost killing of stems, the yield potential of miscanthus is between 30 and 60 dry Mg/ha-yr (Heaton *et al.*, 2008).

Napier grass (*Pennisetum purpureum L Schum*), also known as elephant grass or banagrass, is a perennial warm-season grass native to Africa. It is cultivated for forage in tropical and subtropical regions of the world. It grows to a height of 2–4.5 m (sometimes up to 7.5 m) and has leaves that are 30–120 cm long and 1–5 cm broad. The biomass yield ranges from 7–45 dry Mg/ha-yr, depending on region and climate (Woodard and Prine, 1993).

Energy cane (*Saccharum officinarum L.*) is a crossbreed of sugarcane and a wild grass with a yield potential in excess of 25 dry Mg/ha-yr (Eggleston *et al.*, 2010). Similar to miscanthus, both napier grass and energy cane are established by stem cutting or micropropagation techniques.

Table 10.3 Biomass yield potential of dedicated energy crops and forest biomass

Lignocellulosic feedstocks	Yield, Mg/ha·yr (ton/ace·yr)	References
Switchgrass	5–15 (2.2–6.7)	McLaughlin *et al.*, 2002
Miscanthus	30–60 (13.4–26.7)	Heaton *et al.*, 2008
Napier grass (elephant grass)	7–45 (3.1–20.0)	Woodard and Prine, 1993
Energy cane	8–45 (3.6–20.0)	Knoll *et al.*, 2012
Willow	3–20 (1.3–8.9)	Sevel *et al.*, 2014
Hybrid popular (6-yr cycle)	8–12 (3.6–5.3)	Pearson *et al.*, 2010
Logging residues	7–11 (3.1–4.9)	Baker *et al.*, 2010

Sorghum (*Sorghum bicolor L. Moench*) is a C_4 annual grass (*Poaceae*) with high photosynthetic efficiency. It is grown for either grain, sugar, or forage. It is native to the tropics and considered one of the most drought-resistant crops. It has varied sensitivity to photoperiod and temperature, and has a short life cycle (3–5 months to maturity). Therefore, it can be cultivated multiple times a year in tropical and subtropical climates. Forage sorghum varieties are high biomass crops and are typically taller, leafier, and later maturing than grain varieties. Sweet sorghum produces stalks rich in fermentable sugars (sucrose, glucose, and fructose) for syrup or ethanol production. Sugar yield varies with variety/hybrid, location (soil, water, environment, pests, and diseases), fertilizer input, and production practices. Generally, fresh biomass yields of sweet sorghums can range from 18–107 Mg/ha, with juice comprising 65–80% of the total weight. The total sugar content of the juice varies between 9% and 20%. The bagasse represents about two-thirds of the dry mass (Erickson *et al.*, 2012).

Table 10.3 provides the yield potential of common energy crops adopted in the USA.

10.2.3 Forest Biomass

Forest biomass includes trees, short-rotational woody crops, forest logging residues, and mill residues (Figure 10.4). Generally, forest biomass is classified as softwood (e.g., pine, spruce) and hardwood (e.g., oak, hybrid poplar) species. Softwoods are the major source of lignocellulosic feedstock for paper and pulp industries and bioenergy applications (wood pellets). Trees are grown primarily for lumber and paper products. The USA has approximately 749 million acres of forestland, which is about one-third of the total land area. The total timberland for commercial applications is about 504 million acres, with an average yield potential of 20 ft^3/ac (US DOE, 2011). The US timberlands have a total biomass availability of 368 million dry tons annually, including the current uses of 142 million dry tons. Both hardwood and softwood species are established from seedlings obtained from nurseries and maintained annually for a growth cycle of 15–35 years. An average harvesting age of 25 years is reported for logging pine forestlands. Cutting or logging of roundwood timbers for sawing and pulping applications produces a significant amount of logging residues, consisting of tree tops, limbs, and small branches. Saw mill and pulp mill residues such as sawdust, chips, shavings, and bark are fully utilized for the generation of steam and electricity or for wood pellet production. Short-rotational woody coppices (SRC) such as willow and poplar are grown in a 3–5-year cycle for bioenergy applications, with an average yield of 5–6 dry ton/ac/yr. SRC crops are typically propagated with stem cuttings or plantlets that are planted in double rows, with 125 cm spacing between each set of double rows and 75 cm spacing within each row.

The average yield of common forest biomass is given in Table 10.3. The holocellulose content of woody biomass does not vary much, but the lignin content in softwood is higher than that in hardwood. Forest biomass yield depends on species, tree age, management practices, and climatic

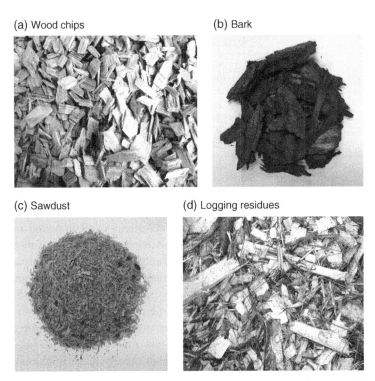

(a) Wood chips

(b) Bark

(c) Sawdust

(d) Logging residues

FIG. 10.4   Forests-based lignocellulosic feedstock. Photos courtesy of Sudhagar Mani, University of Georgia

conditions. It is often related to the diameter of the tree at breast height of 1.3 m. The most commonly studied tree species and their equation parameters are presented in Table 10.4.

Forest biomass estimation

Forest biomass (dry kg), $m = aDBH^b$ (10.6)

$m = a + b \ \log(DBH)$

where m is the amount of dry biomass from a tree in kg; DBH is the tree diameter at breast height in cm; a and b are constants that vary with species.

Natural Resources Canada (NRC) has developed a web-based forest biomass equation that estimates the amount of roundwood, foliage, branches, and bark from a standing tree. The tool can be used for calculating roundwood production and potential logging residues availability from managed forestlands (https://apps-scf-cfs.rncan.gc.ca/calc/en/calculateur-calculator). Similar to crop residues, collection of all logging residues is not recommended for soil conservation and maintaining forest soil health. Sustainable harvesting or removal of logging residues is critical for the long-term accumulation of soil organic matter and soil health.

Table 10.4 Biomass equations for common tree species

Tree types	Species group	Max. DBH (cm)	Equation parameters	
			a	b
Hardwood	Aspen/alder/cottonwood/willow	70	0.1098	2.3867
	Soft maple/birch	66	0.1477	2.3651
	Mixed hardwood	56	0.0837	2.4835
	Hard maple/oak/beech/hickory	73	0.1336	2.4342
Softwood	Cedar/larch	250	0.1309	2.2592
	Douglas fir	210	0.1075	2.4435
	True fir/hemlock	230	0.0790	2.4814
	Pine	180	0.0792	2.4349
	Spruce	250	0.1253	2.3323

Source: Data from Jenkins *et al.*, 2003.

Example 10.2: Estimate the above-ground biomass fractions of lodgepole pine stand with a diameter at breast height (DBH) of 250 cm. Calculate the amount of logging residues generated from this pine stand. If the biomass is removed from the field, how much nutrient (N, P, K) is removed from the field?

Solution: Lodgepole pine yield data obtained from https://apps-scf-cfs.rncan.gc.ca/calc/en/calculateur-calculator is presented in Table 10.5.

Table 10.5 Biomass and nutrient content of tree fractions

	Bark	Branches	Foliage	Stem wood
Biomass (kg)	2,389	3,475	509	87,433
N (kg)	6.5	N/A	5.4	39.6
P (kg)	1.2	N/A	0.6	7.0
K (kg)	4.1	N/A	2.8	35.8

Amount of logging residues generated = branches + foliage = **3, 984 kg**

N/A = not applicable

10.3 Feedstock Logistics

Feedstock logistics is a discrete network of operations to move biomass from a field to an intermediate storage facility or a biorefinery. Any feedstock logistics system consists of three major operations: feedstock harvesting and collection; transport; and storage.

10.3.1 Harvesting and Collection of Crop Residues and Energy Crops

Crop residues are typically harvested just after grain harvesting to prepare land for the next harvesting session. The harvesting window of crop residues typically varies between 30 and 45 days after grain harvesting. Systematic scheduling and efficient collection of agricultural residues are critical to consistently supplying biomass to a biorefinery. Unlike crop residues, energy crops can be harvested after first frosting or at low stem moisture content. Harvesting and collection operations consist of mowing, raking, baling or chopping, and transporting of bales/chops to the roadside for storage. Mowing is a field operation to cut standing crop residues using a shear or impact cutting bars in a mower (Figure 10.5). A mower can be a self-propelled or tractor-mounted unit. The material cut is usually crushed to facilitate field drying or remove stem moisture and left in a row as swaths. Raking is the manipulation or rolling of swaths into narrow rows to facilitate baling or forage chopping (Figure 10.6). Forage harvesters are commonly used to cut and chop grasses or energy crops to directly collect the wet or dry form in forage wagons running in parallel with a harvester (Figure 10.7). Biomass can be collected in five forms after harvesting (Figure 10.8): rectangular bales; round bales; loaves; dry chop piles; and wet chop ensiling.

FIG. 10.5   Mowing operation. Reproduced with permission of John Deere.

FIG. 10.6   Bermuda grass swath during raking before baling/chopping. Photo courtesy of Sudhagar Mani, University of Georgia.

FIG. 10.7   Forage harvester with a chop wagon. Reproduced with permission of John Deere.

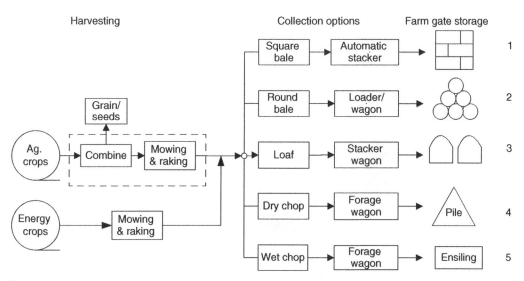

FIG. 10.8   Harvesting and collection options for crop residues and energy crops.

Rectangular and round bales are the most common forms of collecting crop residues and energy crops for bioenergy applications. A baler takes a dry swath (usually less than 15% moisture content, wet basis), compacts it into either round or rectangular bales, and ties it with twine before leaving it in the field (Figure 10.9). The bales are picked up by bale wagons and transported to the roadside. A front-end loader is used to stack the bales for storage. The bale size and density vary with the volume of the bale chamber, chop size, and ram speed. The stacker wagon was the oldest form of collecting forage biomass in wet form, similar to the cotton module builder in the early 1970s developed by John Deere (Figure 10.10).

(a) Round baler

(b) Large rectangular baler

FIG. 10.9   Biomass baling operation. Reproduced with permission of John Deere.

A rectangular bale size of 1.2 × 1.2 × 2.4 m is the most common, with a bulk density ranging from 140–180 kg/m³. Round bales of 1.5 m wide × 1.8 m diameter have an average bulk density of 180 kg/m³. For bioenergy applications, large rectangular bales are recommended for efficient stacking and transport.

FIG. 10.10   Forage stacker wagon. Reproduced with permission of John Deere.

A stacker wagon picks up dry swath and packages it in a high-compression chamber 2.4 m wide, up to 6 m long and 3.6 m high. The material is pressed to increase the bulk density of about 80 kg/m^3 and transported to a storage area or to the roadside. These are no longer in use despite the low cost of collection.

The dry chop collection option requires a forage harvester to pick up dry biomass from a swath, a chopper to cut it into smaller pieces (25–50 mm), and a blower to blow the chops into a forage wagon traveling along with the harvester. The forage wagon transports the chops to the roadside and piles up the material. Similar to the dry chop options, a forage harvester is employed in wet chop operations, but the chopped material is loaded into a concrete pit or a silo for wet storage or ensiling. The entire biomass collection operation requires multiple field machines or passes to collect biomass prior to transport and storage (multi-pass harvesting). Use of multiple field machines increases the cost and energy requirement for biomass collection. Recently, several universities and farm equipment manufacturers have developed single-pass harvesting machines, where one or two connected machines are used to efficiently collect both grain and crop residues at the same time. The collected crop residues are chopped and fed directly to a baler feeder for immediate baling or a wagon for roadside transport and storage using a single field operation.

Agricultural Machinery Performance and Cost

The field performance of any agricultural machinery is evaluated by its capacity. Machinery capacity is defined as the amount of area covered per unit time or the volume of material processed per unit time. It helps in determining the rate at which the material is collected or processed. The field capacity of machinery is reported in ha/hr (ac/hr) and it depends on field machinery speed, width of the machine, and field efficiency.

Field efficiency or machine utilization is the ratio of effective field capacity to the theoretical field capacity of the machinery.

$$\text{Field capacity}, FC = \frac{SW}{10}\eta_f \tag{10.7}$$

where:
FC = Field capacity (ha/hr)
S = Machine speed (km/hr)
W = Machine width (m)
η_f = Field efficiency or machine utilization (%)

Example 10.3: Calculate the bulk density of a large rectangular switchgrass bale with a size of $1.2 \times 1.2 \times 2.4$ m, where the weight of the bale at 12% moisture content is 681 kg (1500 lb).

Solution:

Weight of the bale, m = 681 kg or 1500 lb at 12% moisture content

$$\text{Dry weight of the bale} = 681\frac{(100-12)}{100} = 599.3\,\text{kg}$$

Volume of the bale, V = $1.2 \times 1.2 \times 2.4 = 3.456\,\text{m}^3$

$$\text{Bulk density} = \frac{m}{V} = \frac{599.3}{3.456} = \mathbf{173.4}\,\text{kg/m}^3\,(\text{dry wt.})$$

Field efficiency or machine utilization can never be 100% in the field due to delays in field turning and idle time, implement adjustment time, material clogging time, and fuel filling and lubrication time. Table 10.6 provides a list of common machinery with typical speeds and efficiencies.

Farm equipment ownership cost and material processing cost can be calculated based on the American Society of Biological and Agricultural Engineers standards (ASABE D497.5 & ASABE

Table 10.6 Typical field operating parameters for common farm equipment

Equipment – operations	Range of speed (mph)	Typical speed (mph)	Range of efficiency (%)	Typical efficiency (%)
Combine (sp)	2–5	3	60–85	65
Mower-conditioner	3–6	5	75–85	80
Shred	5–12	7	75–90	80
Rake	4–8	6	70–90	80
Baler – large squares	4–8	5	70–90	75
Baler – round	3–8	5	55–75	65
Forage harvester (sp)	1.5–6	3.5	60–85	70
Windrower (swather)	3–8	5	70–85	80
Stacker wagon (loaf)	4–8	5	60–70	65

Note: sp = self-propelled
Source: Data from ASABE, R2015a, R2015b.

EP496.3). These standards cover detailed procedures to calculate tractor and attachment unit costs for common agricultural equipment. Example 10.4 demonstrates the procedure for estimating machinery cost in the context of a biomass baling operation.

The fuel consumption of agricultural machinery is calculated based on the rated engine power. The lubricant cost is usually assumed as 15% of the fuel cost (ASABE D497.5 standard).

SI unit

$$\text{Machinery fuel consumption rate (L/h)} = 0.305 * P \text{ (for gasoline engine)} \qquad (10.8)$$

$$= 0.223 * P \text{ (for diesel engine)} \qquad (10.9)$$

where P is the rated engine power (kW).

English unit

$$\text{Machinery fuel consumption rate (gal/h)} = 0.0.068 * Q \text{ (for gasoline engine)} \qquad (10.10)$$

$$= 0.0.044 * Q \text{ (for diesel engine)} \qquad (10.11)$$

where Q is the rated engine power (hp).

Example 10.4: Calculate the total cost ($/h) of operating a large round baler attached to a 180 hp diesel tractor. The list prices of a tractor and a large round baler are $150,000 and $40,000, respectively. The size of each bale is 60 × 72 in and each weighs 1000 lb and is tied with twines. Calculate the ownership and operating cost of both a tractor and a baler. Make all necessary assumptions.

Solution: A baling operation needs a baler (attachment) and a power unit (a tractor). The field capacity of the power unit is the same as the attachment. The total cost of the operation will be the sum of the total fixed and operational costs of both a baler and a power unit (Table 10.7)

Table 10.7 Assumptions for the calculations (refer to ASABE Standards D497.5 and EP496.3)

SL No	Parameter	Assumed value/given data
1	Interest rate	8%
2	Diesel price ($/gal)	4.0
3	Hourly labor cost ($/hr)	12
4	Cost of twine (20,000 ft ball), $	26
5	Taxes, insurance, and housing (%)	1% of list price
	Repair and maintenance (%)	24% of list price
	Bale weight (tons)	0.5

(continued overleaf)

Table 10.7 (*continued*)

SL No	Parameter	Assumed value/given data	
		Power unit (tractor)	Attachment (round baler)
6	Horse power (hp)	180	
7	List price ($)	150,000	40,000
8	Bale size		60 in × 72 in
9	Annual use (hr)	600	200
10	Life time (y)	15	10
11	Field capacity (ton/hr)	Depends on/same as attachment	9
12	Remaining salvage value (%)	24	28
	Lubricant cost (%)	15% of fuel cost	

(A) Ownership/fixed cost

(i) Tractor

Purchase price (PP) = 0.85 × list price = 0.85 × $150,000 = $127,000

Salvage value (S) = List price × remaining salvage value (%) = 150,000 × 24/100
$$= \$36,000$$

Depreciation (D) = Purchase price − salvage value = $127,000 − $36,000 = $91,000

$$Capital\ recovery\ factor\ (CRF) = \frac{Interest\ rate\ (1 + interest\ rate)^{year}}{(1 + interest\ rate)^{year} - 1}$$

$$= \frac{0.08}{1 - (1 + 0.08)^{-15}} = 0.117$$

$$Capital\ recovery\ (CR) = Depriciation\ (D) \times Capital\ recovery\ factor\ (CRF)$$
$$+ salvage\ value\ (S) \times interest\ rate$$

$$CR = 91,000 \times 0.0117 + 36,000 \times 0.08 = 13,511.49$$

Taxes, insurance & housing (TIH) $= 0.01 \times$ purchase price (PP)

$$= 0.01 \times 127,000 = 1,270$$

Total ownership or fixed cost (FC) ($)

$$= \text{Capital recovery (CR)} + \text{Taxes, insurance \& housing (TIH)}$$

$$= 13,511.49 + 1,270 = 14,781.49$$

$$Ownership\ cost\ (\$/hr) = \frac{Total\ fixed\ cost}{Annual\ use\ in\ hr} = \frac{14,781.489}{600} = 24.64$$

The hourly ownership cost of the tractor is $24.64.

(ii) Baler

Purchase price $(PP)(\$) = 0.85 \times List\ price = 0.85 \times 40,000 = 34,000$

Salvage value $(S)(\$) = 0.28 \times List\ price = 0.28 \times 40,000 = 11,200$

Depreciation $(D)(\$) = PP - D = 34,000 - 11,200 = 22,800$

Capital recovery $(CR)(\$) = 22,800 \times 0.149 + 11,200 \times 0.08 = 4293.87$

Taxes, insurance & housing $(TIH)(\$) = 0.01 \times PP = 0.01 \times 34{,}000 = 340$

Total ownership or fixed cost $(FC)(\$) = 4{,}293.87 + 340.00 = 4{,}633.87$

Hourly fixed cost $(\$/hr) = 4{,}633.87/200 = 23.17$

The hourly ownership cost of the baler is \$23.17.

(B) Operational cost/variable cost
(i) Tractor

Fuel use $(diesel)$ $(gal/hr) = 0.044 \times Machine\ hp = 0.044 \times 180 = 7.92$

Fuel cost $(\$/hr) = Fuel\ use \times diesel\ price = 7.92 \times 4.00 = 31.68$

Lubricant cost $(\$/hr) = 0.15 \times Fuel\ cost = 0.15 \times 31.68 = 4.75$

Total accumulated machine hour $(hr) = annual\ use \times lifetime = 600 \times 15$
$$= 9{,}000$$

Repair & maintenance cost (RM) $(\$) = 0.24 \times List\ price = 0.24 \times 150{,}000.00$
$$= 36{,}000$$

Repair & maintenance cost (RM) per hour $(\$/hr)$

$$= \frac{Repair\ \&\ maintenance\ cost\ (RM)}{Total\ accumulated\ machine\ hour} = \frac{36{,}000}{9{,}000} = 4.00$$

Machines or equipment need extra labor hours other than actual machine hours for different unavoidable activities associated with different operations. So extra labor hours, about 10–20% of actual working machine hours assumed during calculating the cost of labor. In this case about 10% of extra labor hour is considered for every machine hour.

Hourly labor cost $(\$) = hourly\ labor \times 1.1 \times machine\ hour = 12.00 \times 1.1 \times 1 = 13.2$

Hourly operational cost $(\$) = fuel\ cost + $ Lubricant cost $ + $ RM cost $ + $ labor cost
$$= 31.68 + 4.75 + 4.00 + 13.2 = 53.63$$

Total hourly operational cost of tractor is \$53.63.

(ii) Baler

Total accumulated machine hour $(hr) = 200 \times 10 = 2{,}000$

Repair & maintenance cost (RM) $(\$) = 0.24 \times 40{,}000 = 9{,}600$

Repair & maintenance cost (RM) per hour $(\$/hr) = \dfrac{9{,}600}{2{,}000} = 4.80$

Let us assume a Twine roll is about 20,000 feet long. If the twine is used to tie a round bale (6 × 5 ft) with a twine spacing of 4 inch, the twine length required per bale is

$$= \pi \times d \times \left(\frac{w}{s} + 1 \right)$$

where, d = diameter of a bale (ft); w = width of a bale (ft); s = twine spacing (ft)

$$Twine\ length\ per\ bale = \pi \times d \times \left(\frac{w}{s} + 1\right) = \pi \times 6 \times \left(\frac{5}{0.333} + 1\right) = 302\,ft$$

Number of bale warped / twine roll = $20,000/302 \cong 66$

$$Cost\ of\ twine\ per\ bale\ (\$)$$
$$= Cost\ of\ twine\ roll\,/\,number\ of\ bales\ warped\ by\ a\ twine\ roll$$
$$= 26/66 = 0.39$$

$$Baling\ capacity\ (number\ of\ bale\ per\ hr) = baler\ capacity\ (tons/hr)\,/\,bale\ weight$$
$$Cost\ of\ twine\ per\ hr\ (\$)$$
$$= Baling\ capacity\ (number\ of\ bale\ per\ hr) \times cost\ of\ twine\ per\ bale$$
$$= 9/0.5 \times 0.39 = 7.07$$

$$Hourly\ operational\ cost\ (\$) = RM\ cost + Twine\ cost = 4.80 + 7.07 = 11.87$$

Total hourly operational cost of baler is $ 11.87.

Total hourly cost of baling operation

Total hourly cost of a farm machinery = Hourly fixed cost + Hourly operational cost

$$Total\ hourly\ cost\ of\ power\ unit\ (tractor)(\$/h) = 24.64 + 53.63 = 78.27$$

$$Total\ hourly\ cost\ of\ attachement\ (baler)(\$/h) = 23.17 + 11.87 = 35.04$$

$$Total\ hourly\ cost\ of\ baling\ operation\ (\$/h)$$
$$= Total\ hourly\ cost\ of\ power\ unit + Total\ hourly\ cost\ of\ attachment$$
$$= \mathbf{78.27 + 35.04 = 113.31}$$

10.3.2 Harvesting of Forest Biomass

Harvesting of forest trees and logging residues requires specialized equipment with high machine power. A typical tree-harvesting system is shown in Figure 10.11.

A typical logging operation consists of five major machinery systems, namely, a feller-buncher, a skidder, a log processor, a knuckle boom loader, and a log trailer (Figure 10.12).

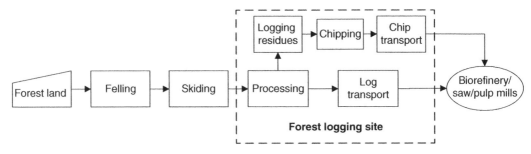

FIG. 10.11 A typical forest biomass harvesting system.

(a) A feller-buncher to saw and fell a tree

(b) A skidder to drag a tree

(c) A log processor to remove tops and branches

(d) Semi-truck log trailer with knuckle boom loader

FIG. 10.12 Typical logging machinery used during tree harvesting. Photo courtesy of Sudhagar Mani, University of Georgia.

Trees are harvested using a feller-buncher, which holds the tree trunk, cuts it using a disk or shear saw, and lays the tree on the ground. A skidder uses a large grapple to pick up one or several cut trees and drag them to a logging site. A logging site has several series of large logging equipment to produce both pulp and saw-wood logs. A log processer takes the skidded trees, cuts limbs, branches, and tops, and measures tree diameter to classify the logs. A log diameter of 6–8 in is piled up for pulp-chip production. Some of the classified pulp logs are currently used for manufacturing wood pellets. A log diameter larger than 8 in is sent for sawmill applications. The classified log piles are loaded onto a semi-truck log trailer using a knuckle boom loader. A front-end loader collects all the tops, limbs, and branches and then piles them up at the logging site for immediate collection or collection after transpirational drying. Sometimes, logging residues are chipped using a whole-tree chipper and residue chips are loaded directly into a chip van for bioenergy applications.

Forest equipment cost estimation is similar to that for agricultural machinery, with some exceptions, and is illustrated in Example 10.5. The detailed cost estimation for forest equipment is described in Brinker *et al.* (2002). The forest equipment is heavier and has a shorter life, as it can be used year round. Unlike crop residues and energy crops, forest biomass can be harvested throughout the year, if weather permits, and requires no long-term storage facility. The machine utilization rate for any forest equipment is about 60–65%, which may increase the material processing cost. Example 10.5 describes the cost calculation for forest equipment, using a chipper.

Example 10.5: A forester purchases a 1,000 hp whole-tree chipper for $750,000. The lifetime of the chipper is 5 years, with a 20% salvage value, and the utilization rate is 60%. Calculate the chipping cost in $/PMH and $/SMH. State all the assumptions. Machine rates for selected harvesting machines can be obtained from Brinker *et al.* (2002).

Solutions: Table 10.8 shows the summary of key assumptions used in this example.

Table 10.8 Assumptions

Parameters	Assumed value/given data
Interest rate	8%
Diesel price ($/gal)	4.0
Hourly labor cost ($/hr)	12
Salvage value (S)	20% of purchase price
Taxes, insurance, and housing (%)	8% of yearly investment cost
Repair and maintenance (%)	80% of depreciation
Lube and oil cost ($/hr)	36.8% of fuel cost
Schedule machine hours @ 8hr/day @ 250day/y (SMH)	2,000
Utilization rate	60%
Purchase price ($)	750,000
Machine life (N) (years)	6
Machine horsepower (hp)	1000

Fixed cost

Productive machine hours (PMH) = *Scheduled machine hours* (SMH) × *Utilization rate*

$$= 2000 \times 0.6 = 1200$$

$$Salvage\ value\ (\$) = 0.2 \times purchase\ price = 0.2 \times 750,000 = 150,000$$

$$Annual\ depreciation\ (AD)(\$) = \frac{Purchase\ price - Salvage\ value}{Machine\ life}$$

$$= \frac{750,000 - 150,000}{6} = 100,000$$

$$Average\ annual\ investment\ (AYI)(\$) = \frac{(Purchase\ price - salvage\ value) \times (Machine\ life + 1)}{2 \times Machine\ life}$$

$$+ Salvage\ value$$

$$= \frac{(750,000 - 150,000) \times (6 + 1)}{2 \times 6} + 150,000 = 500,000$$

$$Annual\ interest\ cost\ (I)(\$) = Interest\ rate \times AYI = 0.08 \times 500,000 = 40,000$$

$$Annual\ insurance\ and\ tax\ (IT)(\$) = 0.08 \times 500,000 = 40,000$$

$$Total\ annual\ fixed\ cost\ (\$) = AD + I + IT = 100,000 + 40,000 + 40,000 = 180,000$$

$$Fixed\ cost / (SMH)(\$) = Total\ annual\ fixed\ cost / Scheduled\ machine\ hours\ (SMH)$$
$$= 180,000/2,000 = 90$$

$$Fixed\ cost / (PMH)(\$) = Total\ annual\ fixed\ cost / Productive\ machine\ hours\ (PMH)$$
$$= 180,000/1,200 = 150.0$$

Operating/Variable cost

$$Fuel\ use / hr = 0.037 \times Machine\ horsepower = 0.037 \times 1,000 = 37$$

$$Fuel\ cost / hr\ (\$) = 37 \times 4.0 = 148$$

$$Lube\ and\ oil\ cost / hr\ (\$) = 0.368 \times Fuel\ cost = 0.368 \times 148 = 54.46$$

$$Repair\ and\ maintenance / hr\ (PMH)(\$) = 0.8 \times AD/PMH = 0.8 \times 100,000/1,200$$
$$= 66.67$$

$$Labor / hr\ (\$) = 1.4 \times labor\ cost / hr \times Machine\ utilization\ rate = 1.4 \times 12 \times 0.6 = 28$$

$$Total\ operating\ cost / hr\ (PMH)\ (\$) = Fuel\ cost + Lube\ \&\ oil\ cost + RM\ cost + labor\ cost$$

$$= 148 + 54.46 + 66.67 + 28 = 297.13$$

$$Total\ operating\ cost\ (\$) / hr\ (SMH) = Total\ operating\ cost\ (\$) / hr\ (PMH)$$
$$\times Machine\ utilization = 297.13 \times 0.6 = 178.28$$

$$Total\ cost\ (\$) / hr\ (PMH) = Fixed\ cost\ (PMH) + variable\ cost\ (PMH)$$
$$= 150 + 297.13 = 447.13$$

$$Total\ cost\ (\$)/hr\ (SMH) = Fixed\ cost\ (SMH)\ +variable\ cost\ (SMH)$$
$$= 90 + 178.28 = 268.28$$

10.3.3 Transportation

Lignocellulosic feedstock is transported from farm or forest gates to a biorefinery or a storage facility, either by truck, train, or a combination. Barges and ships are commonly used for long-distance transport of wood pellets and wood chips from North America to Europe and China, respectively. Figure 10.13 shows the multi-mode transport options available for biomass. Pipeline transport of biomass slurries is used for short-distance transport, similar to a crude oil pipeline. For example, sawmill residues are pneumatically conveyed through pipelines over short distances (less than 5 km). The most common form of transport for any form of biomass is by a trucking system. Rail carts are often used for long-distance transport (>160 km) or wherever it is convenient and economical. Biomass bales are frequently transported using a semi-truck flatbed trailer (Figure 10.14). The

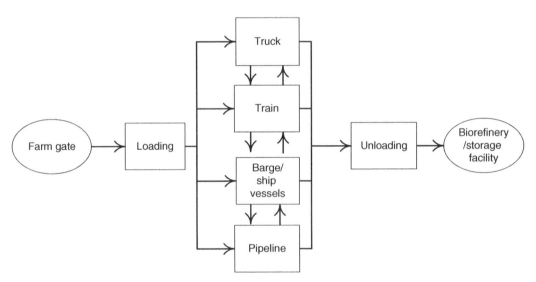

FIG. 10.13   Multi-mode biomass transportation options.

(a) Bale transport by truck   (b) Biomass transport by rail

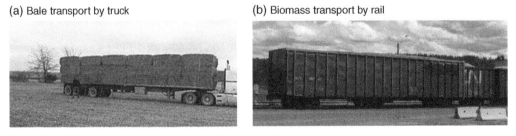

FIG. 10.14   Biomass transport by truck and rail. Photos courtesy of Sudhagar Mani, University of Georgia.

Table 10.9 Bulk density data for different forms of lignocellulosic feedstock

Form of feedstock	Shape and size characteristics	Bulk density (kg/m^3)
Chopped biomass	20–40 mm length	60–80
Ground particle	1.5 mm loose fill	120
Baled biomass	Square/round bales	140–180
Ground particle	1.5 mm pack fill with tapping	200
Cube	33 mm × 33 mm cross-section	400
Granule	2–20 mm	250–500
Pellet	6.24 mm diameter	500–700
Round briquette	50 mm diameter and 300 mm length	~1,000

Source: Data from Mani, 2005.

Table 10.10 Cost and energy consumption equations for transporting feedstocks

Transport mode	Cost ($/Mg)	Energy consumption (MJ/Mg)
Truck	5.7 + 0.1367 D	1.3 D
Rail	17.1 + 0.0277 D	0.68 D
Barge	34.0 + 0.01 D	–
Pipeline	2.67 m$^{-0.87}$ + 0.137 m$^{-0.44}$	160 m$^{-0.87}$ + 22.2 m$^{-0.44}$

D = distance traveled (km); m = weight of biomass transported (Mg)
Source: Data from Sokhansanj *et al.*, 2009.

bulk form of biomass is transported using both intermodal containers and semi-truck and trailer. Wood chips and pellets are often transported to mainland Europe and Scandinavian countries by large ships (Sokhansanj *et al.*, 2009).

All long-haul tractors and trailers in the USA are required to maintain a gross weight of below 36.4 tons. Height, width, and length restrictions are applied in addition to weight. A typical flatbed trailer has an average empty weight of 5 Mg (5.5 tons), with 12.2–16.2 m (40–53 ft) length and 2.4 m (8 ft) width. A typical chip van can have a capacity of 75–80 m^3 with an average weight of 4.5 Mg (5 tons). A larger van can have both a width and height of 2.4 m (8 ft) and a length of 14.2 m (53 ft), along with a maximum road weight restriction of 21 Mg. Biomass should have a theoretical bulk density of approximately 260 kg/m^3. If the bulk density of biomass is less than this threshold limit, it would cost more to transport low-density biomass such as bales and chips. Densification of crop residues, grasses, and wood chips into pellets or briquettes will increase the bulk density up to 650–850 kg/m^3.

Table 10.9 lists the bulk density of biomass in different forms for long-distance transport. The transport cost of biomass depends on fuel prices, haul distance, truck capacity, biomass moisture content, and bulk density. Estimations of cost and energy consumption equations for transporting biomass using truck, rail, and pipeline are given in Table 10.10.

10.3.4 Storage

Safe storage of dry or wet biomass is critical for a consistent supply to a biorefinery. Storage of wet biomass is typically done by an ensiling technique to store corn stover and grasses as silage for animal feed applications. In an ensiling operation, wet biomass is chopped into small pieces, loaded, and compacted into a concrete pit or a tower silo, then tightly covered to avoid air infiltration. This method can

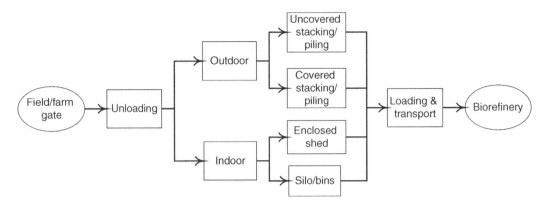

FIG. 10.15   Biomass storage options.

(a) Woodchip piles

(b) Roundwood log piles

(c) Biomass bales

FIG. 10.16   Outdoor storage of lignocellulosic feedstock. Photos courtesy of Sudhagar Mani, University of Georgia.

be used for storing wet biomass for 6–12 months with no or limited dry-matter losses. Figure 10.15 shows various storage options available for both agricultural and forest-based lignocellulosic feedstock.

Outdoor storage of biomass (Figure 10.16) can often be less expensive than indoor storage due to high capital investment. If a large volume of biomass is stored outside for a long period, the losses of

(a) Enclosed bulk storage piles

(b) Bale storage in sheds

(c) Bin storage

FIG. 10.17   Indoor storage of biomass. Photos courtesy of Sudhagar Mani, University of Georgia.

dry matter and key chemical compounds (e.g., sugars, cellulose) can be high due to fungal and microbial decomposition. The storage cost of biomass depends on the type of storage option, costs of dry-matter and quality losses, and storage time. Outdoor storage of bales covered with tarpaulins is recommended as an economical option with limited dry-matter losses. Biomass in granular form such as granules or pellets can be stored in silos or bins, similar to grain storage systems (Figure 10.17).

References

ASABE Standards (R2015a). *D497.7: Agricultural Machinery Management Data*. St. Joseph, MI: American Society of Biological and Agricultural Engineers.

ASABE Standards (R2015b). *EP496.3: Agricultural Machinery Management*. St. Joseph, MI: American Society of Biological and Agricultural Engineers.

Baker, S.A., Westbrook, M.D., Jr., and Greene, W.D. (2010). Evaluation of integrated harvesting systems in pine stands of the Southern US. *Biomass and Bioenergy* 34(5): 720–27.

Brinker, R.W., Kinard, J., Rummer, B., and Lanford, B. (2002). *Machine Rates for Selected Forest Harvesting Machines*. Alabama Agricultural Experiment Station, Circular 296 (revised). Auburn, AL: Auburn University.

Eggleston, G., Grisham, M., and Antoine, A. (2010). Clarification properties of trash and stalk tissues from sugar cane. *Journal of Agricultural and Food Chemistry* 58: 366–73.

El Bassam, N. (1998). *Energy Plant Species: Their Use and Impact on Environment and Development*. London: James and James Science Publishers.

Erickson, J. E., Woodard, K.R., and Sollenberger, L.E. (2012). Optimizing sweet sorghum production for biofuel in the Southeastern US through nitrogen fertilization and top removal. *BioEnergy Research* 5(1): 86–94.

FMO (1987). *Hay and Forage Harvesting: Fundamentals of Machine Operation*. Moline, IL: Deere and Company.

Heaton, E.A., Dohleman, F.G., and Long, S.P. (2008). Meeting U.S. biofuel goals with less land: The potential of *Miscanthus*. *Global Change Biology* 14: 2000–14.

Jenkins, J.C., Chojnacky, D.C., Heath, L.S., and Birdsey, R.A. (2003). National-scale biomass estimators for United States tree species. *Forest Science* 49(1): 12–35.

Knoll, J.E., Anderson, W.F., Strickland, T.C., Hubbard, R.K., and Malik, R. (2012). Low input production of biomass from perennial grasses in the Coastal Plain of Georgia, US. *BioEnergy Research* 5(1): 86–94.

Labrecque, M., and Teodorescu, T.I. (2003). High biomass yield achieved by Salix clones in SRIC following two 3-year coppice rotations on abandoned farmland in southern Quebec, Canada. *Biomass Bioenergy* 25(2): 135–46.

Ladanai, S., and Vinterback, J. (2009). *Global Potential of Sustainable Biomass for Energy*. Report # 013. Uppsala: Swedish University of Agricultural Sciences.

Lewandowski, I., Scurlock, J.M.O., Lindvall, E., and Christou, M. (2003). The development and current status of perennial rhizomatous grasses as energy crops in the US and Europe. *Biomass and Bioenergy* 25(4): 335–61.

Ma, Z., Wood, C.W., and Bransby, D.I. (2000). Soil management impacts on soil carbon sequestration by switchgrass. *Biomass and Bioenergy* 18(6): 469–77.

Mani, S. (2005). A systems analysis of biomass densification technology. PhD dissertation. Department of Chemical & Biological Engineering, University of British Columbia, Canada.

McLaughlin, S.B., De La Torre Ugarte, D.G., Garten, C.T., Jr., *et al.* (2002). High value renewable energy from prairie grasses. *Environmental Science and Technology* 36: 2122–9.

NASS (2013). *Field Crops: Production and Yield Data.* Washington, DC: National Agricultural Statistical Services. www.nass.usda.gov.

NRCS (2010). *Soil Erosions on Croplands.* National Resources Inventory. Washington, DC: Natural Resources Conservation Service. www.nrcs.usda.gov

Pearson, C.H., Halvorson, A.D., Moench, R.D., and Hammon, R.W. (2010). Production of hybrid poplar under short term, intensive culture in Western Colorado. *Industrial Crops and Products* 31: 492–8.

Sanderson, M.A., Reed, R.L., McLaughlin, S.B., *et al.* (1996). Switchgrass as a sustainable bioenergy crop. *Bioresource Technology* 56: 83–93.

Sevel, L., Nord-Larsen, T., Ingerslev, M., Jorgensen, U., and Raulund-Rasmussen, K. (2014). Fertilization of SRC willow I: Biomass production response. *BioEnergy Research* 7: 319–28.

Skytte, K., Meibom, P., and Henriksen, T.C. (2006). Electricity from biomass in the European Union – with or without biomass import. *Biomass and Bioenergy* 30: 385–92.

Sokhansanj, S., Mani, S., Turhollow, A., *et al.* (2009). Large-scale production, harvest and logistics of switchgrass – current technology and envisioning a mature technology. *Biofuels, Bioproducts and Biorefining* 3: 124–41.

Teel, A., Barnhart, S., and Miller, G. (2003). *Management Guide for the Production of Switchgrass for Biomass Fuel in Southern Iowa.* Ames, IA: Iowa State University Extension.

U.S. Department of Energy (2011). *U.S. Billion-Ton Update: Biomass Supply for a Bioenergy and Bioproducts Industry.* R.D. Perlack and B.J. Stokes (eds.), ORNL/TM-2011/224. Oak Ridge, TN: Oak Ridge National Laboratory.

Woodard, K.R., and Prine, G.M. (1993). Regional performance of tall tropical bunchgrasses in the Southeastern USA. *Biomass and Bioenergy* 5(1): 3–21.

Exercise Problems

10.1. Why are all crop residues not available for sustainable harvesting?

10.2. Calculate a harvest index for a winter wheat crop if the grain yield is 50 bu/ac.

10.3. Calculate the total cotton stalk availability from a 250-acre cotton farm in Texas.

10.4. Estimate the above-ground biomass fractions of red oak stand with a diameter at breast height (DBH) of 30 cm. Calculate the amount of logging residues generated from this stand. Calculate the amount of roundwood harvested from a 50-acre forest land with a tree spacing of 3 × 5 m. Also calculate the logging residue that can be available for bioenergy applications.

10.5. Explain various energy crop harvesting and collection operations.

10.6. Calculate the diameter of a corn stover round bale with a bulk density of 180 kg/m^3 and a weight of 1600 lb. Assume the bale width to diameter ratio to be 1.5.

10.7. Calculate the field capacity of a raking machine with a machine width of 8 m and an average speed of 10 km/h. Assume the field efficiency to be 85%. If the machine is used to rake a 250-acre switchgrass field with an average yield of 6 Mg/ac, calculate the time required to rake the entire field. Also calculate the material processing capacity in ton/h.

10.8. Calculate the total cost ($/h) of operating a large rectangular baler attached to a 250 hp diesel tractor. The list price of tractor and baler are $175,000 and $65,000, respectively. The size of each bale is 1.2 × 1.2 × 2.4 m and each bale weighs 1500 lb and is tied with twine. Calculate the ownership and operating cost of both tractor and baler. Make all necessary assumptions.

10.9. If the same rectangular baler as in Problem 10.8 is used for baling a miscanthus field of 400 acres with an average yield of 8 Mg/ac, calculate the baling cost of miscanthus ($/Mg).

10.10. A forester purchases a 500 hp feller-buncher for $400,000. The lifetime of the feller-buncher is 5 years, with a 20% salvage value, and the utilization rate is 60%. Calculate the tree-harvesting cost in $/PMH and $/SMH. State all assumptions.

10.11. Describe various options available for the transportation of lignocellulosic feedstock and their implications.

10.12. Describe various options available for the outdoor and indoor storage of bales, bulk chops, and wood chips.

Algae-Based Feedstocks

Xumeng Ge, Johnathon P. Sheets, Yebo Li, and Sudhagar Mani

What is included in this chapter?

This chapter covers algae biomass production as a bioenergy feedstock. The mechanism of algae growth is discussed, together with algae cultivation, harvesting, drying, and oil extraction technologies.

11.1 Introduction

Algae are a group of plant or plant-like organisms that are commonly present in fresh- and marine-water bodies. You may have seen seaweed (a marine macroalga) on the beach, or algal blooms in eutrophic ponds. If these are your only experiences with algae, you may be surprised to learn that algae have many other functions in our society. Many algae species accumulate lipids or starch that are feedstocks for biofuel production. The algal biomass, or residue following extraction of lipids and/ or starch, can be used to produce biomethane through anaerobic digestion. Some algae species can even produce hydrogen (Pilon and Berberoglu, 2014) and hydrocarbons (Baba and Shiraiwa, 2013). Furthermore, genetically engineered algae have been developed for direct production of ethanol (Enquist-Newman *et al.*, 2014). Algae also play an important role in wastewater remediation, due to their ability to remove pollutants such as nitrogen and phosphorus. Algae are a source of several high-value nutraceuticals, such as protein (20–40%), omega-3 polyunsaturated fatty acids, docosahexaenoic acid (DHA), and astaxanthin (a red pigment carotenoid).

Biodiesel production from *algal-derived lipids* has been extensively studied in recent years. Under ideal conditions, algal cells can accumulate lipid content as high as 75% of the total dry weight (Mata *et al.*, 2010). Estimations have shown that the areal oil yield from algae could reach over 100 times that of traditional oilseed crops, such as soybean (Mata *et al.*, 2010; Pienkos and Darzins, 2009). Moreover, algae cultivation can use non-arable land and saline/brackish water, without competing with food or feed production that uses prime agricultural land and fresh water (Babu and Subramanian, 2013).

Bioenergy: Principles and Applications, First Edition. Edited by Yebo Li and Samir Kumar Khanal.

© 2017 John Wiley & Sons, Inc. Published 2017 by John Wiley & Sons, Inc.

Companion website: www.wiley.com/go/Li/Bioenergy

Algae are a highly diverse group of organisms. In the next section, you will learn basic information pertaining to algae classification, cell structure, and characteristics. Sections 11.3–11.6 cover the mechanisms of algae growth, technologies for algae cultivation, algal biomass harvesting and drying, and lipid extraction from algal biomass. There are three primary types of algae: cyanobacteria, microalgae, and macroalgae. This chapter focuses on microalgae and cyanobacteria, which usually accumulate lipids, although macroalgae can be an alternative feedstock (Gosch *et al.*, 2012; Hu *et al.*, 2008; Rajeshwari and Rajashekhar, 2011). Thus, in the context of this chapter, the term "algae" refers specifically to *microalgae* and *cyanobacteria*.

11.2 Algae Classification, Cell Structure, and Characteristics

According to the definition given by Lee (2008), algae are thallophytes (plants or plant-like organisms lacking roots, stems, and leaves) that "have chlorophyll α as their primary photosynthetic pigment and lack a sterile covering of cells around the reproductive cells." Algae are generally classified into three subgroups: cyanobacteria, microalgae, and macroalgae. *Cyanobacteria* are prokaryotes that lack membrane-bounded organelles. Both *microalgae* and *macroalgae* are eukaryotes that have these organelles in their cells. Cyanobacteria and microalgae are generally unicellular, although some of them can exist as colonies (Schirrmeister *et al.*, 2011). Contrary to cyanobacteria and microalgae, macroalgae are multicellular organisms that resemble the leaves, stems, and roots found in higher plants.

Like other phototrophic organisms, algae (cyanobacteria, microalgae, and macroalgae) also rely on chlorophyll to absorb energy from light. In algal cells, chlorophyll is embedded in a membrane-bound compartment called the thylakoid. In cyanobacteria, the thylakoids are located directly in the cytoplasm. In microalgae and macroalgae, the thylakoids are stacked, and embedded in the well-known chloroplast organelle. Cyanobacteria contain carboxysomes in their cytoplasm for maintaining an adequate CO_2 level in their cells (i.e., the CO_2-concentrating mechanism) for photosynthesis. Similarly, microalgae and macroalgae have pyrenoids in their chloroplasts to mediate the CO_2-concentrating function. Typical cell structures of cyanobacteria, microalgae, and macroalgae are illustrated in Figure 11.1, and their characteristics are compared in Table 11.1.

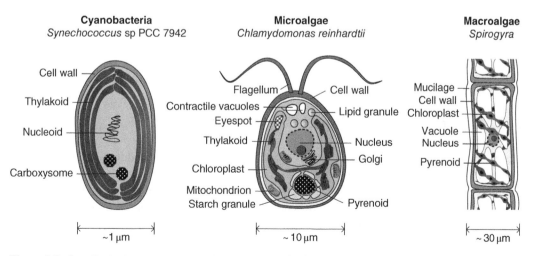

FIG. 11.1   Typical cell structures of cyanobacteria (*Synechococcus*), microalgae (*Chlamydomonas reinhardtii*), and macroalgae (*Spirogyra*).

Table 11.1 Differences between cyanobacteria, microalgae, and macroalgae

Type of algae	Type of cell	Structure	Location of thylakoid	CO_2-concentrating component
Cyanobacteria	Prokaryotic	Unicellular	Cytoplasm	Carboxysome
Microalgae	Eukaryotic	Unicellular	Chloroplast	Pyrenoid
Macroalgae	Eukaryotic	Multicellular	Chloroplast	Pyrenoid

- **Contractile vacuoles**: an organelle that removes excess water out of the cell (i.e., osmoregulation).
- **Eyespot**: a photoreceptive organelle that is used for light direction and irradiance detection, which further induces modification of the swimming behavior of cells.
- **Vacuole**: a membrane-bound enclosed organelle that is filled with water, and inorganic and organic molecules.
- **Golgi**: stacks of membrane-bound structures that modify, sort, and package macromolecules, such as proteins and lipids, for secretion.
- **Mitochondrion**: a membrane-bound organelle that supplies energy (ATP) for eukaryotic cells.
- **Nucleoid**: an irregularly shaped region that contains the chromosome without surrounding membrane.
- **Nucleus**: a membrane-bound structure that contains genetic material.
- **Nucleolus**: a non-membrane-bound structure composed of proteins and nucleic acids within the nucleus.
- *Synechococcus*: a cyanobacterium that is widespread in sea water. Thylakoids are located in the cytoplasm, forming bands close to the cell membrane.
- *Chlamydomonas reinhardtii*: a model microalga that can swim using its two flagella. Inside the cell are a large cup-shaped chloroplast and two contractile vacuoles. An eyespot, located at the outer edge of the chloroplast, detects light. The chloroplast contains a large pyrenoid that is surrounded by starch granules.
- *Spirogyra*: a macroalga that is commonly found in fresh water. Each cell is coated with a mucilaginous layer, which makes the cell surface feel slippery. Large vacuoles occupy most of the volume. The chloroplasts are arranged in a spiral pattern. The cytoplasm forms a thin lining between the cell wall and the organelles.

11.3 Mechanism of Algal Growth

Algae can synthesize carbohydrates from CO_2, and generate O_2 as a byproduct using energy from sunlight. This process is known as *photosynthesis*, and takes place in the cytoplasm (cyanobacteria cells) or chloroplasts (microalgae cells). The process of photosynthesis can be divided into two types of reaction: light-dependent and light-independent. In the light-dependent reaction, thylakoids use light energy to split water into hydrogen ions and oxygen, and release electrons, which are used to generate energy carriers such as ATP and NADH (Figure 11.2). In the light-independent reaction, ATP and NADH are used to incorporate (i.e., fix) CO_2 into carbohydrates via the Calvin cycle, which is a pathway mediated by several enzymes (Figure 11.2). In addition, carboxysomes

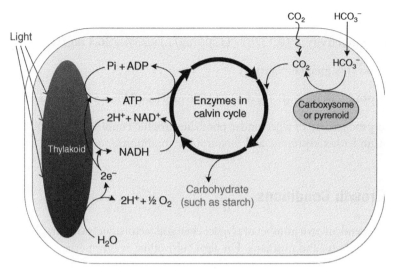

FIG. 11.2   Mechanism of photosynthesis in cyanobacteria or in the chloroplast of microalgae.

Table 11.2 *Chlorella vulgaris*

Conditions	Specific growth rate (1/day)	Algal biomass concentration (mg/L)	Lipid content (%)
Photoautotrophic (light, NO_3^-)	0.04	250	38
Heterotrophic (1% glucose, no light)	0.13	1,206	23
Mixotrophic (1% glucose, light)	0.15	1,696	21

(in cyanobacteria) or pyrenoids (in chloroplasts of microalgae) help maintain a CO_2-rich environment around the Calvin cycle enzymes.

The general equation for photosynthesis is:

$$2n\,CO_2 + 2n\,H_2O + \text{photons} \xrightarrow{\text{Cyanobacteria or Chloroplasts}} 2\,(CH_2O)_n + 2n\,O_2 \qquad (11.1)$$

It should be noted that some microalgal species can also undergo heterotrophic growth through the utilization of organic compounds such as glucose, acetate, ethanol, and glycerol, as carbon and energy sources. Moreover, some microalgae such as *Hematococcus pluvialis* and *Chlorella* spp., exhibit mixotrophic behavior, which is the simultaneous assimilation of CO_2 and organic carbon compounds.

> **Example 11.1:** The specific growth rate, algal biomass yield, and lipid content of *Chlorella vulgaris* cultured under photoautotrophic, heterotrophic, and mixotrophic conditions are shown in Table 11.2. Calculate the lipid productivity for each condition.
>
> *Solution:*
>
> Photoautotrophic (light, NO_3^-)
>
> **Lipid productivity** $= (0.04/d) \times (250\,\text{mg/L}) \times 38\% = \textbf{3.8 mg/(L·day)}$

Heterotrophic (1% glucose, no light)

Lipid productivity $= (0.13/\text{d}) \times (1206\,\text{mg/L}) \times 23\% = \textbf{36.1\,mg/(L·day)}$

Mixotrophic (1% glucose, light)

Lipid productivity $= (0.15/\text{day}) \times (1696\,\text{mg/L}) \times 21\% = \textbf{53.4\,mg/(L·day)}$

The lipid productivity of algae under photoautotrophic conditions is much lower than that under heterotrophic and mixotrophic conditions.

11.4 Algal Growth Conditions

Algal growth is dependent on a number of physicochemical factors, such as light, CO_2 concentration, temperature, pH, salinity, and nutrients. For algal cultivation, it is important to know the optimal conditions and acceptable ranges, and how these factors affect algal growth.

11.4.1 Light

The effects of light are usually based on two parameters: *light irradiance* and light period. The response of the photosynthetic rate to light irradiance shows three major regions at different light intensities (Figure 11.3), namely:

- Light-limited region: at low light intensities, the rate of photosynthesis increases with increasing light irradiance.
- Light-saturated region: after the light irradiance reaches saturation, the rate of photosynthesis stops increasing and is maintained at a constant level with increasing light irradiance.
- Light-inhibited region: following a period of light saturation, the rate of photosynthesis may be reduced with increased light irradiance.

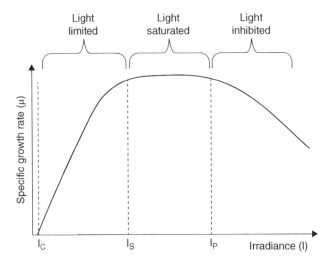

FIG. 11.3   Effect of light irradiance on algal growth. Source: Razeghifard, 2013. Reproduced with permission of John Wiley and Sons, Inc.

The ranges of each region vary with different algal species, but the most widely used light intensities range from 100–200 µE/(m^2·s). Besides light irradiance, the length of the light period can also affect photosynthetic performance (Janssen *et al.*, 2000). Light/dark periods of 24 hr/0 hr, 16 hr/8 hr, and 12 hr/12 hr have commonly been used (Zhao and Su, 2014); however, some algal species may not grow well under continuous light conditions.

11.4.2 CO_2 Concentration

Dissolved CO_2 can react with water to form bicarbonate (HCO_3^-) and hydrogen ions (H^+), thereby decreasing the pH of the medium. Low pH generally inhibits algal growth, especially the activity of carbonic anhydrase, which is the key enzyme in carboxysomes and pyrenoids. Optimal CO_2 concentrations for most algae are 0.038–10% (v/v). Some algae are sensitive to CO_2 concentrations that are higher than 1% (v/v). Conversely, there are a few algal species that can tolerate high CO_2 concentrations of up to 70–100% (v/v; Zhao and Su, 2014).

11.4.3 Temperature, pH, and Salinity

Temperature, pH, and salinity are crucial to maintaining optimal activity of enzymes in algal cells. Most currently used algal species grow well between 16 and 26 °C (Figure 11.4). Low temperatures (<16 °C) can decrease the solubility of CO_2 in the medium, which further slows down algal growth. Very high temperatures (>35 °C) can kill the algae. However, some algal species have shown improved tolerance to high temperatures (up to 40 °C) after an induced acclimation process (Zhao and Su, 2014). The common pH range for algal growth is about 6–9 (Figure 11.5), although some algal species can tolerate a pH below 4 (Zhao and Su, 2014). Optimal salinity levels vary dramatically for different algal species. Low salinity (<1 g/L) is normally required by freshwater algae, while salinities of 20–35 g/L could be optimal for marine algae. Some marine algae, such as *Dunaliella salina*, can tolerate 30–300 g/L salt concentrations (Preetha *et al.*, 2012).

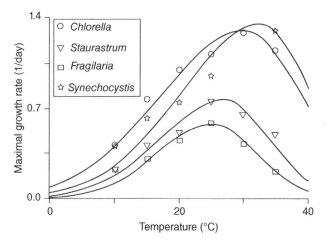

FIG. 11.4   Effect of temperature on algal growth. Source: Dauta, 1990. Reproduced with permission of Springer.

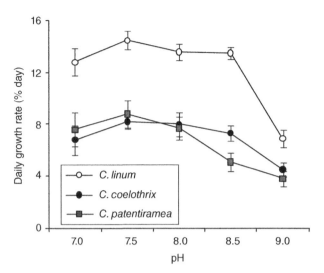

11.4.4 Nutrients

The major nutrients required for photoautotrophic growth are nitrogen (N) and phosphorus (P). Nitrogen uptake by algae is usually in the form of nitrate (NO_3^-) and ammonium (NH_4^+), while algal cells acquire phosphorus in the form of soluble phosphate (PO_4^{3-}). The mechanism of nitrogen assimilation in algal cells is attributed to the enzymatic reduction of NO_3^- to NH_4^+ by a series of enzymes. This is followed by NH_4^+ transport to the chloroplasts, where NH_4^+ is converted into amino acid glutamine by glutamine synthetase (Moberg *et al.*, 2012). Phosphorus is taken up during phosphorylation, where it is primarily converted into adenosine triphosphate (ATP) and adenosine diphosphate (ADP) for energy use (Cai *et al.*, 2013). Other micronutrients (e.g., Fe, K, Mg, Ca, and Na) are also important for the algal photosynthetic process (Cai *et al.*, 2013). Media that have been extensively used for algal growth are Walne's medium, Guillard's F/2 medium, Bold's Basal medium, and the BG11 medium (Table 11.3). The first two are prepared with sea water for marine algae, while the last two are prepared with fresh water for fresh-water algae.

11.5 Steps in Algal-Biodiesel Production

Algal-biodiesel production is a multiple-step process (Figure 11.6) that includes algal cultivation, algal biomass harvesting and drying, lipid extraction, and the conversion of lipids to biodiesel. Generally, algae are cultivated to obtain algal biomass concentrations of about 0.5 g/L (or higher), followed by the harvest of algal cells from a culture to obtain an algal slurry with a total solids (TS) content of about 15–25%. The algal slurry is dried to obtain an algal cake with TS content of about 85–95%, followed by lipid extraction and biodiesel production. In the following subsections, the details of each step are discussed.

Table 11.3 Compositions (mM) of media for algal growth[a]

Components	Marine algae		Fresh-water algae	
	Walne[b]	Guillard f/2[b]	Bold's Basal	BG11
Ca^{2+}	0	0	0.17	0.245
Co^{2+}	8.41×10^{-5}	4.20×10^{-5}	1.68×10^{-4}	1.72×10^{-4}
Cu^{2+}	8.01×10^{-5}	4.20×10^{-5}	6.29×10^{-4}	3.20×10^{-4}
Fe^{3+}	4.81×10^{-3}	1.17×10^{-2}	1.79×10^{-3}	N/A[c]
K^+	0	0	2.20	0.459
Mg^{2+}	0	0	0.304	0.304
Mn^{2+}	1.82×10^{-3}	9.10×10^{-4}	7.28×10^{-4}	9.15×10^{-3}
Na^+	1.29	0.937	3.37	0.180
NH_4^+	4.37×10^{-5}	0	0	N/A[c]
Zn^{2+}	1.54×10^{-4}	7.65×10^{-5}	3.07×10^{-3}	7.65×10^{-4}
BO_3^{3-}	0.543	0	1.85×10^{-2}	4.63×10^{-2}
Citrate	0	0	0	N/A[c]
Cl^-	1.85×10^{-2}	3.69×10^{-2}	0.769	0.508
CO_3^{2-}	0	0	0	0.189
EDTA	0.121	1.12×10^{-2}	1.71×10^{-2}	2.69×10^{-3}
$Mo_7O_{24}^{6-}$	7.28×10^{-6}	0	0	0
MoO_4^{2-}	0	1.40×10^{-5}	4.93×10^{-4}	1.61×10^{-3}
NO_3^-	1.18	0.882	2.94	17.6
PO_4^{3-}	0.112	3.17×10^{-2}	1.72	0.230
SO_4^{2-}	8.01×10^{-5}	1.19×10^{-4}	0.312	0.305
Vitamin B1	3.32×10^{-5}	3.32×10^{-4}	0	0
Vitamin B12	7.44×10^{-6}	3.72×10^{-7}	0	0
Vitamin H	8.19×10^{-7}	2.05×10^{-6}	0	0

[a] Adapted from medium recipes published in http://www.ccap.ac.uk/.

[b] Components from seawater not included.

[c] 6 mg/L of ammonium ferric citrate green ($C_6H_8O_7 \cdot xFe^{3+} \cdot yNH_3$) was added. The mole concentration of these three components could not be calculated based on the formula.

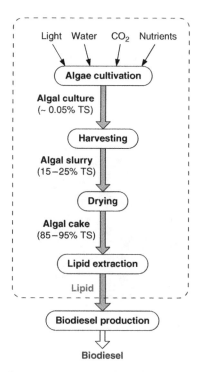

FIG. 11.6 Biodiesel production steps.

11.5.1 Algal Cultivation

Algal Cultivation Systems

There are two systems for algal cultivation: closed (e.g., photobioreactors, PBR) and open (e.g., lakes or ponds).

Closed systems for algal cultivation are carried out in PBRs, which are transparent and compact systems with full control of environmental conditions for enhanced photoautotrophic algae production. PBRs allow for single-species culture with minimal contamination, and can achieve very high algal cell density. Some of the commonly used PBR designs include tubular, flat-plate, and bag reactors (Figure 11.7). PBRs are fabricated from either glass or plastic materials. Most PBRs also include the introduction of mixed air and CO_2 to deliver inorganic carbon, and provide mixing to prevent self-shading. Tubular and flat-plate PBRs reduce the light path in the closed chamber, such that algae receive maximum light exposure. A bag bioreactor system consists of large plastic bags, which is a low-cost design. Internal or external artificial light sources can also be used to provide light energy, but this application greatly increases the cost.

Closed systems are attractive for cultivating algae for high-value nutraceutical and cosmeceutical products, as high cell productivities can be obtained with minimal risk of contamination. High biomass productivities of 0.5–3.0 g/(L·d) have been reported in pilot-scale systems. PBRs also have low evaporation loss, which reduces the volume of water necessary during cultivation. There are drawbacks to PBRs, however, such as overheating and biofouling caused by microbial growth around the wall (Shen *et al.*, 2009). In addition, the algal production cost in a closed system can be an order of magnitude higher than in an open pond system.

In **open systems**, *algal biomass* is grown in a shallow pond about 0.3 m (1 ft) deep, with minimal control of environmental conditions. To promote the optimal photosynthetic rate, CO_2 may be injected into the bottom of the pond. While natural ponds have been used for algal biomass production, the most commonly engineered high-throughput system is an open raceway pond (Figure 11.8).

Open raceway ponds are closed-loop channels constructed with poured concrete or by digging into the ground, and are typically lined with plastic. The raceway ponds are equipped with a paddlewheel to enhance mixing and flow control. Pilot and large-scale systems have algal biomass productivities of 0.05–0.50 g/(L·d). One major merit of an open raceway pond, as compared to closed systems, is the

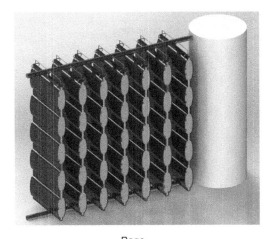

Tubular   Bags   Flat-plates

FIG. 11.7   Common photobioreactor designs.

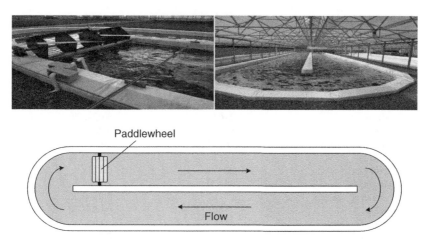

FIG. 11.8   Open raceway ponds. Photos courtesy of Yebo Li, Ohio State University.

Table 11.4 Comparison between open raceway ponds and photobioreactors

Parameters	Open raceway pond	Photobioreactor
Capital costs	Low	High
Operating costs	Low	High
Biomass productivity	Low	High
Carbon dioxide utilization	Low	High
Light source	Natural	Natural/artificial
Large-scale cultivation	Yes	No
Contamination risk	High	Low
Process control	Difficult	Possible
Species control	Not possible	Possible
Mixing	Not uniform	Uniform and complete mix
Footprint	Very large	Very small
Area/volume ratio	Low (5–10, 1/m)	High (20–200, 1/m)
Water losses	Very high	Low
Efficient light utilization	Low	High
Productivity	Low	High (3–5 times)
Biomass concentration	Low	High (3–5 times)
Mass transfer	Low	High

Source: Data from Mata *et al.*, 2010.

lower capital and operating costs due to cheaper construction materials and fewer equipment needs. However, there are several disadvantages of open raceway ponds. These systems are exposed to the natural environment with fluctuations in light irradiance and temperature, which contributes to water loss through evaporation, photo-inhibition, and contamination with other microbes. The low surface area of these systems also limits the mass transfer of CO_2 into the pond and exposure of the algal culture to light. A comparison between open raceway ponds and PBRs is shown in Table 11.4.

Algae Selection

Algal strains are normally selected based on their lipid content and biomass productivity (Table 11.5). However, other characteristics should also be considered, such as nutrient

Table 11.5 Lipid content and productivity of different algae species

Algae species	Lipid content (% dry wt.)	Biomass productivity (g/(L · d))
Ankistrodesmus sp.	24–31	N/A
Botryococcus braunii	25–75	0.02
Chaetoceros sp.	15–40	0.04–0.07
Chlorella sp.	2–63	0.02–7.70
Chlorococcum sp.	~19	0.28
Crypthecodinium cohnii	20–51	10
Dunaliella sp.	6–71	0.09–0.34
Ellipsoidion sp.	27	0.17
Euglena gracilis	14–20	7.70
Haematococcus pluvialis	25	0.05–0.06
Isochrysis sp.	7–40	0.08–1.60
Monodus subterraneus	16	0.19
Monallanthus salina	20–22	0.08
Nannochloris sp.	20–56	0.17–0.51
Nannochloropsis sp.	12–53	0.17–1.43
Neochloris oleoabundans	29–65	N/A
Nitzschia sp.	16–47	N/A
Oocystis pusilla	11	N/A
Pavlova sp.	31–36	0.14–0.16
Phaeodactylum tricornutum	18–57	0.003–1.9
Porphyridium cruentum	9–61	0.36–1.50
Scenedesmus sp.	2–55	0.004–0.74
Spirulina sp.	4–17	0.06–4.3
Thalassiosira pseudonana	21	0.08
Tetraselmis sp.	9–23	0.12–0.32

N/A: not applicable.
Source: Data from Mata *et al.*, 2010.

requirements, resistance to environmental stresses, ease of biomass separation and processing, and the potential for other valuable chemicals.

Site Selection

Resource evaluation and site selection have to be conducted before the establishment of an algal cultivation system. Factors considered for resource evaluation and site selection include climate, water sources, land availability, nutrient supplies, and carbon sources (Zhao and Su, 2014).

Climate. Sunlight and temperature directly affect algal growth rate. For an open pond system, evaporation and precipitation significantly affect the amount of makeup water required, while precipitation and winds could introduce contaminants. For both open pond and closed systems, risks of damage caused by severe weather should also be considered.

Water sources. Adequate water at low cost is required for commercial-scale algal cultivation. Salinity and chemicals in the water should be analyzed to determine their utility for algal cultivation of specific species. For example, seawater and brackish water with high salinity could be used for cultivation of marine algae. Environmental and municipal regulations could make site selection more complicated and difficult.

Land availability. The topography, geology, and land ownership could constrain the installation of algal cultivation systems, especially open pond systems that require large areas. It is economically unfeasible for land with high value to be used for bioenergy production. Land such as national parks, cultural heritage, designated wilderness areas, and military bases that have sensitive environmental or political constraints should also be eliminated from consideration.

Nutrient supplies. Fertilizer-grade nutrients are used to grow algae for high-value products, such as protein supplements, nutraceuticals, and pigments. However, this practice is not feasible for the production of low-value products, like lipids for biodiesel. Therefore, low-cost sources of nutrients, such as wastewater and other waste streams, are commonly considered for commercial-scale algal cultivation. Moreover, the fact that algal cells use the nutrients in wastewaters presents an added benefit in terms of sustainability (Singh and Das, 2014).

CO_2 supplies. Carbon sources can account for up to 40% of energy inputs for algal cultivation. Low-cost CO_2 sources, such as flue gas from power plants, and waste CO_2 gas generated from ethanol fermentation plants could be used for phototrophic growth. Industrial wastewaters that contain organic carbon substrates could be used for heterotrophic algal growth. Thus, proximity to these low-cost carbon sources should be considered when selecting a site for algal cultivation (Singh and Das, 2014).

Example 11.2: Open pond systems are designed with an algal biomass production capacity of 100 kg/d. The areal productivity of biomass is 50 g/(m^2·d). Calculate the number of ponds required assuming that the allowable area for each pond is not to exceed 200 m^2. How much lipid can be produced daily (kg/d) if the biomass lipid content is 40%? Calculate the amount of phosphorus that needs to be added daily (kg/d) if the algal cell composition is $CO_{1.48}H_{1.83}N_{0.11}P_{0.01}$. (Algal biomass is based on dry wt.)

Solution:

Total area required = (100 kg/d)/(0.050 kg/(m^2·d)) = 2,000 m^2
Thus, number of ponds required = 2,000 m^2/200 m^2/pond = **10 ponds**
Lipid produced daily = 100 kg/d × 0.40 = **40 kg/d**
Molecular weight of algal biomass = 12 × 1 + 1.48 × 16 + 1.83 × 1 + 0.11 × 14 + 0.01
$$\times\, 31 = 39.36$$
Phosphorus content in algal biomass = (0.01 × 31)/39.36 * 100% = 0.788%
Therefore, the phosphorus demand for producing 100 kg algal biomass per day is:
100 kg/d × 0.788% = **0.788 kg/d**

Example 11.3: Assume that the algal biomass productivities of a photobioreactor and an open raceway pond are 2.0 g/(L·d) and 0.25 g/(L·d), respectively. Estimate the volumetric requirement (gallons) for the production of 2,000 kg/d of algae biomass for each type of reactor.

Solution:

Photobioreactor (PBR)

$$2,000\,\frac{kg/}{d} \times 1,000\,\frac{g/}{kg} \times \frac{1}{2.0}\,\frac{L \cdot d}{g} \times \frac{1}{3.79}\,\frac{gal/}{L} = \mathbf{263,852\,gal}$$

Open raceway pond

$$2,000\,\frac{kg/}{d} \times 1,000\,\frac{g/}{kg} \times \frac{1}{0.25}\,\frac{L \cdot d}{g} \times \frac{1}{3.79}\,\frac{gal/}{L} = \mathbf{2,110,818\,gal}$$

There is roughly an **eightfold increase** in the volumetric requirement for an open raceway pond compared to a PBR.

Modeling of Algal Growth

As mentioned previously, algal growth is affected by several factors, including light, CO_2, temperature, pH, salinity, and nutrient availability. Some factors, such as pH and salinity, can be easily controlled, while others may vary dramatically, depending on environmental conditions such as sunlight and outdoor air temperature. Furthermore, heterogeneous distribution of substrate and nutrients is commonly observed in large-scale systems. In fact, it is usually unfeasible or even impossible to reach and/or keep the optimal conditions for algal growth in large-scale applications. Therefore, modeling algal growth under different conditions is crucial for predicting the performance of an algal cultivation system in industrial applications.

The Monod model is one of the most commonly used kinetic models that describe the effects of limiting substrates or nutrients on the growth of microorganisms. As discussed in Chapter 13, the Monod model is represented by the following equation:

$$\mu = \frac{\mu_{max} \cdot S}{K_s + S} \tag{11.2}$$

where μ is the specific growth rate (1/d), μ_{max} is the maximum specific growth rate in substrate saturated conditions (1/d), K_s is the half-saturation constant (mg/L), and S is the dissolved concentration of the limiting substrates (such as CO_2) or nutrients (such as nitrogen and phosphorus; mg/L). For modeling of algal growth, S can also be irradiance, and K_s, the half-saturation constant for cell growth, is dependent on average irradiance (W/m^2). The limitation of the Monod model is that the uptake of nutrients, that will be stored and used for later growth, cannot be accounted for. In this case, the quota (Droop) model, which relates the algal growth rate to the intracellular concentration of the limiting nutrient, is more accurate than the Monod model (Flynn, 2003). The cell quota model is described by the following equation:

$$\mu = \mu_{max}' \left(1 - \frac{Q_{min}}{Q}\right) \tag{11.3}$$

where Q is the intracellular nutrient quota (e.g., the amount of nitrogen per unit carbon of biomass; g[N]/g[C]), Q_{min} is the minimal quota (there is no algal growth at this level; g[N]/g[C]), and μ_{max}' is the theoretical growth rate at infinite quota (1/d). One disadvantage of the quota model is that the intracellular nutrient concentration is very difficult to measure.

The effect of temperature can be expressed by assuming an exponential variation due to non-optimal temperature (James and Boriah, 2010):

$$g(T) = e^{-j\left(T - T_{opt}\right)^2} \tag{11.4}$$

where j is the empirical constant for non-optimal temperature ($1/K^2$), T is the temperature (K), and T_{opt} is the optimal temperature for autotrophic growth (K).

Either the Monod or the Droop model can be combined with Equation 11.4, resulting in Equations 11.5 and 11.6, which describe the effects of multiple factors such as CO_2 and nitrogen concentrations, light irradiance, and temperature on algal growth:

$$\mu = \mu_{max} \left(\frac{C}{K_C + C} \right) \left(\frac{N}{K_N + N} \right) \left(\frac{I}{K_I + I} \right) e^{-j\left(T - T_{opt} \right)^2} \tag{11.5}$$

or

$$\mu = \mu_{max}' \left(1 - \frac{Q_{min}}{Q} \right) \left(\frac{C}{K_C + C} \right) \left(\frac{I}{K_I + I} \right) e^{-j\left(T - T_{opt} \right)^2} \tag{11.6}$$

where C is the CO_2 concentration (mg/L), N is the nitrogen concentration (g/L), I is the irradiance (W/m^2), and K_C, K_N, and K_I are the respective half-saturation constants of CO_2, N, and I (mg/L, g/L, and W/m^2, respectively).

During algal cultivation, the conditions in PBR or open pond systems can be more complicated. One issue is that the irradiance has a spatial distribution due to attenuation with depth. Thus, the Lambert–Beer equation (Equation 11.7) is commonly used to describe the reduction of irradiance with depth in algal growth media:

$$I = I_{in} e^{-\alpha X Z} \tag{11.7}$$

where I_{in} is the irradiance at the surface of the growth media (W/m^2), α is the irradiance attenuation coefficient (L/(g·m)), X is the biomass concentration (g/L), Z is the culture depth (m), and I is the light irradiance at depth Z (W/m^2).

Unfortunately, temperature, CO_2, and nutrient concentration distribution profiles of cultivation systems are complicated to model, as their profiles are dependent on factors such as system external temperatures and heat transfer characteristics, media flow patterns, and CO_2 and nutrient supply and consumption rates. Moreover, these factors also have an interactive effect, further complicating the modeling of algal cultivation processes. A promising approach for addressing this issue is to integrate algal growth kinetics with computational fluid dynamics (CFD), which has been employed for solving physical models. A key difficulty with this approach is that integration of CFD with algal growth kinetics can be computationally intensive, due to the large number of kinetic and physical equations that must be formulated and solved. However, increased computational power could overcome this challenge, and studies examining large-scale algal modeling using CFD are being frequently published. More detailed discussion of algal growth models and use of CFD for biological modeling can be found elsewhere (Flynn, 2003; Wu, 2013).

11.5.2 Harvesting

Harvesting is the process of recovering algal cells from the culture medium. The initial total solids (TS) concentration in a typical open raceway pond system is about 0.5 g/L. This suspension must be concentrated to at least 200 g/L before lipid extraction or drying for storage. Algae harvesting is usually carried out in two steps: primary, which increases the biomass concentrations to 10–20 g/L; and secondary, which further increases biomass concentrations to 150–250 g/L. High costs are the main concern with algae harvesting, and can account for up to 20–30% of the total cost of algae biomass production (Gudin and Thepenier, 1986). Algal harvesting technologies include gravity-based sedimentation, filtration with microscreens, ultrasonic vibration, membrane separation (including micro- and ultrafiltration), centrifugation, flocculation, froth flotation, and autoflocculation. Some common harvesting technologies are discussed in what follows.

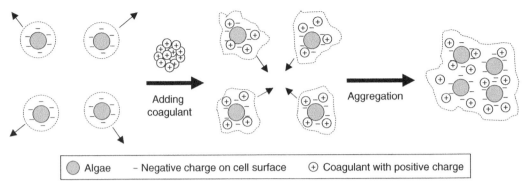

FIG. 11.9   Mechanism of flocculation induced by inorganic coagulants.

Flocculation

Flocculation refers to the aggregation of dispersed particles to form larger groups. Aggregation can be induced by increasing pH leading to autoflocculation, adding chemical coagulants or flocculants, or applying an electrical field, resulting in electrocoagulation or electrolytic flocculation. Flocculation is less energy intensive than mechanical separation. However, the sole use of flocculation is not sufficient to obtain biomass concentrations in the region of 150–250 g/L. Flocculation needs to be combined with other harvesting processes.

During prolonged cultivation under sunlight with limited CO_2 supplies, carbonate, including bicarbonate and/or carbonate, may precipitate with algal cells due to an increase in pH. This process is known as **autoflocculation**, which can be initiated by adding alkaline chemicals (pH >10.0) into the medium.

Flocculation of algal cells can be induced in **chemical coagulation** by adding either *inorganic coagulants* or *organic flocculants* (Figure 11.9). Inorganic flocculants include iron- and aluminum-based chemicals. However, inorganic coagulants require efficient separation and recycling, and are cost prohibitive. Organic flocculants include synthetic polymers, such as polyacrylate, polyacrylamide, polyvinyl alcohol, polystyrene sulfonate, and polyethylene amine, and natural polymers, such as chitosan and alginate. Organic flocculants, such as chitosan, can yield desirable end-product quality at lower dosages, and are less expensive than their inorganic counterparts.

Electrocoagulation and *electrolytic flocculation* are two types of **electrolytic processes** that have been used for algae harvesting. Electrocoagulation is similar to chemical coagulation, except that coagulants are generated from sacrificial electrodes by electrolytic oxidation (Figure 11.10a).

In an *electrolytic flocculation* process, sacrificial electrodes are not required due to the fact that algal cells have negatively charged surfaces. Algal cells migrate to the anode, lose their charge, and naturally form aggregates. The resulting aggregates then rise to the surface, aided by oxygen bubbles produced at the anode, where they can be easily harvested by skimming, as shown in Figure 11.10b. Compared to electrocoagulation, the main advantage of *electrolytic flocculation* is the low risk of contamination from the coagulants. Electrolytic flocculation can recover over 90% of algae with energy consumptions of about 0.3 kWh/m^3 (Poelman *et al.*, 1997).

(a)   (b)

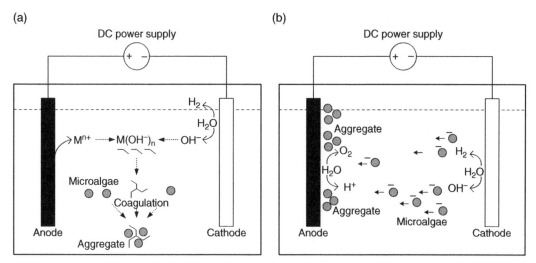

FIG. 11.10   Mechanism of (a) electrocoagulation and (b) electrolytic flocculation.

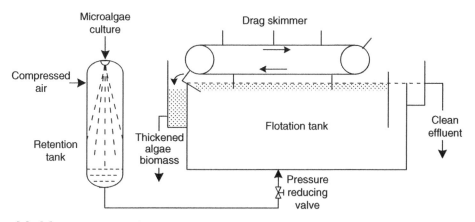

FIG. 11.11   Dissolved air flotation system.

When aluminum (Al) is used as the anode, the following oxidation and reduction reactions occur under DC current:

Oxidation at anode: $Al\,(solid) \rightarrow Al^{3+}\,(aqueous) + 3e^-$

Reduction at cathode: $2H_2O + 2e^- \rightarrow H_2\,(gas) + 2OH^-\,(aqueous)$

Overall reaction: $2Al\,(solid) + 6H_2O \rightarrow 2Al(OH)_3\,(solid) + 3H_2\,(gas)$

The $Al(OH)_3$ serves as a coagulant for the formation of algae aggregates.

Dissolved Air Flotation (DAF)

In the *dissolved air flotation* (DAF) process, algae cultures are mixed with compressed air and sent to a larger flotation tank, resulting in newly formed air bubbles that attach to algal cells and rise to the surface (Figure 11.11). Air bubbles attached on the algal cells result in lowering the overall density of

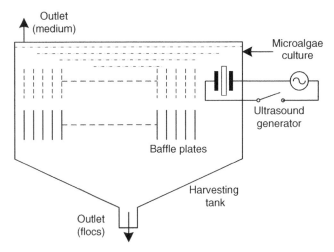

FIG. 11.12   Ultrasound harvesting system.

the algal cells, which start to float on the surface. Algal cells close to the surface are then harvested by a drag skimmer. The generation of air bubbles can be achieved in three ways: saturation of the growth medium at atmospheric pressure along with flotation under vacuum; saturation under a static head with upward flow that results in bubble formation; and saturation at pressures higher than atmospheric pressure, followed by flotation under atmospheric conditions.

The efficiency of separation is usually improved by using flocculants prior to flotation. The typical energy consumption for the DAF process is reported to be 84 kWh/million gallons per day (MGD; Schofield, 2001). The primary problem associated with DAF is formation of oversized bubbles, which can break up the algal flocs and lead to poor separation efficiency.

> A DAF system with a capacity of 5 MGD had been expected to produce an algal biomass slurry with 8% total solids at a cost equivalent to flocculation sedimentation (Benemann and Oswald, 1996). However, this technology has not been demonstrated at a large scale.

Ultrasonic Vibration

The *ultrasound harvesting system* consists of an ultrasonic generator that propagates ultrasonic waves horizontally through the culture medium, causing the flocculation of algal cells (Figure 11.12). Baffle plates are installed below the level of the ultrasound pathway to facilitate vertical settling of aggregates. The apparatus operates in an intermittent mode: when it is on, algal cells flocculate, and when it is turned off, aggregates precipitate through the baffle plates. The algal flocs are removed via an outlet at the bottom of the tank, while the culture medium is removed from an outlet at the top.

Ultrasound harvesting of algae has been applied to the harvesting of *Monodus subterraneus* cultures with 92% efficiency (Bosma *et al.*, 2003). This technology can concentrate an initial algal biomass concentration of about 0.02–0.5 g/L almost twentyfold. However, low harvest rates limit the application of the current technology at large scales (Bosma *et al.*, 2003). High power consumption and

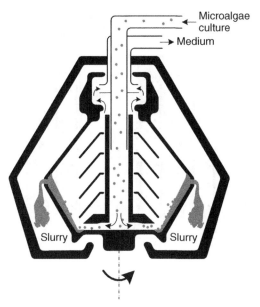

Microalgae
culture
Medium

Slurry   Slurry

FIG. 11.13   Disc-stack centrifuge.

the rate of gravity sedimentation after acoustic agglomeration are also critical issues for scaling up for industrial applications.

Centrifugation

A *centrifuge* includes a rotor that rotates at high speeds around a fixed axis. When the algal suspension is centrifuged in the rotating motor, cells move outward because their density is greater than the medium. Figure 11.13 shows a schematic of a commonly used disc-stack centrifuge.

Filtration

Different filtration techniques, which have been used for algal harvesting with varying degrees of success, include chamber filter pressing, vacuum belt filtration, rotary vacuum filtration, microfiltration, and ultrafiltration.

The **chamber filter press** consists of a series of chambers and is generally operated in a batch mode, because the filter cake needs to be discharged before the next batch (Figure 11.14). Chamber filters with 10–15 mm chamber gaps can be operated at pressures up to 16×10^5 Pa with a filtration cycle of 90–120 min. The diaphragm press, an improved model of the chamber press, utilizes a pumping pressure of 4×10^5 to 6×10^5 Pa to blow off the cake formed in the chamber gaps, thereby reducing the cycle time to 60 min. This method is fairly widely applicable, even to smaller algal species such as *Scenedesmus* (100% recovered; Mohn, 1980) and *Coelastrum proboscideum*, which was concentrated 245 times to a slurry with 27% TS content.

In a **vacuum belt filter**, the filter cloth is moved horizontally while a vacuum tray is fixed underneath (Figure 11.15). The medium is drawn through the filter cloth to the vacuum tray and stored in a reservoir, while the algal slurry is left on the filter cloth and discharged at the end of the horizontal belt.

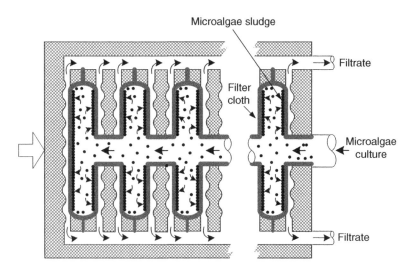

FIG. 11.14   Chamber filter press.

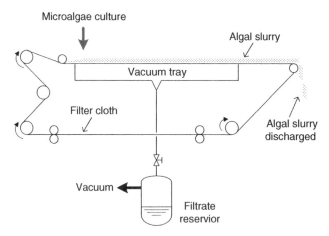

FIG. 11.15   Vacuum belt filter.

Belt filters rated for a minimum of 0.6 metric ton/hr on a dry solids basis are expected to support the harvest of a 160-acre pond with 20 g/(m^2 · d) productivity (Putt, 2007). A key to the successful operation of a belt press is that the feed must be flocculated to avoid blinding of the filter belt and facilitate gravity drainage when the slurry is initially fed to the belt. Other designs with reciprocating vacuum trays have a small capacity (40 m^2 filtration area) and are expensive, but can be operated with a wide range of growth media.

Example 11.4: The harvest efficiency of a commercial centrifuge is 95% at 13,000 × g (earth's gravitational acceleration), 60% at 6,000 × g, and 40% at 1,300 × g. If the maximum inlet flow rate and biomass concentration of algae media are 300 m³/hr and 200 mg/L, respectively, what amount of biomass (kg) can be harvested each day (8 h) at each setting? Assume that the centrifuge energy requirement is 3,000 kWh per metric ton of harvested algae biomass. At the current price ($0.10/kWh), what is the electricity cost per kg of algae biomass ($/kg)?

Solution:

High setting (13,000 × g)

$$300\frac{m^3}{hr} \times \frac{1000\,L}{m^3} \times 200\frac{mg}{L} \times \frac{1kg}{10^6\,mg} \times 8\frac{hr}{d} \times 0.95 = \textbf{456 kg/day}$$

Medium setting (6,000× g)

$$300\frac{m^3}{hr} \times \frac{1000\,L}{m^3} \times 200\frac{mg}{L} \times \frac{kg}{10^6\,mg} \times 8\frac{hr}{day} \times 0.60 = \textbf{288 kg/day}$$

Low setting (1,300 × g)

$$300\frac{m^3}{hr} \times \frac{1000\,L}{m^3} \times 200\frac{mg}{L} \times \frac{kg}{10^6\,mg} \times 8\frac{hr}{day} \times 0.40 = \textbf{192 kg/g}$$

Cost of harvesting

$$3,000\frac{kWh}{metric\,ton} \times \frac{metric\,ton}{1,000\,kg} \times \frac{\$0.10}{kWh} = \textbf{\$0.3/kg}$$

The **rotary vacuum filter** has a drum covered with filter cloth, such as canvas, nylon, dacron, metal, or glass fiber, and rotates at a speed of 0.1–2 rpm while partly submerged in the agitated algal culture (Figure 11.16). The applied vacuum (33.3–66.6 kPa) at the center of the drum draws the growth medium into the drum, leaving the suspended algal cells on the filter medium as a cake (18% TS content). Suitable precoating of the filter medium with potato starch, diatomaceous earth, cellulose, or similar material has been shown to be more efficient in achieving algal cake with a TS content as high as 37% (Gudin and Thepenier, 1986). In fact, a rotary vacuum filter without any precoating was found to be ineffective for pond effluent and is not recommended due to its high energy requirement and filter medium clogging. For example, a rotary vacuum filter without any precoating concentrated *Coelastrum* to a slurry of 18% TS content, but a drastic drop of inlet flow rate (capacity), from 0.4 to 0.8 m/hr, was observed within 15 min.

In cross-flow **membrane filtration** techniques, the bulk flow runs parallel to the filtering membrane and perpendicular to the permeation flux. In spite of the advantages of a faster filtration rate and complete cell removal, cross-flow filtration is capital intensive, and is not suitable for large-scale applications. Membrane fouling is also a major challenge for large-scale applications.

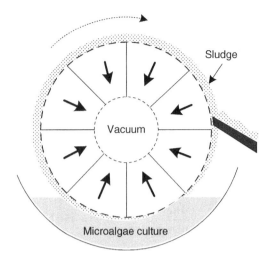

FIG. 11.16   Rotary vacuum filter.

In summary, flocculation, flotation, and ultrasound technologies can only achieve a low TS, although their recovery efficiencies can be up to 96%, and are suitable for primary harvesting. Centrifugation and filtration technologies, which can achieve a high TS, are suitable for secondary harvesting, but have much higher energy consumption.

11.5.3 Drying

The harvested algal biomass may need to be dried to a moisture content of 12–15% (w/w) before oil extraction. *Drying* is also required for storage of algae biomass, which will be essential for continuous operation of an industrial-scale algae biorefinery. The drying process can contribute to as much as 70–80% of the total production cost.

Solar drying is the most inexpensive way of drying algae from about 20% TS content to as high as 90%. Nonetheless, the performance is highly unpredictable due to regional and seasonal variations in climate, loss of product quality, and unpleasant odors. A solar dryer, consisting of a wooden chamber with the inside surface painted black and the top covered with a 2 mm glass plate, could produce an air temperature of 60–65 °C in its interior, which can dry *Spirulina* biomass in 5–6 hr to 92–96% TS content (Shelef *et al.*, 1984).

In a drum drier, the algal slurry is applied to the surface of rotating heated drums as a thin layer and dried quickly. The dry algal biomass is then removed from the surface by a knife (Figure 11.17). During the spray drying process, the slurry is sprayed into a hot air stream, where it is dried into powder (Figure 11.18). Powdered biomass is collected at the bottom. This process needs continuous spraying of slurry at high pressures, which results in high energy consumption. Therefore, spray drying may be a choice for high-value products (>$1,000/metric ton), but is not economically applicable for bioenergy production.

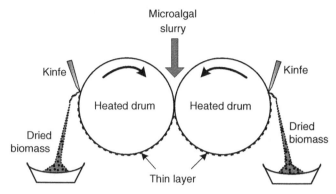

FIG. 11.17   Double drum drier.

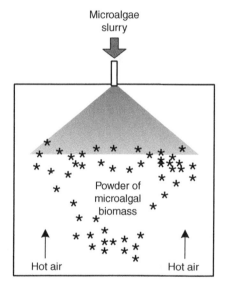

FIG. 11.18   Spray drying.

Example 11.5: About 1,000 kg of algae slurry (with 6% TS) needs to be dried to a final TS content of 85%.
1. Calculate how much water needs to be removed, and the energy required to remove this amount of water.
2. Assuming that the algae slurry is spread to a height of 0.5 cm on a black surface that can receive 480 cal/(cm² · d) radiation, and algal slurry density, ρ, is 1,100 kg/m³, calculate how many days are required to dry the algae slurry to a TS content of 85%.

Solution:

1. Initial amount of algal slurry, m = 1,000 kg
   Initial moisture content of algal slurry, X_i = (100% − % initial TS) = 94%
   Final moisture content after drying, X_f = (100% − % final TS) = 15%

Amount of water that needs to be removed during drying:

$$m_w = m\left[\frac{X_i - X_f(1 - X_i)}{(1 - X_f)}\right]$$

$$= 1000\left[0.94 - 0.15\frac{(1 - 0.94)}{(1 - 0.15)}\right] = \mathbf{929\,kg}$$

Amount of energy required to remove water during drying:

Assume the latent heat of vaporization of water, λ, is 2,257 kJ/kg [from steam table]

$$E = m_w \cdot \lambda = 929\,(kg)\,2,257\left(\frac{kJ}{kg}\right) = \mathbf{2,096,753\,kJ}$$

2. Rate of solar radiation on the drying surface, S = 480 cal/(cm$^2 \cdot$ d) or 20,097 kJ/(m$^2 \cdot$ d)

Algal slurry spreading height, h = 0.5 cm

Algal slurry density, ρ = 1,100 kg/m^3

Area required to spread algae slurry = $A = \dfrac{m}{\rho \cdot h} = \dfrac{1000\,kg}{1100\,\dfrac{kg}{m^3} \times 0.005\,m} = \mathbf{182\,m^2}$

Drying time: $t = \dfrac{E}{SA} = \dfrac{2,096,753}{20,097 \times 182} = \mathbf{0.57\,d}$

11.5.4 Lipid Extraction

There are two types of lipids in algal biomass: neutral and polar. Neutral lipids are mainly used for energy storage, while polar lipids are the main component of bilayer cell membranes. Among neutral lipids, acylglycerols are more desirable than polar lipids for biofuel production due to their high trans-esterification efficacies and the higher oxidation stabilities of the resulting methyl esters. However, some types of neutral lipids, that do not contain fatty acids, such as hydrocarbons (HC), sterols (ST), ketones (K), and pigments (carotenes and chlorophylls), are less desirable. Although these lipid fractions are soluble in organic solvents (hence fitting the definition of lipids), they cannot be converted to biodiesel (Halim *et al.*, 2012). Currently there are no *algal lipid extraction* processes that are established at a commercial scale. However, various laboratory-scale algal lipid extraction technologies have been developed.

Organic Solvent Extraction

The basic mechanism of *organic solvent extraction* (Figure 11.19) includes five key steps (Halim *et al.*, 2012):

1. Penetration of the organic solvent through the cell membrane into the cytoplasm.
2. Interaction of the organic solvent with neutral lipids.
3. Formation of an organic solvent–lipids complex.
4. Diffusion of the organic solvent–lipids complex across the cell membrane.
5. Diffusion of the organic solvent–lipids complex across the static film of organic solvent surrounding the cell.

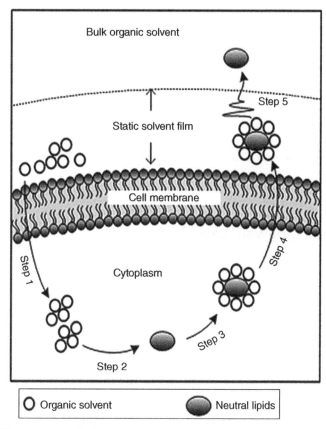

FIG. 11.19   Mechanism of algal lipid extraction with organic solvents. Source: Halim *et al.*, 2012. Reproduced with permission of Elsevier.

Non-polar organic solvents, such as chloroform and hexane, are frequently used to extract neutral lipids from algal cells. However, some neutral lipids coexist with polar lipids in a complex that is linked to membrane proteins via hydrogen bonds. Therefore, polar organic solvents, typically alcohols (such as methanol, ethanol, and isopropanol), are mixed with non-polar organic solvents to facilitate the extraction of membrane-associated neutral lipids.

The most typical organic solvent mixture for lipid extraction is chloroform/methanol (1/2, v/v), which has two main advantages: neutral lipids can be completely extracted, and complete drying of algal biomass is not required. However, the high toxicity of chloroform limits its usage in large quantities. Hexane/isopropanol (3/2, v/v), which has a lower toxicity, can be an alternative to chloroform/methanol. In addition, microwaves and ultrasound have been utilized to quickly disrupt the cellular structure, enhancing lipid extraction by organic solvents. Lipid extraction is usually conducted in batch mode, during which the extraction rate decreases due to lipid mass transfer equilibrium. At laboratory scale, lipids can be extracted with a Soxhlet apparatus.

Supercritical Fluid Extraction

Generally, a substance exists in one of three phases, gas, liquid, or solid, depending on temperature and pressure. However, phase boundaries may disappear when the temperature and pressure reach specific levels (critical values). A *supercritical fluid* is defined as any substance at temperature and pressure above its critical values. Supercritical fluids have three advantages in algal lipid extraction: solvent effectiveness can be varied by adjusting temperature and pressure; it can quickly penetrate cellular structures, which results in a higher extraction yield and shorter extraction time; and extracted lipids are solvent free.

CO_2 is used as the solvent in most *supercritical fluid extractions*, due to its moderate critical pressure (73.8 atm), low critical temperature (31.1 °C), low toxicity, low flammability, and inactivity. Scrubbed flue gas of power stations is an inexpensive source of CO_2.

Cyanotech, Kona, Hawaii: Case Study

Cyanotech Corporation has been in operation since 1984. Currently, it operates a 90-acre algae production facility on the Kona Coast of the Big Island of Hawaii (Figure 11.20). The constant warm temperature permits year-round production of *Spirulina platensis* and *Haematococcus pluvialis*. *Spirulina* is sold as a whole algae nutraceutical supplement in either powder or tablet form under the brand name Hawaiian Spirulina Pacifica™. *Haematococcus* is produced as a source of natural astaxanthin and is sent offsite

FIG. 11.20   Algae production facilities on the Kona Coast of the Big Island, Hawaii, USA.

for supercritical CO_2 extraction, yielding an astaxanthin-rich oleoresin. The oleoresin is further blended and used for the production of softgel capsules, also sold as a nutraceutical supplement, and marketed under the brand name BioAstin™ Hawaiian Natural Astaxanthin. Cyanotech maintains 68 large open culture ponds, which on average are 600 ft long and 50 ft wide, and hold 150,000 gallons of culture medium; 40 of the large culture ponds are used for *Spirulina* production and 28 are used for *Haematococcus* production. *Haematococcus* production is also supported by a series of closed culture systems ranging in size from 20 L, up to 30,000 L.

References

Baba, M., and Shiraiwa, Y. (2013). Biosynthesis of lipids and hydrocarbons in algae. In: Z. Dubinsky (ed.), *Photosynthesis*. Rijeka: InTech, Chapter 13.

Babu, M.G., and Subramanian, K.A. (2013). *Alternative Transportation Fuels: Utilisation in Combustion Engines*. Boca Raton, FL: CRC Press.

Benemann, J.R., and Oswald, W.J. (1996). *Systems and Economic Analysis of Microalgae Ponds for Conversion of CO_2 to Biomass*. Final report. Berkeley, CA: Department of Civil Engineering, University of California, Berkeley.

Bosma, R., van Spronsen, W.A., Tramper, J., and Wijffels, R.H. (2003). Ultrasound, a new separation technique to harvest microalgae. *Journal of Applied Phycology* 15: 143–53.

Cai, T., Park, S.Y., and Li, Y. (2013). Nutrient recovery from wastewater streams by microalgae: Status and prospects. *Renewable and Sustainable Energy Reviews* 19: 360–69.

Dauta, A., Devaux, J., Piquemal, F.O., and Boumnich, L. (1990). Growth rate of four freshwater algae in relation to light and temperature. *Hydrobiologia* 207: 221–6.

de Paula Silva, P.H., Paul, N.A., de Nys, R., and Mata, L. (2013). Enhanced production of green tide algal biomass through additional carbon supply. *PloS One* 8: e81164.

Enquist-Newman, M., Faust, A.M., Bravo, D.D., *et al.* (2014). Efficient ethanol production from brown macroalgae sugars by a synthetic yeast platform. *Nature* 505: 239–43.

Flynn, K.J. (2003). Modelling multi-nutrient interactions in phytoplankton: Balancing simplicity and realism. *Progress in Oceanography* 56: 249–79.

Gosch, B.J., Magnusson, M., Paul, N.A., and Nys, R. (2012). Total lipid and fatty acid composition of seaweeds for the selection of species for oil-based biofuel and bioproducts. *GCB Bioenergy* 4: 919–30.

Gudin, C., and Thepenier, C. (1986). Bioconversion of solar energy into organic chemicals by microalgae. *Advances in Biotechnological Processes* 6: 73–110.

Halim, R., Danquah, M.K., and Webley, P.A. (2012). Extraction of oil from microalgae for biodiesel production: A review. *Biotechnology Advances* 30: 709–32.

Hu, Q., Sommerfeld, M., Jarvis, E., *et al.* (2008). Microalgal triacylglycerols as feedstocks for biofuel production: Perspectives and advances. *The Plant Journal* 54: 621–39.

James, S.C., and Boriah, V. (2010). Modeling algae growth in an open-channel raceway. *Journal of Computational Biology* 17: 895–906.

Janssen, M., de Winter, M., Tramper, J., Mur, L.R., Snel, J., and Wijffels, R.H. (2000). Efficiency of light utilization of *Chlamydomonas reinhardtii* under medium-duration light/dark cycles. *Journal of Biotechnology* 78: 123–37.

Lee, R.E. (2008). *Phycology*. Cambridge: Cambridge University Press.

Mata, T.M., Martins, A.A., and Caetano, N. (2010). Microalgae for biodiesel production and other applications: A review. *Renewable and Sustainable Energy Reviews* 14: 217–32.

Moberg, A.K., Ellem, G.K., Jameson, G.J., and Herbertson, J.G. (2012). Simulated cell trajectories in a stratified gas-liquid flow tubular photobioreactor. *Journal of Applied Phycology* 24: 357–63.

Mohn, F.H. (1980). Experiences and strategies in the recovery of biomass from mass cultures of microalgae. *Algae Biomass: Production and Use* 1980: 548–71.

Pienkos, P.T., and Darzins, A. (2009). The promise and challenges of microalgal-derived biofuels. *Biofuels, Bioproducts and Biorefining* 3: 431–40.

Pilon, L., and Berberoglu, H. (2014). Photobiological hydrogen production. In: S.A. Sherif, D.Y. Goswami, E.K. Stefanakos, and A. Steinfeld (eds.), *Handbook of Hydrogen Energy*. Boca Raton, FL: CRC Press, p. 369.

Poelman, E., De Pauw, N., and Jeurissen, B. (1997). Potential of electrolytic flocculation for recovery of micro-algae. *Resources, Conservation and Recycling* 19: 1–10.

Preetha, K., John, L., Subin, C.S., and Vijayan, K.K. (2012). Phenotypic and genetic characterization of *Dunaliella* (*Chlorophyta*) from Indian salinas and their diversity. *Aquatic Biosystems* 8: 27.

Putt, R. (2007). *Algae as a Biodiesel Feedstock: A Feasibility Assessment*. Auburn, AL: Center for Microfibrous Materials Manufacturing, Department of Chemical Engineering, Auburn University. www.ascension-publishing.com/BIZ/nrelalgae.pdf, accessed April 2016.

Rajeshwari, K.R., and Rajashekhar, M. (2011). Biochemical composition of seven species of cyanobacteria isolated from different aquatic habitats of Western Ghats, Southern India. *Brazilian Archives of Biology and Technology* 54: 849–57.

Razeghifard, R. (2013). *Natural and Artificial Photosynthesis: Solar Power as an Energy Source*. Chichester: John Wiley & Sons Ltd.

Schirrmeister, B.E., Antonelli, A., and Bagheri, H.C. (2011). The origin of multicellularity in cyanobacteria. *BMC Evolutionary Biology* 11: 45.

Schofield, T. (2001). Dissolved air flotation in drinking water production. *Water Science and Technology* 43: 9–18.

Shelef, G., Sukenik, A., and Green, M. (1984). *Microalgae Harvesting and Processing: A Literature Review*. Haifa: Technion Research and Development Foundation.

Shen, Y., Yuan, W., Pei, Z.J., Wu, Q., and Mao, E. (2009). Microalgae mass production methods. *Transactions of the ASABE* 52: 1275–87.

Singh, M., and Das, K.C. (2014). Low cost nutrients for algae cultivation. In: R. Bajpai, A. Prokop, and M. Zappi (eds.), *Algal Biorefineries*. Amsterdam: Springer Netherlands, pp. 69–82.

Wu, B. (2013). Advances in the use of CFD to characterize, design and optimize bioenergy systems. *Computers and Electronics in Agriculture* 93: 195–208.

Zhao, B., and Su, Y. (2014). Process effect of microalgal-carbon dioxide fixation and biomass production: A review. *Renewable and Sustainable Energy Reviews* 31: 121–32.

Exercise Problems

11.1. What are the differences between microalgae and macroalgae? Differentiate between cyanobacteria and microalgae.

11.2. List the advantages and disadvantages of phototrophic, heterotrophic, and mixotrophic cultivation of microalgae.

11.3. Consider the features of two algal cultivation systems: photobioreactors and open ponds. Which one is more suitable for biodiesel production and why? Which is feasible for producing high-value product and why?

11.4. What type of technologies are preferable for primary harvesting, secondary harvesting, and drying, and why?

11.5. List the critical temperatures and pressure values of three different fluids, and discuss which is most suitable for use in supercritical fluid extraction.

11.6. Calculate the lipid productivity of *Nannochloropsis* sp. cultivated in a raceway with a specific growth rate of 0.12 1/d, biomass yield of 1.2 g/L, and lipid content of 45%. Compare the lipid productivity with *Spirulina* sp. with a specific growth rate of 0.25 1/d, biomass yield of 1.8 g/L, and lipid content of 15%.

11.7.   Assume that the biomass productivity of a photobioreactor and an open raceway pond is 2.0 g/(L · d) and 0.25 g/(L · d), respectively. Estimate the volumetric requirement (gallons) for the production of (a) 1 metric ton/day, (b) 10 metric tons/day, and (c) 100 metric tons/day of algae biomass in a raceway pond.

11.8.   If 100 lb of algal biomass is to be produced, how much nitrogen needs to be supplemented based on an algal cell composition of $CO_{1.48}H_{1.83}N_{0.11}P_{0.01}$?

11.9.   Calculate the amount of water that needs to be removed and the energy required to dry 100 tons of algae slurry from an initial TS content of 18% to a final TS content of 90%. Assuming that the algae slurry is spread to a height of 0.4 cm on a black surface that can receive 350 cal/(cm^2·d) of radiation, how long does it take to dry the algae slurry to a solid content of 90%?

11.10. Calculate the amount of water that needs to be removed during harvesting of 5,000 kg of algal medium from 2% TS content to 20% TS content.

Biological Conversion Technologies

Pretreatment of Lignocellulosic Feedstocks

Chang Geun Yoo and Xuejun Pan

What is included in this chapter?

This chapter covers various pretreatment methods for lignocellulosic feedstocks. Discussion of common pretreatment methods, the mechanism, representative conditions, effects on plant cell wall structure and composition, enzymatic digestibility of pretreated substrates, formation of fermentation inhibitors, and co-products potential are covered. Pertinent examples and calculations are also included.

12.1 Introduction

Lignocellulosic feedstocks are the most abundant renewable resource on Earth, with an availability of approximately 200 billion dry metric tons per year (Zhang *et al.*, 2007). Examples of lignocellulosic feedstocks include crop residues, such as corn stover, wheat and rice straws, and other crop stalks; forest residues, such as forest thinnings and sawmill residues; and energy crops, such as switchgrass, miscanthus, sweet sorghum, energy cane, Napier grass, and fast-growing trees (e.g., poplar, eucalyptus, and willow). Compared to starch-rich feedstocks, these bioresources are cost competitive and do not directly compete with food and feed. Lignocellulosic feedstocks have thus been considered the most promising and sustainable resources for bioenergy, chemicals, and other value-added products.

Chemically, lignocellulosic biomass is composed of three major components (cellulose, hemicellulose, and lignin) and minor substances (e.g., extractives and ash). These components coexist in the cell wall of plants. In general, lignocellulosic biomass contains approximately 35–45% cellulose, 25–35% hemicellulose, and 15–30% lignin, depending on plant species, age, and tissue type (Figure 12.1). *Cellulose* is a linear polymer of glucose linked together by β-1,4 glycosidic bonds;

Bioenergy: Principles and Applications, First Edition. Edited by Yebo Li and Samir Kumar Khanal.
© 2017 John Wiley & Sons, Inc. Published 2017 by John Wiley & Sons, Inc.
Companion website: www.wiley.com/go/Li/Bioenergy

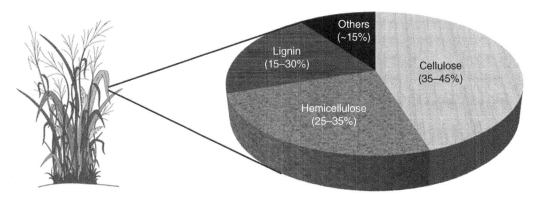

FIG. 12.1   Chemical composition of lignocellulosic biomass.

hemicellulose is a group of branched co-polymers of multiple sugars (e.g., xylose, mannose, arabinose, galactose, glucose, and uronic acids); and *lignin* is a three-dimensional aromatic polymer of phenylpropane units (coniferyl, sinapyl, and *p*-coumaryl alcohols), cross-linked primarily through ether and carbon–carbon bonds (Figure 12.2).

Current bioconversion of lignocellulosic feedstocks to liquid fuels and chemicals is primarily based on a carbohydrate (or sugar) platform, also known as a biochemical platform. Specifically, the sugar (primarily glucose) is first extracted from lignocellulosic biomass by either enzymatic or chemical (acidic) hydrolysis of cellulose, and then fermented into target products such as ethanol, butanol, propanediol, succinic acid, and lactic acid. Therefore, the hydrolysis of cellulose to glucose (saccharification) is considered to be the most critical and important step of the bioconversion process. However, because of the physical and chemical recalcitrance of the biomass to cellulose hydrolytic enzymes (cellulases), effective saccharification of lignocellulosic feedstock continues to be a technical challenge.

12.2 What Does Pretreatment Do?

In the cell wall, cellulose is surrounded by hemicellulose and lignin, so the cellulose is not readily accessible to enzymes or cellulases (Figure 12.2). In addition to acting as a physical barrier, lignin is a natural inhibitor of biodegradation of cell wall components, thus contributing to the self-defense system of plants against attacks by insects, bacteria, and fungi. Minor components of the cell wall, such as extractives, also retard the enzymatic hydrolysis of cellulose. The tough and rigid structure and poor permeability of plant cell walls form another barrier to cellulases. Also, the particle size (surface area) of the processed feedstock, as well as the cellulose's crystallinity and degree of polymerization, limit the accessibility of cellulose to enzymes. All of these barriers contribute to the recalcitrance of lignocellulosic feedstocks toward enzymatic hydrolysis. Before cellulases can effectively hydrolyze cellulose, the recalcitrance of the feedstock must be mitigated or reduced through one or more pretreatment methods. Thus, pretreatment is a critical step in biochemical platforms. In this chapter, representative pretreatment methods are discussed, as well as the mechanisms, conditions, performances, pros and cons, and challenges of these processes.

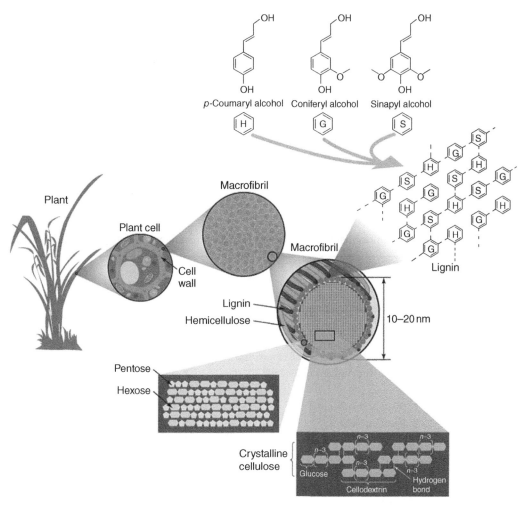

FIG. 12.2   Structural organization of the plant cell wall. Source: Rubin, 2008. Reproduced with permission of Nature Publishing Group.

The objective of *pretreatment* is to enhance the accessibility of cellulose to hydrolytic enzymes. Pretreatment methods are designed to accomplish one or more of the following: remove hemicellulose and/or lignin; modify the lignin structure to reduce negative impacts on enzymes, for example non-productive adsorption; break the matrix structure of the feedstock to reduce particle size, thereby increasing the surface area for the enzymes to access cellulose; destroy the crystalline structure of cellulose; or prehydrolyze cellulose to reduce the degree of cellulose polymerization. Cellulose becomes exposed and more accessible to cellulases after such pretreatment (Figure 12.3a), resulting in pretreated biomass that is more quickly hydrolyzed by cellulases than the unpretreated biomass (Figure 12.3b). Pretreatment can be accomplished by physical, chemical, hydrothermal, biological, and hybrid methods. Many pretreatment technologies have been developed during the last few decades. Typical pretreatment methods and their effects on biomass are summarized in Table 12.1.

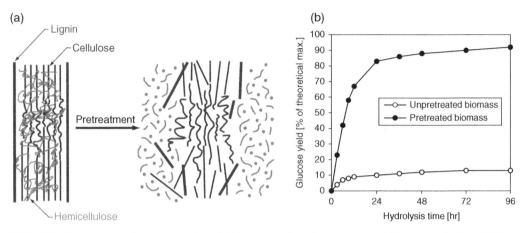

FIG. 12.3   Impact of pretreatment on (a) biomass cell wall structure and (b) enzymatic digestibility.

Table 12.1 Representative pretreatment methods and their effects on biomass

Pretreatment	Examples	Major effects
Physical	• Grinding with hammer mill or knife mill • Milling with ball mill or disk mill	• Size reduction • Surface area development
Thermochemical	• Dilute acid pretreatment • Steam explosion • Hot water pretreatment • AFEX (ammonia fiber expansion) • Organosolv pretreatment • Sulfite pretreatment (SPORL) • Soaking in aqueous ammonia (SAA)	• Hemicellulose removal • Lignin removal/relocation/modification • Size reduction • Surface area/pore size increase • Decrystallization of cellulose • Depolymerization of cellulose • Swelling of cell wall
Biological	• Microbial (fungal) pretreatment	• Delignification (lignin removal)
Other	• Ionic liquid pretreatment • Concentrated phosphoric acid pretreatment • Microwave-enhanced pretreatment • Ultrasound pretreatment	• Dissolution of cellulose • Decrystallization of cellulose • Depolymerization of cellulose • Pore size/surface area increase • Chemical reaction enhancement

Desirable outcomes of pretreatment include the production of readily digestible cellulose substrate, high overall sugar recovery, generation of high-quality lignin for high-value co-products, low capital and operating costs, low energy consumption, minimal formation of fermentation inhibitors, applicability to different types of feedstock, elimination of the need for extensive size reduction (an energy- and cost-intensive operation), and good scalability (Agbor *et al.*, 2011; Lynd *et al.*, 2002).

What are the qualities of the ideal pretreatment process?

- Generating a cellulose substrate with high enzymatic hydrolysis rate and yield, resulting in high yields of fermentable sugars and biofuels

- Minimal degradation of sugars or carbohydrates
- Minimal formation of inhibitory compounds that are toxic to enzymes and microbes in fermentation steps
- Low energy consumption for the process
- Mild process conditions to reduce the capital and operational costs
- Low-cost and recyclable pretreatment chemicals to reduce the operational cost
- Robust and capable of handling a variety of biomass feedstocks
- Co-product generation from lignin and hemicellulose
- Scalable for commercial production
- Minimal generation of toxic wastes and low environmental impact

12.3 Physical Pretreatment

Physical (or mechanical) pretreatment is used to disrupt cellulose crystallinity, reduce particle size, and increase the surface area of the feedstock. In general, physical pretreatment does not affect the chemical composition of the feedstock. No cell walls are degraded, nor do significant structural changes occur to the cellulose, hemicellulose, or lignin components. The improved enzymatic digestibility is primarily attributed to reduced crystallinity, which enhances enzyme accessibility to cellulose, and from the reduced particle size/increased surface area, which improves mass and heat transfer during cellulose hydrolysis. However, physical pretreatment is typically energy intensive and is thereby a cost-intensive operation. Initial particle size, moisture content, biomass species, and types of size-reduction equipment affect energy consumption for biomass grinding. For example, the energy consumption is 1–2 kWh/metric ton for reducing a whole plant to a roughly cut material size of 10–15 cm, and 20–30 kWh/metric ton for reducing precut biomass (1 cm) to fine particles (0.6 mm). Woody biomass usually consumes more energy for size reduction than herbaceous biomass. For example, energy consumptions for corn stover and switchgrass are 11.0 and 27.6 kWh/metric ton, respectively, while those for poplar and pine chips are 85.4 and 118.5 kWh/metric ton, respectively (Esteban and Carrasco, 2006; Mani *et al.*, 2004). Physical pretreatment alone is usually not enough to achieve satisfactory enzymatic digestibility, and it is often combined with thermochemical pretreatment methods. For most thermochemical pretreatment technologies, physical size reduction is essential.

Mechanical pretreatment methods are categorized by the degree of size reduction. Chipping, chopping, grinding, and milling are four common methods. In general, *chipping and chopping* result in a coarse size (>30–50 mm), while grinding and milling are able to achieve smaller particle sizes. Choppers, such as rotary shear shredders, knife cutters, and knife mills, are commonly used for coarse size reduction of baled biomass. A *hammer mill*, in which biomass is crushed by the rotating hammers in a closed chamber, can grind biomass to a wide range of particle sizes, and is normally used for producing fine particles. The particle size is controlled by the size of the screen installed in the mill. Ball

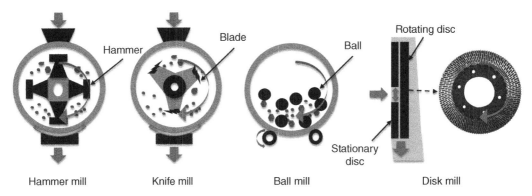

FIG. 12.4 Different types of mill for biomass size reduction.

mills are also widely used to produce fine particles, particularly for use at a laboratory scale. In a ball mill, biomass is crushed by metal or porcelain balls through cascading, shearing, and impacting forces. *Disk mills* have been used in the paper industry for pulping wood chips to generate fibers for paper-making, and for pulp refining to improve the fiber surface for better fiber binding and stronger paper. In a disk mill, wood chips or chopped biomass are crushed and sheared between two counter-rotating disks (plates). The particle (fiber) size is controlled by the clearance between the disks, and by the size and pattern of the bars on the disks. The structures of a hammer mill, knife mill, ball mill, and disk mill are illustrated in Figure 12.4.

Mechanisms of biomass size reduction

Biomass size reduction equipment employs one or more of the following four mechanisms: impact, attrition, shear, and compression. These methods are illustrated in Figure 12.5. *Impact* is the instantaneous collision of one object with another; *attrition* is the scrubbing of material between two hard surfaces; *shear* compresses and cuts the biomass between the edges of two hard surfaces; and *compression* is the force caused by pressing two hard surfaces toward each other.

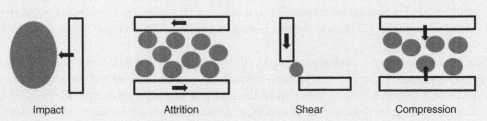

FIG. 12.5 Size reduction mechanisms. Source: Adapted from Womac, 2006.

12.4 Thermochemical Pretreatment

Thermochemical pretreatment refers to methods that use chemical reagent(s) at elevated temperatures to remove/reduce the recalcitrance of the biomass. Thermochemical pretreatments can cause removal of hemicellulose by dissolution or hydrolysis; removal of lignin (delignification) by breaking the structures (depolymerization and dissolution); decrystallization of cellulose; and/or prehydrolysis of cellulose (reduction of cellulose polymerization). However, undesirable side reactions may occur during thermochemical pretreatment. For example, degradation of sugars not only reduces the sugar recovery yield, but leads to the generation of fermentation inhibitors. Also, lignin condensation could retard the removal of lignin and may increase the non-productive adsorption of enzymes on lignin during subsequent enzymatic saccharification. Many chemicals, including acids, bases, and organic solvents, have been studied for the thermochemical pretreatment of biomass.

12.4.1 Acid Pretreatment

Acid pretreatment uses acid to catalyze the hydrolysis of carbohydrates, in particular hemicellulose, from the biomass. Examples of acids that are used include sulfuric acid, hydrochloric acid, phosphoric acid, and sulfur dioxide (SO_2). Among these, sulfuric acid is the most widely used for biomass pretreatment because of its low price and high efficiency. Hardwoods and grasses usually contain 3–5% acetyl groups, which are linked to hemicellulose. The acetyl groups are easily released at elevated temperature by hydrolysis and form acetic acid. The acetic acid can function as a reactive acid in many pretreatment methods, such as hot water, steaming, and steam explosion. Thus, these methods are also considered to be forms of acid pretreatment. A simplified flowchart of acid pretreatment methods is illustrated in Figure 12.6.

Acid pretreatment is typically conducted at elevated temperatures (80–220 °C), with the temperature dependent on the pretreatment method. During acid pretreatment, most hemicellulose is hydrolyzed and dissolved from the cell wall into a liquid hydrolysate comprising oligomeric or monomeric hemicellulosic sugars, which can be fermented to biofuels and biochemicals. These sugars can be further degraded during pretreatment, as shown in Figure 12.7. The removal of hemicellulose creates pores on the cell wall, exposing the cellulose and making it more accessible to enzymes, which

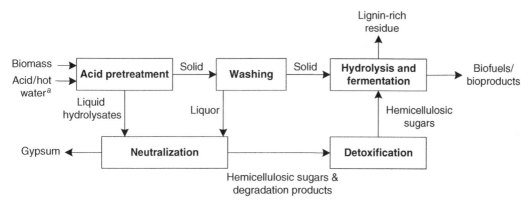

FIG. 12.6   Schematic diagram of acid pretreatment.
[a]Washing and neutralization steps are not required in the case of hot water pretreatment.

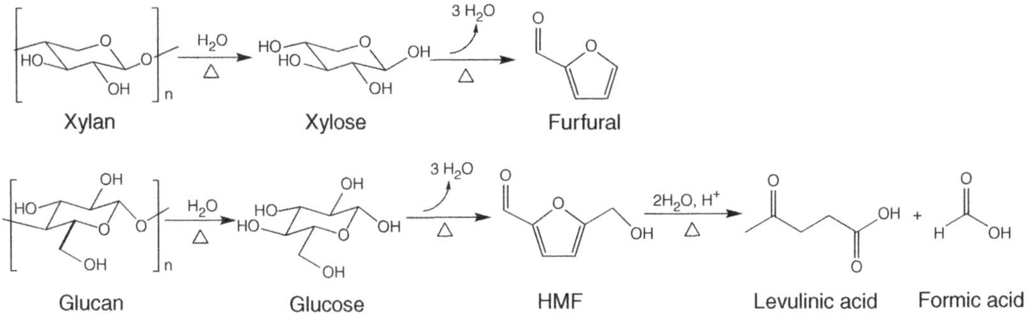

FIG. 12.7   Dehydration pathways of carbohydrates under acidic conditions.

is the primary mechanism by which acid pretreatment improves the enzymatic digestibility of lignocellulosic biomass. Cellulose is partially depolymerized and may even be hydrolyzed to glucose during acid pretreatment, particularly at severe conditions (high temperature and low pH). The depolymerization (prehydrolysis) of cellulose is actually favorable to enzymatic hydrolysis, since it creates new reducing and non-reducing ends for enzymes to attack. In general, there is no significant delignification during acid pretreatment. In fact, in most cases lignin is enriched in the substrate after hemicellulose is dissolved during the acid pretreatment. As a result, the primary components of acid-pretreated solid substrate are cellulose and lignin. The three most widely studied acid pretreatment methods (dilute acid, hot water, and steam explosion) are briefly discussed in what follows.

Dilute acid pretreatment is typically conducted with 0.5–2.0% (concentration) sulfuric acid at 140–180 °C for 10–40 mins. Depending on the types of biomass and pretreatment conditions, this pretreatment method can remove up to 90% of the hemicellulose. Most dissolved hemicellulose is in the form of oligosaccharides or monosaccharides, and part of these sugars is further degraded into furans. Only a small portion of cellulose (up to 10%) and lignin (up to 20%) is dissolved during the pretreatment. Dilute acid pretreatment is effective for grasses and hardwood, but less effective for softwood feedstocks. The major issues and limitations include overall low sugar yield due to sugar degradation during pretreatment; formation of fermentation inhibitors (Figure 12.7), such as furan derivatives (furfural and hydroxymethylfurfural) and organic acids (acetic acid, formic acid, and levulinic acid); acid-induced equipment corrosion; and high downstream costs for acid neutralization and handling of solid waste (e.g., gypsum).

Liquid hot water (LHW) pretreatment utilizes water to treat biomass at elevated temperatures. In this pretreatment, the hydrolysis of polysaccharides, primarily hemicellulose, is catalyzed by the acetic acid produced from the cleavage of O-acetyl groups from hemicellulose. For example, LHW pretreatment of corn stover at 190 °C for 15 min removes 40% of the xylan from the biomass. Since no additional acid is added, formation of fermentation inhibitors is limited, and therefore washing and neutralization may not be necessary prior to enzymatic hydrolysis and fermentation. It was reported that the entire pretreated mixture without washing, including both the liquid hydrolysate and pretreated solid substrate, could be enzymatically hydrolyzed with efficacies of 90% and 80% for cellulose and xylan, respectively (Mosier *et al.*, 2005). However, LHW pretreatment is not as effective as dilute acid pretreatment, particularly for woody biomass.

Steam explosion pretreatment exposes biomass in a reactor to saturated steam at high pressure (0.7–3.4 MPa) and temperature (160–240 °C) for several seconds to a few minutes. The reactor is suddenly depressurized, creating an explosive effect as the materials instantaneously expand. Chemically, the high-temperature steam releases acetic acid from hemicellulose, and the acid catalyzes the hydrolysis of hemicellulose and even cellulose. Acid- and heat-induced sugar degradation occurs as well, generating furans. There is no significant delignification during the steaming, but condensation, relocation, and redistribution of lignin occur. Addition of acid catalysts, such as H_2SO_4 and SO_2, in the steam explosion pretreatment can reduce the reaction time and temperature as well as enhance the removal of hemicellulose. Physically, the steamed biomass experiences intensive shearing, friction, and expansion during the "explosion," which substantially reduces the particle size. Typically, the appearance of steam-exploded biomass changes from a sawdust-like fibrous material to a fine mud with a dark brown color, although it varies with steaming conditions. The color change is caused by lignin condensation. The dissolution of hemicellulose and a substantial size reduction are the major effects of steam explosion to enhance enzymatic digestibility. Steam explosion is an effective pretreatment method for grasses and hardwood, but is less effective for softwood. However, it is an energy-intensive process, and scaleup, in particular in continuous steam explosion, is still challenging.

Degradation of carbohydrates and formation of fermentation inhibitors during acid pretreatment

An undesirable side effect of acid pretreatment is heat- and/or acid-induced sugar degradation. As shown in Figure 12.7, pentoses (5-carbon sugars, such as xylose and arabinose) are dehydrated into furfural, while hexoses (6-carbon sugars, such as mannose, galactose, and glucose) are dehydrated into 5-hydroxymethylfufural (HMF). HMF further decomposes to levulinic acid and formic acid. These sugar degradation products are toxic to fermentation organisms, and are therefore referred to as fermentation inhibitors. Although furfural, HMF, and levulinic acid can be useful platform chemicals for fuels, chemicals, and materials, if fermentation is desired they must be removed via a detoxification operation. Other fermentation inhibitors generated during acid pretreatment include acetic acid from acetyl groups of hemicellulose and phenolic compounds from lignin degradation.

12.4.2 Alkaline Pretreatment

Alkaline pretreatment uses bases, such as sodium hydroxide, ammonia or ammonium hydroxide, calcium hydroxide (lime), and potassium hydroxide, to pretreat lignocellulosic biomass. The bases are able to break ester, ether, and glycosidic bonds to dissolve hemicellulose and lignin (depending on conditions), increase the internal surface area by swelling the cell wall, decrease the degree of polymerization and crystallinity of cellulose, and remove acetyl groups and uronic acids. In general, bases are more effective at delignification than acids because bases, in particular strong bases, can disrupt lignin structures by breaking ether linkages between lignin units. Bases are, however, less effective than acids at removing hemicellulose under typical pretreatment conditions, as base-pretreated substrates retain more hemicellulose than acid-pretreated ones. Thus, if lignin is not extensively dissolved, most base-pretreated substrates usually have poorer enzymatic digestibility than those pretreated with acid.

In general, base pretreatments need moderate reaction temperatures, but have a longer reaction time compared to acid pretreatment. A major issue of base pretreatment is that the dissolved hemicellulose sugars are difficult to use, because either the sugars are decomposed or they are too difficult or expensive to separate and purify from the pretreatment liquor. Another problem of base pretreatment is the handling and treatment of the spent pretreatment liquor, particularly when a significant amount of lignin is dissolved. Flowcharts of different alkaline pretreatment processes are presented in Figure 12.8.

Sodium hydroxide (NaOH) is an effective alkaline pretreatment chemical, in particular for grasses and crop residues, but higher chemical loading and harsher conditions are required for woody biomass. NaOH pretreatment can be conducted at varied conditions, such as sodium hydroxide loading from 1–20% (on dry biomass) and a pretreatment temperature from room temperature to 160 °C. For example, the pretreatment of switchgrass with a low concentration of sodium hydroxide (2%) at 21 °C, 50 °C, and 121 °C results in 63%, 78%, and 86% delignification, respectively. The pretreated biomass has a 3.8-fold higher enzymatic digestibility than untreated material (74% glucan digestibility and 63% xylan digestibility; Xu *et al.*, 2010). The recovery of NaOH after pretreatment is energy and cost intensive, in particular at low concentration, although the technology is available and mature.

Ammonia has high selectivity toward lignin over carbohydrates, high volatility, and a swelling effect on cell walls, and is also able to reduce the crystallinity of cellulose. Moreover, residual ammonia in a pretreated biomass can be used as a nitrogen source for microbial growth in subsequent fermentation. In general, ammonia-based pretreatment is effective for grasses and agricultural residues, but not woody biomass. The recovery of ammonia is relatively easy because

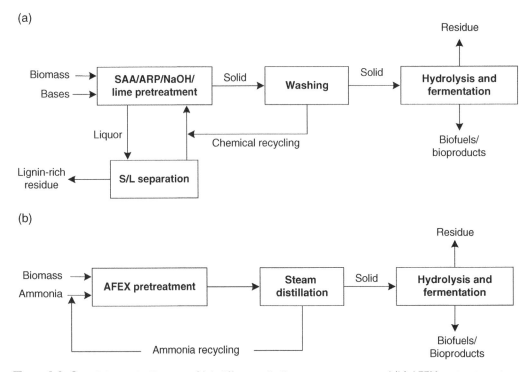

FIG. 12.8   Schematic diagram of (a) different alkaline pretreatments and (b) AFEX pretreatment.

of its high volatility, but high-yield recycling of ammonia is technically and economically challenging at a large scale.

Reaction conditions and outcomes vary for different ammonia-based pretreatments, such as soaking in *aqueous ammonia (SAA)*, *ammonia recycle percolation (ARP)*, *ammonia fiber expansion (AFEX)*, and *low-moisture anhydrous ammonia (LMAA)*. SAA is conducted with 10–15% aqueous ammonia in a batch reactor at moderate temperatures (25–80 °C) for anywhere from 6–24 hrs to 10–60 days. This method preserves most of the cellulose and hemicellulose in the solid substrate. In the ARP process, 5–15% aqueous ammonia percolates through the biomass using a flow-through reactor, at elevated temperatures (170–210 °C) for 10–90 mins. Under these reaction conditions, ARP results in effective delignification (70–90%) and significant removal of hemicellulose (~50%). LMAA pretreatment was developed to reduce energy input and ammonia consumption by using anhydrous ammonia gas. Gaseous ammonia allows rapid distribution of the chemical to the biomass with minimal water loading. AFEX pretreatment exposes biomass to ammonia at moderate temperatures (60–90 °C) with high pressure (above 3 MPa). The pressure is released after being held for a certain time at the target temperature. The quick pressure reduction opens up the biomass structure through expansion. Although there is no significant delignification during AFEX pretreatment, ammonia may modify the lignin structure and thereby reduce its impact on enzyme access to cellulose. AFEX is able to partially decrystalize cellulose as well, while limiting hemicellulose loss. LMAA and AFEX do not need a washing step, unlike the other ammonia pretreatments (SAA and ARP); therefore, the entire carbohydrate fraction can be recovered without loss using these processes.

Pretreatment with **aqueous calcium hydroxide (lime)** at low temperatures (25–55 °C) is another alkaline pretreatment method. Lime pretreatment usually has a long reaction time, but the cost is low because of inexpensive chemicals and mild pretreatment conditions. Increasing the reaction temperature (120 °C) can shorten the lime pretreatment time from weeks to 2–3 hr (Chang *et al.*, 1997). Supplying oxygen during lime pretreatment can enhance the pretreatment effect, but oxygen could also cause cellulose and hemicellulose loss because of the poor selectivity of oxygen. Lime is recyclable, although the removal of lime from the biomass slurry is challenging.

Example 12.1: Corn stover is harvested from Central Iowa for cellulosic ethanol production. The corn stover is composed of 40% cellulose, 30% hemicellulose, 25% lignin, and 5% ash. The company plans to process 100 metric tons of corn stover per day. Two different pretreatment methods (dilute acid and soaking in aqueous ammonia) are evaluated to determine the higher sugar conversion yields. The results of each pretreatment are as in Table 12.2.

Table 12.2 Pretreatment results

Pretreatment method	Delignification (%)	Hemicellulose loss (%)	Cellulose loss (%)	Enzymatic digestibility	
				Glucan (%)	Xylan (%)
Dilute acid	10	80	5	95	70
SAA	70	20	2	90	70

Assumptions:
- Xylan is the only component of hemicellulose.
- Hemicellulose and cellulose lost during pretreatment are not recovered.

- Ash is totally removed during pretreatment.
- The initial biomass is moisture free (completely dry).

1. At the end of each day, how much glucose and xylose are produced and how much residual solid remains after enzymatic hydrolysis for each pretreatment method? (Note: 1 kg cellulose or glucan produces 1.11 kg glucose after complete hydrolysis; and 1 kg xylan generates 1.14 kg xylose after complete hydrolysis.)
2. If the capacity of the pretreatment reactor is 15 metric tons per run, which pretreatment method is more productive in terms of total sugar yield? The reaction time of dilute acid and SAA pretreatments are 30 min and 6 hr, respectively.

Solution:

1. With dilute acid pretreatment
Since the pretreatment keeps 95% cellulose and 20% hemicellulose of the corn stover in the pretreated substrate, and 95% cellulose and 70% hemicellulose in the substrate are hydrolyzed to sugars by enzymes, the amount of sugars released is:

Glucose: 100 metric tons $\times$ (40% $\times$ 95% $\times$ 95%) $\times$ 1.11 = **40.1 metric tons**

Xylose: 100 metric tons $\times$ (30% $\times$ 20% $\times$ 70%) $\times$ 1.14 = **4.8 metric tons**

All ash and 10% lignin were removed during the pretreatment, and also 95% cellulose and 70% hemicellulose in the pretreated biomass are hydrolyzed; therefore, **the residual solid** is:

100 metric tons $\times$ {25% $\times$ (100% – 10%) + [40% $\times$ 95% $\times$ (100% – 95%)
    + 30% $\times$ 20% $\times$ (100% – 70%)]} = **26.2 metric tons**

With SAA pretreatment
By the samz. calculation methods, the amount of sugars released after SAA pretreatment and enzymatic hydrolysis is:

Glucose: 100 metric tons $\times$ (40% $\times$ 98% $\times$ 90%) $\times$ 1.11 = **39.2 metric tons**

Xylose: 100 metric tons $\times$ (30% $\times$ 80% $\times$ 70%) $\times$ 1.14 = **19.2 metric tons**

The residual solid is:

100 metric tons $\times$ {25% $\times$ (100% – 70%) + [40% $\times$ 98% $\times$ (100% – 90%)
    + 30% $\times$ 80% $\times$ (100% – 70%)]} = **18.6 metric tons**

2. To process 100 metric tons of corn stover, dilute acid pretreatment takes:

100 metric tons / 15 metric tons $\times$ 0.5 hr = 3.33 hr
Sugar productivity: $(40.1 + 4.8)/3.33$ = **13.5 metric tons/hr**

Since SAA pretreatment takes 6 hr per run, the time to process 100 metric tons of corn stover is:

100 metric tons / 15 metric tons × 6 hr = 40 hr

Sugar productivity: $(39.2 + 19.2)/40 = $ **1.5 metric tons/hr**

Based on the sugar production for each pretreatment, dilute acid pretreatment has a higher productivity than SAA pretreatment.

12.4.3 Organosolv Pretreatment

Organosolv pretreatment refers to the use of organic solvents with or without catalysts (bases, acids, and salts). The solvents used include alcohols (methanol, ethanol, ethylene glycol, glycerol, tetrahydrofurfuryl alcohol, and butanol); organic acids (formic and acetic); acetone or other ketones; phenol or cresols; and dimethyl sulfoxide. Among these, ethanol is considered to be the most promising for organosolv pretreatment because of its low cost and availability. More importantly, when the pretreatment is for cellulosic ethanol production, ethanol has a unique advantage over other solvents because there is no need to purchase ethanol from an external supplier. Moreover, as ethanol concentration after fermentation, and ethanol recycling after pretreatment, can share the same distillation facility, capital costs are greatly reduced (Pan, 2013).

Organosolv pretreatment is able to selectively extract and dissolve lignin from biomass by cleaving ether linkages between lignin units. The extracted lignin is called organosolv lignin. Organosolv lignin is considered to be the best lignin for co-product development because of its high purity, high reactivity toward modification, and low and uniform molecular weight. With an acid catalyst, almost all hemicelluloses are hydrolyzed into sugars, with some of the sugars being degraded to furfural, HMF, levulinic acid, and formic acid. Depending on the pretreatment conditions, cellulose is partially depolymerized (prehydrolyzed) during the organosolv pretreatment. With the removal of lignin and hemicellulose, and the prehydrolysis of cellulose, the organosolv substrate (cellulose fraction) is readily digestible, and can be easily hydrolyzed into glucose within 24 hr with low enzyme loading.

Organosolv pretreatment is conducted within a wide range of temperatures ($100–250\,°C$), depending on the solvent used. Organosolv ethanol pretreatment is usually operated at $160–180\,°C$. One of the issues for organosolv pretreatment is solvent recycling. Although the organic solvents are relatively easy to recover and reuse, the distillation operation is energy and cost intensive. Also, the solvents are more volatile and create a higher vapor pressure than water at the same temperature; therefore, a high-pressure reactor is required for organosolv pretreatment. Figure 12.9 illustrates the organosolv pretreatment process.

12.4.4 Sulfite-Based Pretreatment

Sulfite-based pretreatment refers to the method using aqueous sulfite or bisulfite solution. A sulfite pretreatment technology that has been extensively investigated is SPORL (sulfite pretreatment to overcome recalcitrance of lignocellulose). SPORL uses sulfite or bisulfite with a small amount of acid (to maintain pH at about 2.0) to pretreat the biomass. Representative conditions for SPORL pretreatment are 2–6% sulfuric acid, 2–10% sulfite, $150–180\,°C$, and 10–30 min, depending on the biomass species (Zhu *et al.*, 2009). All chemical loading rates are based on dry biomass.

During SPORL pretreatment, lignin reacts with sulfite to become sulfonated, which causes part of the lignin (20–40% of total lignin) to be dissolved as lignosulfonate. Lignosulfonate is a valuable lignin product, and has been widely used as a surfactant and adhesive. The lignin retained in the

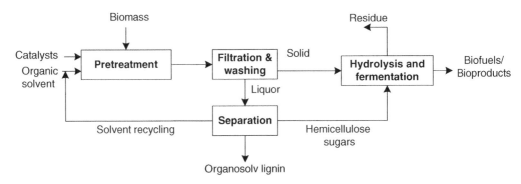

FIG. 12.9   Schematic diagram of organosolv pretreatment process.

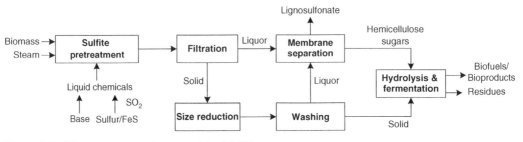

FIG. 12.10   Schematic diagram of the SPORL process.

solid substrate is partially sulfonated as well, which reduces the hydrophobicity of lignin. The reduction of hydrophobicity weakens the lignin's impact on cellulases by reducing non-productive hydrophobic adsorption. This is one of the reasons why SPORL-pretreated biomass has better enzymatic digestibility compared to many of the other pretreatment processes.

Most hemicellulose is dissolved by acidic hydrolysis during SPORL pretreatment, which is another reason for the excellent enzymatic digestibility of the SPORL substrate. For sufficient removal of hemicellulose, maintaining a moderately low pH (~2.0) is critical, which is why hemicellulose-rich feedstocks, such as hardwood and grasses, require more acid catalyst. The low pH of SPORL pretreatment partially depolymerizes cellulose, which improves the digestibility of the SPORL substrate as well.

In addition to better digestibility, SPORL pretreatment results in higher total sugar recovery. As a result of the buffer function of sulfite, the pH of the SPORL pretreatment liquor is higher than that of the corresponding dilute acid pretreatment at the same acid loading. The slightly higher pH prevents sugars from acid-induced degradation. Accordingly, SPORL pretreatment generates less fermentation inhibitors than the corresponding acid pretreatment. The SPORL process is shown in Figure 12.10.

Another feature of SPORL pretreatment is that wood chips (e.g., 50 mm × 30 mm × 5 mm) can be directly treated without size reduction, which can be conducted after the chemical sulfite treatment. This arrangement significantly reduces the energy consumed for size reduction. For instance, the energy consumption for a typical size reduction of wood chips to fiber was about 0.2–0.6 kWh/kg of wood, but it dropped to 0.05 kWh/kg of wood when the size was reduced with a disk mill after chemical treatment (Zhu *et al.*, 2009). SPORL is an effective pretreatment for high-lignin woody biomass, including softwood. Compared to acid and base, the recovery of sulfite is more complicated and expensive, in particular at low concentrations, although there is mature technology available for sulfite recovery in the paper industry.

12.4.5 The Combined Severity (CS) Factor

Thermochemical pretreatments involve many process parameters, such as chemical dosage (concentration), pretreatment temperature, reaction time, and pH of the pretreatment liquor. The values for these parameters reflect the pretreatment severity. However, it is difficult to compare the severity of different pretreatments under a variety of conditions. For this purpose, a single variable, known as *Combined Severity* (CS), which integrates reaction time, reaction temperature, and pH of the liquid, is used. A higher CS value corresponds to a more severe process:

$$CS = \log(R_0) - pH \tag{12.1}$$

$$R_0 = \int_0^t \left[\frac{(T_H - T_R)}{14.75} \right] dt \tag{12.2}$$

where t is the reaction time (min), T_H is the reaction temperature (°C), T_R is the reference temperature (100 °C), and pH refers to the pH of the pretreatment liquid.

> **Example 12.2:** Wood chips harvested from North Carolina are pretreated with two different dilute acid pretreatment methods. The first method is a two-stage dilute acid pretreatment. 1 metric ton of wood chips are pretreated in the first stage with 10 metric tons of 0.1 M H_2SO_4 at 150 °C for 20 mins, and then the wood chips are treated in the second stage with 5 metric tons of 0.1 M H_2SO_4 at 190 °C for 10 min. The second method is a single-stage acid pretreatment process using 10 metric tons of 0.2 M HCl at 170 °C for 30 min. If all other reaction conditions are the same, which pretreatment method is more severe? K_a (acid dissociation constant) of HSO_4^- is 1.2×10^{-2}.

Solution:

Two-stage dilute sulfuric acid pretreatment
Dissociation of sulfuric acid in water (concentration 0.1 M):

Step 1: $H_2SO_4 \rightarrow HSO_4^- + H^+$ Complete
Step 2: $HSO_4^- \rightarrow SO_4^{-2} + H^+$ Incomplete

Table 12.3 ICE (Initial, change, and equilibrium) table of Step 2

	HSO_4^-	SO_4^{-2}	H^+
Initial	0.1	0	0.1
Change	−x	x	x
Equilibrium	0.1−x	x	0.1+x

$$K_a = \left[SO_4^{-2} \right] \left[H^+ \right] / \left[HSO_4^- \right] = x(0.1 + x)/(0.1 - x) = 0.012$$

x = 0.01 M

$$[H^+] = 0.1 + 0.01 = 0.11\,M$$

$$pH = -\log[H^+] = -\log(0.11) = 0.96$$

Therefore, $R_0 = 20 \times \left[\exp\dfrac{(150-100)}{14.75}\right] + 10 \times \left[\exp\dfrac{(190-100)}{14.75}\right] = 5059$

$$CS = \log(5059) - 0.96 = 2.74$$

One-stage dilute hydrochloric acid pretreatment
Hydrochloric acid is a strong acid and dissociates completely in water.

$$[H^+] = HCl\,concentration = 0.2\,M$$

$$pH = -\log[H^+] = -\log(0.2) = 0.70$$

Therefore, $R_0 = 30 \times \left[\exp\dfrac{(170-100)}{14.75}\right] = 3{,}452$

$$CS = \log(3{,}452) - 0.70 = 2.84$$

From the results, it is apparent that the single-stage hydrochloric acid pretreatment has a higher severity value. In other words, the hydrochloric acid pretreatment might have a slightly more severe pretreatment effect than the two-stage dilute sulfuric acid pretreatment.

12.5 Other Pretreatments

Other pretreatment methods, such as cellulose solvent-based pretreatment; biological pretreatment; and high-energy radiation pretreatment methods that use ultrasound, UV, γ-ray, pulsed electrical field, or microwaves, have also been investigated for enhancing the enzymatic hydrolysis of lignocellulosic biomass. In this section, cellulose solvent-based, biological, ultrasonic, and microwave pretreatments are briefly introduced.

12.5.1 Cellulose Solvent-Based Pretreatment

Cellulose solvent-based pretreatment uses cellulose solvents (e.g., ionic liquid) to fractionate cell wall components (cellulose, hemicellulose, and lignin) by dissolving cellulose and, in some cases, all of the biomass. Compared to traditional pretreatment methods, this process is biomass independent as it works with all lignocellulosic feedstocks, including agricultural residues, bioenergy crops, and woody biomass. It also requires only modest reaction conditions and enables fractionation of the biomass components for improved utilization. The high price of the solvents, solvent recycling, and high capital costs are the major challenges of cellulose solvent-based pretreatment technologies (Sathitsuksanoh *et al.*, 2013).

Biomass pretreatment using **ionic liquids** (IL), which can dissolve the cellulose and lignocellulose components of biomass, has been extensively investigated. Ionic liquids are organic salts that consist

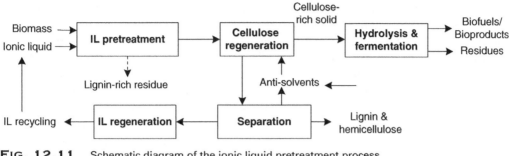

FIG. 12.11   Schematic diagram of the ionic liquid pretreatment process.

of a large organic cation, such as alkylimidazolium, alkylpyridinium, tetraalkylammonium, or tetra-alkylphosphonium, and an organic/inorganic anion, such as Cl^-, Br^-, PF_6^-, SCN^-, $CF_3SO_3^-$, or AcO^-. They exist as liquids at relatively low temperatures (below 100 °C). The dissolution of cellulose primarily results from the breakage of the inter- and intramolecular hydrogen bonds, caused by the interaction between cellulose hydroxyl groups and anions of the ionic liquid. Hemicellulose and lignin can also be dissolved in some ionic liquids. Ionic liquid pretreatments are usually conducted at mild reaction temperatures (70–140 °C). After the pretreatment, most cellulosic content is precipitated and recovered by adding anti-solvents, such as acetone and water. The pretreated material is significantly less crystalline, has more surface area, and is readily susceptible to enzymatic hydrolysis. However, the recovery cost for the regeneration of ionic liquids remains a major challenge. Figure 12.11 shows the ionic liquid pretreatment process.

Phosphoric acid with a concentration of 85% at room temperature is able to disrupt hydrogen bonds of the crystalline cellulose network and completely solubilize cellulose. The dissolved cellulose can be regenerated by adding water to yield readily digestible amorphous cellulose. For example, bamboo, corn stover, giant reed, sugarcane, miscanthus, and poplar pretreated with concentrated phosphoric acid had >87% enzymatic digestibility (Sathitsuksanoh *et al.*, 2013). The cost, recovery of phosphoric acid, and the fractionation of cellulose, hemicellulose, and lignin after the pretreatment are the major challenges of this process.

N-methylmorpholine-N-oxide (NMMO) and **NaOH/urea** are two additional cellulose solvents for biomass pretreatment. NMMO has highly polar N-O bonds that can form strong hydrogen bonds with cellulose hydroxyl groups, thereby destroying the intra- and intermolecular hydrogen bonds and the crystalline structure of cellulose. For example, oak and spruce pretreated with 85% NMMO at 90–130 °C for 1–3 hr had 65% and 84% enzymatic digestibility, respectively (Shafiei *et al.*, 2010). NaOH/urea can dissolve cellulose at subzero temperatures and has been shown to be an effective biomass pretreatment method. For example, NaOH/urea pretreated spruce significantly improved glucan digestibility (87–96%). However, this method is costly due to the need for a high NaOH concentration and a subzero (-15 to -20 °C) operating temperature (Kuo and Lee, 2009).

12.5.2 Biological Pretreatment

Biological pretreatment employs microorganisms such as white-, brown-, and soft-rot fungi and bacteria to treat biomass. These microorganisms reduce biomass recalcitrance by modifying or degrading lignin and hemicellulose. White-rot fungi have been considered the most effective biocatalysts for biomass degradation due to their unique suite of lignin-modifying enzymes. Many white-rot fungi, including *Phanerochaete chrysosporium*, *Pleurotus ostreatus*, *Trametes versicolor*, *Cyathus stercoreus*, *Ceriporiopsis subvermispora*, and *Irpex lactues* (among others), have been evaluated for biomass pretreatment. Unfortunately, most of the white-rot fungi degrade not only lignin but also carbohydrates (cellulose and hemicellulose); therefore, the overall cellulose and hemicellulose recovery is low. However, it was found that *Ceriporiopsis subvermispora* has highly selective delignification, and is considered to be a promising strain for biomass pretreatment. A mutant strain of *Trametes versicolor* incapable of cellulose hydrolysis has also been described that may limit cellulose loss during pretreatment (Canam *et al.*, 2011; Wan and Li, 2011). Biological pretreatment technology has many advantages, including no chemical use, low energy inputs, and mild and environment-friendly reaction conditions. However, long reaction times of weeks to several months, fastidious environmental requirements for microorganism growth, and the need for a large pretreatment space are the major limitations of biological pretreatment.

12.5.3 Ultrasonic Pretreatment

Ultrasonic pretreatment uses ultrasound, which is defined as sound waves with frequencies higher than the human hearing range (>20 kHz), to treat biomass. Ultrasound creates pressure waves in a liquid medium that enhance physical and chemical effects. When used in biomass pretreatment, mechano-acoustic effects caused by the pressure waves improve the accessibility of biomass through microjet erosion, cell wall collapse, and mass transfer augmentation. Microjet forms when the ultrasound-induced cavitation bubbles collapse near a solid boundary. The produced microjet is able to erode cell walls and increase the surface area. In addition, ultrasound augments chemical reactions by improving mass and heat transfer; therefore, it results in faster reactions at lower temperatures with reduced chemical loading. Ultrasound is able to improve delignification by cleavage of lignin macromolecules, and by the linkages between lignin and hemicellulose (Bussemaker and Zhang, 2013). For example, ultrasonic-assisted ionic liquid pretreatment reduced the dissolution time of holocellulose (the mixture of cellulose and hemicellulose left when lignin is dissolved) by up to 52%, compared to the same pretreatment without ultrasound (Yue *et al.*, 2012). Although ultrasonic-enhanced pretreatment improves biomass conversion, it is energy intensive and expensive; therefore, it is not currently feasible for industrial applications.

12.5.4 Microwave Pretreatment

Microwaves are electromagnetic waves that range from 300 MHz to 300 GHz. They are reflected by metals, absorbed by some dielectric materials, and transmitted through other materials without a significant loss of energy. For example, water, foods with a high water content, and some organic solvents are good microwave absorbers; whereas ceramics, quartz glass, and most thermoplastic materials only absorb microwaves slightly. When a material absorbs microwaves, the interactions between the molecular dipoles and the irradiation cause heating via dipolar polarization and ionic conduction.

Microwaves have been shown to provide rapid, internal heating and enhance chemical reactions. When applied to biomass, microwaves are absorbed by water, polar components of biomass, and pretreatment solvents and chemicals, causing rapid heating. In addition, "hot spots" created in the heterogeneous material improve the disruption of biomass structures, and the microwave radiation generates a changing magnetic field, resulting in the vibration of polar bonds, which accelerates destruction of the crystal structure (Hu and Wen, 2008). However, microwave pretreatment is energy intensive and expensive. It is also challenging to pretreat biomass with microwaves at an industrial scale.

12.6 Co-products from Lignocellulosic Feedstock Pretreatment

Current bioconversion is primarily focused on converting cellulose to biofuels and biochemicals, and insufficient attention has been paid to the non-cellulose components that account for 60–70% of biomass. Conversion of these non-cellulose components to value-added co-products would more completely utilize the biomass and offset the production costs of the main product (e.g., biofuel), improving the overall economics of the bioconversion process. Many pretreatment methods are able to fractionate hemicellulose and/or lignin from the biomass while preparing the cellulose fraction for subsequent enzymatic hydrolysis. Potential uses for hemicellulose and lignin are discussed in the following sections.

12.6.1 Hemicellulosic Sugars

Hemicellulose is the second largest fraction of polysaccharides in lignocellulosic biomass, and is composed of a group of heteropolymers of hexoses (such as galactose, mannose, and glucose), pentoses (xylose and arabinose), and uronic acids. Some of the sugars are acetylated. As already discussed, the release of acetyl groups forms acetic acid that catalyzes the hydrolysis of hemicellulose during some forms of pretreatment, such as hot water and steam explosion. The composition and structure of hemicellulose vary depending on the species of biomass. For instance, xylan is the main component of herbaceous biomass and hardwood hemicellulose, while glucomannan is the major component of softwood hemicellulose.

Hemicellulose is dissolved from biomass during pretreatment by hydrolysis, in particular during acid pretreatment. Depending on the pretreatment method and conditions, the dissolved hemicellulosic sugars are typically in the form of monomeric sugars (hexoses and pentoses) and oligomeric sugars. The dissolved sugars can be dehydrated into furans (see the following section) and further degraded to organic acids. Hemicellulosic sugars have the potential for producing valuable co-products. For example, hexoses can be directly fermented to ethanol or other products. Pentoses are not as easily fermentable as hexoses, but some genetically modified microorganisms have the ability to ferment pentoses into ethanol. Oligomeric hemicellulosic sugars have been successfully used as animal feed additives and as plant growth regulators. In addition, hemicellulosic sugars can be chemically or biologically converted into many platform chemicals.

12.6.2 Furans (Furfural and HMF)

Furans such as *furfural* and *5-hydroxymethylfurfural (HMF)* are byproducts formed during the acidic pretreatment of lignocellulosic biomass. Furfural is generated from the dehydration of pentoses, such as xylose and arabinose. HMF is the product of the dehydration of hexoses, such as glucose. The dehydration pathways of biomass sugars to furfural and HMF under acidic conditions are summarized in Figure 12.7. In brief, polysaccharides (cellulose and hemicellulose) are first hydrolyzed into monomeric sugars (e.g., glucose and xylose), then these sugars are dehydrated to HMF and furfural, respectively. HMF can be further decomposed to levulinic acid (LA) and formic acid. Furfural and HMF are fermentation inhibitors because they have toxic effects on microorganisms, but they are useful platform chemicals for many products. For example, furfural is a starting material for many chemicals and materials such as nylon and tetrahydrofuran (THF). HMF can be used to derive a spectrum of useful chemicals, including dimethylfuran (DMF), 2,5-furandicarboxylic acid (FDA), 2,5-diformylfuran (DFF), dihydroxymethylfuran, and 5-hydroxy-4-keto-2-pentenoic acid (Rosatella *et al.*, 2011). In addition, both furfural and HMF can be used as feedstocks or intermediates for producing hydrocarbon fuels.

12.6.3 Lignin

After polysaccharides, *lignin* is the next largest component in lignocellulosic biomass. As discussed earlier, lignin is a three-dimensional phenolic polymer of phenylpropane units (coniferyl, sinapyl, and *p*-coumaryl alcohols). In nature, lignin provides structural integrity to plants and protects them against attack from microorganisms; therefore, it is a major barrier and inhibitor to the bioconversion of biomass. For this reason, many pretreatment methods aim to remove or modify lignin to enhance enzymatic hydrolysis of the biomass. Some of the dissolved lignins have great potential for high-value lignin products. For example, the lignin from organosolv pretreatment has been evaluated for applications involving adhesives and polymeric composite materials, because of the high purity and reactivity of the organosolv lignin moieties. Lignosulfonate from the sulfite-based pretreatment is a good surfactant, and it has been widely used in adhesives, concrete admixtures, emulsifiers, road binders, and pesticide dispersants. Lignin is also a potential feedstock for producing aromatic compounds and liquid fuels through pyrolysis, oxidation, reduction, and hydrolysis. However, because of the heterogeneous physical and chemical properties of lignin, the development of high-quality and high-value lignin products remains challenging. The most common use of lignin today is as a combustible fuel for heat and power production.

> **Example 12.3:** Switchgrass (100 kg dry matter) is pretreated with dilute sulfuric acid. The biomass contains 40% cellulose, 40% hemicellulose, and 20% lignin. Half of the hemicellulose is xylan and the rest is arabinan. During pretreatment, 5% of cellulose, 20% of xylan, and 40% of arabinan are decomposed to furans (HMF and furfural). Assume that there is no further decomposition of furans. Calculate the amounts of HMF and furfural formed during the acid pretreatment.

> *Solution:*
>
> **Hydrolysis and dehydration reactions from cellulose to HMF**
>   Cellulose to glucose: $(C_6H_{10}O_5)_n + n(H_2O) \rightarrow n(C_6H_{12}O_6)$
>   Glucose to HMF: $C_6H_{12}O_6 \rightarrow C_6H_6O_3 + 3(H_2O)$
> $100 \, \text{kg} \times 40\% \times 5\% \times 180/162 \times 126/180 = \textbf{1.56\,kg}$
>
> **Hydrolysis and dehydration reactions from hemicellulose to furfural**
>   Pentosan to pentose: $(C_5H_8O_4)_n + n(H_2O) \rightarrow n(C_5H_{10}O_5)$
>   Pentose to furfural: $C_5H_{10}O_5 \rightarrow C_5H_4O_2 + 3(H_2O))$
> $100 \, \text{kg} \times (20\% \times 20\% + 20\% x \, 40\%) \times 150/132 \times 96/150 = \textbf{8.73\,kg}$
>
> According to the results, 1.56 kg of HMF and 8.73 kg of furfural are produced from 100 kg of switchgrass during the dilute sulfuric acid pretreatment.

References

Agbor, V.B., Cicek, N., Sparling, R., Berlin, A., and Levin, D.B. (2011). Biomass pretreatment: Fundamentals toward application. *Biotechnology Advances* 29: 675–85.

Bussemaker, M.J., and Zhang, D. (2013). Effect of ultrasound on lignocellulosic biomass as a pretreatment for biorefinery and biofuel applications. *Industrial and Engineering Chemistry Research* 52: 3563–80.

Canam, T., Town, J.R., Tsang, A., McAllister, T.A., and Dumonceaux, T.J. (2011). Biological pretreatment with a cellobiose dehydrogenase-deficient strain of Trametes versicolor enhances the biofuel potential of canola straw. *Bioresource Technology* 102: 10020–27.

Chang, V.S., Burr, B., and Holtzapple, M.T. (1997). Lime pretreatment of switchgrass. *Applied Biochemistry and Biotechnology* 63–65: 3–19.

Esteban, L.S., and Carrasco, J.E. (2006). Evaluation of different strategies for pulverization of forest biomass. *Powder Technology* 166: 139–51.

Hu, Z., and Wen, Z. (2008). Enhancing enzymatic digestibility of switchgrass by microwave-assisted alkali pretreatment. *Biochemical Engineering Journal* 38: 369–78.

Kuo, C.H., and Lee, C.K. (2009). Enhancement of enzymatic saccharification of cellulose by cellulose dissolution pretreatments. *Carbohydrate Polymers* 77: 41–6.

Lynd, L.R., Weimer, P.J., Zyl, W.H., and Pretorius, I.S. (2002). Microbial cellulose utilization: Fundamentals and biotechnology. *Microbiology and Molecular Biology Reviews* 66: 506–77.

Mani, S., Tabil, L.G., and Sokhansanj, S. (2004). Grinding performance and physical properties of wheat and barley straws, corn stover and switchgrass. *Biomass and Bioenergy* 27: 339–52.

Mosier, N., Hendrickson, R., Ho, N., Sedlak, M., and Ladisch, M.R. (2005). Optimization of pH controlled liquid hot water pretreatment of corn stover. *Bioresource Technology* 96: 1986–93.

Pan, X.J. (2013). Organosolv biorefining platform for producing chemicals, fuels, and materials from lignocellulose. In: H. Xie and N. Gathergood (ed.), *The Role of Green Chemistry in Biomass Processing and Conversion*. Hoboken, NJ: John Wiley & Sons, Inc.

Rosatella, A.A., Simeonov, S.P., Frade, R.F.M., and Afonso, C.A.M. (2011). 5-Hydroxymethylfurfural (HMF) as a building block platform: Biological properties, synthesis and synthetic applications. *Green Chemistry* 13: 754–93.

Rubin, E.M. (2008). Genomics of cellulosic biofuels. *Nature* 454: 841–5.

Sathitsuksanoh, N., George, A., and Zhang, Y.-H.P. (2013). New lignocellulose pretreatments using cellulose solvents: A review. *Journal of Chemical Technology and Biotechnology* 88: 169–80.

Shafiei, M., Karimi, K., and Taherzadeh, M.J. (2010). Pretreatment of spruce and oak by N-methylmorpholine-N-oxide (NMMO) for efficient conversion of their cellulose to ethanol. *Bioresource Technology* 101: 4914–18.

Wan, C., and Li, Y. (2011). Effectiveness of microbial pretreatment by *Ceriporiopsis subvermispora* on different biomass feedstocks. *Bioresource Technology* 102: 7507–12.

Womac, A.R. (2006). *Size reduction actions*. Biomass Processing. Knoxville, TN: University of Tennessee. http://www.biomassprocessing.org/grinding_sizereduction_actionspage.htm, accessed April 2016.

Xu, J., Cheng, J.J., Sharma-Shivappa, R.R., and Burns, J.C. (2010). Sodium hydroxide pretreatment of switchgrass for ethanol production. *Energy Fuels* 24: 2113–19.

Yue, F., Lan, W., Zhang, A., Liu, C., Sun, R., and Ye, J. (2012). Dissolution of holocellulose in ionic liquid assisted with ball-milling pretreatment and ultrasonic irradiation. *BioResources* 7: 2199–208.

Zhang, Y.-H.P., Ding, S.-Y., Mielenz, J.R., *et al.* (2007). Fractionating recalcitrant lignocellulose at modest reaction conditions. *Biotechnology and Bioengineering* 97: 214–23.

Zhu, J.Y., Pan, X.J., Wang, G.S., and Gleisner, R. (2009). Sulfite pretreatment (SPORL) for robust enzymatic saccharification of spruce and red pine. *Bioresource Technology* 100: 2411–18.

Exercise Problems

12.1. Explain the objectives of pretreatment for the bioconversion of lignocellulosic biomass.

12.2. Switchgrass is used for ethanol production. The composition of the switchgrass is 37% cellulose, 24% xylan, 3% galactan, 4% arabinan, 20% lignin, 7% extractives, and 5% ash. A dilute acid pretreatment method is applied to the switchgrass before enzymatic hydrolysis and fermentation. The pretreatment hydrolyzes 10% hexosan and 90% pentosan into monomeric sugars. Approximately 30% of the hydrolyzed pentoses are decomposed to furfural. Assume that there is no decomposition of the hydrolyzed hexoses. How much of each lignocellulosic sugar (glucose, xylose, galactose, and arabinose) is produced when pretreating 1,000 kg (dry matter) switchgrass?

12.3. In Problem 12.2, how much water is consumed/produced during hydrolysis and dehydration reactions in the pretreatment? Assume that lignin, extractives, and ash do not change during the pretreatment.

12.4. Develop an overall mass balance for Problem 12.2. Assume that 75% lignin and 60% ash are left in the biomass while all the extractives are dissolved in the liquid.

12.5. The capacity of reactors is limited in a cellulosic ethanol plant. To improve process efficiency, engineers are trying to develop a pretreatment method that has a short pretreatment time and high delignification. Among three candidate pretreatment methods are dilute sulfuric acid pretreatment, ammonia recycle percolation (ARP) pretreatment, and ammonia fiber expansion (AFEX) pretreatment. Which method would meet the requirements, and why?

12.6. Calculate the combined severity factor of the pretreatment system in Figure 12.12.

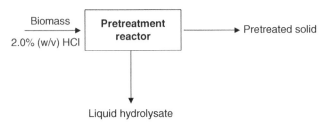

FIG. 12.12   Pretreatment system.

The pretreatment conditions are 10 min pretreatment time and 190 °C pretreatment temperature.

12.7. Red oak has 42% cellulose, 28% hemicellulose (22% xylan and 6% arabinan), 24% lignin, 5% extractives, and 1% ash. In an acid hydrolysis experiment, 100 kg of red oak sawdust is hydrolyzed. The results show that the cellulose and hemicellulose are completely hydrolyzed to monomeric sugars. In addition, some of the hydrolyzed sugars (30% glucose, 50% xylose, and 50% arabinose) are dehydrated to HMF and furfural. If there is no further decomposition or condensation of the furans generated, what is the quantity of furans (furfural and HMF) produced in this process?

12.8. A company plans to produce cellulosic ethanol from corn stover. The harvested corn stover has 5% moisture content, and the composition of the dry biomass is 38% glucan, 22% xylan, 6% galactan, 4% arabinan, 18% lignin, 5% ash, and 7% extractives. Liquid hot water (LHW) pretreatment is used to pretreat 10 metric tons of corn stover with a water flow rate of 50 L/min at 200 °C for 30 min. During the pretreatment, 2% glucan, 90% xylan, and all other hemicellulose are hydrolyzed into monomeric sugars without the formation of furans. If 95% glucan and 80% xylan in the pretreated corn stover are hydrolyzed into glucose and xylose, respectively, in the following enzymatic hydrolysis stage, what is the quantity of fermentable sugars (monosaccharides) that are produced in both pretreatment and enzymatic hydrolysis stages?

12.9. If a catalyst (either 1% sulfuric acid or 10% aqueous ammonia) is added to the LHW pretreatment in Problem 12.8, how would each of the catalysts influence the pretreatment? Predict the changes in the composition of pretreated substrate with each catalyst.

12.10. A biorefinery is designed to produce cellulosic ethanol from corn stover with a capacity of 10 million gallons per year. Corn stover is pretreated with dilute sulfuric acid, and a simultaneous saccharification and co-fermentation (SSCF) process is used to convert cellulose and hemicellulose in the pretreated corn stover into ethanol.

 (a) How much corn stover is needed per year by the biorefinery? Assume that 3% cellulose and 25% hemicellulose of the corn stover are lost during the pretreatment due to acid-induced degradation; 86% cellulose and 92% hemicellulose in the pretreated corn stover are hydrolyzed into glucose and xylose, respectively; and 90% glucose and xylose are fermented to ethanol at 90% of their theoretical yields. The moisture content of the corn stover is 16%. The corn stover contains 40% cellulose and 30% hemicellulose on dry corn stover.

 (b) Assume that 40% of the land surrounding the biorefinery is used to grow corn. The corn stover yield is approximately 7.41 metric tons/ha (dry). All farmers are contracted to provide corn stover to the biorefinery. What is the collection radius of the corn stover for the biorefinery?

 (c) If the volume of the batch reactors used in the biorefinery for corn stover pretreatment is 300 m^3, sulfuric acid concentration for the pretreatment is 2%, the reactor is loaded up to 85% of its volume, and the ratio of the acid solution to the corn stover is 3.5:1 (v/w), for each batch, how much corn stover can be pretreated? How much concentrated (98%) sulfuric acid (density 1.84 g/cm^3) is needed? How much water is required? Since the acid concentration is low, the density of 2% sulfuric acid can be considered as 1.0 g/cm^3. The particle density of the corn stover is 250 kg/m^3 at 16% moisture.

Enzymatic Hydrolysis

David Hodge and Wei Liao

What is included in this chapter?

This chapter covers the fundamentals of enzyme kinetics. Different types of hydrolytic enzymes (hydrolases) and their applications in carbohydrate depolymerization for biofuel production are discussed. Factors affecting enzyme hydrolysis, enzyme inhibition, and pertinent examples are also considered.

13.1 Introduction

Enzymes are biological catalysts (biocatalysts) and are known to catalyze a large number of reactions in biological systems, such as the synthesis of virtually all organic compounds produced by living organisms and the biological degradation of other organic compounds. For instance, amylase enzymes in our saliva convert starch in food to glucose that provides the energy we need. Enzymes have diverse applications ranging from food processing, pharmaceuticals, nutraceuticals, environmental remediation, and bioenergy production. Compared to chemical catalysts, biocatalysts have several important and unique characteristics, such as typically milder reaction conditions, greater reaction specificity, and the capacity for regulation (Voet and Voet, 2013). Biochemical reactions catalyzed by enzymes are called *enzymatic reactions*. Enzymatic reactions can be classified according to six general reaction types: oxidation and reduction; transfer reaction; hydrolysis; formation or removal of a double bond; isomerization; and bond formation with ATP hydrolysis (Table 13.1). Among these, the hydrolysis reactions account for more than 60% of all enzymatic reactions in industrial applications (Faber, 1997). The enzymes catalyzing hydrolysis reactions (i.e., breaking a covalent bond by addition of water) are classified as hydrolases.

Humans have used hydrolases for thousands of years in food preparation and brewing. One of the oldest examples is the malting process for fermented beverage production, where cereal grains were allowed to begin germination in order to activate enzymes to hydrolyze starch to dextrins and glucose.

Bioenergy: Principles and Applications, First Edition. Edited by Yebo Li and Samir Kumar Khanal.
© 2017 John Wiley & Sons, Inc. Published 2017 by John Wiley & Sons, Inc.
Companion website: www.wiley.com/go/Li/Bioenergy

Table 13.1 Enzyme classification according to reaction type

Classification	Type of reaction	Examples
Oxidoreductases	Oxidation and reduction	Pyruvate dyhydrogenase that converts pyruvate to acetyl CoA in the tricarboxylic acid (TCA) cycle
Transferases	Transfer of functional groups	Lipid kinases that phosphorylate lipids in the cell to change the reactivity and localization of the lipid
Hydrolases	Hydrolysis reactions	Xylanases that depolymerize xylan (a homopolymer of xylose) by hydrolysis of glycosidic bonds
Lyases	Group elimination to form double bonds	Pyruvate decarboxylase that catalyzes the decarboxylation of pyruvic acid to acetaldehyde and carbon dioxide
Isomerases	Isomerization	Ribose-5-phosphate isomerase that interconverts aldoses and ketoses
Ligases	Bond formation coupled with ATP hydrolysis	DNA ligase that forms phosphodiester bonds to join DNA strands together

In the late 17th and early 18th centuries, starch degradation by saliva and meat digestion from stomach secretions was recognized. The first hydrolase, diastase (α-, β- or γ-amylase), was discovered by the French chemists Anselme Payen and Jean-Francois Persoz in 1833 (Seetharaman and Bertoft, 2012). With the protein nature of enzymes being established in the early 20th century (Sumner, 1926), more hydrolases, including glycoside hydrolases, esterases, and proteases, were discovered and purified. Advances in microbial fermentation for hydrolase production further catalyzed the commercial production of hydrolases. They are currently being used extensively for industrial, agricultural, and human health applications. In the bioenergy industry, hydrolases play a critical role in the biochemical production of biofuels from starch and cellulose. They depolymerize starch and cellulose into monomeric sugars that can be readily fermented by microorganisms for biofuel production.

13.2 Nomenclature and Classification of Hydrolases

Generally, enzyme names end with "*–ase*" and are named after the substrate (reactant) whose transformation is catalyzed by the enzymes, such as amylase, cellulase, or xylanase for the transformation of amylose, cellulose, or xylan, respectively. Each enzyme also has its own systematic name and classification number according to the enzyme nomenclature of the International Union of Biochemistry and Molecular Biology (IUBMB) Enzyme Commission (EC). The systematic name shows the exact action of an enzyme, and the classification number assigns a unique code to it. For instance, the systematic name of a starch hydrolase, α-amylase, is 1,4-α-D-glucan glucanohydrolase, and the classification number is EC3.2.1.1. In the classification number of α-amylase, the first number, 3, indicates that the class of amylase is hydrolase; the second number, 2, shows its subclass of glycosylase; the third number, 1, designates its sub-subclass of hydrolyzing *O*- and *S*-glycosyl compounds; and the fourth number, 1, is the arbitrarily assigned sub-subclass number.

As already mentioned, 3 as the first number in the EC enzyme nomenclature system is assigned to the class of hydrolases. Some selected hydrolases and their functions are listed in Table 13.2. Lipases (EC3.1) catalyze the breakdown of triglycerides into three fatty acids and one glycerol; nucleases (EC3.1) cleave the phosphodiester bonds to release nucleotides from nucleic acids; glycosylases (EC3.2) catalyze the depolymerization of polysaccharides (starch, cellulose, hemicellulose, and pectin) into simple sugars; epoxide hydrolases (EC3.3) act on ether bonds to convert epoxides into trans-dihydrodiols; and peptidases and proteases (EC3.4) hydrolyze proteins into amino acids and short peptides. The glycosylases as an example are discussed in detail in Section 13.4.

Table 13.2 Selected hydrolases

EC number	Types of bond hydrolyzed	Examples
3.1	Ester	
3.1.1	Carboxylic acid esters	Triacylglycerol lipase, acetylcholine esterase, phospholipase A1, gluconolactonase, lipoprotein lipase
3.1.3–4	Phosphoric acid mono- or diesters	Phospholipase, nucleotidase
3.2	Glycosidic	
3.2.1	O- and S-glycosides	α-amylase, oligo-1,6-glucosidase, lysozyme, α-glucosidase, β-galactosidase, xylan 1,3-β-xylosidase
3.2.2	N-glycosides	DNA-deoxyinosine glycosylase, NAD+ nucleosidase
3.3	Ether	
3.3.1	Thioether	Thioether hydrolases
3.3.2	Eposides	Epoxide hydrolase
3.4	Peptide	
3.4.11	Leucine residues at N-terminus of peptides	Leucine aminopeptidase
3.4.16	Peptide bond	Serine type carboxypeptidases

Proteins and enzymes

Proteins are polymers composed of amino acids that are covalently linked in peptide bonds. They are critical in performing a diverse range of functions, including most of the catalytic, transport, structural, and regulatory functions of the cell. In order to better understand the functions of proteins, a three-dimensional relationship has been used to describe the structure of proteins. The structure includes four levels (primary, secondary, tertiary, and quaternary; Voet and Voet, 2013). The primary structure is the linear sequence of amino acids; the secondary structure is the local spatial arrangement of amino acids at various segments of the polypeptide chain; the tertiary structure refers to the three-dimensional arrangement of combined secondary structures of a polypeptide; and the quaternary structure is the spatial arrangement of two or more polypeptides with non-covalent interactions. Most enzymes are globular, water-soluble proteins where the tertiary structure gives the molecules their three-dimensional structure, which influences their catalytic function. In nature, these enzymes are coupled in multiple reactions or metabolic pathways that are responsible for either metabolism of a carbon source for the generation of energy (e.g., ATP) and/or reducing equivalents (e.g., NADH), or the generation of all the organic compounds required for cell function.

13.3 Enzyme Kinetics

Enzyme kinetics refers to the rates of enzyme-catalyzed reactions. The study of enzyme kinetics is important, because the kinetics can provide insight into the catalytic mechanism of the enzyme, delineate the relationship between enzyme and reaction rate, and identify superior enzymes and reaction conditions for a selected enzyme-catalyzed transformation. The basic principles of enzyme kinetics are discussed in this section.

13.3.1 Fundamentals of Reaction Rate: Transition State Theory

The objective of kinetic analysis is to describe reaction rates with regard to the properties of reactants and products. In order to achieve this objective, transition state theory was developed in the 1930s by Henry Eyring (1935). Transition state theory establishes the relationship between reactants and activated transition state complexes.

Given a reaction with substrate A and product B:

$$A \rightarrow B \tag{13.1}$$

In order to produce product B, a high-energy (unstable) intermediate called the activated complex (A∗) must be formed from the substrate A. Equation 13.1 can be rewritten as follows:

$$A \leftrightarrow A^* \rightarrow B \tag{13.2}$$

This two-step model of the reaction can be described in a transition state diagram (Figure 13.1). The point of forming A^* is called the transition state. The energy input must reach the transition state in order to complete the reaction. The lower the free energy the activated complex has, the less stable the transition state would be and the faster the reaction would occur.

13.3.2 Reaction Rate and Reaction Orders

A general chemical or enzymatic reaction equation can be given as follows:

$$aA_1 + bA_2 + cA_3 + \cdots \rightarrow B \tag{13.3}$$

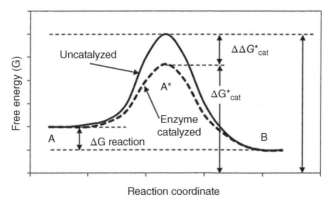

FIG. 13.1   Transition state diagram of an enzymatic reaction.

Notes:

1. ΔG reaction is the Gibbs free energy of the reaction; ΔG^*_{cat} is the Gibbs free energy of the activated complex.

2. Reaction coordinate is the minimum energy pathway of a reaction.

3. $\Delta\Delta G^*_{Cat}$ is the reduction by the enzyme. $\Delta\Delta G^*_{Cat} = \Delta G^*_{Uncat} - \Delta G^*_{Cat}$. The enzymatic reaction leads to an activated transition state complex with lower Gibbs free energy than the reaction without enzyme. Lower Gibbs free energy means a lower activation barrier, so that the rate of reaction for enzymatic reaction is enhanced.

where A_1, A_2, and A_3 are the substrates; B is the product; and a, b, and c are the coefficients.

The rate of this reaction is proportional to the product of the substrate concentrations raised to the power of each substrate's coefficient. The rate equation can be expressed as follows:

$$r = kC_{A1}^a C_{A2}^b C_{A3}^c \ \cdots\cdots \tag{13.4}$$

where r is the overall reaction rate, k is the reaction rate constant, and C is the concentration of reactants (g/L).

The order of the reaction is defined as the sum of the stoichiometric coefficients of substrates, $a + b + c + \cdots$. Thus, when the sum of the stoichiometric coefficients is 0, 1, or 2, the corresponding reactions are termed zero-order, first-order, or second-order reactions.

A basic reaction rate expression just considering the formation of primary product from primary reactant (Equation 13.5) can be used to discuss individual kinetics:

$$nA \xrightarrow{k} B \tag{13.5}$$

The reaction rate (r) is defined as a function of time and concentration of reactant or product; k is the reaction rate constant; and n is the coefficient of reactant A:

$$r = -\frac{dC_A}{dt} = \frac{dC_B}{dt} \tag{13.6}$$

where C_A is the concentration of reactant A (g/L), C_B is the concentration of product B (g/L), and t is the time (hr).

Zero-order, first-order, and second-order kinetics for the reaction considering the formation of primary product from primary reactant can then be derived using the definition of reaction rate, and the relationship between the rates and substrate concentrations.

Zero-Order Kinetics

Considering $0A \rightarrow B$, a *zero-order reaction*, its reaction rate can be expressed as follows:

$$r = k_0 C_A^0 = -\frac{dC_A}{dt} \rightarrow r = k_0 = -\frac{dC_A}{dt} \tag{13.7}$$

where k_0 is the zero-order rate constant (g/(L·hr)).

Equation 13.7 can be rearranged as follows:

$$dC_A = -k_0 dt \tag{13.8}$$

With the initial condition of $C_A = C_{A0}$ at $t = 0$, the expression can be integrated:

$$\int_{C_{A0}}^{C_A} dC_A = -k_0 \int_0^t dt \tag{13.9}$$

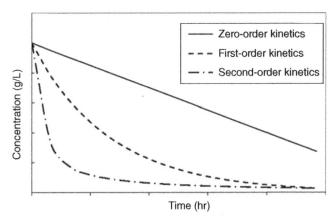

FIG. 13.2   Substrate concentration versus time for zero-, first-, and second-order kinetics.

Zero-order kinetics gives the following relationship between substrate concentration and reaction time:

$$C_A = C_{A0} - k_0 t \tag{13.10}$$

A plot of C_A vs. time gives a straight line with slope $-k_0$ (Figure 13.2).

First-Order Kinetics

Considering $A \to B$, a *first-order reaction*, its reaction rate can be expressed as follows:

$$r = k_1 C_A = -\frac{dC_A}{dt} \tag{13.11}$$

where k_1 is the first-order volumetric rate constant (1/hr).

Solving Equation 13.11 with the initial condition of $C_A = C_{A0}$ at $t = 0$, the first-order kinetics gives the following relationship between substrate concentration and reaction time:

$$\ln C_A - \ln C_{A0} = -k_1 t \tag{13.12}$$

The plot of C_A vs. time is presented in Figure 13.2.

Second-Order Kinetics

Considering $2A \to B$, a *second-order reaction*, its reaction rate can be expressed as follows:

$$r = k_2 C_A^2 = -\frac{dC_A}{dt} \tag{13.13}$$

where C_A is the concentration of reactant A (g/L) and k_2 is the second-order volumetric rate constant ((hr·L)/g).

Similarly, with the initial condition of $C_A = C_{A0}$ at $t = 0$, the second-order kinetics gives the following relationship between substrate concentration and reaction time:

$$\frac{1}{C_A} = \frac{1}{C_{A0}} + k_2 t \qquad (13.14)$$

The plot of C_A vs. time is presented in Figure 13.2.

Example 13.1: Three different α-amylases (A, B, and C) were tested in order to compare their hydrolysis performance on corn starch. A starch concentration of 10 g/L was used for the hydrolysis. The starch concentrations (g/L) during hydrolysis were monitored and are presented in Table 13.3. The experiment was conducted at 50 °C. The reaction time was 3 hr. Determine the rate constants for the hydrolysis with each enzyme.

Table 13.3 Starch concentration in enzymatic hydrolysis

	Starch concentration (g/L)		
Time (hr)	Enzyme A	Enzyme B	Enzyme C
0	10.0	10.0	10.0
0.5	8.4	7.8	5.7
1.0	6.8	6.1	4.0
1.5	5.2	4.7	3.1
2.0	3.6	3.7	2.5
2.5	2.0	2.9	2.1
3.0	0.4	2.2	1.8

Solution: Starch concentrations were plotted as a function of time for each individual enzyme (Figure 13.3). According to the derivation in this section, zero-order, first-order, and second-order kinetics fit the data for Enzymes A, B, and C, respectively.

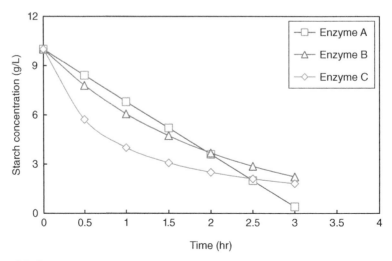

FIG. 13.3   Starch concentration vs. time.

In order to find out rate constants from these equations, the new plots are made according to Equations 13.10, 13.12, and 13.14 (Figure 13.4).

From the slopes of the kinetic equation plots (Figure 13.4), the rate constants are 3.2 g/(L·hr), 0.5 1/hr, and 0.15 (hr·L)/g for enzymes A, B, and C, respectively.

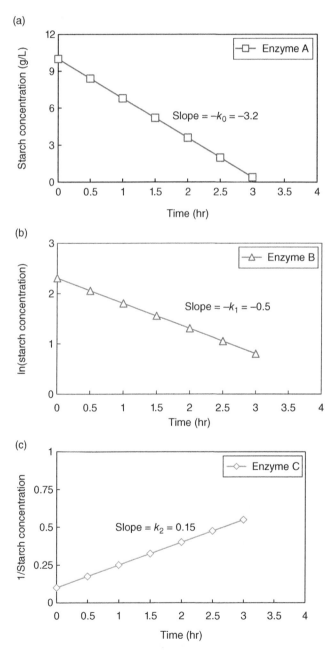

FIG. 13.4   Plots of kinetic equations: (a) Zero-order kinetics of starch concentration vs. time; (b) first-order kinetics of ln(starch concentration) vs. time; and (c) second-order kinetics of 1/(starch concentration) vs. time.

13.3.3 Michaelis–Menten Kinetics

The German biochemist Leonor Michaelis and the Canadian biochemist Maud L. Menten first proposed a mechanism that elucidates the dependence of the initial rate of enzyme-catalyzed reactions on concentration in 1913 (Michaelis and Menten, 1913). Thus, the kinetic equation was named in their honor and is popularly known as Michaelis–Menten kinetics, which is derived by combining first-order kinetics with an additional equilibrium expression for the enzyme–substrate binding. A two-step reaction in which the substrate first forms a complex with the enzyme and then the complex is converted into product and enzyme is given as follows:

$$E + S \underset{k_{-1}}{\overset{k_1}{\leftrightarrow}} ES \overset{k_2}{\to} P + E \tag{13.15}$$

where E, S, ES, and P represent the enzyme, substrate, enzyme–substrate complex, and product, respectively. The constants k_1 and k_{-1} are the reaction rate constants for the forward and reverse reactions of E + S and ES, respectively, and k_2 is the reaction rate for the reaction of ES to P + E. In most enzyme reactions, the substrate concentration is much higher than the enzyme concentration, and all enzymes are quickly converted into enzyme–substrate complex. So the second step of the reaction becomes rate limiting and the overall reaction rate is related to the second reaction.

Thus, the reaction rate of the two-step enzyme reaction can be expressed as follows:

$$r = \frac{dC_P}{dt} = k_2 C_{ES} \tag{13.16}$$

where C_P and C_{ES} are the respective concentrations of product (P) and enzyme–substrate complex (ES) (g/L), and k_2 is the rate constant of the second step reaction (1/hr).

The concentration of ES will change with time, though, and that change can also be expressed as the sum of three reactions of ES production ($k_1 C_E C_S$), ES conversion to product ($k_2 C_{ES}$), and ES dissociation ($k_{-1} C_{ES}$):

$$\frac{dC_{ES}}{dt} = k_1 C_E C_S - k_{-1} C_{ES} - k_2 C_{ES} \tag{13.17}$$

where C_S and C_E are the concentrations of substrate and enzyme (g/L), respectively.

In order to integrate Equation 13.17, a steady-state condition is assumed. With the exception of the starting phase of the reaction, ES remains constant until the substrate is completely exhausted. This is referred to as a rapid equilibrium assumption. Therefore, ES is considered to be at steady state during the reaction.

$$\frac{dC_{ES}}{dt} = 0 \tag{13.18}$$

Thus, the expression of ES change can be rearranged as follows:

$$k_1 C_E C_S = k_{-1} C_{ES} + k_2 C_{ES} \tag{13.19}$$

Since C_E and C_{ES} are impossible to measure at a specific moment in a reaction, the total enzyme concentration (C_{ET}) is used to establish the relationship between C_E and C_{ES}:

$$C_E = C_{ET} - C_{ES} \tag{13.20}$$

C_{ES} can then be found by solving Equations 13.19 and 13.20:

$$C_{ES} = \frac{C_{ET}C_S}{\frac{k_{-1}+k_2}{k_1} + C_S} \tag{13.21}$$

$\frac{k_{-1}+k_2}{k_1}$ is defined as the Michaelis constant (K_m) with unit of concentration (g/L), and Equation 13.20 is simplified to:

$$C_{ES} = \frac{C_{ET}C_S}{K_m + C_S} \tag{13.22}$$

The reaction rate, r, can be expressed as:

$$r = k_2 C_{ES} = \frac{k_2 C_{ET} C_S}{K_m + C_S} \tag{13.23}$$

Since the maximum reaction rate (v_{max}) (g/(L·hr)) occurs when the enzyme is entirely in the form of ES:

$$v_{max} = k_2 C_{ET} \tag{13.24}$$

Thus, the reaction rate can be expressed as follows:

$$r = \frac{v_{max} C_S}{K_m + C_S} \tag{13.25}$$

Equation 13.24 is known as the Michaelis–Menten equation and v_{max} is the maximum initial reaction rate (g/(L·hr)). The value of K_m is equivalent to the substrate concentration at the half-maximum initial reaction rate.

Since r can also be expressed by the substrate concentration ($r = -\frac{dC_S}{dt}$), Equation 13.25 can be integrated with the initial condition of $C_S = C_{S0}$ at $t = 0$:

$$\int_{C_{S0}}^{C_S} \left(\frac{K_m + C_S}{C_S} \right) dC_S = -v_{max} \int_0^t dt \tag{13.26}$$

The rate expression for Michaelis–Menten kinetics yields the following relationship between substrate concentration and reaction time:

$$K_m \ln\frac{C_S}{C_{S0}} + (C_S - C_{S0}) = -v_{max}t \tag{13.27}$$

The plot of C_S vs. time (t) is presented in Figure 13.5.

K_m and υ_{max} need to be determined in order to use the Michaelis–Menten equation. Before the availability of computers, these parameters were determined by transforming the Michaelis–Menten equation into a straight-line plot popularly known as the *Lineweaver–Burk plot* and the *Eadie–Hofstee diagram*, which are still in use. Now, computer-based software is also being used widely for the determination of these parameters. The Lineweaver–Burk plot approach for determining K_m and υ_{max} is discussed in the following section. The Matlab® approach is given in Example 13.2.

The Lineweaver–Burk plot can be obtained by taking the reciprocal of Equation 13.25:

$$\frac{1}{r} = \left(\frac{K_m}{\upsilon_{max}}\right)\frac{1}{C_S} + \frac{1}{\upsilon_{max}} \tag{13.28}$$

A plot of $1/r$ versus $1/C_S$ (Lineweaver–Burk or *double-reciprocal plot*) gives a straight line with an intercept of $1/\upsilon_{max}$ and a slope of (K_m/υ_{max}) (Figure 13.6). The intercept on the *x*-axis is $-1/K_m$. Thus, the values of K_m and υ_{max} can be obtained from the plot.

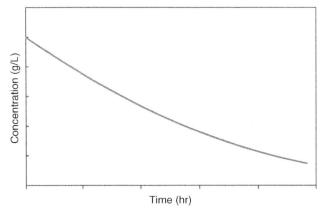

FIG. 13.5   Substrate concentration vs. time for Michaelis–Menten kinetics.

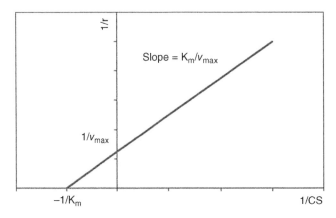

FIG. 13.6   Lineweaver–Burk plot for the determination of K_m and υ_{max}.

The dual nature of Michaelis–Menten kinetics

Figure 13.7 demonstrates the relationship between Michaelis–Menten kinetics and zero-order, first-order, and second-order kinetics. Michaelis–Menten kinetics is a combination of zero- and first-order kinetics. When substrate concentration (C_S) is much higher than the Michaelis constant (K_m), the Michaelis–Menten equation becomes:

$$r = \frac{v_{max}C_S}{K_m + C_S} \approx \frac{v_{max}C_S}{C_S} = v_{max} \tag{13.29}$$

Since v_{max} is a constant, the Michaelis–Menten equation for the case of $C_S \gg K_m$ is simplified to zero-order kinetics and a reaction rate that is nearly independent of substrate concentration.

On the other end, at low substrate concentrations, when K_m is much higher than C_S, the Michaelis–Menten equation becomes:

$$r = \frac{v_{max}C_S}{K_m + C_S} \approx \frac{v_{max}C_S}{K_m} = \frac{v_{max}}{K_m}C_S \tag{13.30}$$

Since v_{max} and K_m are constants, Michaelis–Menten kinetics for the case where $K_m \gg C_S$ represents first-order kinetics.

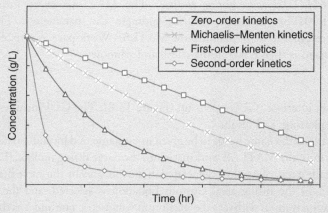

FIG. 13.7   Substrate concentration versus time for Michaelis–Menten kinetics compared with zero-, first-, and second-order kinetics.

Example 13.2: Enzymatic hydrolysis of potato starch was performed in order to quantify the hydrolysis kinetics of α-amylase from a bacterium, *Bacillus subtilis*. The α-amylase attacks the 1,4-linkages of amylose in a random fashion, and the hydrolysis releases maltosaccharide fragments with one reducing end. The starch concentration (g/L) and volumetric reaction rate (g/(L·hr)) were measured and are presented in Table 13.4.

1. Estimate the maximum volumetric reaction rate (v_{max}) (g/(L·hr)) and the Michaelis constant (K_m) (g/L).
2. Determine the time course of the enzymatic reaction in a time interval of 2.5 hours at an initial substrate concentration of 10 g/L.

Table 13.4 Substrate concentration and volumetric reaction rate of enzymatic hydrolysis

Substrate concentration (C_S) (g/L)	Volumetric reaction rate (r) (g/(L·hr))
0	0
1	2.64
2	4.22
3	4.84
4	5.02
5	4.98
6	5.30
7	5.52
8	5.60
9	5.70
10	5.64
11	5.82
12	5.90
13	5.80
14	5.98

Solution:

1. Parameter estimation

Method 1. MATLAB curve fitting toolbox The MATLAB curve fitting toolbox (MATLAB R2012b) has been used to estimate the parameters. The data in Table 13.4 need first to be loaded into the MATLAB Workspace as variables. The following command is then typed in the command window:

>> cftool

A graphic user interface (GUI) opens up (Figure 13.8). The GUI includes four areas: data, fitting, plotting, and results.

In the data area, the name needs to be given in Fit name. *X* data and *Y* data are used to load the data from the MATLAB workspace into the curve fitting toolbox (Figure 13.8). After loading the data, a data plot is generated in the Plot area by the curve fitting toolbox (Figure 13.9). In the fitting area, Custom Equation is selected, and the Michaelis–Menten equation is input with *a* as the maximum volumetric rate and *b* as the Michaelis constant (Figure 13.9).

The results from the curve fitting are as follows:

$$v_{max} = 6.44, \ K_m = 1.20$$

The R-square for the curve fitting is 0.99.

Method 2. Lineweaver–Burk plot A Lineweaver–Burk plot is used to obtain v_{max} and K_m from the intercept ($1/v_{max}$) and slope (K_m/v_{max}) of Equation 13.28 (also see Figure 13.10).

From the plot, two relationships can be obtained:

$$\frac{1}{v_{max}} = 0.15 \text{ and } \frac{K_m}{v_{max}} = 0.22$$

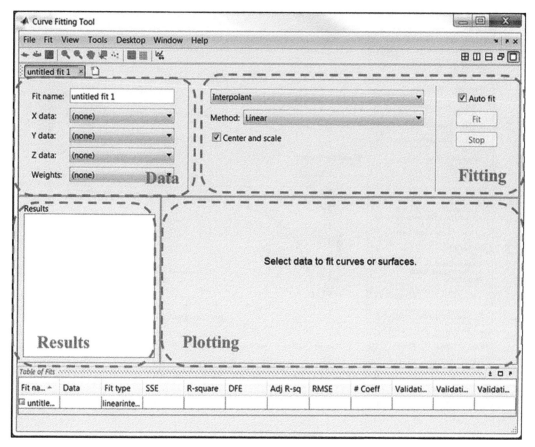

The results from the plot are:

$$v_{max} = 6.66,\ K_m = 1.47$$

R-square is 0.97, which is slightly lower than in the MATLAB method. Therefore, the results from the MATLAB method are used for the analysis of enzyme kinetics.

2. Enzyme kinetics

Substituting the values of v_{max} and K_m from Step 1, and C_{S0} with 10 g/L in Equation 13.27, we have:

$$1.20\ln\frac{C_S}{10} + (C_S - 10) = -6.44t$$

The equation can be simplified to:

$$1.20\ln C_S + C_S = -6.44t + 12.77$$

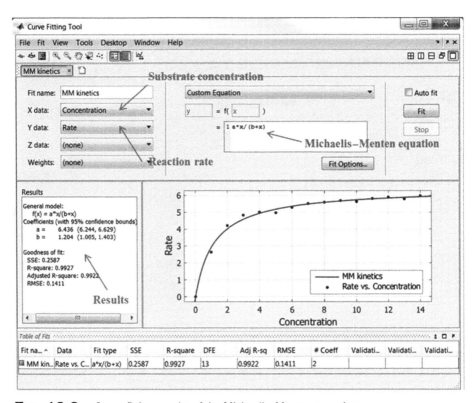

FIG. 13.9 Curve fitting results of the Michaelis–Menten equation.

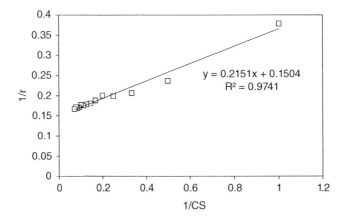

FIG. 13.10 Lineweaver–Burk plot.

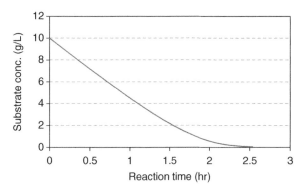

FIG. 13.11   Calculated starch hydrolysis kinetics.

Plotting this equation, the enzyme kinetics for 2.5 hr of hydrolysis is shown in Figure 13.11.

13.3.4 Enzyme Inhibition

Many substances present in the medium inhibit enzymatic reactions. Compounds that reduce the reaction rate are known as inhibitors. There are several mechanisms to describe *enzyme inhibition*. The three major inhibition models, namely competitive, uncompetitive, and mixed inhibition (noncompetitive inhibition), are discussed here. In *competitive inhibition*, an inhibitor competes directly with a substrate for an enzymatic binding site and inhibits the reaction. Competitive inhibitors influence the reaction by reducing the concentration of free enzyme available for substrate binding. In *uncompetitive inhibition*, the inhibitors directly bind to the enzyme–substrate complex instead of the free enzyme, but still prevent catalysis. Uncompetitive inhibitors function by distorting the structure of the active site and making the enzyme inactive. The *mixed inhibition* model describes when an inhibitor can bind to either free enzyme or the enzyme–substrate complex and includes both competitive inhibition and uncompetitive inhibition. The differences among these three inhibitions are presented in Figure 13.12.

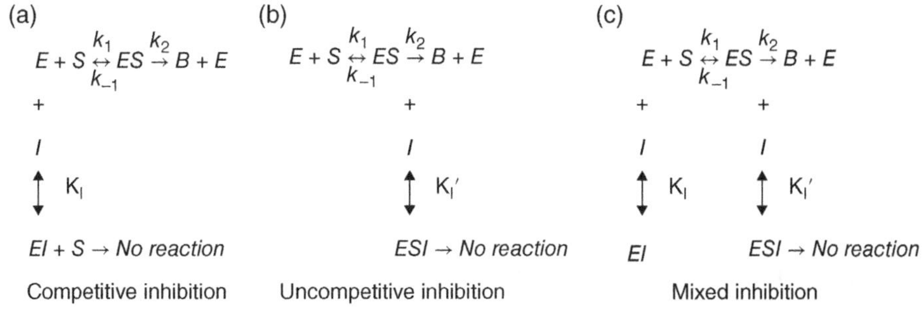

(a)

$$E + S \underset{k_{-1}}{\overset{k_1}{\rightleftarrows}} ES \overset{k_2}{\rightarrow} B + E$$

$+$

I

$\updownarrow K_I$

$EI + S \rightarrow$ No reaction

Competitive inhibition

(b)

$$E + S \underset{k_{-1}}{\overset{k_1}{\rightleftarrows}} ES \overset{k_2}{\rightarrow} B + E$$

$+$

I

$\updownarrow K_I{}'$

$ESI \rightarrow$ No reaction

Uncompetitive inhibition

(c)

$$E + S \underset{k_{-1}}{\overset{k_1}{\rightleftarrows}} ES \overset{k_2}{\rightarrow} B + E$$

$+$ $+$

I I

$\updownarrow K_I$ $\updownarrow K_I{}'$

EI $ESI \rightarrow$ No reaction

Mixed inhibition

FIG. 13.12   Schematics of inhibition models.
Notes: K_I and $K_I{}'$ are dissociation constants for competitive inhibition and uncompetitive inhibition, respectively. EI is the enzyme–inhibitor complex. ESI is the enzyme–substrate–inhibitor complex.

Table 13.5 The effects of inhibitors on the parameters of the Michaelis–Menten equation

Type of inhibition	υ_{max}	K_m
None	υ_{max}	K_m
Competitive	υ_{max}	αK_m
Uncompetitive	υ_{max}/α'	K_m/α'
Mixed	υ_{max}/α'	$\alpha K_m/\alpha'$

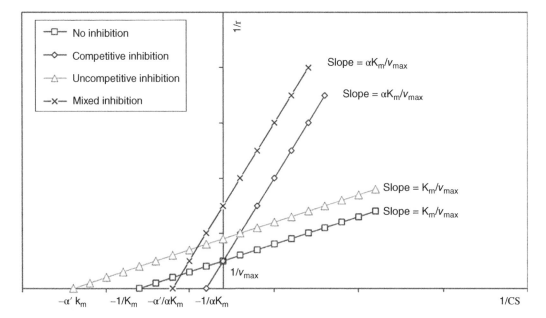

FIG. 13.13 Lineweaver–Burk plot of Michaelis–Menten equations of competitive, uncompetitive, and mixed inhibitions.

The derivation of Michaelis–Menten equations for different types of inhibition is similar to the derivation of the general Michaelis–Menten equation in Section 13.3.3. The expression of the parameters of υ_{max} and K_m is listed in Table 13.5.

In Table 13.5, α and α' are inhibition coefficients for competitive inhibition and uncompetitive inhibition, respectively, and are defined as $\alpha = 1 + \frac{C_I}{K_I}$ and $\alpha' = 1 + \frac{C_I}{K_I'}$; and C_I is the concentration of inhibitor, $K_I = [E][I]/[EI]$ and $K_I' = [ES][I]/[ESI]$.

Differences in Michaelis–Menten equations for competitive, uncompetitive, and mixed inhibitions can be delineated by a Lineweaver–Burk plot (Figure 13.13).

13.4 Enzymatic Hydrolysis of Carbohydrates

13.4.1 Carbohydrate Structure

"Sugar," "carbohydrate," "glycan," and terms containing "-saccharide" or "glyco-" are all labels that are applied to the class of compounds that share a number of related features, and are of significant

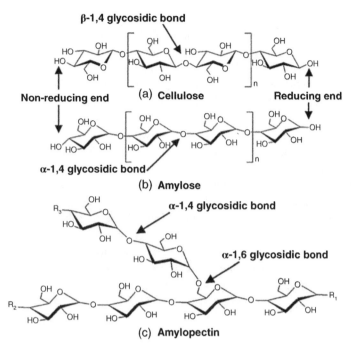

FIG. 13.14   Representative polymers of glucose contained in (a) cellulose, (b) amylose, and (c) amylopectin. Amylose and amylopectin are the two polysaccharide components of starch.

importance for bioenergy and biobased product generation. *Monosaccharides* are the simplest type of carbohydrate and are typically represented by the chemical formula $(CH_2O)_n$ (although exceptions include deoxy sugars and uronic acids), with n ranging from 3 to 7. Monosaccharides, single sugar molecules, can be linked together into polysaccharides by a type of ether linkage known in carbohydrate chemistry as a glycosidic linkage to form glycosides. These orientations are locked into the polymer and determine its structure. Examples of these structures in cellulose (containing only β-1,4 glycosidic bonds) and starch (containing α-1,4 and α-1,6 glycosidic bonds) are presented in Figure 13.14.

The depolymerization of starch and cell wall polysaccharides (cellulose and hemicellulose) requires breaking a glycosidic bond (Figure 13.14). Chemically, one molecule of water is consumed for every glycosidic bond cleaved, because one H+ and one OH- ion are incorporated into the sugar molecule at each hydrolysis point. Consequently, enzymes that catalyze the hydrolysis of a glycosidic bond are known as glycosidases. Since biological depolymerization of starches, cellulose, hemicelluloses, and pectins (among others) are dependent on glycosidases, these are clearly important enzymes. Starch-hydrolyzing enzymes (amylases) are critical for the production of fermentable sugars from starch-based feedstocks, while cellulose-hydrolyzing enzymes (cellulases) are central for the hydrolysis of cellulosic-based feedstocks for monomeric sugar production.

13.4.2 Starch Depolymerization

As covered in Chapter 24, dry grinding is the dominant process for corn-based ethanol production. The two important enzymatic steps during the hydrolysis of starch to fermentable sugars are *liquefaction* and *saccharification*. Prior to treatment with enzymes, the corn is first hammer-milled, mashed, and cooked/heated for starch gelatinization. *Gelatinization* is the process of starch swelling with water at an elevated temperature. The temperature at which this occurs is a function of starch properties such as the degree of branching and solvent properties such as pH. Within the starch granules, the starch contains tightly packed crystalline regions stabilized by H-bonding, which is disrupted by infiltrating water during gelatinization. This has two important outcomes. The first is that the viscosity of the mash is substantially increased and the mash exhibits "shear thickening" behavior. The second important outcome is that water can infiltrate between individual starch polymer chains, which consequently allows hydrolytic enzymes access to more potential sites for hydrolysis.

The liquefaction of starch is often performed in combination with "jet cooking" using bacterial thermophilic α-amylases, or starch-depolymerizing enzymes with optimum activities at temperatures as high as 105 °C due to the improved flowability of gelatinized starch slurries at these temperatures. These enzymes hydrolyze internal α-1,4 glycosidic bonds within the polymer to break the starch into smaller fragments known as dextrins, and significantly decrease the viscosity of the starch slurry. This is followed by a separate hydrolysis stage known as saccharification or the conversion of dextrin oligomers to glucose. The enzymes include glucoamylase, which releases glucose from dextrins by hydrolyzing an α-1,4 bond at the polymer's reducing end, and pullulanases that debranch amylopectin by breaking α-1,6 glycosidic bonds (Figure 13.15). Additionally, protease enzymes are added during the saccharification stage in some dry-grind mills to break down protein from the corn into smaller peptides that can be utilized as a nutrient source by the yeast during fermentation. The saccharification is typically performed concurrently with fermentation and is known as simultaneous saccharification and fermentation (SSF), as discussed in Chapter 14. Advantages of SSF include eliminating the feedback inhibition of glucoamylase by glucose as it is released by the enzymes, decreasing the opportunity for bacterial inhibition, and reducing such capital requirements as the reactor residence time for both saccharification and fermentation is shortened.

An alternative adopted by many dry-grind mills over the last several years is to employ *"granular starch-hydrolyzing enzymes"* during SSF. This approach employs an α-amylase that allows the direct conversion of granular starch during SSF to glucose, eliminating the gelatinization, jet-cooking, and liquefaction steps. Some of the merits include low initial sugar concentrations during SSF, lower energy consumption, and the potential to operate at higher solids loading (resulting in higher final ethanol concentrations), since the solution is not viscosity limited.

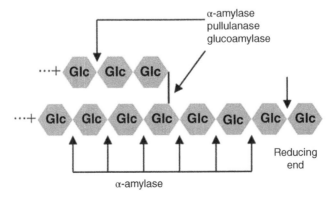

FIG. 13.15 Illustration of enzymes for complete hydrolysis of starch including endo-acting α-amylase, debranching pullulanase, and glucose-liberating glucoamylase.

Commercial starch-hydrolyzing enzymes are typically derived from fungi (e.g., *Aspergillus* spp.) and bacteria (e.g., *Bacillus* spp.). Enzyme loadings of α-amylase and glucoamylase are around 100–200 enzyme units/g corn starch and 120 enzyme units/g corn starch, respectively, for starch ethanol production. These enzymes contribute approximately $0.035 to the cost of a gallon of starch fuel ethanol, which is about 3% of the total production cost (McAloon *et al.*, 2000).

13.4.3 Cellulose Hydrolysis

Many ongoing commercialization efforts for full-scale lignocellulosic biofuel plants plan to employ a biochemical conversion route. Biochemical production of lignocellulosic biofuels is possible at prices cheaper than for thermochemical routes. Although either enzymes or acid can be used for the release of monomeric sugars from cellulosic biomass, the former is the preferred method due to higher sugar yields and lower chemical inputs. Thus, enzymatic hydrolysis is considered to be an integral part of biochemical methods for lignocellulosic biofuel production. A combination of cellulase enzymes (endo-β-1,4-glucanases, exo-β-1,4-glucanases, and β-1,4-glucosidases) is utilized for the hydrolysis of cellulose to monomeric units.

Cellulases

The hydrolysis of cellulose requires a cocktail of many enzymes, collectively referred to as *cellulases*, for the effective conversion of cellulose into fermentable sugars. These are often used in combination with "accessory enzymes," which may include xylanases and pectinases. Virtually all commercial enzyme cocktails for plant cell wall deconstruction are fungal enzymes derived from *Trichoderma* and *Aspergillus* spp., while some cocktails applied primarily in food processing may derive from *Humicola insolens* and *Bacillus* spp. of bacteria (Banerjee *et al.*, 2010).

In nature, *cellulases* are employed as either complexed or non-complexed enzyme systems. Complexed cellulase systems are bound together on a scaffold protein attached to the cell membrane and are collectively referred to as a cellulosome. These systems are found in anaerobic bacteria, including the *Clostridia*. Non-complexed (free enzyme) cellulase systems are found in aerobic and anaerobic fungi and bacteria. In these systems, the enzymes are secreted by the organism into its surroundings, either to generate sugars for uptake by the cell or, in the case of some rumen bacteria, to remove hemicelluloses to gain access to cellulose. Non-complexed fungal and bacterial glycosidases that act on cellulose and hemicelluloses typically contain a catalytic core protein where the hydrolysis reaction takes place and a non-catalytic binding domain consisting of a carbohydrate binding module (CBM). CBMs are necessary to bring the protein glycoside hydrolases into the proximity of the substrate and orienting it for catalysis. CBM binding is thought to disrupt the local H-bonding network in the region of the cellulose surrounding the binding site, allowing the catalytic site access to the polymer, while the introduction of hydrolytic chain breaks results in further disruption.

Enzyme Components of Cellulases

Although only β-1,4 glycosidic bonds are required to be hydrolyzed in order to depolymerize cellulose completely, several distinct classes of enzymes working in synergy are required for complete and efficient depolymerization to glucose. This is due to the crystalline structure of cellulose, whereby the packing of individual polymer chains severely limits the access of enzymes and, unlike starch, cellulose crystallinity is resistant to disruption by water. The enzymes in each of these cellulase classes breaking a β-1,4 glycosidic bond include exo-β-1,4-glucanases or exoglucanases (also known as cellobiohydrolases because they release cellobiose); endo-β-1,4-glucanases or endoglucanases;

β-glucosidases; and the recently recognized "lytic polysaccharide monooxygenases." A generalized overview of these enzymes acting synergistically for cellulose hydrolysis to monomeric glucose is given in Figure 13.16.

Exo-β-1,4-glucanases are "exo" acting because they work at the ends of the polymer. Individual cellulose polymers are directional because they have a distinct "reducing" end and a "non-reducing" polysaccharide end that influences their reactivity to hydrolytic enzymes. These exoglucanases are also known as cellobiohydrolases because they release cellobiose (a disaccharide) from the chain end (Figure 13.16). In the filamentous fungus *T. reesei*, there are two major exoglucanases that are active on either a reducing or a non-reducing end of cellulose. The enzyme specific for cellulose reducing ends is known as cellobiohydrolase I (CBHI). The enzyme responsible for acting on non-reducing ends is known as cellobiohydrolase II (CBHII). These enzymes are also processive, which means that they work their way down a chain releasing cellobiose. Cellobiose released by these enzymes can be hydrolyzed to glucose by β-glucosidase.

Endoglucanases are so called because they cleave a cellulose chain in the middle of a polymer. These "nicks" within the cellulose chain generated by the action of endoglucanases result in the formation of a new reducing end and a non-reducing end with every hydrolytic event, which in turn are able to serve as binding sites for exoglucanases. As such, the cooperative combination of endo-acting and exo-acting enzymes causes a synergetic effect that results in higher overall hydrolysis rates.

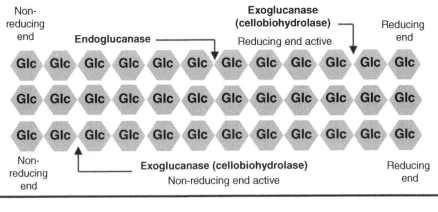

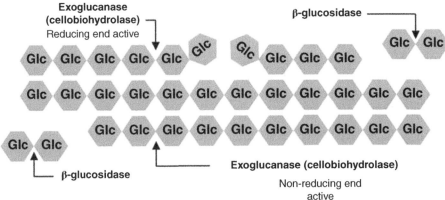

FIG. 13.16   Illustration of enzymes for complete hydrolysis of crystalline cellulose by cellulases showing cooperativity and synergy between endo- and exo-acting hydrolases.

"Lytic polysaccharide monooxygenase" (LPMO) is a recently identified enzyme that is an important component of many fungal and bacterial cellulolytic enzyme systems. This enzyme is not a glycosidase, but rather belongs to the general enzyme class known as oxidoreductases. It introduces an oxidative chain break comparable to an endo-β-1,4-glucanase, with the difference that cellulose is oxidized to form a δ-gluconolactone or gluconic acid. Recent optimization of LPMO enzyme activity in commercial enzyme cocktails has resulted in substantial improvements in decreasing the protein loadings required to achieve comparable yields for a target hydrolysis.

13.4.4 Hemicellulose Hydrolysis

Besides cellulose hydrolysis, the hydrolysis of hemicelluloses may be necessary for some pretreatments that retain a significant fraction of the hemicellulose in polymeric or oligomeric form. In particular, hemicelluloses are known to form strong associations with cellulose microfibril surfaces and to block the enzyme's access to the cellulose. Thus, the supplementation of hemicellulases results in improved cellulose hydrolysis. One challenge to the hydrolysis of hemicelluloses is that not only do these polysaccharides comprise several classes of polymers (e.g., glucomannans, xylans, xyloglucans, and β-glucans), but each of these polymers requires several enzymes for complete depolymerization. For example, the complete depolymerization of glucuronoarabinoxylan in energy crops may require as many as five individual enzyme activities: acetyl esterase, endoxylanase, α-glucuronidase, arabinosylfuranosidase, and feruloyl esterases. Because the relative abundance of the hemicelluloses varies depending on the plant type (softwoods, hardwoods, forbs, or grasses) and the pretreatment methods, the optimal enzyme cocktail for hemicellulose depolymerization varies.

13.4.5 Key Factors Affecting the Enzymatic Hydrolysis of Lignocellulosic Feedstocks

The *enzymatic hydrolysis* of lignocellulosic biomass could be affected by various factors such as the physical properties of the substrate (composition, crystallinity, and degree of polymerization), enzyme synergy (origin and composition), mass transfer (substrate adsorption and bulk and pore diffusion), and intrinsic kinetics. It is widely accepted that the most important biomass properties influencing its enzymatic hydrolysis are the degree of crystallinity and the nature and amount of associated constituents (lignin and hemicelluloses), which make lignocellulose virtually water insoluble. Like the hydrolysis of gelatinized starch, the hydrolysis of pretreated biomass undergoes liquefaction as insoluble cell wall polymers (cellulose and hemicellulose) are converted into soluble monomers and oligomers. This liquefaction results in a substantial decrease in the water-insoluble solids content as hydrolysis proceeds. However, unlike starch liquefaction where the gelling capacity of the polymer is disrupted by its depolymerization, the liquefaction of pretreated biomass is primarily due to the solubilization of insoluble cell wall polysaccharides. The solubilization requires water to remove polysaccharides from the cell, which presents a problem in achieving a high sugar concentration. High sugar concentrations need the hydrolysis to be performed at a high solid content, while high solid hydrolysis results in little free water in the reaction for solubilizing polysaccharides and reduces hydrolysis efficiency.

Besides water insolubility, a number of additional factors that influence the rate and extent of cellulose hydrolysis are related to the complicated structures of the plant cell wall. These can be considered factors that relate primarily to the accessibility of the cellulose to hydrolytic enzymes. At the polymer level, cellulose crystallinity effectively limits the accessibility of cellulose polymers relative to other amorphous (non-crystalline) polysaccharides. This is a consequence of multiple cellulose polymers packed into a tight crystalline microfibrillar structure that excludes water and enzymes from internally packed cellulose chains. Hemicelluloses such as xylans and glucomannans are also known to sheath the surfaces of cellulose microfibrils and further limit access to the cellulose. One of the most important limiting factors is lignin, which acts as a highly effective barrier to exclude water from the cell wall and limit access to the polysaccharides. In addition to its role in excluding enzymes from penetrating into the cell wall, lignin (as well as other polyphenolic compounds such as tannins) is very effective at adsorbing and precipitating proteins, including cellulolytic enzymes.

Because of these structural barriers, cellulose hydrolysis requires significantly higher enzyme loadings compared to starch hydrolysis. For example, starch–ethanol processes typically employ total enzyme loadings (α-amylase and glucoamylase) of 0.7–1.2 mg protein per g of starch. By comparison, pretreated lignocellulose requires total cellulase enzyme loadings of 5.0–20 mg protein per g of cellulose to achieve the expected sugar yields within a reasonable time, and, as such, represent a large contribution to the costs of cellulosic biofuels. It has been reported that lignocellulose enzymes contribute approximately \$0.30–0.50 per gallon of cellulosic fuel ethanol, which is about 10% of the total production cost (McMillan *et al.*, 2011).

Example 13.3: An enzymatic hydrolysis was performed beginning with 100 g of pretreated lignocellulose per kg slurry with an initial cellulose content of 52% by mass. After 48 hours, the insoluble solids content reaches 62.5 g per kg slurry and the cellulose content of the insoluble solids is reduced to 23.5%. Calculate the estimated cellulose conversion (i.e., the percentage of the cellulose that is reacted).

Solution:

1. **The cellulose conversion ratio** is $\dfrac{Cellulose\ Reacted}{Initial\ Cellulose}$ or $\dfrac{Initial\ Cellulose - Final\ Cellulose}{Initial\ Cellulose}$ or $1 - \dfrac{Final\ Cellulose}{Initial\ Cellulose}$ Both terms in this expression need to be determined.

2. **Calculate the initial cellulose concentration** Initial cellulose = 100 $\dfrac{g\ pretreated\ biomass}{kg\ slurry} \cdot 0.52 \dfrac{g\ cellulose}{g\ pretreated\ biomass} = 52 \dfrac{g\ cellulose}{kg\ slurry}$

3. **Determine the cellulose concentration after 48 hr** Final cellulose = $62.5 \dfrac{g\ insoluble\ solids}{kg\ slurry} \cdot 0.235 \dfrac{g\ cellulose}{g\ insoluble\ solids} = 14.7 \dfrac{g\ cellulose}{kg\ slurry}$

4. **Calculate the cellulose conversion** $1 - \dfrac{14.7}{52} = 0.718$ or **72%**

References

Banerjee, G., Scott-Craig, J.S., and Walton, J.D. (2010). Improving enzymes for biomass conversion: A basic research perspective. *BioEnergy Research* 3(1): 82–92.

Eyring, H. (1935). The activated complex in chemical reactions. *Journal of Chemical Physics* 3(2): 107–15.

Faber, K. (1997). *Biotransformations in Organic Chemistry: A Textbook*. Berlin: Springer.

McAloon, A., Taylor, F., Yee, W., Ibsen, K., and Wooley, R. (2000). *Determining the Cost of Producing Ethanol from Corn Starch and Lignocellulosic Feedstocks*. Springfield, VA: U.S. Department of Agriculture/U.S. Department of Energy.

McMillan, J.D., Jennings, E.W., Mohagheghi, A., and Zuccarello, M. (2011). Comparative performance of precommercial cellulases hydrolyzing pretreated corn stover. *Biotechnology for Biofuels* 4: 29.

Michaelis, L., and Menten, M.L. (1913). The kenetics of the inversion effect. *Biochemische Zeitschrift* 49: 333–69.

Seetharaman, K., and Bertoft, E. (2012). Perspectives on the history of research on starch. Part II: On the discovery of the constitution of diastase. *Starch-Starke* 64(10): 765–9.

Sumner, J.B. (1926). The isolation and crystallization of the enzyme urease. Preliminary paper. *Journal of Biological Chemistry* 69(2): 435–41.

Voet, D., and Voet, J. (2013). *Biochemistry*, 4th edn. Chichester: John Wiley & Sons Ltd.

Exercise Problems

13.1. What are hydrolases? Describe the catalytic function of three types of hydrolases. In which types of conversions are hydrolases employed for biofuels processes?

13.2. What is the transition state theory?

13.3. Discuss the relationship between Michaelis–Menten kinetics, zero-order kinetics, and first-order kinetics.

13.4. The initial rate of reaction for the enzymatic hydrolysis of deoxyribose-phosphate was measured as a function of initial substrate concentration. The substrate concentration and initial reaction rate of enzymatic hydrolysis of deoxyribose-phosphate are as in Table 13.6.

Table 13.6 Initial reaction rate and substrate concentration

Substrate concentration (μmol/L)	Initial reaction rate (μmol/(L·min))
6.6	0.30
3.4	0.24
1.7	0.15

(a) Estimate the Michaelis–Menten kinetic parameters for this reaction.

(b) When an inhibitor was added, the initial reaction rate decreased. The substrate and inhibitor concentrations and initial reaction rate of enzymatic hydrolysis of deoxyribose-phosphate are presented in Table 13.7. Is this an example of competitive, non-competitive, or mixed inhibition? Use a graph to justify your answer.

Table 13.7 Initial reaction rate, inhibitor concentration, and substrate concentration

Substrate concentration (μmol/L)	Inhibitor concentration (μmol/L)	Initial reaction rate (μmol/(L·min))
6.6	145	0.10
3.4	145	0.07
1.7	145	0.05

13.5. An enzymatic hydrolysis of native corn starch was performed in order to quantify the hydrolysis kinetics of an amylase from a fungus, *Aspergillus oryzae*. The starch concentration (g/L) and volumetric reaction rate (g/(L·hr)) were measured and are presented in Table 13.8.

(a) Estimate the values for the maximum volumetric reaction rate (g/(L·hr)) and the Michaelis constant (g/L).

(b) Demonstrate the time course of substrate concentration change in a time interval of 1 hr with an initial substrate concentration of 10 g/L.

Table 13.8 Volumetric reaction rate and substrate concentration

Substrate concentration (C_A) (g/L)	Volumetric reaction rate (r) (g/(L·hr))
0	0
1	1.3
2	2.0
3	2.4
4	2.4
5	2.5
6	2.6
7	2.8
8	2.8
9	2.8
10	2.8
11	2.9
12	3.0
13	2.9
14	3.0

13.6. A reaction was carried out by immobilized enzyme in a 5 L bioreactor. The inlet substrate concentration was 100 mol/L and the flow rate was set at 1 L/hr. After reaching a steady state, the outlet substrate concentration was 10 mol/L.

(a) What is the reaction rate in the reactor?

(b) The steady-state outlet substrate concentration was measured as a function of the inlet flow rate. The flow rate and outlet substrate concentration of the bioreactor were as in Table 13.9. Estimate Michaelis–Menten kinetic parameters by using a plotting technique.

Table 13.9 Outlet substrate concentration and flow rate

Flow rate (L/hr)	Outlet substrate concentration (mol/L)
0.7	4.0
0.8	7.0
1.0	10.0
1.2	14.0

13.7. The initial reaction rate of acetylcholine hydrolysis by an animal serum (enzyme source) was measured with and without prostigmine (inhibitor). The inhibitor concentration is 1.5×10^{-7} mol/L. The substrate concentration and initial reaction rate (mol/(L·min)) with and without inhibitor were as listed in Table 13.10. Determine the type of inhibition for prostigmine, and calculate the Michaelis–Menten kinetic parameters in the presence of prostigmine.

Table 13.10 Initial reaction rate and substrate concentration

	Initial reaction rate (mol/(L·min))	
Substrate concentration (µmol/L)	Without prostigmine	With prostigmine
0.0032	0.111	0.059
0.0049	0.148	0.071
0.0062	0.143	0.091
0.0080	0.166	0.111
0.0095	0.200	0.125

13.8. Describe the difference in complexed versus non-complexed enzyme systems for polysaccharide depolymerization.

13.9. Describe what is meant by "synergy" between cellulose-hydrolyzing enzymes.

13.10. In what ways to do the higher-order structure of lignocellulosic feedstocks affect the depolymerization of polysaccharides contained within the cell walls of these feedstocks?

13.11. If 100 g of a pretreated lignocellulose has a mass composition of 35% lignin, 40% cellulose, 15% xylan, and 10% "other," what is the composition of the solids following hydrolysis conversion of 75% of the cellulose and 80% of the xylan? Assume that the "other" fraction is unaffected by the hydrolysis.

Ethanol Fermentation

Saoharit Nitayavardhana and Samir Kumar Khanal

What is included in this chapter?

This chapter covers biochemical pathways for ethanol production from both hexose and pentose sugars. Microbial cultures and factors affecting ethanol fermentation are discussed. Current ethanol fermentation and recovery technologies are also included.

14.1 Introduction

The art of *fermentation* has been known to humans since the Neolithic (New Stone) era (~10,000–2,000 BC). Fermented food, beer, and wine are quite possibly the earliest examples where fermentation was deliberately (or perhaps inadvertently) applied. In ancient Egypt, Mesopotamia, China, Greece, and Rome, alcoholic beverages were commonly used for palliative care as well as in religious rituals and celebration. The earliest archeological evidence of fermentation dates back 9,000 years from alcoholic beverage residue found in pottery in China's Yellow River Valley. The oldest known wine vessel, 7,500 years old, was discovered in Hajji Firuz, Iran. Exactly how the concept of fermentation entered human society remains a mystery, but the methods and techniques for producing fermented beverages were broadly similar worldwide. In the modern era, fermentation is used extensively for the production of bread, various dairy products, pickles, tempeh, kimchi, sauerkraut, vinegar, and, of course, a plethora of enjoyable beverages. At an industrial scale, vitamins (B2, B12), proteins, nutraceuticals, pharmaceuticals, and enzymes, among many others, are all produced by specialized fermentation techniques.

Bioenergy: Principles and Applications, First Edition. Edited by Yebo Li and Samir Kumar Khanal.
© 2017 John Wiley & Sons, Inc. Published 2017 by John Wiley & Sons, Inc.
Companion website: www.wiley.com/go/Li/Bioenergy

What is fermentation?

Fermentation is the biochemical transformation of organic substances into simpler compounds by the action of enzymes produced by microorganisms. The word "fermentation" is derived from the Latin word meaning "to boil," since the bubbling and foaming of early fermented beverages closely resembled boiling. French microbiologist Louis Pasteur (Figure 14.1a) is credited for his contributions in helping to understand the fundamental principles governing fermentation. He was the first to point out that fermentation is caused by microbial growth. He carried out the fermentation experiment using the set-up similar to the one shown in Figure 14.1b. He further showed that bacterial contamination results in wine/beer spoilage. Soon after, he proposed heat treatment to decontaminate undesirable microbes, known as *pasteurization*, which is a well-established method for sterilization.

(a)   (b)

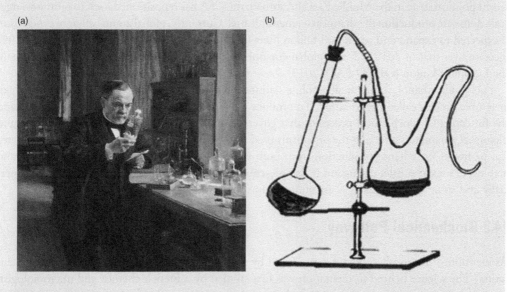

FIG. 14.1   (a) Louis Pasteur, 1885, and (b) his experiment to produce beer. Source: https://en. wikipedia.org/wiki/Louis_Pasteur#/media/File:Albert_Edelfelt_-_Louis_Pasteur_-_1885.jpg. Public domain.

The use of ethanol (consumable alcohol) as a fuel source has a long and interesting history. From as early as the 1850s, ethanol was used to light street and household lamps across the USA. The practice was short-lived, however, as federal taxes levied on industrial alcohol during the Civil War drove up prices, consequently paving the way for cheaper petroleum-based alternatives. Not long after, in 1906, ethanol taxes were reduced and the very first production-line-assembled automobile, the Model T, was introduced in 1908 by the Ford Motor Company, using ethanol to power its internal combustion engine. As automobile production increased, interest began to grow in a cheaper fossil-derived fuel source generated as waste in oil refineries. The product, named gasoline, was a hydrophobic, energy-dense liquid that could be piped over great distances without absorbing water, consequently rendering obsolete the expensive rail system that connected the nation. With the advent of better processing and improved infrastructure, the lower cost of gasoline displaced ethanol as the primary fuel for transportation, and for decades the USA and the world have exploited use of the finite resource.

Throughout the 19th century, the USA was a relatively young country, rich in oil and coal buried beneath its soil. In the years leading up to the 1970s, however, the production of oil (and subsequently gasoline) started to peak, and the USA began to import fossil fuels from foreign nations to meet its increasing oil demands. After heated international conflicts, the USA found itself with a scarcity of transportation fuel during the Oil Embargo of 1973. Ethanol, a domestically produced resource, began to regain appeal, and significant research efforts were launched on its production and development as a renewable fuel. The US federal government made significant efforts to energize the widespread use of ethanol as a fuel alternative to gasoline through the implementation and enactment of campaigns, laws, incentives, and mandates. Globally, the USA and Brazil worked to establish some of the very first models for successfully incorporating ethanol into the transportation sector. During the 1980s, interest in biofuels (primarily ethanol) declined again in response to a fall in gasoline prices, but, following the recent political strife in the Middle East and abroad, the USA has repositioned itself to promote large-scale domestic production of ethanol as a renewable fuel. Currently, global annual ethanol production is expected to expand and reach 168 billion liters (44.4 billion gallons) by 2022, more than a 70% increase over that of 2010–12, and global ethanol markets are dominated by the USA, Brazil, and the European Union (OECD-FAO, 2013).

Fuel-grade ethanol, found at all gasoline stations today, can be produced biologically through an anaerobic process called *fermentation*. To produce ethanol efficiently, it is important to understand the fundamental biochemical processes that govern fermentation, and the physical, chemical, and biological parameters affecting the organisms used in the process. This chapter covers the biochemical pathways for ethanol production from both six-carbon (glucose) and five-carbon (xylose and arabinose) sugars. A brief discussion is also presented on the various microorganisms used commercially and some of the challenges encountered in industrial scale-up.

14.2 Biochemical Pathway

As mentioned previously, ethanol is produced by means of a biological process known as fermentation. The science behind fermentation was first studied by a French chemist and microbiologist, Louis Pasteur, who discovered that alcohol generation was the result of microbial activity in which organic substances were metabolized by microbes in the absence of oxygen. More specifically, ethanol (i.e., biofuel) is generated as the final product from the fermentation of soluble sugar molecules.

Ethanol fermentation is one of the anaerobic fermentation processes that naturally break down organic matter (i.e., sugars) under anaerobic conditions. Therefore, the process does not require oxygen. However, it is important to understand that ethanol can be produced under both anaerobic and anoxic (oxygen-limiting) conditions.

Anaerobic processes

Anaerobic processes are classified as either anaerobic fermentation or anaerobic respiration (Figure 14.2), depending on the type of electron acceptor(s). Anaerobic respiration requires external electron acceptors in the electron-transport chain, whereas organic matter acts as both electron donor and acceptor in anaerobic fermentation. In anaerobic respiration, energy is produced through both substrate-level phosphorylation and oxidative phosphorylation when electrons are transferred in the electron-transport chain. In anaerobic fermentation, substrate-level phosphorylation, in which ATP (adenosine triphosphate) is produced by transferring a high-energy phosphate bond from high-energy intermediates to ADP (adenosine diphosphate), is the major energy-generating process. Common anaerobic fermentation processes are lactic acid production and ethanol production. Examples of anaerobic respiration processes include denitrification (NO_3^- to N_2), sulfate reduction (SO_4^{2-} to H_2S), and methanogenesis (CO_2 to CH_4).

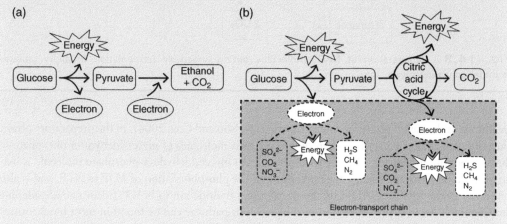

FIG. 14.2   (a) Anaerobic fermentation of glucose to ethanol, (b) anaerobic respiration of glucose.

14.2.1 Hexose Fermentation

Although the overall process of ethanol production is commonly called fermentation, it involves two major metabolic pathways, glycolysis and fermentation. *Glycolysis* is a near-universal carbohydrate catabolic pathway that occurs with or without the presence of oxygen. Hexose is degraded into pyruvate via glycolysis. The pyruvate is further degraded into ethanol and carbon dioxide–regenerating coenzyme (NAD^+) for glycolysis (Figure 14.3).

Glycolysis

Glycolysis is a ubiquitous catabolic process in which the glucose molecule is broken down to pyruvates while conserving energy as ATP and NADH. The term *glycolysis* is derived from the Greek words *glykus*, "sweet," and *lysis*, "loosening." It is a series of enzymatic reactions, 10 successive steps that occur under both aerobic and anaerobic conditions. The general equation of this life-sustaining process is as follows:

$$\text{Glucose} + 2NAD^+ + 2ADP + 2P_i \rightarrow 2\text{Pyruvate} + 2NADH + 2H^+ + 2ATP + 2H_2O$$

$$(14.1)$$

In glycolysis, one molecule of glucose is converted into two molecules of pyruvate with the simultaneous generation of two ATPs. The details of glucose catabolism in glycolysis can be illustrated in 10 steps. Steps 1–5 are known as the preparatory phase (or energy investment), whereas Steps 6–10

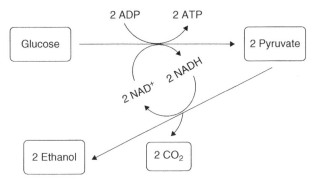

FIG. 14.3 Regeneration of NAD+ molecules during anaerobic fermentation to allow glucose degradation via glycolysis.

are known as the payoff phase (or energy recovery; Nelson and Cox, 2004). In the preparatory phase, glucose (a six-carbon sugar) is broken down into two molecules of glyceraldehyde-3-phosphate (a three-carbon intermediate). In the payoff phase, each glyceraldehyde-3-phosphate molecule is oxidized, yielding pyruvate. Energy is gained through the phosphorylation of ADP to ATP, and is also conserved in the form of NADH by the transfer of a hydride ion to NAD+ (nicotinamide adenine dinucleotide). An in-depth discussion of the glycolysis pathway can be found in most biochemistry textbooks. In brief, the 10 successive steps are illustrated in this section.

Step 1: Phosphorylation of glucose. The first irreversible reaction starts when glucose enters the cell.

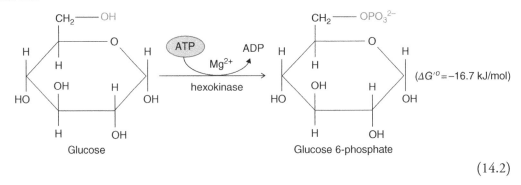

$$(14.2)$$

Step 2: Conversion of glucose 6-phosphate to fructose 6-phosphate

$$(14.3)$$

Step 3: Phosphorylation of fructose 6-phosphate to fructose 1,6-phosphate

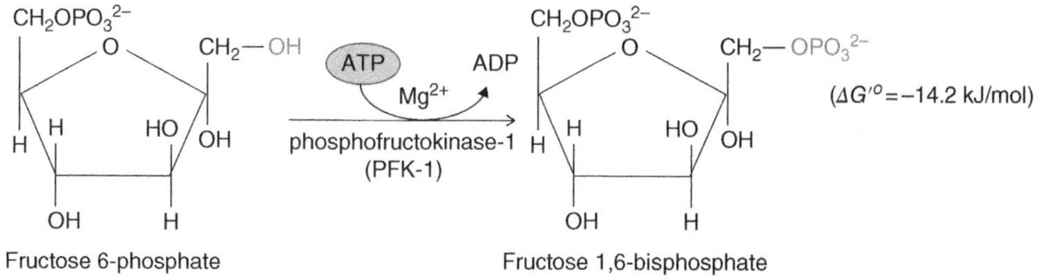

Fructose 6-phosphate

Fructose 1,6-bisphosphate

$(\Delta G'^O = -14.2 \text{ kJ/mol})$

$$(14.4)$$

Step 4: Cleavage of fructose 1,6-bisphophate. In this "lysis" step, the six-carbon molecule is cleaved to yield two different three-carbon molecules, giving its pathway the name "glycolysis."

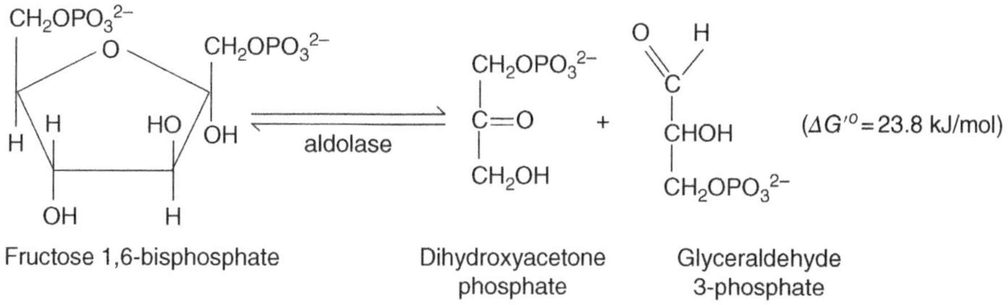

Fructose 1,6-bisphosphate

Dihydroxyacetone phosphate

Glyceraldehyde 3-phosphate

$(\Delta G'^O = 23.8 \text{ kJ/mol})$

$$(14.5)$$

Step 5: Interconversion of the triose phosphates. The dihydroxyacetone phosphate is rapidly and reversibly isomerized to glyceraldehyde 3-phosphate, because only glyceraldehyde 3-phosphate continues through the glycolysis pathway.

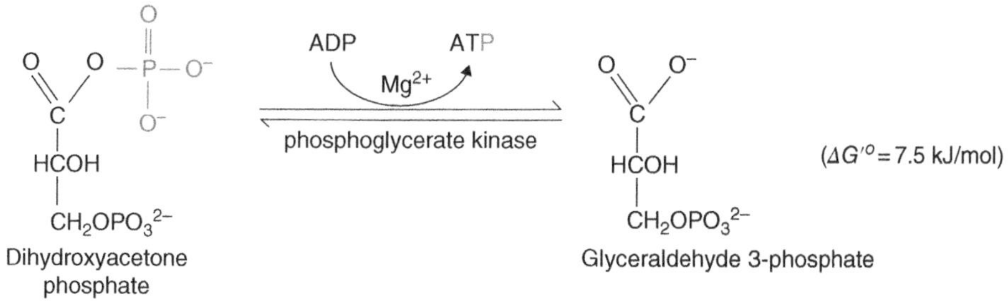

Dihydroxyacetone phosphate

Glyceraldehyde 3-phosphate

$(\Delta G'^O = 7.5 \text{ kJ/mol})$

$$(14.6)$$

In summary, the preparatory phase of glycolysis allows the phosphorylation of the glucose molecule at C-1 and C-6. The phosphorylated molecule then splits into two molecules of glyceraldehyde 3-phosphate. Two molecules of ATP are invested in the phosphorylation reactions. Thus, this stage is also known as energy investment.

The next five steps are part of the payoff phase (energy recovery), where each molecule of glyceraldehyde 3-phosphate is oxidized to pyruvate, and the energy is conserved in ATP and NADH.

Step 6: Oxidation and phosphorylation of glyceraldehyde 3-phosphate to 1,3-bisphosphoglycerate. Energy is conserved in NADH by transferring the hydride ion to NAD⁺, and it is also conserved in the acyl phosphate group at C-1 of 1,3-bisphosphoglycerate.

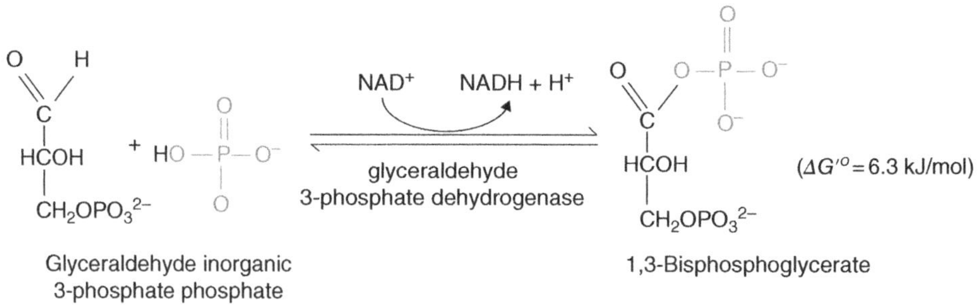

Glyceraldehyde inorganic
3-phosphate phosphate

glyceraldehyde
3-phosphate dehydrogenase

1,3-Bisphosphoglycerate

$(\Delta G'^{o} = 6.3 \text{ kJ/mol})$

$$(14.7)$$

Step 7: Transfer of the phosphoryl group from 1,3-bisphosphoglycerate to ADP. This step is referred to as substrate-level phosphorylation, in which ATP formation takes place through the phosphoryl group transfer from a substrate.

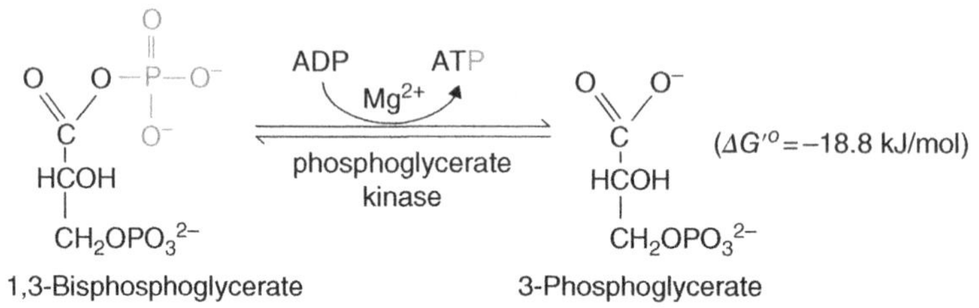

1,3-Bisphosphoglycerate

phosphoglycerate
kinase

3-Phosphoglycerate

$(\Delta G'^{o} = -18.8 \text{ kJ/mol})$

$$(14.8)$$

Steps 6 and 7 together are an energy-coupling process. The energy released during the oxidation of the aldehyde to the carboxylate group is conserved by the formation of ATP from ADP and P_i.

Step 8: Conversion of 3-phosphoglycerate to 2-phosphoglycerate

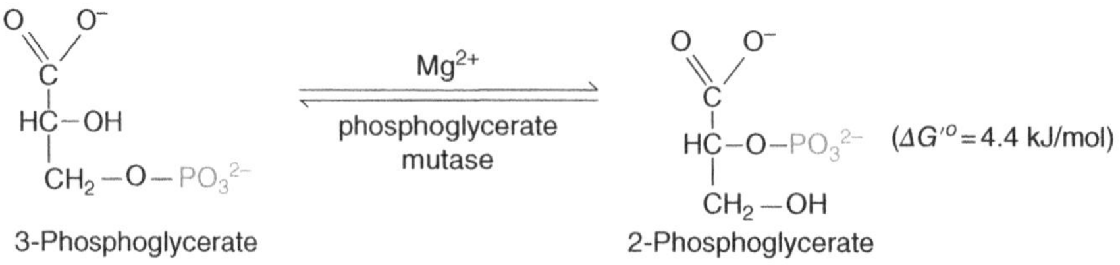

3-Phosphoglycerate

phosphoglycerate
mutase

2-Phosphoglycerate

$(\Delta G'^{o} = 4.4 \text{ kJ/mol})$

$$(14.9)$$

Step 9: Dehydration of 2-phosphoglycerate

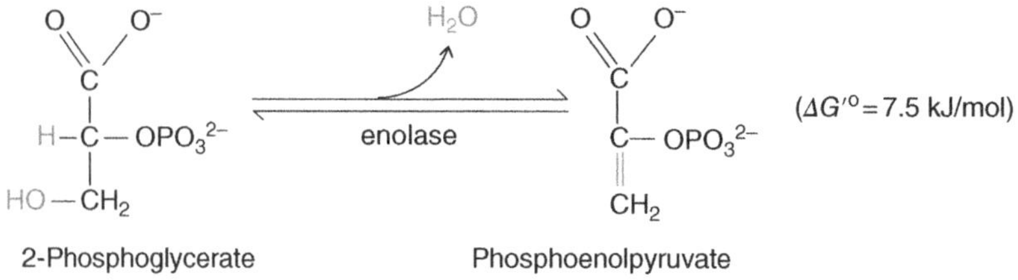

2-Phosphoglycerate

enolase

Phosphoenolpyruvate

$(\Delta G'^{o} = 7.5 \text{ kJ/mol})$

$$(14.10)$$

Step 10: Phosphoryl transfer from phosphoenolpyruvate to ADP. This final step is substrate-level phosphorylation to generate ATP.

$$\text{Phosphoenolpyruvate} \xrightarrow[\text{pyruvate kinase}]{\substack{\text{ADP} \quad\quad \text{ATP} \\ \text{Mg}^{2+}, \text{K}^+}} \text{Pyruvate} \quad (\Delta G'^0 = -31.4 \text{ kJ/mol}) \qquad (14.11)$$

In the payoff phase of glycolysis, two molecules of pyruvate (obtained from one molecule of glucose) yield four molecules of ATP and two molecules of NADH. Because two molecules of ATP are invested in the preparatory phase, the net ATP derived from the complete degradation of one glucose molecule via glycolysis is 2, as shown in Equation 14.1.

Three major chemical reactions in glycolysis

- Degradation of glucose (six-carbon sugar) to pyruvate (three-carbon molecule)
- Phosphorylation of ADP to ATP by high-energy phosphate compounds
- Hydride ion transfer to NAD^+ to form NADH

In normal cells under aerobic conditions, pyruvate is further oxidized to acetate. Acetate then enters the citric acid cycle (also known as the tricarboxylic acid [TCA] cycle or the Krebs cycle) and is further oxidized to CO_2 and H_2O. The reduced electron carrier NADH, produced during glucose degradation, is reoxidized to NAD^+ by transferring its electrons to O_2 (the terminal electron acceptor) in the aerobic electron transport chain (Equation 14.12). Recall that NAD^+ acts as an electron acceptor in glycolysis, allowing for the completion of the next cycle of glycolysis.

$$2NADH + 2H^+ + 2O_2 \rightarrow 2NAD^+ + 2H_2O \qquad (14.12)$$

Failure to regenerate NAD^+ could cease glycolysis (and the energy-generating process critical to sustaining life). Therefore, under anaerobic conditions, organisms must regenerate NAD^+ by other means due to the lack of O_2 as the terminal electron acceptor in the electron transport chain. Under an anaerobic environment, yeasts implement fermentation pathways for regenerating NAD^+ (Figure 14.3). In ethanol fermentation, yeasts transfer electrons from NADH to an internal electron acceptor (an organic intermediate compound) to form a reduced end product, ethanol, and NAD^+ is regenerated in the fermentation process.

Ethanol Fermentation

On closer inspection of Figure 14.3, ethanol fermentation involves a two-step enzymatic reaction, as in Equation 14.13:

$$\text{Pyruvate} \xrightarrow[\substack{\text{pyruvate} \\ \text{decarboxylase}}]{CO_2} \text{Acetaldehyde} \xrightarrow[\substack{\text{alcohol} \\ \text{dehydrogenase}}]{\substack{\text{NADH} \quad \text{NAD}^+}} \text{Ethanol} \qquad (14.13)$$

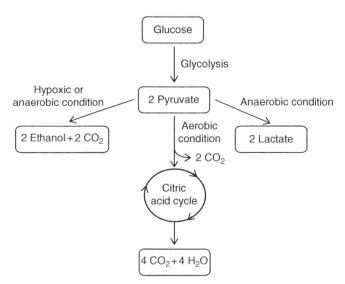

FIG. 14.4 Fates of pyruvate under aerobic and anaerobic conditions. Courtesy of Saoharit Nitayavardhana, Chiang Mai University.

Pyruvate is first decarboxylated to acetaldehyde, and acetaldehyde is then reduced to ethanol by NADH, consequently regenerating NAD^+ for glycolysis (Figure 14.3). The overall equation for ethanol production from glucose is given by Equation 14.14:

$$Glucose + 2ADP + 2P_i \rightarrow 2Ethanol + 2CO_2 + 2ATP + 2H_2O \tag{14.14}$$

Aerobically, a complete degradation of glucose by glycolysis follows the citric acid cycle (Kreb cycle), with oxygen as a terminal electron acceptor in the electron-transport chain, which results in the generation of a significant amount of energy (ATPs). The glucose molecule is completely oxidized to CO_2 and H_2O, as shown by Equation 14.15:

$$Glucose + 6O_2 + 36ADP + 36P_i \rightarrow 6CO_2 + 36ATP + 6H_2O \tag{14.15}$$

The energy generated by aerobic respiration is considerably higher (36 ATPs) than that generated by anaerobic fermentation (2 ATPs). Therefore, in the presence of oxygen, glucose is broken down to pyruvate via glycolysis, which then enters the citric acid cycle, oxidizing pyruvate to CO_2 and H_2O. Aerobically, yeast utilizes the excess energy (ATP) to produce more yeast cells. The metabolic pathway switches to anaerobic fermentation when dissolved oxygen is exhausted. From this point, yeast cells start to ferment sugars into ethanol.

It is important to mention that pyruvate, the product of glycolysis, can also go through a different fermentation route to produce lactic acid (Equation 14.16). This is common in active skeletal muscles when the oxygen supply is insufficient or in cells that have no mitochondria (erythrocytes) for cellular respiration. The regeneration of NAD^+ is critical to allow for the perpetuation of the glycolysis pathway (as already mentioned). Figure 14.4 summarizes the three different possible catabolic fates of pyruvate. The overall steps for ethanol production in yeast cell are illustrated in Figure 14.5.

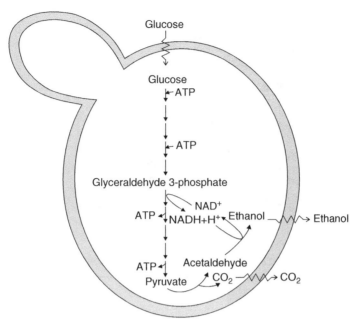

FIG. 14.5   Overall processes for ethanol production in a yeast cell.

$$(14.16)$$

From the equations already given, carbon dioxide (CO_2) is co-generated during anaerobic fermentation of sugars. Theoretically, one mole of CO_2 is generated per mole of ethanol produced. The CO_2 in the gaseous phase is commonly recovered, and can be used in the food and beverage industries.

Rapid consumption of glucose by yeast, *Saccharomyces cerevisiae*, with simultaneous production of ethanol during batch fermentation, is shown in Figure 14.6. The initial slow uptake of glucose is mainly due to the acclimation period of yeast in the fermentation media. Once acclimatized, yeast start to ferment the glucose rapidly, with almost exponential growth of yeast until the glucose is exhausted from the fermentation broth. Although ethanol is the major product of sugar fermentation, at the end of the fermentation period when the yeast cells reach the stationary growth phase, generation of other undesirable products such glycerol, and acetic and lactic acids, has been found during batch fermentation. During batch fermentation, ethanol concentration reaches as high as 13% (v/v). At higher ethanol concentrations, yeasts are inhibited by osmotic pressure.

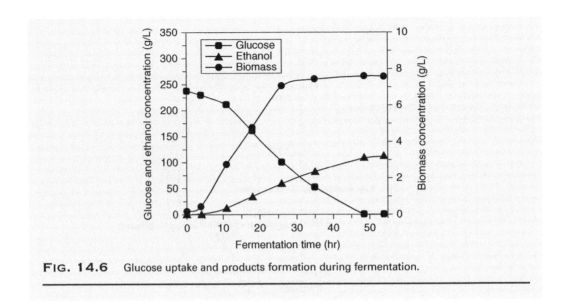

FIG. 14.6   Glucose uptake and products formation during fermentation.

14.2.2 Pentose Fermentation

As mentioned earlier, glycolysis is capable of degrading glucose (C-6 sugar) or other sugars that produce any of its intermediates. However, the traditional microorganism for commercial ethanol production (from C-6 sugars), *Saccharomyces cerevisiae*, is unable to utilize pentose (C-5 sugars). It lacks enzymes that degrade pentose into intermediates for further degradation via glycolysis or other pathways. Other microorganisms, however, can utilize pentose (e.g., arabinose and xylose), and thus could eventually produce ethanol. Pentose metabolism varies depending on the type of microbes. The metabolic pathways for L-arabinose and D-xylose degradation in both bacteria and fungi are discussed in the following section. The resulting intermediates, including D-xylulose 5-phosphate (from D-xylose catabolism in bacteria and fungi, and from L-arabinose catabolism in fungi) and L-ribulose 5-phosphate (from L-arabinose degradation in bacteria), enter the pentose phosphate pathway (PPP) for conversion into a number of other compounds. The ethanol fermentation is then carried out using the same biochemical pathway as elucidated in the previous section.

Although bacteria and fungi use different metabolic pathways to utilize pentose, both of them convert D-xylose to D-xylulose 5-phosphate. In bacteria, D-xylose is isomerized to form D-xylulose (Equation 14.17). The D-xylulose is then phosphorylated by transferring a phosphoryl group from ATP to form D-xylulose 5-phosphate (Equation 14.18).

$$\begin{array}{ccc}
\underset{\text{D-xylose}}{\begin{array}{c} H \quad O \\ \diagdown \diagup \\ C \\ | \\ H-C-OH \\ | \\ HO-C-H \\ | \\ H-C-OH \\ | \\ CH_2OH \end{array}} & \underset{\text{xylose isomerase}}{\rightleftharpoons} & \underset{\text{D-xylulose}}{\begin{array}{c} CH_2OH \\ | \\ C=O \\ | \\ HO-C-H \\ | \\ H-C-OH \\ | \\ CH_2OH \end{array}}
\end{array} \qquad (14.17)$$

$$
\begin{array}{ccc}
\begin{array}{l}
CH_2OH \\
| \\
C{=}O \\
| \\
HO{-}C{-}H \\
| \\
H{-}C{-}OH \\
| \\
CH_2OH \\
\text{D-xylulose}
\end{array}
&
\xrightarrow[\text{xylulokinase}]{\quad ATP \quad\quad ADP\quad}
&
\begin{array}{l}
CH_2OH \\
| \\
C{=}O \\
| \\
HO{-}C{-}H \\
| \\
H{-}C{-}OH \\
| \\
CH_2O{-}\textcircled{P} \\
\text{D-xylulose 5-phosphate}
\end{array}
\end{array}
\tag{14.18}
$$

L-arabinose catabolism in bacteria starts from the reversible reaction converting L-arabinose to L-ribulose by enzyme arabinose isomerase (Equation 14.19). The L-ribulose is then phosphorylated to form L-ribulose 5-phosphate (Equation 14.20). Both D-xylulose 5-phosphate (from D-xylose degradation) and L-ribulose 5-phosphate (from L-arabinose degradation) pass through the pentose phosphate pathway to produce the intermediate compound (glyceraldehyde 3-phosphate), which enters the glycolysis pathway, is catabolized, and is then fermented to the final product, ethanol.

$$
\begin{array}{ccc}
\begin{array}{l}
H\quad O \\
\diagdown\!\!\diagup \\
C \\
| \\
H{-}C{-}OH \\
| \\
HO{-}C{-}H \\
| \\
HO{-}C{-}H \\
| \\
CH_2OH \\
\text{L-arabinose}
\end{array}
&
\underset{\text{arabinose isomerase}}{\rightleftharpoons}
&
\begin{array}{l}
CH_2OH \\
| \\
C{=}O \\
| \\
HO{-}C{-}H \\
| \\
HO{-}C{-}OH \\
| \\
CH_2OH \\
\text{L-ribulose}
\end{array}
\end{array}
\tag{14.19}
$$

$$
\begin{array}{ccc}
\begin{array}{l}
CH_2OH \\
| \\
C{=}O \\
| \\
HO{-}C{-}H \\
| \\
HO{-}C{-}OH \\
| \\
CH_2OH \\
\text{L-ribulose}
\end{array}
&
\xrightarrow[\text{ribulokinase}]{\quad ATP \quad\quad ADP\quad}
&
\begin{array}{l}
CH_2OH \\
| \\
C{=}O \\
| \\
HO{-}C{-}H \\
| \\
HO{-}C{-}OH \\
| \\
CH_2O{-}\textcircled{P} \\
\text{L-ribulose 5-phosphate}
\end{array}
\end{array}
\tag{14.20}
$$

In fungi, a common step of pentose catabolism is the oxidation-reduction reaction of both sugars to form xylitol. Xylitol is converted to D-xylulose and continues through a reaction shown previously in Equation 14.18.

D-xylose is reduced to xylitol by xylose reductase (Equation 14.21). Xylitol dehydrogenase catalyzes the oxidation of xylitol to yield D-xylulose (Equation 14.22).

$$
\begin{array}{ccc}
\begin{array}{l}
H\quad O \\
\diagdown\!\!\diagup \\
C \\
| \\
H{-}C{-}OH \\
| \\
HO{-}C{-}H \\
| \\
H{-}C{-}OH \\
| \\
CH_2OH \\
\text{D-xylose}
\end{array}
&
\underset{\text{xylose reductase}}{\overset{NAD(P)H \quad NAD(P)^+}{\rightleftharpoons}}
&
\begin{array}{l}
CH_2OH \\
| \\
H{-}C{-}OH \\
| \\
HO{-}C{-}H \\
| \\
H{-}C{-}OH \\
| \\
CH_2OH \\
\text{Xylitol}
\end{array}
\end{array}
\tag{14.21}
$$

$$
\begin{array}{c}
\text{CH}_2\text{OH} \\
| \\
\text{H}-\text{C}-\text{OH} \\
| \\
\text{HO}-\text{C}-\text{H} \\
| \\
\text{H}-\text{C}-\text{OH} \\
| \\
\text{CH}_2\text{OH} \\
\text{Xylitol}
\end{array}
\quad
\xrightarrow[\text{xylitol dehydrogenase}]{\text{NAD}^+ \quad \text{NADH}}
\quad
\begin{array}{c}
\text{CH}_2\text{OH} \\
| \\
\text{C}=\text{O} \\
| \\
\text{HO}-\text{C}-\text{H} \\
| \\
\text{H}-\text{C}-\text{OH} \\
| \\
\text{CH}_2\text{OH} \\
\text{D-xylulose}
\end{array}
\qquad (14.22)
$$

Similarly, L-arabinose is first reduced by xylose reductase to form L-arabitol (Equation 14.23). L-arabitol is oxidized to L-xylulose using the catalytic activity of the enzyme arabitol dehydrogenase (Equation 14.24). Xylitol is formed after the reduction of L-xylulose by xylulose reductase (Equation 14.25).

$$
\begin{array}{c}
\text{H} \quad \text{O} \\
\diagdown \diagup \\
\text{C} \\
| \\
\text{H}-\text{C}-\text{OH} \\
| \\
\text{HO}-\text{C}-\text{H} \\
| \\
\text{HO}-\text{C}-\text{H} \\
| \\
\text{CH}_2\text{OH} \\
\text{L-arabinose}
\end{array}
\quad
\xrightleftharpoons[\text{xylose reductase}]{\text{NAD(P)H} \quad \text{NAD(P)}^+}
\quad
\begin{array}{c}
\text{CH}_2\text{OH} \\
| \\
\text{H}-\text{C}-\text{OH} \\
| \\
\text{HO}-\text{C}-\text{H} \\
| \\
\text{HO}-\text{C}-\text{H} \\
| \\
\text{CH}_2\text{OH} \\
\text{L-arabitol}
\end{array}
\qquad (14.23)
$$

$$
\begin{array}{c}
\text{CH}_2\text{OH} \\
| \\
\text{H}-\text{C}-\text{OH} \\
| \\
\text{HO}-\text{C}-\text{H} \\
| \\
\text{HO}-\text{C}-\text{H} \\
| \\
\text{CH}_2\text{OH} \\
\text{L-arabitol}
\end{array}
\quad
\xrightleftharpoons[\text{arabitol dehydrogenase}]{\text{NAD}^+ \quad \text{NADH}}
\quad
\begin{array}{c}
\text{CH}_2\text{OH} \\
| \\
\text{C}=\text{O} \\
| \\
\text{H}-\text{C}-\text{OH} \\
| \\
\text{HO}-\text{C}-\text{H} \\
| \\
\text{CH}_2\text{OH} \\
\text{L-xylulose}
\end{array}
\qquad (14.24)
$$

$$
\begin{array}{c}
\text{CH}_2\text{OH} \\
| \\
\text{C}=\text{O} \\
| \\
\text{H}-\text{C}-\text{OH} \\
| \\
\text{HO}-\text{C}-\text{H} \\
| \\
\text{CH}_2\text{OH} \\
\text{L-xylulose}
\end{array}
\quad
\xrightleftharpoons[\text{xylulose reductase}]{\text{NAD(P)H} \quad \text{NAD(P)}^+}
\quad
\begin{array}{c}
\text{CH}_2\text{OH} \\
| \\
\text{H}-\text{C}-\text{OH} \\
| \\
\text{HO}-\text{C}-\text{H} \\
| \\
\text{H}-\text{C}-\text{OH} \\
| \\
\text{CH}_2\text{OH} \\
\text{Xylitol}
\end{array}
\qquad (14.25)
$$

As mentioned previously, xylitol is converted and phosphorylated to D-xylulose 5-phosphate, and enters the pentose phosphate pathway for subsequent ethanol fermentation. The schematic diagram of pentose metabolism in fungi (solid line) and bacteria (dashed line) is shown in Figure 14.7.

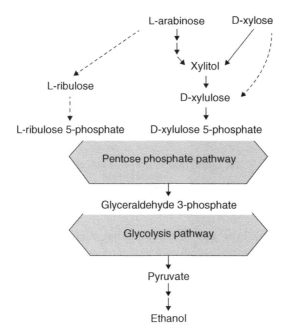

FIG. 14.7   Pentose catabolism in fungi (solid line) and bacteria (dashed line).

Intermediate compounds, L-ribulose 5-phosphate and D-xylulose 5-phosphate, are derived from L-arabinose and D-xylose catabolism. These compounds are further catabolized via the pentose phosphate pathway using the catalytic activity of several enzymes (transaldolase, transketolase, and epimerase) to produce glyceraldehyde 3-phosphate. Glyceraldehyde 3-phosphate can enter glycolysis (Step 6) to generate a three-carbon compound, pyruvate, which is then fermented to ethanol as the final product (see Section 14.2.1).

14.3 Byproducts Formation during Ethanol Fermentation

In the actual ethanol fermentation process, other byproducts are also generated, resulting in lower ethanol yields. At the initial stage of fermentation, when the amount of acetaldehyde is limited, yeasts utilize different pathways for NAD$^+$ regeneration to allow for further breakdown of glucose via glycolysis. Glycerol, for example, is produced under such conditions. When increased concentrations of acetaldehyde are supplied in the cell, yeasts switch their metabolic pathway to ethanol production (Figure 14.3). The overall glycerol production pathway is illustrated in Equation 14.26. Normal glycerol concentration in the fermentation broth is about 1.0% (w/v).

$$
\begin{array}{c}
\text{NADH} \qquad \text{NAD}^+ \\
\text{Dihydroxyacetone} \xrightarrow{} \text{Glycerol 3-phosphate} \xrightarrow{} \text{Glycerol} \\
\text{phosphate} \qquad \text{glycerol 3-phosphate} \qquad\qquad \text{glycerol} \\
\text{dehydrogenase} \qquad\qquad\qquad \text{3-phosphatase}
\end{array}
\qquad (14.26)
$$

When yeasts are cultivated at pH 7.0 or above, glycerol and acetic acid are produced. The formation of acetic acid results from the increased activity of the enzyme acetaldehyde dehydrogenase (optimal pH of around 8.75), which catalyzes acetic acid production from acetaldehyde (Equation 14.27). NADH generated in this step must be reoxidized; therefore, in the absence of oxygen, dihydroxyacetone phosphate is reduced to form glycerol 3-phosphate, and later glycerol.

$$\text{Acetaldehyde} \xrightarrow[\substack{\text{acetaldehyde} \\ \text{dehydrogenase}}]{\text{NAD}^+ \quad \text{NADH}} \text{Acetic acid} \qquad (14.27)$$

Another source of undesirable byproduct formation during ethanol fermentation is due to microbial contamination. Microbes consume sugar during fermentation, and compete for substrate and nutrients during yeast fermentation, thereby limiting the ethanol yield. They can also degrade ethanol through specific enzyme activities like those found in humans. Bacteria (namely, lactic acid bacteria, acetobacter, and *Clostridium* sp.) and wild-type yeasts are the most common contaminating microbes observed during ethanol fermentation.

Lactic acid bacteria (LAB; *Lactobacillus* sp.) are the primary contaminant of concern in industrial ethanol fermentation. LAB are quite robust microbes that can survive at high ethanol concentrations, low pH, and low oxygen concentrations. Lactobacilli contamination causes a significant loss of ethanol productivity (up to 22%) due to lactic and acetic acid production, and the inhibition of yeasts (i.e., *S. cerevisiae*). Another main contaminant causing a persistent problem in ethanol fermentation is the growth of wild yeasts, such as *Dekkera bruxellensis*. *D. bruxellensis* utilizes ethanol as a substrate for acetic acid production when other carbon sources are in short supply under aerobic conditions. Similarly, acetobacter also oxidizes ethanol into acetic acid. Although some acetogenic bacteria can degrade ethanol to acetic acid anaerobically, their activity is limited by a relatively high ethanol concentration in the fermentation broth. Moreover, anaerobically, glucose can be oxidized through multiple biochemical reactions to pyruvate and eventually forms butyric acid, which is a main byproduct due to *Clostridium* sp. contamination.

To control microbial contamination in ethanol fermentation, antibiotics including virginiamycin and penicillin are commonly used. However, for unwanted yeast contamination, the use of fungicide has to be implemented under caution, as it can also have detrimental effect on the yeast population for ethanol production. One common fungicide, targeting *D. bruxellensis* without harming *S. cerevisiae*, is polyhexamethyl biguanide (PHMB).

14.4 Microbial Cultures

The most common microbial culture used for industrial fuel ethanol production is the yeast *S. cerevisiae*. The yeast converts sugars (hexoses) into ethanol using the biochemical pathways discussed earlier. It is critical to maintain optimal conditions for ethanol fermentation in the broth throughout the fermentation process, and to limit contamination from byproduct generation and unwanted microbial contamination.

14.4.1 Yeast Culture for Hexose Fermentation

Yeasts, the eukaryotic microorganisms, are classified in the kingdom Fungi. As heterotrophic osmotrophs, yeasts require an external nutrient source. They excrete digestive enzymes to break down complex nutrients, and absorb digested soluble nutrients through their cell walls via active and passive transport. The yeast species *S. cerevisiae* is a unicellular organism that reproduces asexually by budding. *S. cerevisiae* cells are commonly ellipsoidal, with a size ranging from 5–10 μm and 1–7 μm along the long and short axes, respectively. Yeast cells can adopt an elongated form during nitrogen-limited conditions. The size of yeast cells differs greatly depending on the strain and growth state.

Figure 14.8 shows a typical yeast cell structure. In brief, as a eukaryote, yeast contains a cell wall, plasma membrane, cytoplasm, and membrane-bound organelles, including a nucleus, endoplasmic reticulum (ER), Golgi apparatus, lysosome, mitochondrion, and vacuole. Biochemical macromolecules of yeast are mainly proteins (structural, hormones, and enzymes), glycoproteins (cell wall components and enzymes), polysaccharides (cell wall components, capsular components, and carbon storage), phosphates (storage), lipids (structural, storage, and functional), and nucleic acids (DNA and RNA).

The yeast cell is surrounded by a rigid cell wall and develops bud scars during cell division. The yeast cell envelope consists of three main constituents: from the inside out, plasma membrane, periplasmic space, and cell wall. The cell wall stabilizes osmotic homeostasis in the cell, protects the cell against physical stress or mechanical damage, maintains cell shape, and acts as a scaffold for proteins, thereby limiting the permeability of the cell wall for macromolecules. The cell wall, however, is flexible, allowing the cell to respond quickly to rapid cell volume changes when a sudden shift in osmotic pressure occurs. The yeast cell wall significantly varies in composition during the natural growth

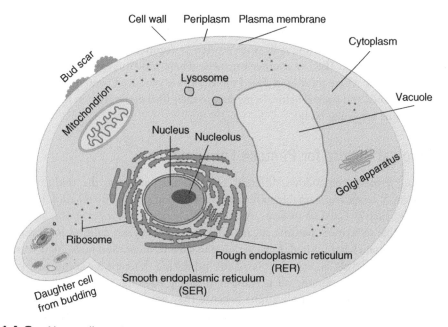

FIG. 14.8   Yeast cell structure.

cycle, and is mainly composed of 80–90% polysaccharides (mannans and glucans with small amounts of chitin) and small amounts of protein. In other words, the yeast cell wall is primarily made up of glycoproteins (one or several oligosaccharides joined covalently to a protein). The periplasm (cell wall–associated region located between the cell wall and the plasma membrane) consists of secreted proteins (mainly enzymes) that are essential for hydrolysis reactions. These secreted proteins cannot permeate through the plasma membrane; therefore, they catalyze the hydrolysis of substrates before entering the cell plasma membrane. Invertase, an enzyme converting sucrose into glucose and fructose, is located in the yeast periplasm, allowing the yeast to break down the disaccharide into monosaccharides, facilitating transport into the cell for further catabolism. The inner structure of the cell envelope is the plasma membrane, a lipid bilayer with embedded protein. The plasma membrane provides selective permeability of certain substances in and out of the cytoplasm, regulating nutrient uptake and toxin extrusion.

The cytoplasm, a gel-like fluid material, contains low and intermediate molecular weight compounds; proteins, glycogen, and other soluble macromolecules; and membrane-bound organelles. Thus, the cell metabolic pathways (such as the glycolytic pathway and fatty acid, enzyme, and protein biosynthesis) take place in this region. The yeast cytoplasm has freely suspended ribosomes that aid in protein synthesis, and lysosomes that help break down waste and cellular debris.

There are several membrane-bound organelles located in the cytoplasm. The vacuole is likely formed under stressful conditions, as it stores and isolates toxic ions that could be harmful to the cell. It is also the location for enzyme, amino acid, and polyphosphate storage. The nucleus is an organelle enclosed by an impermeable double membrane with nuclear pores; DNA (deoxyribonucleic acid) synthesis is its major function. Thus, the nucleus is responsible for containing all of the genetic material (DNA) and protein (histone). DNA and histone proteins in the nucleus are organized in a complex structure called a chromosome. Inside the nucleus, there is a non-membrane-bound structure called the nucleolus where rRNA (ribosomal ribonucleic acid) is transcribed, processed, and assembled. Similar to the nucleus, the mitochondrion, a site where energy (ATP) is generated through oxidative phosphorylation during aerobic conditions (cellular respiration), is a double membrane-bound organelle. However, during anaerobic fermentation, mitochondria have no energy-generation capabilities in the cell, since oxygen is required as the final electron acceptor. The endoplasmic reticulum (ER) is another organelle where protein synthesis takes place (rough ER), and is also the site where most lipid-based molecules are produced (smooth ER). The proteins and lipids are processed and packed before being sent to their final destination in the organelle, called the Golgi apparatus (or Golgi body).

14.4.2 Microbial Culture for Pentose Fermentation

Because *pentose fermentation* follows different biochemical pathways, the traditional ethanol-fermenting yeast, *S. cerevisiae*, is unable to ferment five-carbon sugars into ethanol. However, several yeast species, such as *Candida shehetae*, *Kluveromyces marxianus*, *Pichia stipitis*, and *Pachysolen tennophilus*, are able to ferment pentose sugars. The use of these yeast cultures for industrial ethanol production is somewhat limited, however, due to slow fermentation rates, high sensitivity to inhibitors and oxygen levels, and product (ethanol) inhibition. Bacteria are commonly capable of fermenting pentose sugars. Among all pentose-fermenting bacteria, thermophiles are the most promising for commercial-scale applications because they are less vulnerable to contamination during fermentation, consume a wide range of substrates, and produce less cell biomass. However, these cultures, including *Thermoanaerobacterium saccharolyticum*, *Clostridium thermohydrosulfurium*, *Clostridium thermosaccharolyticum*, *Clostridium thermosulfurogenes*, *Clostridium tetani*, and

Thermoanaerobacter ethanolicus, usually achieve low ethanol concentrations because of low product tolerance and byproduct formation during fermentation. Filamentous fungi, *Monilia* sp., *Neocallimastix* sp., *Trichoderma reesei*, and *Fusarium oxysporum*, can directly convert cellulose/hemicellulose to ethanol in a single step, and are commonly used in the biological pretreatment of lignocellulosic biomass. Drawbacks of such organisms include long fermentation periods, byproduct formation, and high viscosities in the fermentation broth, which limit their use in industrial ethanol production.

Presently, research efforts focus heavily on the molecular level to develop suitable microbial strains that can ferment both hexose and pentose sugars with high ethanol yields. Metabolic engineering techniques have gained considerable attention in recent years. Specific genes for pentose metabolism from pentose-fermenting microbes are engineered into host microorganisms like *S. cerevisiae*. There are several advantages of using *S. cerevisiae* over bacteria. This commercial ethanol-fermenting yeast has a high resistance to hydrolysate inhibitors. It can also grow at a low pH and requires lower amounts of nutrients. Genetically modified *S. cerevisiae,* which are inserted with genes for xylose reductase and xylitol dehydrogenase production from *P. stipitis*, have been shown to produce ethanol from xylose.

14.5 Environmental Factors Affecting Ethanol Fermentation

14.5.1 Nutrients

The major carbon source for yeast growth and ethanol production comes from sugar molecules. *S. cerevisiae* can use a wide variety of sugars, such as monosaccharides (glucose, mannose, fructose, and galactose), disaccharides (sucrose, maltose, and trehalose), and trisaccharide (maltotriose and raffinose), but cannot grow on lactose and cellobiose (disaccharides) or pentose (monosaccharide). Yeast can utilize both organic (yeast extract, peptone, urea, and nucleotide bases) and inorganic (ammonium) nitrogen (N) sources. However, they lack the ability to use nitrate and nitrite. Typically, nutrient utilization is contingent on growth conditions. For example, yeast cannot use proline (amino acid) as an N source under anaerobic conditions, but it can utilize all amino acids under aerobic conditions. Phosphorus (P) is another key macronutrient required for producing energy to support cell growth, and for the synthesis of material for cell reproduction such as protein (enzymes), genetic material (DNA), and cell membrane (unsaturated fatty acid and sterols). Phosphorus is usually assimilated from a phosphate ion. Sulfur (S) is also essential and can be assimilated from a sulfate ion or amino acids (methionine and cysteine). Some other micronutrients (K, Mg, Na, and Ca) and trace elements (Al, B, Co, Fe, Mn, and Zn) are also necessary to support yeast metabolic pathways. These nutrients are required in small amounts, which are usually sufficient in the raw material for ethanol fermentation. However, very high or very low sodium ion concentrations can have detrimental effects on the ethanol yield. Often, vitamins such as biotin, pantothenic acid, inositol, thiamin, pyridoxine, and nicotinic acid are also needed in modest quantities.

14.5.2 pH

pH can limit the growth rate of yeasts by altering enzymatic activity, cell permeability, and metal ion availability. The yeast can survive in a wide range of pH (2.0–8.0), but the optimal pH for its growth is 4.8–5.0, which is slightly acidic. A pH between 4.0 and 5.0 has no significant effect on the ethanol yield during fermentation, but a longer fermentation time has been observed when fermentation was carried out at a lower pH. Ethanol fermentation at a pH above 5.0 can lower the ethanol yield.

14.5.3 Temperature

Temperature fluctuations affect the growth rate, nutritional requirements, cell permeability, enzyme activity, and metabolism. The major mechanisms affected by temperature are the structure and composition of cytoplasmic membranes, which determine the rate of substrate utilization. *S. cerevisiae* can grow in a temperature range of 5–38 °C; but fermentation usually takes place at a temperature of about 30 °C, since it is thermodynamically favorable for producing ethanol. To expedite growth at the initial stage, the temperature can be set at near 35 °C. However, at a later stage when high concentrations of ethanol are present in the fermentation broth, higher temperatures have a significant negative effect on yeast activity, and result in a higher risk of heat-resistant *Lactobacillus* sp. contamination.

14.5.4 Others

Some other factors affecting ethanol fermentation are ethanol and organic acid (among a few others). Ethanol, the end product of fermentation, can inhibit yeast activity through a process known as feedback inhibition. Therefore, the normal concentration of ethanol at the end of the fermentation period is usually fixed at around 10–15% (v/v), although a higher ethanol concentration (18–20% (v/v)) can be obtained using some strains that can tolerate high ethanol concentrations of around 20% (v/v). Moreover, organic acids (lactic and acetic acids) produced from undesired microbial contamination are detrimental to yeast, resulting in a lower ethanol yield.

14.6 Industrial Fuel-Grade Ethanol Production

Ethanol production nowadays is carried out in a large, closed, and sterile bioreactor in which temperature, pH, and agitation are maintained at specific levels for optimal ethanol fermentation performance. Figure 14.9 is an example of a sugarcane-ethanol facility in Brazil. For fuel applications, ethanol can be produced from many substrates other than fermentable sugars, but the complex feedstocks typically used, such as polysaccharides derived from starch and lignocellulosic materials, need first to be broken down into fermentable sugars. In general, the feedstocks for bioethanol production are broadly classified into three categories: sugar, starch, and lignocellulosic-based feedstocks. Sugar-based feedstocks (sugarcane, molasses, sugar beet, and sweet sorghum) and starch-based feedstocks (corn, sorghum grains, and cassava) are considered to be first-generation feedstocks. Sugar-based feedstocks are mainly composed of carbohydrates in the form of simple (soluble) sugars that are readily fermentable by microbes. Starch-based feedstocks, on the other hand, are composed of slightly more complex carbohydrates, which need to be broken down into simple sugars by the use of specific enzymes (e.g., amylases) before they can be fermented.

Ethanol is currently produced from first-generation feedstocks, such as corn (USA), sugarcane (Brazil), sugar beet (European Union), and cassava (Thailand). Although ethanol is commercially produced from first-generation feedstocks, there are many concerns over a food/feed versus fuel issue. Non-edible lignocellulosic feedstocks (agricultural and forestry residues, and dedicated energy crops known as second-generation feedstocks) are ideal low-cost substrates for bioethanol production. Cellulose, hemicellulose, and lignin are the main constituents of lignocellulosic biomass; but only the

FIG. 14.9 Sugarcane-ethanol plant in São Paulo, Brazil. Photo courtesy of Saoharit Nitayavardhana, Chiang Mai University.

carbohydrates (cellulose and hemicellulose) can be broken down into simple sugars for ethanol production. The recalcitrant nature of lignocellulosic biomass requires extensive pretreatment (see Chapter 12) to disrupt the biomass structure, thereby allowing enzymes to access cellulose and/or hemicellulose in the subsequent enzymatic hydrolysis (see Chapter 13). More detail about ethanol production processes from these feedstocks is given in Chapter 23 (Sugar-Based Biorefinery), Chapter 24 (Starch-Based Biorefinery), and Chapter 25 (Lignocellulosic-Based Biorefinery). This chapter will only discuss the ethanol fermentation process after fermentable sugars are obtained from several pretreatment steps.

14.6.1 Seed Culture Preparation

At a commercial scale, *S. cerevisiae* is first propagated in a series of laboratory-scale steps. The original yeast stock culture, usually frozen or freeze-dried, is reactivated in an enriched medium such as yeast extract peptone dextrose (YEPD or YPD). The reactivation step can be conducted in a test tube (10 ml) and incubated at an optimal temperature ($\sim$30 °C) for about 24 hr. In the yeast propagation step, the activated culture is transferred aseptically and cultivated in YPD. The culture is then transferred and cultivated on saccharified product derived from the desired feedstock used for ethanol production. The yeast culture is propagated in a series of steps to increase the volume of the saccharified product and produce a seed culture for the industrial-scale operation (Figure 14.10). An amplification ratio of 1:10–15 is used for the yeast propagation steps. All the cultivations are conducted under controlled temperatures of around 30 °C for 15–20 hr. Depending on the volume of the fermentation broth, the final yeast inoculum for ethanol fermentation is approximately 5–10% of the

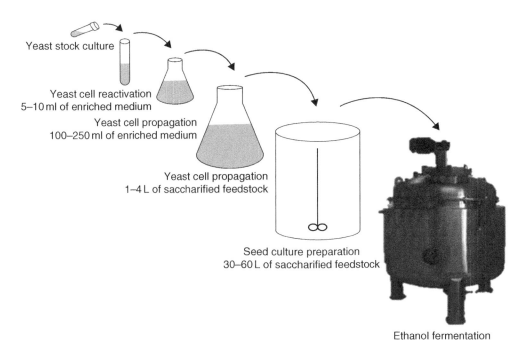

Ethanol fermentation

FIG. 14.10   Seed culture preparation for ethanol fermentation.

medium (sugar solution). Yeast propagation is critical in industrial-scale ethanol fermentation, as a pure yeast culture is required for high ethanol productivity. Therefore, during this step, aseptic techniques are important to limit undesirable microbial contamination and to maintain a pure culture.

14.6.2 Industrial Ethanol Fermentation

There are two common types of fermenters, batch and continuous, used in commercial ethanol production. Batch fermentation is simple and easy to control, with less chance of undesirable microbial contamination. This system is generally used in ethanol industries. Batch ethanol fermentation normally starts with fermenter disinfection, followed by substrate/nutrient loading and yeast culture inoculation. Mass production of an active yeast culture is followed by ethanol production, and the fermentation broth is harvested at the end of a predetermined period. The first stage to establish active yeast cells (6–8 hr) requires oxygenation for energy generation through oxidative phosphorylation (cellular respiration). A sufficient oxygen supply in the culture is also needed to synthesize other compounds necessary for cellular reproduction. Sugars are mainly used to support the growth of yeast cells. Later, when sufficient yeast cells are generated, oxygenation is no longer required. The yeasts ferment sugars to ethanol anaerobically when dissolved oxygen is exhausted. Rapid ethanol, CO_2, and heat generation during the active fermentation period can occur and last for about 12 hr. After that, biochemical reactions become limited due to the availability of substrate (sugars), and the fermentation rate slows down rapidly. High ethanol concentrations in the fermentation broth also inhibit fermentative metabolism in the yeast cells.

Normally, it takes about 72 hr for complete ethanol fermentation from glucose, and the final ethanol concentration in the fermentation broth reaches around 10–15% (v/v). However, because yeast cell populations experience different growth stages in batch fermentation, the ethanol production efficiency is relatively low. Moreover, byproducts such as glycerol formation are inevitable during the stationary growth phase of yeast cells. Therefore, continuous fermentation systems, which maintain an active fermentation stage, could be one avenue to overcome these issues. Continuous fermentation includes two stages: start-up and steady-state operations. Start-up, similar to batch ethanol fermentation, occurs at the beginning (12–18 hr) to establish an active yeast culture inoculum for ethanol fermentation. Later, substrate and nutrients are continuously fed into the fermenter, while the fermentation broth or beer is continuously siphoned out of the fermenter under steady-state conditions. Suspended yeast cells flowing out with the beer can be separated and recycled back to the fermenter. The ethanol fermentation efficiency and production rate are high in continuous systems, as the yeast cells are in their active fermentation stage. However, it is difficult to control contamination in continuous fermentation systems because of the perpetual inflow and outflow of substrate and effluent, respectively.

> **Example 14.1:** A fermenter is designed to ferment 5,000 gal/day of sugarcane juice with a sucrose concentration of 60 g/L at 35 °C.
>
> 1. Calculate the gallons of ethanol produced daily.
> 2. If 10% of the sugars are consumed in yeast cell production, how much ethanol will be produced?

Solution:

1. **When all sugars are converted to ethanol (See Section 23.2 in Chapter 23 for more details)**

$$C_{12}H_{22}O_{11} \rightarrow 4C_2H_5OH + 4CO_2$$

342 g 4×46

1 g $4 \times 46/342 = 0.538$ g

Or 1 kg sucrose will produce 0.538 kg ethanol.
   Total ethanol produced:

$$= (5,000 \, \text{gal/day}) \times (3.785 \, \text{L/gal}) \times (60 \times 10^{-3} \text{kg sucrose/L}) \times (0.538 \, \text{kg ethanol/} \\ \text{kg sucrose}) \times (1/0.790 \, \text{kg/L})$$
$$= \mathbf{773.3 \, L/day \, (204.3 \, gal/day)}$$

2. **When 10% of sugar is consumed by the yeast cells, sugar available**

   Amount of sucrose available:
$$= [(5,000 \, \text{gal/day}) \times (3.785 \, \text{L/gal}) \times (60 \times 10^{-3} \text{kg sucrose/L}) \times (0.90)]$$
$$= 1,022 \, \text{kg/day}$$

Total ethanol produced:

$$= [1,022 \, \text{kg/day} \times (0.538 \, \text{kg ethanol/kg sucrose})] \times (1/0.790 \, \text{kg/L})$$
$$= \mathbf{696 \, L/day \, (184 \, gal/day)}$$

Simultaneous-saccharification and fermentation (SSF)

In commercial starch-based ethanol production, SSF is employed such that saccharification and fermentation are combined in a single fermenter. SSF allows fermentable sugars produced in saccharification processes to be consumed immediately, thereby maintaining low sugar concentrations in the fermentation broth. This way, yeast cells are maintained at the active fermentation stage during the SSF process, and saccharification can occur without product inhibition (feedback inhibition) resulting from high concentrations of sugar. Because of low sugar concentrations in the fermentation medium, SSF requires a lower amount of enzyme (glucoamylase) for the saccharification process and shortens the fermentation period.

14.6.3 Ethanol Recovery

The ethanol recovery process is critical for the industry as it determines the quantity and quality of the final product. Commonly, the ethanol–water mixture (fermented broth or beer) derived from the fermentation process (10–15% (v/v) ethanol) is separated by traditional distillation technology. However, due to the azeotropic, non-ideal behavior of the ethanol–water mixture, the maximum ethanol concentration that can be achieved by distillation is 84% mol/mol (93% (v/v)). A further separation process, dehydration, must be employed to remove water to produce fuel-grade ethanol (>99% (v/v) or 200 proof ethanol).

Distillation is used to separate various components of a liquid mixture based on their distribution in the liquid and vapor phases. This technology can be used to separate only a non-ideal mixed solution, as it requires the compositions of the components in the two phases to remain different when the mixture is in equilibrium; in other words, the boiling temperature of the two (or more) substances must be notably different.

In industrial practice, ethanol is normally recovered by rectification (fractionation) or stage distillation (Figure 14.11). The distillation column is equipped with a series of plates or trays serving effectively as a series of miniature distillation stages. Under operation, the temperature is at the mixture boiling point (or between the bubble and dew points) throughout the distillation column. Taking a closer look at each stage, liquid flows downward to the stage immediately below, while vapor flows upward to the stage immediately above. Theoretically speaking, the vapor stream and liquid stream entering and leaving each stage are in equilibrium. The distillation column is divided into two sections: a rectifying section (upper portion) and a stripping section (lower portion), located by the feeding location, usually found near the middle of the column. Vapor enters the trays from the bottom and continues rising to the upper stage with the more volatile component; that is, ethanol. Consequently, the ethanol content in the vapor increases as it gets closer to the top of the distillation column. The vapor stream at the top, with the highest ethanol concentration, finally passes through a condenser to recover the overhead product (distillate) containing about 90% (v/v) of ethanol, while a portion of the condensed liquid is sent back to the top of a distillation column (reflux), serving as the source of a downflow liquid stream. Similarly, at the bottom of the distillation column, the liquid stream containing the lowest ethanol content is partially vaporized in a reboiler. The vapor is sent back to the bottom of the column, while the remaining liquid (bottom product or residue) containing about 1% (v/v) ethanol flows out.

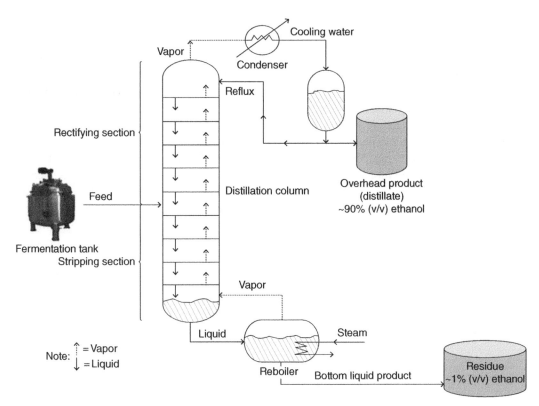

FIG. 14.11 A typical fractionating distillation column for an ethanol separation process.

Typically, fractional distillation columns range from 0.65–6.0 m in diameter and 6–60 m in height. The distillation columns are commonly classified into two categories: packed and stage (tray or plate) columns. In a packed column, packing materials, which have a high surface area and high porosity (void ratio), are placed inside the column, serving as media for mass transfer between liquid and vapor phases throughout the packed column. Liquid flows down on the surface of the packing material, while vapor rises through the void space in the column. However, in large-scale distillation, the packed column requires high fabrication costs and specialized installation to prevent vapor/liquid bypassing and reductions in efficiency. Consequently, this technology is not cost effective and is limited to relatively low-volume and high-value product processing. The other distillation setup is the stage (plate or tray) column, which is commonly used in ethanol industries. Perforated plates (or sieve trays) with a slight slope are installed in the column, allowing liquid to flow down to the lower stage. The small openings on the plates allow vapor to rise through a plate to an upper stage. As mentioned earlier, the degree of ethanol–water separation is greater with the higher number of stages (plates or trays). Therefore, the number of plates in the column is designed specifically for a given separation system (calculations for this phenomenon are beyond the scope of this book).

Distillate derived from distillation columns contains about 90% (v/v) ethanol; therefore, a further **dehydration** process is needed to produce fuel-grade anhydrous ethanol (>99% (v/v)). Commonly, molecular sieve adsorption technology is used in the ethanol industry. Molecular sieve adsorption can separate a mixture based on differences in the size of molecules. The solid adsorbent contains pores, in which one component could be adsorbed and separated from the other components. Molecular sieves used to separate water in final dehydration processes are porous crystalline aluminosilicates (zeolites). The zeolites form an open crystal lattice containing precisely uniform pores with a diameter of 0.30–0.35 nm. Hydrous alcohol from the distillation column is preheated and vaporized before entering the molecular sieve unit. Under pressure, the vapor is passed through a molecular sieve packed column where the water (a smaller molecule, 0.28 nm) is adsorbed, consequently allowing the bigger molecule, ethanol (0.32 nm), to pass through as an anhydrous vapor (<1% (v/v) water). The anhydrous ethanol vapor is then condensed and cooled down to generate the final fuel-grade ethanol. The molecular sieves can be reused for several years, but regeneration is needed when the water adsorption capacity reaches its saturation state. The water accumulated in molecular sieves is removed by evaporation under vacuum. Typically, multiple dehydration columns are used in ethanol plants. The dehydration and regeneration processes are sequentially repeated for continuous operation. A molecular sieve requires low energy input. Because no chemicals are used, it is an environmentally friendly technology that could generate nearly pure ethanol (>99.9% or ~200 proof). This technology is also beneficial for the distillation process, as the inlet hydrous vapor can contain a wide range of ethanol contents (80–95% (v/v)), rendering strict quality control of the distillate unnecessary. Further, molecular sieves have a very long life of over 5 years with proper handling.

References

Abbott, D.A., and Ingledew, W.M. (2005). The importance of aeration strategy in fuel alcohol fermentations contaminated with *Dekkera/Brettanomyces* yeasts. *Applied Microbiology and Biotechnology* 69: 16–21.

Bai, F.W., Anderson, W.A., and Moo-Young, M. (2008). Ethanol fermentation technologies from sugar and starch feedstocks. *Biotechnology Advances* 26: 89–105.

Beckner, M., Ivey, M.L., and Phister, T.G. (2011). Microbial contamination of fuel ethanol fermentations. *Letters in Applied Microbiology* 53: 387–94.

Berry, D.R., and Slaughter, J.C. (2003). Alcoholic beverage fermentation. In: G.H. Andrew and J.R. Piggott (eds.), *Fermented Beverage Production*, 2nd edn. (ed.). New York: Plenum, pp. 25–40.

Bischoff, K.M., Liu, S., Leathers, T.D., Worthington, R.E., and Rich, J.O. (2008). Modeling bacteria contamination of fuel ethanol fermentation. *Biotechnology and Bioengineering* 103: 117–22.

BNDES (2008). *Sugarcane-Based Bioethanol: Energy for Sustainable Development/Coordination*. Rio de Janeiro: BNDES/CGEE.

Brown, E., Cory, K., and Arent, D. (2007). *Understanding and Informing the Policy Environment: State-Level Renewable Fuels Standards*. Technical report NREL/TP-640-41075. Golden, CO: National Renewable Energy Laboratory.

Cheng, J.J. (2010). Biological process for ethanol production. In: J. Cheng (ed.), *Biomass to Renewable Energy Processes*. Bristol, PA: CRC Press, pp. 209–69.

Cronwright, G.R., Rohwer, J.M., and Prior, B.A. (2002). Metabolic control analysis of glycerol synthesis in *Saccharomyces cerevisiae*. *Applied and Environmental Microbiology* 68: 4448–56.

de Souza Liberal, A., Basílio, A., do Monte Resende, A., *et al.* (2007). Identification of *Dekkera bruxellensis* as a major contaminant yeast in continuous fuel ethanol fermentation. *Journal of Applied Microbiology* 102: 538–47.

Dias, L., Pereira da Silva, S., Tavares, M., Malfeito Ferreira, M., and Loureiro, V. (2003). Factors affecting the production of 4-ethylphenol by the yeast *Dekkera bruxellensis* in enological conditions. *Food Microbiology* 20: 377–84.

EIA (2013). *Today in Energy*. Washington, DC: U.S. Energy Information Administration. http://www.eia.gov/todayinenergy/detail.cfm?id=9791, accessed March 2013.

Elsztein, C., Menezes, J.A.S., and de Morais, M.A. (2008). Polyhexamethyl biguanide can eliminate contaminant yeasts from fuel-ethanol fermentation process. *Journal of Industrial Microbiology and Biotechnology* 35: 967–73.

Geankoplis, C.J. (2003). Vapor-liquid separation processes. In: C.J. Geankoplis (ed.), *Transport Process and Separation Process Principles (Includes Unit Operation)*, 4th ed. Upper Saddle River, NJ: Pearson Education, pp. 696–759.

Katzen, R., Madson, P.W., and Moon, G.D., Jr. (1997). Ethanol distillation: The fundamentals. In: K.A. Jacques, T. P. Lyons, and D.R. Kelsall (eds.), *The Alcohol Textbook*. Nottingham: Nottingham University Press, pp. 269–88.

Madigan, M.T., Martinko, J.M., and Parker, J. (2000). *Biology of Microorganisms*, 10th edn. London: Prentice Hall International.

Matsuoka, S., Ferro, J., and Arruda, P. (2009). The Brazilian experience of sugarcane ethanol industry. *In Vitro Cellular and Developmental Biology – Plant* 45: 372–81.

McGovern, P.E. (2009). Alcoholic beverages: Whence and whither? In: P.E. McGovern, *Uncorking the Past: The Quest for Wine, Beer, and Other Alcoholic Beverages*. Berkeley, CA: University of California Press, pp. 266–82.

Melo, T.C.C., Machado, G.B., Belchior, C.R.P., *et al.* (2012). Hydrous ethanol-gasoline blends –Combustion and emission investigations on a Flex-Fuel engine. *Fuel* 97: 796–804.

Narendranath, N. (2003). Bacterial contamination and control in ethanol production. In: K.A. Jacques, T.P. Lyons, and D.R. Kelsall (eds.), *The Alcohol Textbook*. Nottingham: Nottingham University Press, pp. 287–98.

Nelson, D.L., and Cox, M.M. (2004). Glycolysis. In: D.L. Nelson and M.E. Cox, *Lehninger Principles of Biochemistry*, 4th edn. New York: W.H. Freeman, pp. 521–43.

OECD-FAO (2013). *Agricultural Outlook 2013*. Washington, DC: OECD/FAO. http://www.fao.org/fileadmin/templates/est/COMM_MARKETS_MONITORING/Oilcrops/Documents/OECD_Reports/OECD_2013_22_biofuels_proj.pdf, accessed April 2016.

Renewable Fuels Association (2014). *Industry Statistics*. Washington, DC: Renewable Fuels Association. http://www.ethanolrfa.org/pages/statistics, accessed February 2014.

Rovere, E.L.L., Pereira, A.S., and Simões, A.F. (2011). Biofuels and sustainable energy development in Brazil. *World Development* 39: 1026–36.

Thomas, K.C., Hynes, S.H., and Ingledew, W.M. (2001). Effect of lactobacilli on yeast growth, viability and batch and semi-continuous alcoholic fermentation of corn mash. *Journal of Applied Microbiology* 90: 819–28.

Wang, Z.-X., Zhuge, J., Fang, H., and Prior, B.A. (2001). Glycerol production by microbial fermentation: A review. *Biotechnology Advances* 19: 201–3.

Exercise Problems

14.1. What are the main differences between anaerobic respiration and anaerobic fermentation? Why is ethanol fermentation started with an air supply during the beginning of ethanol fermentation?

14.2. Name the main biochemical pathways for glucose conversion into ethanol. Also explain the importance of the ethanol fermentation process in relation to glucose degradation via glycolysis.

14.3. Lignocellulosic ethanol is becoming more attractive, but the conversion of its sugar component is challenging. What are the three main components in lignocellulosic biomass structure and what component(s) can contribute sugars as a fermentation substrate?

14.4. Lignocellulosic biomass has a different sugar composition to that of sugar- and starch-based feedstocks. What is/are the sugar(s) in lignocellulosic material that could not be obtained from sugar- and starch-based feedstocks?

14.5. Discuss the biochemical pathways for five-carbon sugar degradation and fermentation to produce ethanol.

14.6. Explain batch fermentation in terms of yeast growth and biochemical pathway, and discuss the importance of an oxygen-free and undesired microbial-free environment in ethanol fermentation.

14.7. What is simultaneous saccharification and fermentation (SSF)? Why is SSF important in the starch-to-ethanol process?

14.8. Provide the names of microorganisms commonly used in ethanol production from both five- and six-carbon sugars.

14.9. You are assigned to operate a lignocellulosic ethanol plant using yeast, *Saccharomyces cerevisiae*. Your substrate is hydrolysate containing both five- and six-carbon sugars. Explain your strategy for how to use *S. cerevisiae* for lignocellulosic ethanol production from mixed sugars in one fermentation tank.

14.10. Explain the ethanol recovery processes commonly used in industrial ethanol production. What are the advantages and limitations of these processes?

Butanol Fermentation

Victor Ujor and Thaddeus Chukwuemeka Ezeji

What is included in this chapter?

This chapter is an overview of acetone-butanol-ethanol (ABE) fermentation, its biochemistry, microbiology, and stoichiometry. Bioreactor design and downstream processing of butanol are included, along with worked examples.

15.1 Introduction

Butanol (1-butanol) is a colorless, four-carbon, flammable liquid with a characteristic banana-like odor, which was first reported in 1862 by Louis Pasteur (Dürre, 2008, 2011; Ni and Sun, 2009). Pasteur named the producing microorganism (possibly a mixed culture), which bore a resemblance to *Clostridium butyricum*, "Vibrion butyrique." Albert Fitz was perhaps the first microbiologist to obtain a pure culture of a butanol-producing (solventogenic) microorganism in 1876, which he named *Bacillus butylicus*. Butanol fermentation is also referred to as acetone-butanol-ethanol (ABE) fermentation due to the co-production of acetone and ethanol. ABE fermentation gained biotechnological and political significance during World War I, when it served as the major source of acetone, the critical bulk chemical for the production of the smokeless ammunition cordite (Dürre, 2008, 2011; Ni and Sun, 2009). A strain of *Clostridium acetobutylicum* with superior product formation capacity, isolated by Chaim Weizman, was the microorganism of choice for the production of acetone during World War I (Dürre, 2008; Ni and Sun, 2009). However, after the war butanol assumed a central role among the three products of ABE fermentation. This was stimulated by Henry Ford's introduction of the motor assembly line, which led to a significant increase in automobile production, during which butanol was identified as an excellent precursor for the synthesis of butyl acetate. Butyl acetate is used to produce quick-drying lacquers employed in car

---

Bioenergy: Principles and Applications, First Edition. Edited by Yebo Li and Samir Kumar Khanal.
© 2017 John Wiley & Sons, Inc. Published 2017 by John Wiley & Sons, Inc.
Companion website: www.wiley.com/go/Li/Bioenergy

manufacturing. As a result, butanol production by ABE fermentation remained fairly popular until the 1960s. During this time, there were plants in the UK, the USA, South Africa and China (Dürre, 2008; Ni and Sun, 2009).

Owing to the low cost of crude petroleum in the 1960s through to the end of the 20th century, coupled with the rising cost of molasses, a major substrate for ABE fermentation, industrial ABE fermentation became economically unfeasible worldwide (Dürre, 2008, 2011; Ni and Sun, 2009). Consequently, petroleum became the preferred feedstock for the synthesis of n-butanol. Petroleum-derived propylene is converted to butyraldehyde by an oxo reaction, which is then reduced to n-butanol by hydrogenation (Dürre, 2008). Butanol is an essential bulk chemical with diverse industrial applications. It is used mainly as a precursor of a catalogue of esters, ethers, butyl acetate, butyl amines, and butyl glycol, among other derivatives, which find application in industry as super absorbents, elastomers, adhesives, fibers, oil additives, in paper finishing, and as additives in ink and leather manufacturing (Dürre, 2008). Other applications of butanol or its derivatives include synthetic fruit flavorings, plasticizers, paint thinners, detergents, intermediates in the production of alkaloids, camphor, hormones, antibiotics, and perfumes, and as a solvent in chromatography (Dürre, 2008).

More recently, biobutanol (butanol from microbial fermentation) has been attracting tremendous attention as a source of renewable fuel. It is expected that the production of butanol by microbial fermentation from renewable resources will lead to reduced emission of greenhouse gases, as well as reduce dependence on crude oil (Ezeji *et al.*, 2007). Ethanol is currently used as a blend in transport fuel. Most gasoline sold in the USA is blended with at least 10% ethanol, commonly known as E10. Ethanol is commercially produced through the fermentation of sugars by the yeast *S. cerevisiae* (Chapter 14). Butanol can be produced by microbial fermentation using a number of microorganisms, generally referred to as solventogenic clostridia. Examples include *Clostridium beijerinckii* (Figure 15.1) and *Clostridium acetobutylicum*. Compared to ethanol, butanol has a number of desirable properties that make it an ideal biofuel (Table 15.1).

ABE fermentation, which was a large biotechnology industry during World War I, declined due to increasing substrate costs in comparison to petrochemically derived butanol. Interest in ABE fermentation has resurged in recent years, however. This chapter discusses the fundamentals of ABE

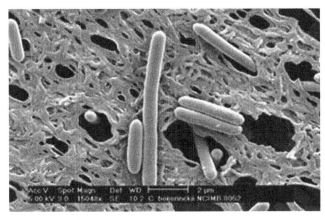

FIG. 15.1 Scanning electron micrograph (SEM) of *C. beijerinckii* NCIMB 8052 used for ABE fermentation. Scale 25,600 X. Source: http://genome.jgi-psf.org/clobe/clobe.home.html. Public domain.

Table 15.1 Advantages of butanol over ethanol

Ethanol	Butanol
Reduced mileage per gallon	More mileage per gallon, comparable to gasoline
Lower energy density (76,000 Btu/gallon)	Higher energy density (110,000 Btu/gallon)
Higher heat of vaporization (840 KJ/Kg)	Lower heat of vaporization (582 KJ/Kg), closer to that of gasoline (305 KJ/Kg)
Higher hygroscopicity, hence corrosive to Pipelines	Suitable to existing pipelines due to low hygroscopicity

fermentation, including biochemistry and stoichiometry, the kinetics of product yield, factors affecting ABE fermentation, substrates, bioreactor design, and downstream processing.

15.2 Butanol Fermentation

15.2.1 Acetone-Butanol-Ethanol (ABE) Fermentation

The major products of ABE fermentation include butyrate, acetate, butanol, acetone, ethanol, hydrogen, and carbon dioxide (Bahl *et al.*, 1986; Ezeji *et al.*, 2010). The growth profile of solventogenic clostridia is biphasic. During the first (exponential) growth phase, referred to as the acidogenic phase (acidogenesis), cells undergo rapid growth with a concomitant secretion of acetate and butyrate, which consequently lowers the pH of the growth medium (Bahl *et al.*, 1986; Bryant and Blaschek, 1988; Chen and Blaschek, 1999). At the end of the exponential growth phase, clostridia transition to the second stationary phase, known as the solventogenic phase (solventogenesis). During solventogenesis, clostridia reabsorb the secreted acids for the biosynthesis of acetone, butanol, and ethanol. The typical mass ratio of acetone, butanol, and ethanol is 3:6:1. The shift from acidogenesis to solventogenesis is triggered by the impending risk of cell death due to low pH (Costa and Moreira, 1983).

Increase in acid concentration in the medium elevates proton levels in the immediate environment external to the cytoplasm. Anaerobic bacteria must maintain cytoplasmic pH within a narrow range, hence the pH of the cytoplasm decreases dramatically parallel to the external environment during acidogenesis. This is because undissociated (unionized) acids diffuse into the cytoplasm due to the pH gradient between the cytoplasm and the growth medium (Ezeji *et al.*, 2010). In the cytoplasm with higher pH relative to the immediate environment, the acids dissociate, and this lowers cytoplasmic pH. That leads to the collapse of the membrane proton gradient and, consequently, cell death (Welker and Papoutsakis, 1999). Hence, reabsorption and conversion of acids to neutral compounds are an adaptive response by ABE-producing bacteria. Such an adaptive response prolongs the metabolic life of solventogenic clostridia, and provides them with a significant ecological advantage over non-acid-tolerant/metabolizing species. On the switch from acidogenesis to solventogenesis, additional carbon and electrons are channeled toward the formation of solvents. The accumulation of solvents (ABE), particularly butanol, can exceed a toxic threshold, which consequently halts the metabolism by fluidizing and disrupting membrane function. The final concentration of solvents at the end of fermentation is approximately 2% (v/v). Solventogenic *Clostridium* species sporulate during butanol production and the sporulation increases when the butanol concentration in the fermentation medium nears the toxic threshold (5 g/L) to avoid butanol toxicity.

Example 15.1: A novel strain of *Clostridium beijerinckii* genetically modified to abolish acetone production resulted in enhanced synthesis of butanol, but only when grown in a mixed-sugar medium. This was accompanied by an unwanted accumulation of acetone to a final concentration of 45% of the amount of butanol produced. Calculate the total product yield to substrates, taking into account the acetone levels if fermentation resulted in 17 g/L butanol and 3.2 g/L ethanol in a medium originally containing 36 g/L glucose, 18.5 g/L xylose, and 4.75 g/L arabinose, where the residual sugars were 14%, 24%, and 22% of the original concentrations, respectively.

Solution:

Step A. Find the total concentration of utilized substrates (sugars)

Glucose: $(100 - 14\%) \times 36$ g/L $= 31$ g/L

Xylose: $(100 - 24\%) \times 18.5$ g/L $= 14$ g/L

Arabinose: $(100 - 22\%) \times 4.75$ g/L $= 3.7$ g/L

Total   48.7 g/L

Step B. Find the total product concentration

17 g/L butanol + 3.2 g/L ethanol + 7.65 g/L acetone
(equivalent to 45% of 17 g/L butanol) = 27.85 g/L

Step C. $Yield_{product} = $ Total product concentration/Concentration of consumed substrate

$\rightarrow 27.85$ (g/L)$/48.7$ (g/L) $= 0.57$ g/g or 57%

15.2.2 Biochemical Pathway

ABE producers catabolize hexoses (six-carbon sugars) via the Embden–Meyerhof–Parnas pathway, where one molecule of hexose is converted to two molecules of pyruvate with the net generation of two molecules of both ATP and NADH (Girbal and Soucialle, 1994a, b; Girbal *et al.*, 1995; Gheshlaghi *et al.*, 2009; Cooksley *et al.*, 2012). Conversely, pentoses (five-carbon sugars) are metabolized through the phosphoketolase and pentose phosphate pathways (Gheshlaghi *et al.*, 2009; Cooksley *et al.*, 2012). The pentoses are converted into pentose 5-phosphate and catabolized by the transketolase–transaldolase sequence, thereby producing fructose 6-phopshate and glyceraldehydes 3-phosphate, which consequently enter the glycolytic pathway. Dissimilation of three molecules of pentose generates five molecules each of pyruvate, ATP, and NADH. Butanol biosynthesis in solventogenic clostridia is catalyzed by a series of enzymes, as depicted in Figure 15.2. These enzymes include acetoacetyl-CoA:acetate/butyrate-coenzyme A transferase (abridged as CoA transferase), butyraldehyde and butanol dehydrogenases, thiolase, 3-hydroxybutyryl-CoA dehydrogenase, and Crotonase.

During primary metabolism, solventogenic clostridia convert sugars to pyruvate, ATP, and NADH. Pyruvate is oxidatively decarboxylated to acetyl-CoA by pyruvate-ferredoxin oxidoreductase. The reducing power generated by the decarboxylation of pyruvate is in part dissipated by a

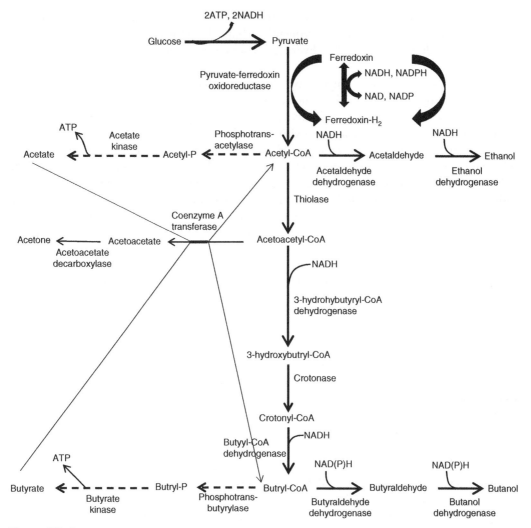

FIG. 15.2   Metabolic pathways and enzymes employed by solvent-producing clostridia for the production of acetone, butanol, and ethanol from pyruvate. Source: Ezeji *et al.*, 2010. Reproduced with permission of Springer.

hydrogenase in the form of hydrogen gas. This reaction is an important factor in the alkalinization of the cytoplasm (Cooksley *et al.*, 2012). During acidogenesis, hydrogen serves as an electron sink for both protons and electrons, which results in an increase in hydrogen generation. However, during solventogenesis, hydrogen production reduces as excess reducing equivalents (NADH, NADPH) are oxidized during the production of butanol and ethanol.

Acetyl-CoA is the precursor of organic acids (acetate and butyrate) and butanol (as well as acetone and ethanol). Acetyl-CoA metabolism branches into acetate via phosphotransacetylase and acetate kinase, while in the central pathway it is converted to acetoacetyl-CoA, the precursor of four-carbon solvents in a condensation reaction catalyzed by thiolase. The conversion of acetyl-CoA to acetoacetyl-CoA is critical to the ratio of two-carbon (acetate and ethanol), three-carbon (acetone), and four-carbon (butanol and butyrate) products. *In vivo* activity of thiolase is under the regulation of the coenzyme A/acetyl-CoA ratio. Acetoacetyl-CoA is converted to 3-hydroxybutyryl-CoA, which is then converted to crotonyl-CoA, and subsequently to butyryl-CoA in a series of reduction reactions catalyzed by 3-hydroxybutryryl-CoA dehydrogenase, crotonase, and butyryl-CoA dehydrogenase, respectively.

Butyryl-CoA is converted to butyrate by phosphotransbutyrylase and butyrate kinase, while butyraldehyde dehydrogenase and butanol dehydrogenase are responsible for the conversion of the former to butanol. The acetate and butyrate reabsorption by coenzyme A transferase serves as a detoxification strategy for reducing the toxicity of acids to cells. This function is the rate-limiting step in the ABE pathway. In *Clostridium acetobutylicum*, the most studied solvent-producing clostridia, the enzymes involved in butanol synthesis are encoded in different operons: *bdhB*, *bdhA*, *adhE2*, and *sol*. *bdhB* and *bdhA* are organized in contiguous, monocistronic operons. The *sol* operon consists of a bifunctional butyraldehyde/butanol dehydrogenase (adhE), and the two subunits of CoA transferase (ctfA and ctfB). The transcriptional factor, SpoA, functions as the master regulator of sporulation and as an initiator of solvent production by the activation of adc (acetoacetate decarboxylase), adhE, and ctfAB in *C. acetobutylicum* and *C. beijerinckii*.

Electron flux, hence the redox state of the cell, is central to the product profile and yield by clostridia (Welker and Papoutsakis, 1999). It is also involved in regulation of the switch from acidogenesis to solventogenesis. Consequently, pyruvate-ferredoxin oxidoreductase is a key determinant of the metabolic state of ABE-producing clostridia. The cleavage of pyruvate by pyruvate-ferredoxin oxidoreductase in the presence of coenzyme A to produce carbon dioxide and acetyl-CoA is accompanied by the conversion of oxidized ferredoxin to its reduced state. Clostridial ferredoxin yields one molecule of each of acetyl-CoA and carbon dioxide per molecule of ferredoxin reduced, with the concomitant transfer of two electrons. Clostridial pyruvate-ferredoxin oxidoreductase contains an iron-sulfur-chromophor in the redox center, which transports electron flow from pyruvate to ferredoxin. The pyruvate-ferredoxin oxidoreductase of clostridia is highly unstable and hyper-oxygen sensitive. It loses 50% of its activity following 1 hr exposure to oxygen.

Although lactate is rarely a major product of ABE fermentation, it can be accumulated appreciably by solventogenic clostridia under stressful conditions such as iron and sulfur limitation, or carbon monoxide gassing at neutral pH (Zhou and Yang, 2004; Bahl *et al.*, 1986). Thus, lactate production is triggered when the conversion of pyruvate to acetyl-CoA is moderately blocked. This block increases the intracellular levels of fructose-1,6-diphosphate, which therefore activates lactate dehydrogenase. Under stress conditions such as iron limitation or a high pH, pyruvate may be converted to lactate via the action of lactate dehydrogenase, probably to speed up the regeneration of NAD+ necessary for the continuation of glycolysis.

15.2.3 Stoichiometry and Product Yield

The stoichiometry of ABE fermentation is an important aspect of butanol production. Theoretical yields of the various products of ABE fermentation are presented in Table 15.2, which can be summarized in the overall stoichiometric equation of glucose conversion to solvents and acids:

$$5C_6H_{12}O_6 + 3H_2O \rightarrow C_4H_9OH + CH_3-CO-CH_3 + 2C_2H_5OH + 2CH_3-COOH$$
$$+ C_3H_7-COOH + 11CO_2 + 10H_2 + H_2O \tag{15.1}$$

Theoretically, 5 moles of glucose is expected to yield 1 mole of acetone, 1 mole of butanol, 2 moles of ethanol, 2 moles of acetic acid, and 1 mole of butyric acid, with the evolution of 11 moles and

Table 15.2 The stoichiometry of ABE fermentation and theoretical product yield

Reactions	Product yield (Y) (g/g)
$C_6H_{12}O_6 + H_2O \rightarrow CH_3COCH_3\,(\textbf{Acetone}) + 3CO_2 + 4H_2$	0.32
$C_6H_{12}O_6 + H_2O \rightarrow C_4H_9OH\,(\textbf{Butanol}) + 2CO_2 + 2H_2O$	0.41
$C_6H_{12}O_6 + H_2O \rightarrow 2C_2H_5OH\,(\textbf{Ethanol}) + 2CO_2 + H_2O$	0.51
$C_6H_{12}O_6 + 2H_2O \rightarrow 2CH_3COOH\,(\textbf{Acetic acid}) + 2CO_2 + 4H_2$	0.67
$C_6H_{12}O_6 + H_2O \rightarrow C_3H_7COOH\,(\textbf{Butyric acid}) + 2CO_2 + 2H_2 + H_2O$	0.49

Note: Product yield (Y): Yield of ethanol, butanol, acetone, butyric acid, or acetic acid per gram of glucose consumed.

10 moles of carbon dioxide and hydrogen, respectively (Dai *et al.*, 2012; Papoutsakis, 1984). Hence, the theoretical yield of butanol is estimated to be up to 0.94 mol/mol of glucose in traditional ABE fermentation or 1.33 mol/mol of glucose in mixotrophic fermentation, where glucose and the CO_2 generated during fermentation are maximally utilized in the presence of sufficient H_2 (Tracy, 2012). These translate into a total solvent mass yield of approximately 33–40%. However, in practice yields around 37% are rarely sustained, a factor that currently impedes attempts at industrializing ABE fermentation (Gapes, 2000; Dai *et al.*, 2012; Papoutsakis, 1984). Butanol toxicity to fermenting cells and the production of acetone and CO_2 are major limitations to butanol yield during ABE fermentation. These have sparked tremendous research interests and efforts at constructing more robust strains capable of producing and tolerating higher butanol titers. It is thought that increasing total solvent titer to about 28 g/L in a batch fermentation time of about 40–60 hr would be viable for commercialization.

The biochemical reactions of the ABE fermentation pathway are stoichiometrically represented by the following equations (Dai *et al.*, 2012; Papoutsakis, 1984):

Glucose + 0.873NADH + 14.85ATP $\leftrightarrow$ 6 biomass (cell dry weight)

Glucose $\rightarrow$ 2 pyruvate + 2NADH + 2ATP

Pyruvate $\rightarrow$ acetyl-CoA + CO_2 + reduced ferredoxin

Pyruvate $\rightarrow$ acetoin + 2CO_2

Acetyl-CoA + 2NADH $\rightarrow$ ethanol

Acetyl-CoA $\leftrightarrow$ acetate + ATP

Acetyl-CoA $\leftrightarrow$ acetoacetyl-CoA

Acetoacetyl-CoA + acetate $\rightarrow$ acetone + CO_2 + acetyl-CoA

Acetoacetyl-CoA + butyrate $\rightarrow$ acetone + CO_2 + butyryl-CoA

Acetoacetyl-CoA + 2NADH $\leftrightarrow$ butyryl-CoA

Butyryl-CoA $\leftrightarrow$ butyrate + ATP

Butyryl-CoA + 2NADH $\rightarrow$ butanol

Reduced ferredoxin $\rightarrow$ H_2

Reduced ferredoxin $\leftrightarrow$ NADH

15.2.4 Microbiology

Native ABE producers, generally referred to as solventogenic clostridia, are hetero-fermentative, obligately anaerobic Gram-positive, spore-forming bacteria. They are mesophilic, motile, rod-shaped bacteria with oval, sub-terminal spores and peritrichous flagella. During the early exponential phase, the cells are predominantly long, filamentous, and highly motile. During the solventogenic stage (stationary phase), cells shorten and become rounder in shape with reduced motility (Figure 15.1). They are ubiquitous in soil, lake sediments, well water, and animal gut. Until recently, the identities of solvent-producing species/strains of clostridia were muddled up (Keis *et al.*, 2001).

Following the use of molecular methods, including polymerase chain reaction (PCR) amplification of 16S rRNA, sequencing, DNA fingerprinting, and DNA–DNA hybridization (Keis *et al.*, 2001), ABE-producing clostridia have been reclassified into four broad species: *Clostridium acetobutylicum*, *C. beijerinckii*, *C. saccharobutylicum*, and *C. saccharoperbutylacetonicum* (Table 15.3). Different strains of *C. beijerinckiis* produce 2-propanol at varying concentrations. Some strains, such as B-593, convert the entire acetone produced to 2-propanol by means of a secondary alcohol dehydrogenase, while others produce varying amounts of both acetone and 2-propanol. *C. pasteurianum* exhibits ABE-fermentative capacity when grown on glycerol as the sole carbon source. *C. tetanomorphum* MG1 and *C.*

Table 15.3 Solvent-producing clostridia

Species	Strain(s)	Products	Remarks
C. acetobutylicum	ATCC 824 DMZ 1733	Acetone Butanol Ethanol	Co-metabolizes glycerol and sugars
C. beijerinckii	NCIMB 8052	Acetone Butanol Ethanol	Co-metabolizes glycerol and sugars
C. beijerinckii	B-593	2-propanol Butanol Ethanol	Co-metabolizes glycerol and sugars
C. pasteurianum	ATCC 6013 and I-53	Acetone Butanol Ethanol	Utilizes glycerol as the sole carbon source
C. tetanomorphum	MG1	Butanol Ethanol	N/A
*C. thermosaccharolyticum	ATCC 31960	Butanol Ethanol	Co-metabolizes glycerol and sugars
C. thermocellum	ATCC 27405	Ethanol	N/A
C. cellulolyticum	ATCC 35319	Ethanol	
C. butyricum	ATCC 859 and 19398	1,3-propanediol	Utilizes glycerol as the sole carbon source
C. saccharobutylicum	DSM 13864	Acetone Butanol Ethanol	Co-metabolizes glycerol and sugars
C. saccharoperbutylacetonicum	N1-4	Acetone Butanol Ethanol	Co-metabolizes glycerol and sugars

NA: not available;
*Reclassified as *Thermoanaerobacterium thermosaccharolyticum* (Keis *et al.*, 2001)

thermosaccharolyticum (saccharolytic, thermophilic) do not produce acetone; however, they produce butanol and ethanol in equimolar ratios. Thermophilic solventogenic clostridia such as *C. thermocellum* and *C. cellulolyticum*, which are capable of cellulose utilization, produce ethanol alone. *C. butyricum* is a potent producer of 1,3-popanediol when grown on glycerol as the sole source of carbon.

The quest for robust ABE-producing organisms able to produce solvent titers high enough for commercial-scale production has led to cloning and heterologous expression of the ABE pathway in non-native producers such as *Escherichia coli*, *Pseudomonas putida*, *Bacillus subtilis*, *lactobacillus brevis*, and *Saccharomyces cerevisiae* (Dürre, 2011; Berezina *et al.*, 2010; Nielsen *et al.*, 2009). However, while butanol was produced by the resulting strains, titers were extremely low. The highest concentration was observed with *E. coli*, which produced 1.2 g/L butanol, which is only 10–11% of the levels produced by native producers (solventogenic clostridia). Poor expression of some of the enzymes of the ABE pathway and butanol toxicity are some of the factors underlying the low productivity observed in engineered non-native producers.

15.3 Factors Affecting Butanol Fermentation

The metabolism of ABE producers is complex, which in part explains the relatively low levels of success despite extensive efforts at constructing homo-fermentative or overproducing strains by genetic manipulation. Intra- and extracellular pH, acid concentrations and their cellular reabsorption capacity, availability of reducing equivalents (NADH/NADPH) and the ATP pool, nutritional composition of the fermentation broth, and feedback (product) inhibition are major factors that affect ABE fermentation.

15.3.1 pH

Change in the pH of the broth during fermentation and the butyric acid concentration of the fermentation broth are key factors that trigger the onset of solventogenesis (Bryant and Blaschek, 1988; Chen and Blaschek, 1999; Costa and Moreira, 1983; Guo *et al.*, 2012). Compared to acetic acid, butyric acid has a lower toxicity threshold on solventogenic *Clostridium species*; hence, a slight increase in concentration initiates reassimilation (to stave off the impending toxicity) and consequent conversion to solvents. In the pH range of 5.0–6.8, cell growth is favored, with optimal growth occurring at about pH 6.0. Conversely, a lower pH in the range of 4.7–5.5 favors solvent accumulation. As fermentation progresses, organic acid accumulation increases, with an increase in cell biomass resulting in a pH drop. Although other factors contribute to the dynamics of the switch from acidogenesis to solventogenesis, a drop in pH to below 5.5 amplifies acid toxicity, which leads to reassimilation of the accumulated acids and their conversion to neutral solvents (acetone, butanol, and ethanol) to relieve acid stress. Consequently, maintaining a medium pH at higher levels by buffering reduces the rate at which transition of acidogenesis to solventogenesis occurs, and increases the levels of acid tolerated by *Clostridium* species.

15.3.2 Availability of Co-factors (NADH)

NADH supplies the reducing power for bioreductive reactions as an electron donor. Four molecules of NADH are required for the biosynthesis of one mole of butanol (Gheshlaghi *et al.*, 2009; Ezeji *et al.*, 2010; Cooksley *et al.*, 2012). Under pH-controlled conditions where a higher pH (5.5) is maintained during the exponential growth phase (to favor cell growth) before a subsequent switch of pH to 4.9–5.2 (which supports higher solvent production), a lower NADH/NAD⁺ ratio is obtained compared to non-pH-controlled fermentation (Guo *et al.*, 2012). This indicates that solvent biosynthesis is accompanied by rapid consumption of NADH. Growth under conditions that promote NADH generation increases flux toward butanol and ethanol (alcohologenesis), with concomitant increases in the activities of NADH-dependent alcohol and butyraldehyde dehydrogenases, and ferredoxin:NAD(P)$^+$ reductases (Gheshlaghi *et al.*, 2009; Ezeji *et al.*, 2010; Cooksley *et al.*, 2012; Welker and Papoutsakis, 1999; Girbal and Soucialle, 1994a, b; Girbal *et al.*, 1995). Such conditions include growth in glucose-glycerol- or glucose-glycerol-pyruvate-based media, addition of neutral red (an artificial electron carrier), and reduction of intracellular hydrogenase activity by addition of methyl viologen or by gassing the bioreactor/culture medium with carbon monoxide (Girbal and Soucialle, 1994a, b; Girbal *et al.*, 1995). Conversely, typical solventogenesis in ABE fermentation is associated with increases in the expression of NADPH-dependent alcohol and butyraldehyde dehydrogenases and of the enzymes that catalyze acetone formation.

15.3.3 Medium Composition

Carbon is the major metabolic precursor of ABE, and solventogenic clostridia have a broad carbon substrate utilization spectrum (Section 15.5). However, they exhibit varying growth and solvent profiles relative to available carbon sources, with glucose being the most favored substrate. Metal ions such as Mg^{2+}, Mn^{2+}, Fe^{2+}, and Ca^{2+} play critical roles in the metabolism of solvent-producing clostridia (Bahl *et al.*, 1986; Han *et al.*, 2013; Vasileva *et al.*, 2012). These ions function largely as co-factors of key solventogenic enzymes, and also participate in metabolic and genetic regulation and modulation. Fe^{2+} is particularly crucial to ABE fermentation. In addition to its role as a key co-factor in solventogenic clostridia, Fe^{2+} plays a central role in redox homeostasis. Furthermore, principal aspects of the metabolism of *C. acetobutylicum* and *C. beijerinckii* are iron dependent owing to the involvement of iron-rich proteins. For instance, the iron–sulfur-containing pyruvate:ferredoxin oxidoreductase along with its electron acceptor, ferredoxin, is vital to solventogenic metabolism by virtue of its role in the extraction of reducing equivalents by decarboxylation (Bahl *et al.*, 1986). Calcium ion (Ca^{2+}), on the other hand, exerts wider pleiotropic effects on solvent-producing clostridia by enhancing cell growth, sugar transport and utilization, and solvent production and tolerance (Han *et al.*, 2013).

> **Example 15.2:** In a fermentation using thermochemically pretreated corn stover that originally contained 45 g/L glucose, 17.5 g/L xylose, and 6 g/L arabinose as substrate, *C. acetobutylicum* produced 4 g/L butanol, 1.3 g/L acetone, and 0.15 g/L ethanol after 72 hr fermentation. If the residual sugar concentrations in the bioreactor were 24 g/L, 15 g/L, and 4.9 g/L glucose, xylose, and arabinose, respectively:
>
> 1. Calculate the butanol yield.
> 2. Explain the reason for the abysmally low solvent titer and yield.
> 3. Outline possible remedies to the problems in (2).
>
> ***Solution:***
>
> 1. Find the total concentration of utilized substrates (sugars):
>
>   Glucose : $45 - 24$ = 21 g/L
>
>   Xylose : $17.5 - 15$ = 2.5 g/L
>
>   Arabinose : $6 - 4.9$ = 1.1 g/L
>
>   Total   24.6 g/L
>
>   Butanol yield = 4.0/24.6 = 0.16 g butanol/g sugar

2. Generation of microbial inhibitory compounds such as furfural, 5-hydroxymethyl furfural, syringaldehyde, furoic acid, and acetic acid, among several other compounds that arise from dehydration/degradation reactions of sugars released during dilute acid pretreatment at elevated temperatures, and products of lignin degradation, account for poor growth and consequently poor butanol production in lignocellulosic biomass hydrolysates such as corn stover hydrolysates.

3. Currently, biological, chemical, and physical methods such as overliming, ion-exchange chromatography, and enzymatic treatment are available for the removal of lignocellulose-derived microbial inhibitory compounds. However, these methods add to the cost of butanol production. Consequently, tremendous research efforts are currently being devoted to the genetic construction of more robust strains that are capable of simultaneous utilization of biomass-derived sugars and detoxification of the co-produced inhibitors. Other possibilities include design of novel technologies for lignocellulose hydrolysis, which are capable of reduced generation of inhibitors and efficient release of fermentable sugars.

15.3.4 Product Inhibition

Another factor that influences the outcome of ABE fermentation is product inhibition, also known as feedback inhibition (Ezeji *et al.*, 2010). The products of ABE fermentation, particularly butanol, are severely toxic to vegetative cells. Butanol permealizes the bacterial cell membrane and damages macromolecular structures, thereby causing cell death. Consequently, higher concentrations of butanol, often around 7–15 g/L, lead to increased stress on cells, culminating in shutdown of metabolism. Consequently, this results in the onset of sporulation to evade butanol toxicity. Sporulation is a major factor hampering yield and productivity, and, thus, the commercialization of ABE fermentation.

15.4 Substrates for Butanol Fermentation

Solvent-producing clostridia can ferment six-carbon sugars (hexoses: e.g., glucose, galactose, mannose, and fructose), five-carbon sugars (pentoses: e.g., xylose and arabinose), disaccharides (lactose, cellobiose, and sucrose), and starch; however, glucose is often the preferred substrate (Bahl *et al.* 1986; da Silva *et al.*, 2009; Dürre, 2011; Ezeji *et al.*, 2007). Inexpensive carbon sources – namely, glycerol, lactose, pentoses (from lignocellulosic biomass), or inulin from plant wastes (biodiesel byproduct, agro-wastes, etc.) – can also be used for butanol fermentation (Ujor *et al.*, 2014a; da Silva et al., 2009; Dürre, 2011; Ezeji *et al.*, 2007). A major challenge to the use of these cheaply available substrates is low butanol yield and productivity. Although the butanol titer on lactose is approximately 7.3 g/L, 13.8 g/L on inulin, and 17 g/L on glycerol using *C. pasteurianum*, productivity in each case is low for commercialization (Ujor *et al.*, 2014b; Marchal *et al.*, 1985; Sabra *et al.*, 2014). Ongoing metabolic engineering efforts continue to target these areas for improvement, with the intention of constructing more robust strains proficient at utilizing these cheap substrates. Among these substrates, lignocellulosic biomass is the most abundant and renewable source of fermentable sugars (hexoses and pentoses).

15.5 Advanced Butanol Fermentation Techniques and Downstream Processing

Butanol fermentation is product limiting owing to butanol toxicity to metabolizing microbial cells. The toxicity of butanol is the major reason for the accumulation of low concentrations of butanol in fermentation broth using conventional batch fermentation (ca. 1.5% w/w), unlike ethanol fermentation where concentrations exceeding 15% (w/w) are commonly obtained from commercial batch fermentations. To realize the full potential of butanol-producing microorganisms with respect to butanol concentration and productivity, various simultaneous butanol fermentation and recovery systems are commonly employed at laboratory scale to remove butanol from the bioreactor during fermentation. By keeping the concentration of butanol in the bioreactor below the threshold of toxicity, the growth rates of butanol-producing microorganisms, rates of substrate utilization, and butanol productivity could be remarkably enhanced. Notably, *in situ* butanol recovery techniques during fermentation should be microbial friendly, scalable, and non-fouling, and should enhance butanol productivity.

15.5.1 Gas Stripping

Gas stripping is a simple technique that allows for the selective removal of volatiles from fermentation broth, and does not involve the use of membranes. It can be integrated with butanol fermentation to recover butanol simultaneously during fermentation. Given that butanol fermentation is product limiting, this technique is useful for keeping butanol concentration in the bioreactor below the toxic threshold, such that fermentation can proceed unimpeded until all the sugars in the medium are fully converted to butanol.

Integrated butanol fermentation and recovery by gas stripping entail bubbling oxygen-free nitrogen or fermentation gases (CO_2 and H_2) through the fermentation broth (Figure 15.3). As the gas is bubbled through the bioreactor, it captures butanol (enriched gas), which is condensed at the cold trap equipped with a condenser and a butanol collector. The butanol-free gas is recycled back to the bioreactor to capture more butanol. The recovered butanol is pumped out of the collecting unit to avoid its recirculation through the loop. Performance of a gas-stripping process can be evaluated by its selectivity, which is the weight ratio of the target product (butanol) and other products (e.g., water) in the condensate to the weight ratio of the target product and other products (water, etc) in the fermentation broth (bioreactor; Ezeji *et al.*, 2003, 2004, 2005). Using butanol as an example, selectivity is represented by the α symbol and is defined mathematically as:

$$\alpha = \frac{(Y/(1-Y))}{(X/(1-X))} \tag{15.2}$$

where Y is butanol concentration in % in the condensate; X is butanol concentration in % in the fermentation broth; and gas stripping has a butanol selectivity range of 2–35 depending on the concentration of butanol in the bioreactor. At low butanol selectivity (<10), the gas-stripping process removes a large amount of water from the bioreactor and generates highly dilute butanol in the recovered stream, which will require a significant amount of energy for its concentration and purification.

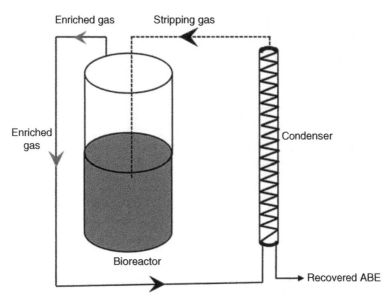

FIG. 15.3   Schematic of gas-stripping system for butanol recovery.

Example 15.3: Integrated ABE fermentation and simultaneous *in situ* butanol recovery by gas stripping were operated for 60 hr. The condensate contained 18.74% butanol, while the residual butanol in the bioreactor was 1.26%. Calculate the butanol selectivity of the process.

Solution:

Using Equation 15.2, $\alpha = \dfrac{Y}{1-Y} \div \dfrac{X}{1-X}$

Therefore $\alpha = \dfrac{Y}{1-Y} \times \dfrac{1-X}{X}$

$$\alpha = \left(\frac{0.187}{1-0.187}\right) \times \left(\frac{1-0.0126}{0.0126}\right) = 0.23 \times 78.11 = 18.0$$

15.5.2 Vacuum Fermentation

Vacuum fermentation is a technique in which fermentation is conducted under vacuum such that the volatile products of interest are vaporized and removed at low pressure. In integrated butanol fermentation and *in situ* product recovery by vacuum fermentation, broth boils at a low temperature to facilitate instant evaporation and recovery of butanol as it is formed (Figure 15.4). Simultaneous butanol fermentation and recovery by vacuum is an advanced fermentation process that requires no membrane, sparger, or agitation, and is microbe friendly. Vacuum-assisted product recovery offers some invaluable advantages such as the utilization of concentrated feed during butanol fermentation, high bioreactor productivity, absence of membrane requirement, and reduced susceptibility to fouling or clogging (Mariano *et al.*, 2011).

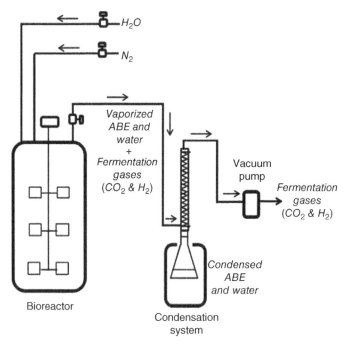

FIG. 15.4 Schematics of vacuum fermentation system for butanol recovery.

The vacuum fermentation technique is not new for ethanol fermentation, as it has long been demonstrated as an effective method for the continuous recovery of ethanol from fermentation beer (Cysewski and Wilke, 1977).

Although vacuum technology is suitable for recovering most products that are more volatile than water, its use has been limited to ethanol and acetaldehyde recovery (Roffler *et al.*, 1984), while excluding butanol, which has a higher boiling point (118 °C) than water. However, the realization that butanol and water form a heteroazeotropic mixture, which at concentrations below 77 g/L boils at a lower temperature than the boiling point of either butanol or water, led to the successful application of this technology to simultaneous butanol fermentation and recovery (Mariano *et al.*, 2011, 2012).

15.5.3 Liquid–Liquid Extraction

Liquid–liquid extraction is a technique in which the headspace of a bioreactor is partially or completely filled with a water-insoluble organic extractant, resulting in direct contact with the fermentation broth (Figure 15.5). Given that butanol is more soluble in the organic (extractant) phase than in the aqueous (fermentation broth) phase, butanol selectively concentrates in the organic phase, thereby lowering its concentration in the fermentation broth below the threshold of toxicity to butanol-producing microbial cells (Ezeji *et al.*, 2007). Following fermentation, the concentrated butanol in the extraction solvent is recovered by back extraction into another extraction solvent or by distillation. The extractant of choice for *in situ* butanol recovery is oleyl alcohol, because it is considerably non-toxic to microorganisms, as well as being a high-quality extractant. Nonetheless, this advanced fermentation system has some technical challenges associated with its operation, which

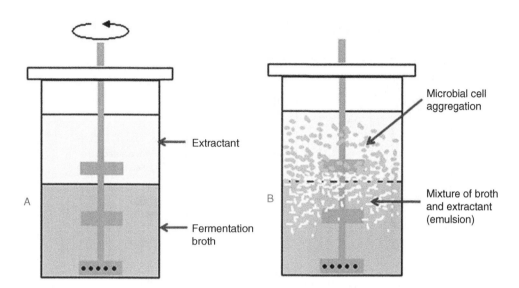

FIG. 15.5   Schematic diagram showing addition of extractant (e.g., oleyl alcohol) to the fermentation broth (Bioreactor A) and the formation of emulsions and cell aggregation at the liquid–liquid interface (rag layers), extending to the whole extracting solvent during fermentation (Bioreactor B).

include extraction solvent toxicity to butanol-producing microbial cells, formation of an emulsion, loss of extraction solvent, and the accumulation of microbial cells at the extractant and fermentation broth interphase, forming a rag layer (Ezeji *et al.*, 2007, Karcher *et al.*, 2005).

15.5.4 Pervaporation

Pervaporation is a technique in which volatile products are selectively removed from broths during fermentation using a low vacuum and a membrane. A pervaporative fermentation system is generally composed of bioreactor, feed pump, membrane module, condenser, recirculation tube, and vacuum pump (Figure 15.6). This technique has been used extensively to maintain the butanol concentration in the bioreactor below the toxicity threshold during laboratory-scale butanol fermentation (Qureshi *et al.*, 1992). In this system, butanol diffuses through the membrane as a vapor, followed by recovery by condensation. Membranes used in pervaporation are either hydrophilic or hydrophobic, depending on the characteristics of the target products for recovery.

Mass transfer of butanol in a pervaporative butanol fermentation system involves three steps: adsorption of butanol into the upstream surface of the membrane in the pervaporation module; diffusion of dissolved butanol through the membrane; and absorption of dissolved butanol into permeate vapor at the downstream surface of the membrane (Shao and Huang, 2007; Ezeji *et al.*, 2010). Polydimethylsiloxane (PDMS) is the standard pervaporative membrane material mostly employed in ethanol and butanol fermentations because the ethanol–water separation factor for PDMS membranes ranged from 4.4–10.8, while the reported butanol–water separation factor for PDMS ranged from 40–60, which is 6–10 times more than that of ethanol–water (Vane, 2005). Application of pervaporation to batch and fed-batch butanol fermentation systems has been demonstrated elsewhere (Qureshi *et al.*, 1992; Vane, 2005).

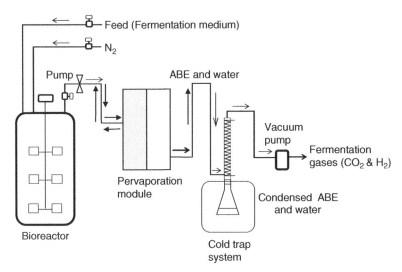

FIG. 15.6 Schematic diagram of pervaporative fermentation applied to butanol production. ABE stands for acetone, butanol, and ethanol, respectively. The butanol-free broth containing residual substrates and microbial cells, also called retentate, is recycled back to the bioreactor, where these residual substrates are converted to ABE.

References

Bahl, H., Gottwald, M., Kuhn, A., Rale, V., Andersch, W., and Gottschalk, G. (1986). Nutritional factors affecting the ratio of solvents produced by *Clostridium acetobutylicum*. *Applied and Environmental Microbiology* 52: 169–72.

Berezina, O.V., Zakharova, N.V., Brandt, A., Yarotsky, S.V., Schwarz, W.H., and Zverlov, V.V. (2010). Reconstructing the clostridial n-butanol metabolic pathway in *Lactobacillus brevis*. *Applied Microbiology and Biotechnology* 87: 635–46.

Bryant, D.L., and Blaschek, H.P. (1988). Buffering as a means for increasing the growth and butanol production of *Clostridium acetobutylicum*. *Journal of Industrial Microbiology* 3: 49–55.

Chen, C.K., and Blaschek, H.P. (1999). Effect of acetate on molecular physiological aspects of *Clostridium beijerinckii* NCIMB 8052 solvent production and strain degeneration. *Applied and Environmental Microbiology* 65: 499–505.

Cooksley, C.M., Zhang, Y., Wang, H., Redl, S., Winzer, K., and Minton, N.P. (2012). Targeted mutagenesis of the *Clostridium acetobutylicum* acetone-butanol-ethanol fermentation pathway. *Metabolic Engineering* 14: 630–41.

Costa, J.M., and Moreira, A.R. (1983). Growth inhibition kinetics of the acetone-butanol-ethanol fermentation. In: H.W. Blanch, E.T. Papoutsakis, and G. Stephanopoulos (eds.), *Foundations of Biochemical Engineering*. Washington, DC: American Chemical Society, pp. 501–2.

Cysewski, G.R., and Wilke, C.R. (1977). Rapid ethanol fermentations using vacuum and cell recycle. *Biotechnology and Bioengineering* 19: 1125–43.

Dai, Z., Dong, H., Zhu, Y., Zhang, Y., Li, Y., and Ma, Y. (2012). Introducing a single secondary alcohol dehydrogenase into butanol-tolerant *Clostridium acetobutylicum* Rh8 switches ABE fermentation to high level IBE fermentation. *Biotechnology for Biofuels* 5: 44–54.

da Silva, G.P., Mack, M., and Contiero, J. (2009). Glycerol: A promising and abundant carbon source for industrial microbiology. *Biotechnology Advances* 27: 30–39.

Dürre, P. (2008). Fermentative butanol production: Bulk chemical and biofuel. *Annals of the New York Academy of Sciences* 1125: 353–62.

Dürre, P. (2011). Fermentative production of butanol: The academic perspective. *Current Opinion in Biotechnology* 22: 1–6.

Ezeji, T.C., Qureshi. N., and Blaschek, H.P. (2003). Production of butanol by *Clostridium beijerinckii* BA101 and *in-situ* recovery by gas stripping. *World Journal of Microbiology Biotechnology* 19: 595–603.

Ezeji, T.C., Qureshi, N., and Blaschek, H.P. (2004). Acetone-butanol-ethanol (ABE) production from concentrated substrate: Reduction in substrate inhibition by fed-batch technique and product inhibition by gas stripping. *Applied Microbiology and Biotechnology* 63: 653–8.

Ezeji, T.C., Karcher, P.M., Qureshi, N., and Blaschek, H.P. (2005). Improving the performance of a gas stripping-based recovery system to remove butanol from *Clostridium beijerinckii* fermentation. *Bioprocess and Biosystems Engineering* 27: 207–14.

Ezeji, T.C., Qureshi, N., and Blaschek, H.P. (2007). Bioproduction of butanol from biomass: From genes to bioreactors. *Current Opinion in Biotechnology* 18: 220–27.

Ezeji, T., Milne, C., Price, N.D., and Blaschek, H.P. (2010). Achievements and perspectives to overcome the poor solvent resistance in acetone and butanol-producing microorganisms. *Applied Microbiology and Biotechnology* 85: 1697–712.

Gapes, J.R. (2000). The economics of acetone-butanol fermentation: Theoretical and market considerations. *Journal of Molecular Microbiology and Biotechnology* 2: 27–32.

Gheshlaghi, R., Scharer, J.M., Moo-Young, M., and Chou, C.P. (2009). Metabolic pathways of clostridia for producing butanol. *Biotechnology Advances* 27: 764–81.

Girbal, L., and Soucaille, P. (1994a). Regulation of *Clostridium acetobutylicum* metabolism as revealed by mixed-substrate-steady continuous cultures: Role of NADH/NAD ratio and ATP pool. *Journal of Bacteriology* 176: 6433–8.

Girbal, L., and Soucaille, P. (1994b). Regulation of solvent production in *Clostridium acetobutylicum*. *Trends in Biotechnology* 16: 11–16.

Girbal, L., Croux, C., Vasconcelos, S., and Soucaille, P. (1995). Regulation of metabolic shifts in *Clostridium acetobutylicum* ATCC 824. *FEMS Microbiology Reviews* 17: 287–97.

Guo, T., Sun, B., Jiang, M., *et al.* (2012). Enhancement of butanol production and reducing power using a two-stage controlled-pH strategy in batch culture of *Clostridium acetobutylicum* XY16. *World Journal of Microbiology and Biotechnology* 28: 2551–8.

Han, B., Ujor, V., Lai, L.B., Gopalan, V., and Ezeji, T.C. (2013). Use of proteomic analysis to elucidate the role of calcium in acetone-butanol-ethanol fermentation by *Clostridium beijerinckii* NCIMB 8052. *Applied and Environmental Microbiology* 79: 282–93.

Karcher, P.M., Ezeji, T.C., Qureshi, N., and Blaschek, H.P. (2005). Microbial production of butanol: Product recovery by extraction. In: T. Satyanarayann and B.N. Johri (eds.), *Microbial Diversity: Current Prospective and Potential Applications*. New Delhi: I.K. International, pp. 865–80.

Keis, S., Shaheen, R., and Jones D.T. (2001). Emended description of *Clostridium acetobutylicum* and *Clostridium beijerinckii*, and descriptions of *Clostridium saccharoperbutylicum* sp. nov. and *Clostridium saccharobutylicum* sp. nov. *International Journal of Systematic and Evolutionary Microbiology* 51: 2095–103.

Marchal, R., Blanchet, D., and Vandecasteele, J.P. (1985). Industrial optimization of acetone-butanol-ethanol fermentation: A study of the utilization of Jerusalem artichoke. *Applied Microbiology and Biotechnology* 23: 92–8.

Mariano, A.P., Qureshi, N., Filho, R.M., and Ezeji, T.C. (2011). Bioproduction of butanol in bioreactors: New insights from simultaneous *in situ* butanol recovery to eliminate product toxicity. *Biotechnology and Bioengineering* 108: 1757–65.

Mariano, A.P., Qureshi, N., Filho, R.M., and Ezeji, T.C. (2012). Assessment of in situ butanol recovery by vacuum during acetone butanol ethanol (ABE) fermentation. *Journal of Chemical Technology and Biotechnology* 87: 334–40 .

Ni, Y., and Sun, Z. (2009). Recent progress in industrial fermentative production of acetone-butanol-ethanol by *Clostridium acetobutylicum* in China. *Applied Microbiology and Biotechnology* 83: 415–23.

Nielsen, D.R., Leonard, E., Yoon, S.-H., Tseng, H.-C., Yuan, C., and Prather, K.L.J. (2009). Engineering alternative butanol production platforms in heterologous bacteria. *Metabolic Engineering* 11: 262–73.

Papoutsakis, E.T. (1984). Equations and calculations for fermentations of butyric acid bacteria. *Biotechnology and Bioengineering* 26: 174–87.

Qureshi, N., Maddox, I.S., and Friedl, A. (1992). Application of continuous substrate feeding to the ABE fermentation: Relief of product inhibition using extraction, perstraction, stripping and pervaporation. *Biotechnology Progress* 8: 382–90.

Roffler, S.R., Blanch, H.W., and Wilke, C.R. (1984). *In situ* recovery of fermentation products. *Trends in Biotechnology* 2: 129–36.

Sabra, W., Groeger, C., Sharma, P.N., and Zeng, A.-P. (2014). Improved n-butanol production by a non-acetone producing *Clostridium pasteurianum* DSMZ 525 in mixed substrate fermentation. *Applied Microbiology and Biotechnology* 98: 4267–76.

Shao, P., and Huang, R.Y.M. (2007). Polymeric membrane pervaporation. *Journal of Membrane Science* 287: 162–79.

Tracy, B.P. (2012). Improving butanol fermentation to enter the advanced biofuel market. *mBio* 3(6): 1–3.

Ujor, V., Bharathidasan A.K., Cornish, K. and Ezeji, T.C. (2014a). Feasibility of producing butanol from industrial starchy food wastes. *Applied Energy* 136: 590–98.

Ujor, V., Bharathidasan A.K., Cornish, K., and Ezeji, T.C. (2014). Evaluation of industrial diary waste (milk dust powder) for acetone-butanol-ethanol production by solventogenic *clostridium* species. *SpringerPlus* 3: 387.

Vane, A. (2005). Review of pervaporation for product recovery from biomass fermentation processes. *Journal of Chemical Technology and Biotechnology* 80: 603–29.

Vasileva, D., Janssen, H., Honicke, D., Ehrenreich, A., and Bahl, H. (2012). Effect of iron limitation and *fur* gene inactivation on the transcriptional profile of the strict anaerobe *Clostridium acetobutylicum*. *Microbiology* 158: 18–29.

Welker, N.E., and Papoutsakis, E.T. (1999). Metabolic flux analysis elucidates the importance of the acid-formation pathways in regulating solvent production by *Clostridium acetobutylicum*. *Metabolic Engineering* 1: 206–13.

Zhou, Y., and Yang, S.-T. (2004). Effect of pH on metabolic pathway shift in fermentation of xylose by *Clostridium tyrobutyricum*. *Journal of Biotechnology* 110: 143–57.

Exercise Problems

15.1. What is biphasic fermentation? Discuss its biotechnological significance.

15.2. What are the two major acids produced by solventogenic *Clostridium* species?

15.3. How do solventogenic *Clostridium* species mitigate acid build-up in the bioreactor during ABE fermentation?

15.4. Discuss the reactions in the ABE biosynthetic pathway where ATPs are generated (exclude the glycolytic pathway).

15.5. Discuss the reactions in the ABE biosynthetic pathway where nicotinamide adenine dinucleotide is reduced and oxidized (NADH, NAD$^+$).

15.6. Briefly discuss the major role of Mg^{2+}, Mn^{2+}, Fe^{2+}, and Ca^{2+} in the metabolism of solvent-producing clostridia.

15.7. Discuss the conditions under which lactate is significantly produced during ABE fermentation.

15.8. How is vacuum fermentation different from gas stripping–based fermentation? Discuss the potential advantages of vacuum fermentation over gas stripping–based fermentation.

15.9. Integrated ABE fermentation and simultaneous *in situ* butanol recovery by vacuum was operated for 72 hr during which 880 g of glucose was utilized. Condensate containing 250.5 g/L butanol was obtained while the concentration of residual butanol in the bioreactor was 4.6 g/L. If the fermentation was conducted using a 5 L bioreactor of which the fermentation/reaction volume is 3 L, calculate (a) the total butanol yield, and (b) the butanol selectivity of the vacuum recovery process.

15.10. Describe processes commonly applied to relieve butanol toxicity to solventogenic *Clostridium* species during ABE fermentation.

15.11. Enumerate the technical challenges associated with the operation of liquid–liquid extractive fermentation.

15.12. A novel strain of *Clostridium beijerinckii* was genetically modified to enhance the biosynthesis of butanol, by disrupting acetate production. However, this resulted in poor acid reassimilation, hence poor ABE production during fermentation on 60 g/L glucose. At the end of fermentation, butanol was accumulated to a final concentration of 3.2 g/L, while acetone and ethanol reached 1.6 and 0.2 g/L respectively. If acetate and butyrate were accumulated to final concentrations of 3.9 and 2.6 g/L, respectively, calculate:

(a) Butanol yield and productivity, if the fermentation lasted for 48 hours and residual glucose concentration in the bioreactor was 40 g/L.
(b) ABE yield and productivity.
(c) Total yield (of acids and solvents).
(d) Suggest a possible solution to the problem of excess acid accumulation in this scenario.

15.13. A newly isolated strain of *Clostridium saccharobutylicum* exhibits robust solvent production and tolerance. If this strain produces 14.5, 8, and 3.4 g/L of butanol, acetone, and ethanol, respectively, on corn stover hydrolysate containing 74%, 19%, and 7% glucose, xylose, and arabinose respectively, all of which make up 68 g/L of total sugars, calculate:

(a) Butanol yield.
(b) ABE yield.

Syngas Fermentation

Mark R. Wilkins, Hasan K. Atiyeh, and Samir Kumar Khanal

What is included in this chapter?

This chapter covers the production of fuels and chemicals from syngas using bacteria. The process description of syngas fermentation, stoichiometry, reactor design, and the generation of different co-products as well as mass balance are discussed. Relevant examples and calculations are also included.

16.1 Introduction

This book has already discussed how liquid biofuels can be produced through two major pathways, biochemical and thermochemical. In biochemical conversion, the feedstocks are converted into simple sugars following pretreatment and with the use of enzymes (no external enzyme is needed for sugar-based feedstocks). The sugars produced are subsequently fermented into biofuels (ethanol/butanol) using microbial catalysts. Thermochemical conversion involves the gasification and pyrolysis of feedstocks (Chapters 21 and 22). Gasification mainly converts lignocellulosic biomass or other reduced-carbon feedstocks into synthesis gas, also known as syngas (a mixture of CO_2, H_2, and CO). As shown in Figure 16.1, syngas can be converted into biofuels by either using chemical catalysts (through the Fischer–Tropsch [FT] process) or microbial catalysts (through syngas fermentation). Syngas fermentation has several advantages over other biofuel-production technologies, such as utilization of all organic biomass including lignin; elimination of complex pretreatment steps and costly enzymes; higher specificity of the biocatalysts than FT catalysts; independence of the H_2:CO ratio for bioconversion; operation of the bioreactor at ambient conditions; and no issue of noble metal poisoning.

In the late 1980s, Professor J.L. Gaddy of the University of Arkansas isolated *Clostridium ljungdahlii* from chicken waste, and this was able to metabolize syngas components into ethanol (Barik *et al.*, 1988). The discovery was part of the early stages of syngas fermentation research for biofuel production.

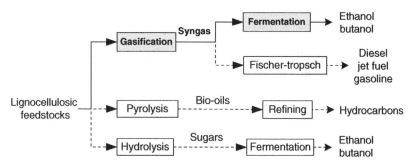

FIG. 16.1   Pathways for liquid biofuels production from lignocellulosic feedstock.

In this chapter, the bacteria responsible for syngas fermentation, stoichiometry, and the metabolic pathway of syngas fermentation are discussed. The factors affecting syngas fermentation, especially mass transfer and bioreactor configurations, are covered. Moreover, product separation and recovery are also discussed.

16.2 Stoichiometry

Syngas-fermenting bacteria typically produce alcohols and organic acids from syngas. There are several biochemical reactions involved in producing these products, and those reactions depend on the availability of CO and H_2 to the bacteria. The main alcohols formed by syngas-fermenting bacteria are ethanol and n-butanol. The primary organic acids formed by syngas-fermenting bacteria are acetic acid and n-butyric acid, as illustrated in the box.

Reactions for ethanol and n-butanol

$$6CO + 3H_2O \rightarrow \underset{\text{Ethanol}}{CH_3CH_2OH} + 4CO_2 \quad \Delta G° = -216\,kJ/mol \tag{16.1}$$

$$2CO_2 + 6H_2 \rightarrow CH_3CH_2OH + 3H_2O \quad \Delta G° = -97\,kJ/mol \tag{16.2}$$

Combining Equations 16.1 and 16.2, we have:

$$3CO + 3H_2 \rightarrow CH_3CH_2OH + CO_2 \quad \Delta G° = -157kJ/mol \tag{16.3}$$

$$12CO + 5H_2O \rightarrow \underset{\text{n-Butanol}}{CH_3CH_2CH_2CH_2OH} + 8CO_2 \quad \Delta G° = -484kJ/mol \tag{16.4}$$

$$4CO_2 + 12H_2 \rightarrow CH_3CH_2CH_2CH_2OH + 7H_2O \quad \Delta G° = -245\,kJ/mol \tag{16.5}$$

Combining Equations 16.4 and 16.5, we have:

$$6CO + 6H_2 \rightarrow CH_3CH_2CH_2CH_2OH + H_2O + 2CO_2 \quad \Delta G° = -365\,kJ/mol \tag{16.6}$$

Reactions for acetic acid and n-butyric acid

$$4CO + 2H_2O \rightarrow \underset{\text{Acetic acid}}{CH_3COOH} + 2CO_2 \quad \Delta G° = -155 kJ/mol \tag{16.7}$$

$$2CO_2 + 4H_2 \rightarrow CH_3COOH + 2H_2O \quad \Delta G° = -76 kJ/mol \tag{16.8}$$

Combining Equations 16.7 and 16.8, we have

$$2CO + 2H_2 \rightarrow CH_3COOH \quad \Delta G° = -115 kJ/mol \tag{16.9}$$

$$10CO + 4H_2O \rightarrow \underset{\text{n-Butyric acid}}{CH_3CH_2CH_2COOH} + 6CO_2 \quad \Delta G° = -511 kJ/mol \tag{16.10}$$

$$4CO_2 + 10H_2 \rightarrow CH_3CH_2CH_2COOH + 6H_2O \quad \Delta G° = -312 kJ/mol \tag{16.11}$$

Combining Equations 16.10 and 16.11, we have:

$$5CO + 5H_2 \rightarrow CH_3CH_2CH_2COOH + H_2O + CO_2 \quad \Delta G° = -412 kJ/mol \tag{16.12}$$

CO can be used by the bacteria as either a source of carbon or a source of electrons. H_2 is only a source of electrons, whereas CO_2 is only a carbon source.

Example 16.1: A gasifier produces 2 metric tons of syngas daily containing 50% CO and 50% H_2 (w/w). What is the maximum volume of ethanol (L) that can be produced stoichiometrically? Assume that the density of ethanol is 0.79 L/kg.

Solution:

1. Mass of CO and H_2

$$(0.5 \times 2 (\text{metric ton}) \times 1,000 (\text{kg/metric ton}) \times (\text{kgmol CO}/28 \text{kg})) = 35.7 \text{kgmol CO}$$

$$(0.5 \times 2 (\text{metric ton}) \times 1,000 (\text{kg/metric ton}) \times (\text{kgmol H}_2/2 \text{kg})) = 500.0 \text{kgmol H}_2$$

2. Volume of ethanol (from Equation 16.3)

$$3CO \quad + \quad 3H_2 \quad \rightarrow \quad CH_3CH_2OH + CO_2$$
3 kgmol   3 kgmol   1 kgmol

CO in the syngas is limiting.

Volume of ethanol = (35.7 (kgmol CO) × (kgmol ethanol/3 kgmol CO)
× 46 (kg/kgmol ethanol))/(0.79 (kg/L)) = **693 L/d (184 gallon/d)**

16.3 Syngas-Fermenting Bacteria

There are relatively few bacteria that have been discovered to produce alcohols using syngas. The bacteria that can produce alcohols from syngas are anaerobic bacteria. Syngas-fermenting bacteria are also classified as chemoautotrophs, since these microbes use inorganic compounds, such as CO and H_2, as energy and electron sources. Syngas-fermenting bacteria include *Clostridium ljungdahlii*, *Clostridium carboxidivorans*, *Clostridium ragsdalei*, *Alkalibaculum bacchi*, *Clostridium*

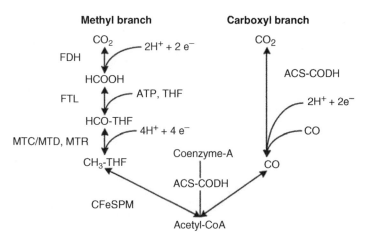

FIG. 16.2   The acetyl-CoA pathway for acetyl-CoA production in syngas-fermenting bacteria with the associated enzymes presented. ACS-CODH: actetyl-CoA synthase–carbon monoxide dehydrogenase; CFeSPM: corrinoid iron sulfur protein methyltransferase; FDH: formate dehydrogenase; FTL: formate-tetrahydrofolate ligase; MTC/MTD: bifunctional methenyl-tetrahydrofolate cyclohydrolase/methylene-tetrahydrofolate dehydrogenase; MTR: 5,10-methylene-tetrahydrofolate reductase. Source: Adapted from Bruant *et al.*, 2010.

---

autoethanogenum, Eubacterium limosum (also known as *Butyribacterium methylotrophicum*; Wilkins and Atiyeh, 2011), and *Clostridium coskatii* (Zahn and Saxena, 2011).

16.3.1 Biochemical Pathway

Syngas-fermenting bacteria use a common metabolic pathway commonly known as the Wood–Ljungdahl pathway or the acetyl-CoA pathway to utilize CO, CO_2, and H_2 (Figure 16.2; Wood *et al.*, 1986). The Wood–Ljungdahl pathway consists of two branches, one that produces the methyl group (CH_3-) in acetyl-CoA and one that provides the carbonyl group (CO) in acetyl-CoA. The methyl group is formed through a series of enzymatic reactions. First, CO_2 is converted to formic acid by reduction with two electrons. Next, formic acid combines with tetrahydrofolate (THF) with the consumption of one adenosine triphosphate (ATP). The formate-THF compound is reduced with four electrons to form methyl-THF. The methyl group is then combined with a cobalt center in a corrinoid protein and transferred to the enzyme carbon monoxide dehydrogenase (CODH). The carbonyl group is produced by either reduction of CO_2 to CO by two electrons or incorporation of a CO molecule. CO is combined with the methyl group, produced as described earlier, and coenzyme-A (CoA) to produce acetyl-CoA.

The electrons required to produce acetyl-CoA can come from either H_2 or CO. Electrons from H_2 are transferred to an electron carrier called ferredoxin by the enzyme hydrogenase (-Equation 16.13). CO is first converted to H_2 through the water gas shift reaction, and then electrons are transferred to ferredoxin by hydrogenase (Equation 16.14):

$$H_2 \rightarrow 2H^+ + 2e^- \tag{16.13}$$

$$CO + H_2O \rightarrow H_2 + CO_2 \tag{16.14}$$

Once acetyl-CoA is formed, it can be used in anabolic processes to synthesize bacterial cells, converted to acetic acid to generate one ATP, reduced to form ethanol, or combined with another

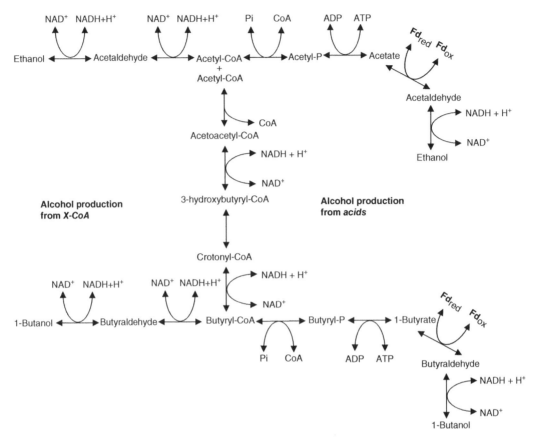

FIG. 16.3 Metabolic pathway showing genes responsible for the production of different acids and alcohols from acetyl-CoA in syngas-fermenting bacteria.

acetyl-CoA molecule to form the precursors to n-butanol and n-butyric acid (Figure 16.3). In addition, many syngas-fermenting bacteria have an enzyme called alcohol oxidoreductase that can reduce the organic acids, acetic acid and n-butyric acid, to ethanol and n-butanol, respectively.

C. carboxidivorans, C. ragsdalei, and *E. limosum* have been reported to produce n-butanol from CO, CO_2, and H_2 (Ukpong *et al.*, 2012; Maddipati *et al.*, 2011; Shen *et al.*, 1999). *C. ljungdahlii* and *C. ragsdalei* can also produce alcohols from organic acids and/or ketones, and H_2 (Perez *et al.*, 2013; Ramachandriya *et al.*, 2011), as shown in Figure 16.3 (right hand side).

$$CH_3COOH + 2H_2 \rightarrow CH_3CH_2OH + H_2O \quad \Delta G^\circ = -22\,kJ/mol \quad (16.15)$$

Acetic acid   Ethanol

$$CH_3COCH_3 + 2H_2 \rightarrow CH_3CHOHCH_3 + H_2O \quad \Delta G^\circ = -253\,kJ/mol \quad (16.16)$$

Acetone   Isopropanol

16.3.2 Genetic Transformation of Syngas-Fermenting Bacteria

Much work has been done in recent years on characterizing and manipulating genetic pathways in syngas-fermenting bacteria. Bruant *et al.* (2010) carried out an analysis of the genes present in

C. carboxidivorans P7 and identified the genes typically found in bacteria that produce acetone-butanol-ethanol from sugars, with the exception of a lack of acetone-production genes. Köpke *et al.* (2010) successfully transformed *C. ljungdahlii* to produce n-butanol by inserting genes for butanol production. Researchers at Syngas Biofuels Energy, Inc. have modified various *Clostridium* species to produce n-butanol and acetone, and found increased ethanol over wild type by deleting genes in the acetate and ethanol pathways, and inserting acetone and butanol production pathway genes (Berzin *et al.*, 2013, 2012a, b). New tools for the transformation of syngas-fermenting bacteria have also been developed that will foster more genetic work in these organisms (Banerjee *et al.*, 2014; Leang *et al.*, 2013).

16.3.3 Microbial Kinetics

Syngas-fermentation kinetics should be developed to facilitate process design and scaleup. Microbial kinetics should include the effect of process parameters such as gas composition, pH, temperature, and pressure on the efficacy of syngas fermentation. Microbial kinetics allow the estimation of cell growth, CO and H_2 consumption, product formation rates, and overall product yields. Cell and product concentrations during syngas fermentation can be predicted using mathematical models. The specific growth rate can be estimated from Equation 16.17:

$$\mu = \frac{1}{X}\frac{dX}{dt} \tag{16.17}$$

Integrating this equation allows estimation of the specific growth rate, μ (1/hr), of the bacterium:

$$\mu = \frac{\ln(X/X_0)}{t} \tag{16.18}$$

where X is cell mass concentration (g/L) at time t (hr) and X_o is the initial cell mass concentration (g/L). The value of μ is estimated from the slope of the linear portion of the growth curve in the ln (X) vs. time (t) plot. Another way to estimate microbial growth rate is using the doubling time (t_d), which equals the time required for microbial cell concentration to double:

$$t_d = \frac{\ln 2}{\mu} \tag{16.19}$$

> **Example 16.2:** A syngas-fermenting bacterium is grown in a batch reactor. It has a constant growth rate and a doubling time of 4.5 hr. If the reactor is initially inoculated with 0.05 g cells/L, find the cell concentration after 24 hr.
>
> **Solution:**
>
> $\mu = \ln(2)/4.5 = 0.154 \ 1/\text{hr}$
> $X = X_0 e^{(\mu \cdot t)} = 0.05 \times e^{(0.154 \times 24)} = \mathbf{2.01\,g/L}$

The microbial cell yield ($Y_{x/s}$) is another important kinetic parameter that provides information on how much syngas (carbon) is diverted to cell synthesis. The cell mass yield is calculated as:

$$\text{Cell mass yield, } Y_{X/S} = \frac{X_t - X_0}{N_{CO}} \tag{16.20}$$

where $Y_{x/s}$ is the cell mass yield (g cell mass increase/mole CO consumed), X_t is the cell mass concentration at time t (g cells/L), X_0 is the initial cell mass concentration (g cells/L), and N_{co} is the CO consumed (moles/L) during the fermentation. Typically, ethanol yield is reported per moles of CO consumed and compared to the stoichiometric ethanol, which can be obtained from CO as follows:

$$\text{Maximum ethanol yield, } \% = \frac{\frac{(C_{e,t} - C_{e,0}) \cdot V}{46 \cdot N_{CO}}}{1/6} \times 100 \tag{16.21}$$

where $C_{e,t}$ is the ethanol concentration at time t (g ethanol/L), $C_{e,0}$ is the initial ethanol concentration (g ethanol/L), 46 is the molecular weight of ethanol (g/mol), V is the reactor working volume (L), N_{co} is the moles of CO consumed, and 1/6 is the maximum stoichiometric conversion factor (mol ethanol/mol CO). The percentages of CO and H_2 utilization are calculated based on the mass balance calculation between gas consumed and gas supplied to the bacterium, as follows:

$$\text{CO utilization, } \% = \frac{N_{CO}}{\dot{N}_{CO} \cdot t} \times 100 \tag{16.22}$$

$$\text{H}_2 \text{ utilization, } \% = \frac{N_{H_2}}{\dot{N}_{H_2} \cdot t} \times 100 \tag{16.23}$$

where N_{CO} is the moles of CO consumed, $\dot{N}_{CO}$ is the rate CO supplied to the reactor (mol CO/hr), N_{H2} is the moles of H_2 consumed, $\dot{N}_{H2}$ is the rate of H_2 supplied to the reactor (mol H_2/hr), and t is the fermentation time (hr). The specific uptake rates of CO or H_2 by the bacterium are calculated based on microbial cells present in the fermentation medium, as follows:

$$q_{CO} = -\frac{1}{X \cdot V} \frac{dN_{CO}}{dt} \tag{16.24}$$

$$q_{H_2} = -\frac{1}{X \cdot V} \frac{dN_{H_2}}{dt} \tag{16.25}$$

where q_{co} and q_{H2} are the specific uptake rates of CO and H_2 in mol CO/(g cell· hr) and mol H_2/(g cell · hr), respectively, X is cell mass concentration (g/L), V is the reactor working volume (L), dN_{CO}/dt is the change in moles of CO over time (mol H_2/hr), and dN_{H2}/dt is the change in moles of H_2 over time (mol H_2/hr).

Specific growth rates of syngas-fermenting bacteria are given in Table 16.1.

Table 16.1 Specific growth rates of syngas-fermenting bacteria

Microorganisms	Substrates	μ (1/hr)	Reference
A. bacchi CP15	$CO/CO_2/H_2$(40 %/30 %/30 %)	0.12	Liu *et al.*, 2012
C. ljungdahlii	H_2/CO_2	0.25	Tanner *et al.*, 1993
C. ljungdahlii	CO/CO_2(80 %/20 %)	0.055	Phillips *et al.*, 1994
C. ljungdahlii	H_2/CO_2(75 %/25 %)	0.033	Phillips *et al.*, 1994
C. carboxidivorans	H_2/CO_2(80 %/20 %)	0.12	Liou *et al.*, 2005
C. carboxidivorans	$CO/CO_2/N_2$(70 %/6 %/24 %)	0.16	Liou *et al.*, 2005
C. carboxidivorans	$CO/CO_2/H_2/N_2$(20 %/15 %/5 %/60 %)	0.04	Datar *et al.*, 2004
C. ragsdalei P11	CO	0.17	Huhnke *et al.*, 2010
E. limosum	H_2/CO_2	0.036	Zeikus *et al.*, 1980

16.4 Factors Affecting Syngas Fermentation

16.4.1 Medium Composition

Bacteria synthesize all structural and functional components required to operate, maintain, and reproduce cells. These components include fats, lipids, nucleic acids, polysaccharides, and proteins, which, along with trace elements, are essential to build new cells. It is important to formulate a medium that contains trace elements, vitamins, minerals, and other nutrients in sufficient amounts to support cell growth during fermentation. For example, typical medium compositions reported for syngas-fermenting bacteria contained minerals, metals, and small amounts of the vitamins pantothenate, biotin, and thiamine for growth (Phillips *et al.*, 1993). The composition of a medium to support cell growth can be formulated differently than a medium to promote the production of products such as alcohols. A readily available, low-cost fermentation medium is critical for efficient syngas fermentation (Gao *et al.*, 2013). Strategies to reduce medium cost, such as reducing concentrations of yeast extract, using inexpensive complex nutrient sources such as cotton seed extract and corn steep liquor, and eliminating costly buffers, have also been shown to increase ethanol production by syngas-fermenting bacteria (Gao *et al.*, 2013; Kundiyana *et al.*, 2011, 2010; Maddipati *et al.*, 2011; Phillips *et al.*, 1993). The amount of different nutrients required for cell synthesis can be calculated if the composition of the cell is known (Gao *et al.*, 2013; Phillips *et al.*, 1993). Example 16.3 describes the approach of determining the nutrient requirements for syngas fermentation.

16.4.2 pH

The pH of the syngas fermentation medium is an important factor that affects both growth and product formation. The optimum pH range for the growth of most *Clostridium* species utilizing CO and H_2 is between 5.0 and 6.0. However, for *A. bacchi*, a moderately alkaliphilic microorganism utilizing H_2 and CO, the optimum pH is between 8.0 and 8.5 (Allen *et al.*, 2010). During syngas fermentation acetic acid is produced, which decreases the pH of the medium and induces ethanol production. The switch by *Clostridium* species from acetic acid production to ethanol production usually occurs at a pH between 4.4 and 5.5. However, *A. bacchi* switched from acetic acid production to ethanol production at a pH between 6.5 and 7.0 (Liu *et al.*, 2012). This clearly shows that the pH of the syngas-fermentation medium plays an important role in growth and product formation.

16.4.3 Temperature

The typical temperature for syngas fermentation is from 30–37 °C, since all organisms reported in the literature at this time that perform syngas fermentation are mesophiles (they grow well at temperatures ranging from 20–40 °C). Temperatures greater than 40 °C will result in cell death and temperatures lower than 30 °C will result in decreased metabolic activity. In addition to cell behavior, temperature affects the solubility of gases in the aqueous medium. CO and H_2 solubility in the medium increases as temperature decreases, which results in greater mass transfer in the reactor. Therefore, a balance between increased solubility and decreased metabolic activity of the bacteria at lower temperatures must be achieved for optimum alcohol production.

Example 16.3: A syngas-fermenting bioreactor is operating at mesophilic conditions with total cell concentration of 50 g/L. Nutrient medium is supplemented daily at a liquid flow rate of 100 L/day. How much nitrogen, phosphorus, and sulfur needs to be supplemented daily? Assume that the cell composition of anaerobic cells is $C_5H_7O_2NP_{0.06}S_{0.1}$.

Solution:

Total cells generated in the fermenter $= 50\,g/L \times 100\,L/day = 5\,kg/day$

Molecular weight of cells $= (5 \times 12\,kg/kmol\,C) + (7 \times 1\,kg/kmol\,H) + (2 \times 16\,kg/kmol\,O) + (1 \times 14\,kg/kmol\,N) + (0.06 \times 31\,kg/kmol\,P) + (0.1 \times 32\,kg/kmol\,S) = 118.06\,kg/kmol\,cells$

Based on the cell composition, 118.06 kg of cells has: 14 kg Nitrogen (N)

1.86 kg Phosphorus (P)

3.2 kg Sulfur (S)

Thus, the nutrients required for producing 5 kg/day of cells are:

Nitrogen requirement: $(14/118.06) \times 5 = 0.593\,kg/day = $ **593 g/day**

Phosphorus requirement: $(1.86/118.06) \times 5 = 0.079\,kg/day = $ **79 g/day**

Sulfur requirement: $(3.2/118.06) \times 5 = 0.136\,kg/day = $ **136 g/day**

It is important to note that other micronutrients and trace elements are required for cell synthesis. Complete compositional analysis of the cell will provide such information.

16.4.4 Mass Transfer

Mass transfer of sparingly soluble gaseous substrates, such as CO and H_2, in the aqueous phase is typically a rate-limiting step and a potential bottleneck in syngas fermentation. During syngas fermentation, the gaseous substrate is first absorbed at the gas–liquid interface and then diffused through the culture medium to the cells. Microbes consume the diffused substrates as their carbon and energy sources, and produce metabolites such as alcohols and organic acids. There are several intermediate steps involved in transporting gaseous substrate into the microbial cells. These steps include diffusion through the bulk gas phase to the gas–liquid interface, moving across the gas–liquid interface, transport into the bulk liquid phase surrounding the microbial cells, and diffusive transport through the liquid–solid boundary. Of these, the gas–liquid interface mass transfer is the main resistance for gaseous substrate diffusion (Klasson *et al.*, 1992). Poor solubility of a gaseous substrate in the culture medium results in low substrate availability to microorganisms, and thus leads to low productivity.

The effectiveness of various reactor configurations is usually quantified and compared by the volumetric mass transfer coefficient (k_La), which represents the hydrodynamic condition in the reactor. Klasson *et al.* (1992) proposed the following equation to calculate the volumetric mass transfer coefficient (k_La) in the liquid phase:

$$\frac{dN_S^G}{V_L\,dt} = \frac{k_L a}{H}\left(P_S^G - P_S^L\right) \tag{16.26}$$

where N_S^G is the molar substrate transferred from the gas phase (mol), V_L is the volume of the reactor (L), P_S^G and P_S^L are the partial pressures of the gaseous substrate in gas and the liquid phases, respectively (atm), H is Henry's law constant ((L·atm)/mol), and a is the gas-liquid interfacial area per unit volume (m²/L).

The difference in the partial pressures of the gaseous substrate $\left(P_S^G - P_S^L\right)$ is the driving force for mass transfer, and thus controls the solubility of the substrate. High-pressure operation improves the solubility of the gas in the aqueous phase. However, at higher concentrations of gaseous substrates, especially CO, anaerobic microorganisms are inhibited. Therefore, determination of a correlation between substrate diffusion and the specific substrate uptake rate, q_S (1/hr), is important in order to evaluate process kinetics (Equation 16.27):

$$q_S = \frac{q_m \times P_S^L}{K'_p + P_S^L + \left(P_S^L\right)^2/W'} \tag{16.27}$$

where q_m (1/hr), W' (atm), and K'_p (atm) are empirical constants. Furthermore, Q_S, the substrate uptake rate (mg/(L·hr)), can be written as $Q_S = q_S \cdot X$, where X is the microbial cell concentration, (mg/L). By comparing Equations 16.26 and 16.27, it is evident that the difference between the partial pressures of the gaseous substrates and the cell concentration of the reactor is directly proportional to the volumetric mass transfer coefficient (Vega et al., 1990). Various approaches such as high gas and liquid flow rates, large specific gas–liquid interfacial areas, increased pressure, different reactor configurations, innovative impeller designs, modified fluid flow patterns, varying mixing times and speeds, and the use of microbubble dispersers have been examined to enhance gas solubility in the liquid phase. Many of these approaches increase the agitator's power input to volume ratio, which facilitates bubble breakup and increases the interfacial surface area available for mass transfer. This approach, however, is not economically attractive for commercial-scale syngas fermentation due to high energy costs. Additionally, higher agitation speeds and vigorous mixing create a higher shear force, which is known to damage the microorganisms in the culture medium. In order to achieve energy-efficient mass transfer, alternative bioreactor configurations such as trickling beds, airlift reactors, and membrane bioreactors have been examined for syngas fermentation.

Younesi et al. (2008) investigated the relationship between agitation speed and $k_L a$ in a stirred-tank reactor using Phodospirillum rubrum as the microbial culture. The authors reported a $k_L a$ value of 72.8 1/hr at an agitation speed of 500 rpm. The results obtained further revealed that mass transfer improves with an increase in agitation speed. Yang et al. (2001) reported a slightly higher $k_L a$ value (119 1/hr) for CO gas in a slurry bubble column operated at a temperature of 20 °C and a pressure of 10 bar. In a separate study, Bredwell et al. (1999) reported a maximum $k_L a$ of 190 and 75 1/hr for H_2 gas in a stirred-tank reactor at a mixing speed of 300 rpm with and without microbubble sparging, respectively. The authors used a mixed culture of sulfate-reducing bacteria (SRB) in their study. Munasinghe and Khanal (2010) reported $k_L a$ values for CO ranging from 0.4 to 91 1/hr for eight different reactor configurations, including a submerged composite hollow fiber membrane (CHFM) reactor. However, caution should be used when comparing $k_L a$ values between reactors because of the assumptions made in the analysis of each reactor and the variability in the active mass transfer and system volumes for each reactor, as discussed in Orgill *et al.* (2013). In addition, k_La values are affected by gas flow rate, sparger and impeller type, and the material of the CHFM. Orgill *et al.* (2013) reported that a non-porous poly dimethyl siloxane (PDMS) CHFM provided the highest k_La (1,062 1/hr), followed by the trickle-bed reactor (421 1/hr), and then the stirred tank reactor (114 1/hr).

16.4.5 Bioreactor Configurations

Bioreactor configuration is closely related to effective gas–liquid mass transfer; thus, innovative reactor design plays an important role in syngas fermentation. High mass transfer rates, high cell densities, low operation and maintenance costs, simple design, and easy scaleup are some of the key parameters for designing an efficient bioreactor system. Similarly, the bioreactor size greatly depends on the rate of mass transfer for sparingly soluble gases (Vega *et al.*, 1990). Some of the commonly used reactor configurations are discussed in this section.

Continuous-stirred tank reactors (CSTR) are the most commonly employed bioreactors in syngas fermentation (Figure 16.4a). In a CSTR, gas is continuously supplied into the liquid phase, while an agitator controls the gas–liquid mass transfer (Vega *et al.*, 1990). Higher agitation speeds lead to a higher mass transfer rate between the substrate gases and the microbes. CSTR is probably one of the simplest bioreactor configurations for syngas fermentation. However, in industrial-scale fermenters, higher agitation speeds increase the agitator's power consumption, thus increasing the operational cost of the plant. Power consumption should be considered while designing a CSTR system.

The **bubble column** consists of a glass column mounted on a steel base (Figure 16.4b). The substrate gases are introduced into the reactor through a fritted glass disk with a pore size of 4–6 microns. These reactors are mainly designed for industrial applications with large reactor volumes. A high mass transfer rate and relatively low operational and maintenance costs are among the merits of this type of reactor, while back-mixing and coalescence are common drawbacks of the system (Datar *et al.*, 2004).

A **trickle-bed reactor** is a slender column with a packing medium (Figure 16.4c). The microbial culture medium continuously flows down and the gaseous substrate flows either upward (counter-current) or downward (co-current), depending on the application. Trickle-bed reactors are operated under atmospheric pressure and no agitation is necessary; therefore, trickle-bed reactors consume less energy than a conventional CSTR (Bredwell *et al.*, 1999). However, the gas–liquid mass transfer in trickle-bed reactors largely depends on the packing medium, gas flow rate, and liquid recirculation flow rate. In trickle-bed reactors, liquid recirculation is the most energy-intensive unit operation.

A **microbubble sparged reactor** is a combination of a CSTR reactor and a microbubble disperser (Figure 16.4d). In this reactor configuration, the microfiltration (MF) unit is used to filter the microbial cells, which are recycled back into the reactor. The filtered fermentation medium is mixed with syngas, which enters into the microbubble generator, consisting of a high-speed spinning disk. Syngas-rich culture medium is then pumped into the fermentation vessel using a peristaltic pump. In this reactor configuration, the large specific interfacial area of the gas bubbles and the longer retention time stimulate high gas–liquid mass transfer. However, the use of an agitator to mix the liquid culture increases energy consumption.

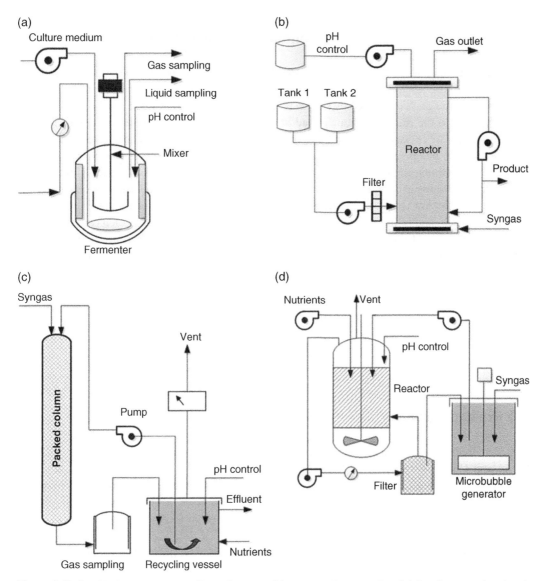

FIG. 16.4   Various reactor configurations used in syngas fermentation (a) Continuous stirred-tank reactor (CSTR); (b) bubble column reactor; (c) trickle-bed reactor; and (d) microbubble sparged reactor.

Composite **hollow fiber membranes** (CHFM) have been proposed as a technologically and economically feasible method for syngas fermentation (Tsai *et al.*, 2009). Even though the technology is yet to be adopted for industrial-scale syngas fermentation, it has been widely studied in water and wastewater treatment (Lee and Rittmann, 2002; Nerenberg and Rittmann, 2004). In CHFM reactors, the inside-out configuration (gas flows inside the fiber lumens) is the most commonly used operational mode. This novel CHFM system offers several advantages such as higher microbial cell retention, higher yield, and higher tolerance to toxic compounds (tar, acetylene, NO_x, etc.). The high capital cost is the major drawback of a CHFM system.

Batch, semi-continuous, and continuous fermentation

Syngas fermentation can be carried out in batch, semi-continuous, or continuous mode. In the batch mode, the reactor is pressurized with syngas to a certain pressure. The fermentation ceases when the gas is consumed by the bacteria. In semi-continuous mode, the gas is continuously fed into the reactor without the addition of liquid medium. The fermentation stops when cell activity ceases due to nutrient limitation or product inhibition. The semi-continuous mode can also be operated by continuous feeding of gas with intermittent feeding of liquid medium containing the nutrients into the reactor. However, the process is typically difficult to control, resulting in cell lysis when nutrients are depleted. Although the batch process is easy to control with a low chance of microbial contamination, it is time and labor intensive, requires pressure vessels, and is limited by the amount of gas that can be processed in each batch due to the low solubility of the gaseous substrates (CO and H_2). The merits of continuous syngas fermentation are the potential to operate at a high cell concentration with cell recycling and efficient control of fermentation parameters such as gas and liquid medium feed rates, pH, and agitation speed. The continuous supply of fresh medium and syngas ensures cell activity for extended operation and can increase the overall productivity of the syngas-fermentation process.

16.5 Product Recovery

A detailed discussion of alcohol recovery, such as ethanol and butanol recovery, is presented in Chapters 14 and 15, respectively.

Organic acids, such as acetic acid, are recovered from the fermentation medium by precipitation or solvent extraction (Rogers et al., 2006). Solvent extraction is based on the relative solubility of the compound of interest in two different immiscible solvents, which form two phases.

Example 16.4: In a CO mass transfer experiment, the results in Table 16.2 were obtained.

Table 16.2 CO mass transfer experiment

Partial pressure in gas phase (P_{CO} gas) (atm)	0.17	0.33	0.53	0.69	0.84
dN_{CO}/dt (mmol/hr)	15	30	45	60	75

The liquid volume of the reactor is 1.5 L and the operating temperature of the reactor is 25 °C. The partial pressure of CO in the liquid phase is assumed to be zero (mass transfer limited region). What is the volumetric mass transfer coefficient ($k_L a$) for this reactor?

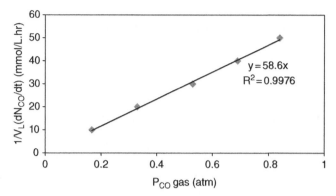

FIG. 16.5   Variation between 1/V$_L$(dN$_{CO}$/dt) and P$_{CO}$ gas.

Solution: Plot the variation between 1/V$_L$(dN$_{CO}$/dt) and P$_{CO}$ gas (Figure 16.5). Using Equation 16.15, in this case:

$$P_S^L = 0$$

Therefore, Equation 16.15 becomes:

$$\frac{dN_S^G}{V_L\,dt} = \frac{k_L a}{H}\left(P_S^G\right)$$

The gradient of the plot gives:

$$k_L a/H = 58.6\,\text{mmol}/(\text{L}\cdot\text{hr}\cdot\text{atm})$$

Henry's constant for CO at 25 °C

$$H = 1052.6\,(\text{atm}\cdot\text{L})/\text{mol}$$

$$\text{Therefore, } k_L a = \left(58.6\,\text{mmol}/(\text{L}\cdot\text{hr}\cdot\text{atm})\right) \times \left(1052.6\,(\text{atm}\cdot\text{L})/\text{mol}\right)$$

$$= \mathbf{61.7\ 1/hr}$$

References

Allen, T.D., Caldwell, M.E., Lawson, P.A., Huhnke, R.L., and Tanner, R.S. (2010). *Alkalibaculum bacchi* gen. nov., sp. nov., a CO-oxidizing, ethanol-producing acetogen isolated from livestock-impacted soil. *International Journal of Systematic and Evolutionary Microbiology* 60: 2483.

Banerjee, A., Leang, C., Ueki, T., Nevin, K.P., and Lovley, D.R. (2014). Lactose-inducible system for metabolic engineering of *Clostridium ljungdahlii*. *Applied and Environmental Microbiology* 80: 2410–16.

Barik, S., Prieto, S., Harrison, S.B., Clausen, E.C., and Gaddy, J.L. (1988). Biological production of alcohols from coal through indirect liquefaction. *Applied Biochemistry and Biotechnology* 18: 363–78.

Berzin, V., Kiriukhin, M., and Tyurin, M. (2012a). Elimination of acetate production to improve ethanol yield during continuous synthesis gas fermentation by engineered biocatalyst *Clostridium* sp MTEtOH550. *Applied Biochemistry and Biotechnology* 167: 338–47.

Berzin, V., Kiriukhin, M., and Tyurin, M. (2012b). Selective production of acetone during continuous synthesis gas fermentation by engineered biocatalyst *Clostridium* sp MAceT113. *Letters in Applied Microbiology* 55: 149–54.

Berzin, V., Tyurin, M., and Kiriukhin, M. (2013). Selective n-Butanol production by *Clostridium* sp MTBu-tOH1365 during continuous synthesis gas fermentation due to expression of synthetic thiolase, 3-hydroxy butyryl-CoA dehydrogenase, crotonase, butyryl-CoA dehydrogenase, butyraldehyde dehydrogenase, and NAD-dependent butanol dehydrogenase. *Applied Biochemistry and Biotechnology* 169: 950–59.

Bredwell, M.D., Srivastava, P., and Worden, R.M. (1999). Reactor design issues for synthesis gas fermentations. *Biotechnology Progress* 15: 834–44.

Bruant, G., Lévesque, M.J., Peter, C., Guiot, S.R., and Masson, L. (2010). Genomic analysis of carbon monoxide utilization and butanol production by *Clostridium carboxidivorans* strain P7T. *PLoS One* 5: e13033.

Datar, R.P., Shenkman, R.M., Cateni, B.G., Huhnke, R.L., abd Lewis, R.S. (2004). Fermentation of biomass-generated producer gas to ethanol. *Biotechnology Engineering* 86: 587–94.

Gao, J., Atiyeh, H.K., Phillips, J.R., Wilkins, M.R., and Huhnke, R.L. (2013). Development of low cost medium for ethanol production from syngas by *Clostridium ragsdalei*. *Bioresource Technology* 147: 508–15.

Huhnke, R.L., Lewis, R.S., and Tanner, R.S. (2010). Isolation and characterization of novel clostridial species. *US Patent No.* 7,704,723.

Klasson, K.T., Ackerson, C.M.D., Clausen, E.C., and Gaddy, J.L. (1992). Biological conversion of synthesis gas into fuels. *International Journal of Hydrogen Energy* 17: 281–8.

Köpke, M., Held, C., Hujer, S., *et al.* (2010). *Clostridium ljungdahlii* represents a microbial production platform based on syngas. *Proceedings of the National Academy of Sciences* 107: 13087.

Kundiyana, D.K., Huhnke, R.L., Maddipati, P., Atiyeh, H.K., and Wilkins, M.R. (2010). Feasibility of incorporating cotton seed extract in *Clostridium* strain P11 fermentation medium during synthesis gas fermentation. *Bioresource Technology* 101(24): 9673–80.

Kundiyana, D.K., Wilkins, M.R., Maddipati, P., and Huhnke, R.L. (2011). Effect of temperature, pH and buffer presence on ethanol production from synthesis gas by "Clostridium ragsdalei." *Bioresource Technology* 102: 5794–9.

Leang, C., Ueki, T., Nevin, K.P., and Lovley, D.R. (2013). A genetic system for *Clostridium ljungdahlii*: A chassis for autotrophic production of biocommodities and a model homoacetogen. *Applied and Environmental Microbiology* 79: 1102–9.

Lee, K.C., and Rittmann, B.E. (2002). Applying a novel autohydrogenotrophic hollow-fiber membrane biofilm reactor for denitrification of drinking water. *Water Research* 36: 2040–52.

Liou, J.S.C., Balkwill, D.L., Drake, G.R., and Tanner, R.S. (2005). *Clostridium carboxidivorans* sp nov., a solvent-producing clostridium isolated from an agricultural settling lagoon, and reclassification of the acetogen *Clostridium scatologenes* strain SL1 as *Clostridium drakei* sp nov. *International Journal of Systematic and Evolutionary Microbiology* 55: 2085–91.

Liu, K., Atiyeh, H.K., Tanner, R.S., Wilkins, M.R., and Huhnke, R.L. (2012). Fermentative production of ethanol from syngas using novel moderately alkaliphilic strains of *Alkalibaculum bacchi*. *Bioresource Technology* 104: 336–41.

Maddipati, P., Atiyeh, H.K., Bellmer, D.D., and Huhnke, R.L. (2011). Ethanol production from syngas by *Clostridium strain* P11 using corn steep liquor as a nutrient replacement to yeast extract. *Bioresource Technology* 102: 6494–501.

Munasinghe, P.C., and Khanal, S.K. (2010). Syngas fermentation to biofuel: Evaluation of carbon monoxide mass transfer coefficient (kLa) in different reactor configurations. *Biotechnology Progress* 26: 1616–21.

Nerenberg, R., and Rittmann, B. (2004). Hydrogen-based, hollow-fiber membrane biofilm reactor for reduction of perchlorate and other oxidized contaminants. *Water Science and Technology* 49: 223–30.

Orgill, J.J., Atiyeh, H.K., Devarapalli, M., Phillips, J.R., Lewis, R.S., and Huhnke, R.L. (2013). A comparison of mass transfer coefficients between trickle-bed, hollow fiber membrane and stirred tank reactors. *Bioresource Technology* 133: 340–46.

Perez, J.M., Richter, H., Loftus, S.E., and Angenent, L.T. (2013). Biocatalytic reduction of short-chain carboxylic acids into their corresponding alcohols with syngas fermentation. *Biotechnology and Bioengineering* 110: 1066–77.

Phillips, J., Klasson, K., Clausen, E., and Gaddy, J. (1993). Biological production of ethanol from coal synthesis gas. *Applied Biochemistry and Biotechnology* 39–40: 559–71.

Phillips, J.R., Clausen, E.C., and Gaddy, J.L. (1994). Synthesis gas as substrate for the biological production of fuels and chemicals. *Applied Microbiology and Biotechnology*, 45/46: 145–57.

Ramachandriya, K.D., Wilkins, M.R., Delorme, M.J.M., *et al.* (2011). Reduction of acetone to isopropanol using producer gas fermenting microbes. *Biotechnology and Bioengineering* 108: 2330–38.

Rogers, P., Chen, J.-S., and Zidwick, M.J. (2006). Organic acid and solvent production. Part I: Acetic, lactic, gluconic, succinic and polyhydroxyalkanoic acids. *Prokaryotes* 1: 511–755.

Shen, G.J., Shieh, J.S., Grethlein, A.J., Jain, M.K., and Zeikus, J.G. (1999). Biochemical basis for carbon monoxide tolerance and butanol production by *Butyribacterium methylotrophicum*. *Applied Microbiology and Biotechnology* 51: 827–32.

Tanner, R.S., Miller, L.M., and Yang, D. (1993). *Clostridium ljungdahlii* sp. nov., an acetogenic species in clostridial rRNA Homology Group I. *International Journal of Systematic and Evolutionary Microbiology* 43(2): 232–6.

Tchobanoglous, G., Theisen, H., Vigil, S.A. (1993). *Integrated Solid Waste Management: Engineering Principles and Management Issues*. New York: McGraw-Hill.

Tsai, S.-P., Datta, R., Basu, R., and Yoon, S.-H. (2009). Syngas conversion system using asymmetric membrane and anaerobic microorganism. US Patent Application 12/036,007.

Ukpong, M.N., Atiyeh, H.K., De Lorme, M.J.M., *et al.* (2012). Physiological response of *Clostridium carboxidivorans* during conversion of synthesis gas to solvents in a gas-fed bioreactor. *Biotechnology Bioengineering* 109: 2720–28.

Vega, J., Clausen, E., and Gaddy, J. (1990). Design of bioreactors for coal synthesis gas fermentations. *Resources, Conservation and Recycling* 3: 149–60.

Wilkins, M.R., and Atiyeh, H.K. (2011). Microbial production of ethanol from carbon monoxide. *Current Opinion in Biotechnology* 22: 326–30.

Wood, H.G., Ragsdale, S.W., and Pezacka, E. (1986). The acetyl-CoA pathway of autotrophic growth. *FEMS Microbiology Letters* 39: 345–62.

Yang, W.G., Wang, J.F., and Jin, Y. (2001). Mass transfer characteristics of syngas components in slurry system at industrial conditions. *Chemical Engineering and Technology* 24: 651–7.

Younesi, H., Najafpour, G., Ku Ismail, K.S., Mohamed, A.R., and Kamaruddin, A.H. (2008). Biohydrogen production in a continuous stirred tank bioreactor from synthesis gas by anaerobic photosynthetic bacterium: *Rhodopirillum rubrum*. *Bioresource Technology* 99: 2612–19.

Zahn, J.A., and Saxena, J. (2011). Novel ethanologenic *Clostridium* species, *Clostridium coskatii*. US Patent Application 2011/0229947 A1.

Zeikus, J.G., Lynd, L.H., Thompson, T.E., Krzycki, J.A., Weimer, P.J., and Hegge, P.W. (1980). Isolation and characterization of a new, methylotrophic, acidogenic anaerobe, the Marburg strain. *Current Microbiology* 3 (6): 381–6.

Exercise Problems

16.1. You have a syngas mixture that contains 40% CO, 30% H_2, and 30% CO_2 by volume. Assuming that the organism uses the reaction in Equation 16.6 first, how much butanol in liters can be produced from the fermentation of 1,000 kg of syngas?

16.2. You have a stream containing 100 g/L of acetic acid. If you assume that you have a bacterium that can convert all of the acetic acid to ethanol using H_2, how many kg of H_2 would it take to convert all of the acetic acid in 8,000 L of your stream to ethanol?

16.3. A syngas-fermenting culture is grown in a CSTR. It has a constant growth rate and a doubling time of 3 hr. If it is initially inoculated with 0.10 g cells/L, find the cell concentration after 36 hr.

16.4. Briefly discuss the opportunities and challenges of syngas-fermentation technology for the production of biofuels and biobased chemicals.

16.5. Process parameters such as the composition of gas mixture and medium, pH, and culture kinetics greatly affect syngas fermentation. Briefly explain how gas composition affects the product profile during syngas fermentation.

16.6. Mass transfer of syngas into aqueous phase is critical for the commercialization of syngas-fermentation technology. What are the factors that affect the mass transfer?

16.7. As a biological engineering student, you are asked to design a bioreactor system that could maintain a minimum of 4 mg CO/L for syngas fermentation. CO is supplied by diffusing biomass-derived syngas consisting of 50% CO, 10% H_2, and 40% N_2 (by volume) at the bottom of the bioreactor. Assume that the bioreactor is operated at a constant temperature of 20 °C and 1 atm pressure.

(a) Is the CO concentration sufficient in the aqueous phase?

(b) Draw a hypothetical CO profile at different temperatures, pressures, and cell concentrations.

16.8. Based on a lab-scale experiment, the mass transfer coefficient for CO was observed to be 61.7 1/hr (see Example 16.3). What volume of industrial-scale bioreactor for syngas fermentation is needed to handle a CO gas flow rate of up to 10 m^3/hr? Assume a temperature of 37 °C and 1 atm pressure and that the gas mix entering the bioreactor is 50% CO (by volume).

16.9. In designing a syngas-fermenting bioreactor, what are the major considerations for improving syngas mass transfer? Discuss briefly.

16.10. Discuss briefly the methods of recovering ethanol and organic acids from syngas-fermentation medium.

Fundamentals of Anaerobic Digestion

Samir Kumar Khanal and Yebo Li

What is included in this chapter?

This chapter provides the basics of anaerobic digestion, including microbiology, biochemical pathway, stoichiometry, and theoretical yield calculation for bioenergy production. Important considerations in anaerobic digestion and a brief overview of Anaerobic Digestion Model No. 1 (ADM1) are also discussed.

17.1 Introduction

An anaerobic environment is common in nature, such as swamp/marsh land or sediments from lakes/ponds, which are often devoid of oxygen. Such an environment also exists in human and animal guts, and in the guts of several insects (e.g., termites and cockroaches). In an *anaerobic digester*, organic matter (e.g., industrial wastewaters, crop residues, animal manures, sewage sludge, food wastes, and energy crops) is decomposed by diverse microbial communities through a series of metabolic stages, and produce the final gaseous product: biogas. An anaerobic environment can be man-made, such as a covered lagoon or tanks commonly known as digesters, where environmental conditions are closely controlled to maximize waste stabilization and/or biogas yield. *Biogas* is primarily a mixture of methane (CH_4) and carbon dioxide (CO_2), with trace amounts of nitrogen (N_2), hydrogen (H_2), hydrogen sulfide (H_2S), ammonia (NH_3), and oxygen (O_2), as shown in Table 17.1. Biogas can be used as a renewable energy source for cooking, generating heat and electricity (through a combined heat and power [CHP] unit), or it can be upgraded into biomethane for using as a transportation fuel or for injecting into a gas grid.

Bioenergy: Principles and Applications, First Edition. Edited by Yebo Li and Samir Kumar Khanal.
© 2017 John Wiley & Sons, Inc. Published 2017 by John Wiley & Sons, Inc.
Companion website: www.wiley.com/go/Li/Bioenergy

Table 17.1 Typical composition of biogas

Component	Content (% by volume)
Methane (CH_4)	50–75%
Carbon dioxide (CO_2)	25–40%
Nitrogen (N_2)	<5%
Hydrogen (H_2)	<1%
Oxygen (O_2)	<1%
Hydrogen sulfide (H_2S)	50–5,000 ppm

There is historical evidence that the Assyrians utilized anaerobic processes to produce biogas for heating bath water during the 10th century (Verma, 2002). It was the Italian physicist Alessandro Volta (Figure 17.1) who discovered CH_4 gas at Lake Maggiore while on summer vacation in 1776. The first full-scale anaerobic treatment of domestic wastewater in an air-tight chamber known as the *Mouras Automatic Scavenger* was published in the French journal *Cosmos* in 1881. Donald Cameron built a septic tank modeled on the principle of Mouras Automatic Scavenger in the city of Exeter, England in 1895. This was designed to collect biogas for heating and lighting. In 1897, waste-disposal tanks at a leper colony in Matunga, Mumbai, India were reported to have been designed with a biogas-collection system, and the biogas was used to drive a gas engine (Bushwell and Hatfield, 1938).

FIG. 17.1 Alessandro Volta.

Two-stage systems known as the Travis tank (1904) and the Imhoff tank (1905) were developed primarily for settled sludge (solids) treatment in Germany. One of the first sludge-heating apparatuses with separate digestion of sludge was reported at the Essen-Rellinghausen Plant, Germany in

1927 (Imhoff, 1938). Separate sludge digestion became immensely popular in larger cities, and the importance of biogas generation was widely recognized. Besides heating of digesters, biogas was collected and delivered to municipal gas systems, and was used for power generation for operating biological wastewater-treatment systems. Today, *anaerobic digestion* (*AD*) has been widely adopted globally for the stabilization of sewage sludge and recovery of bioenergy.

One significant development in AD was the better understanding of the importance of uncoupling the *hydraulic retention time* (*HRT*) from the solids retention time (SRT) in the 1950s (see Section 17.4.7). This led to high-rate anaerobic reactors such as the clarigester (1950), the anaerobic contact process (1955), the anaerobic filter (1969), and the upflow anaerobic sludge blanket (UASB) reactor (1979), among others (see Chapter 18). These high-rate anaerobic reactors are able to maintain a very long SRT, of the order of 100 days or longer (thus retaining slow-growing methanogens in the reactor), irrespective of an HRT of as low as 4 hours. It is also important to point out that high-rate reactors are mainly suitable for wastewaters with a very low suspended solids concentration. High-solids feedstocks are still digested in a completely mixed digester using a suspended growth system.

Many of the early applications of AD primarily focused on wastewater treatment and sewage sludge stabilization. Following the energy crisis of the 1970s, there was significant interest in AD for producing renewable bioenergy (i.e., biogas) from organic wastes (e.g., food wastes, animal manures, organic fraction of municipal solid waste (OFMSW), municipal and industrial wastewaters, etc.). With falling energy prices in the 1980s and 1990s, AD technology did not see significant growth, especially for bioenergy production. In recent years, rising energy prices coupled with growing concern about climate change due to excessive use of fossil-derived energy have drawn significant interest in AD technology for bioenergy production. For example, in Europe alone there are 9,766 agricultural-based biogas plants, and Germany is the global leader with 7,515 operating plants in 2012 (European Biogas Association, 2013). These biogas plants are mainly fed with maize silage, sugar beet, lawn grass, crop residues, animal manure, and dedicated energy crops.

In developing countries, especially in rural areas, household digesters have been playing an even more important role, because they provide biogas for cooking and lighting. In China, the government has been providing strong support for the installation of household digesters, with over 40 million total household digesters reported in 2012. Nepal has a successful biogas development program and currently has over 268,000 household digesters in operation (Surendra *et al.*, 2014).

This chapter provides the basics of AD, including microbiology, biochemical pathway, stoichiometry, and theoretical yield calculation for bioenergy production. Important considerations in AD and an overview of Anaerobic Digestion Model No. 1 (ADM1) are also discussed.

17.2 Organic Conversion in an Anaerobic Process

The breakdown of complex organic matter in an anaerobic process involves multiple steps, which are carried out by several groups of microorganisms. The end product of anaerobic degradation of organic compounds is biogas, an energy-rich gas mixture consisting of mainly CH_4 and CO_2. Figure 17.2 shows the schematics of various steps and microbial groups involved during AD.

Hydrolysis

Complex organic compounds such as proteins, carbohydrates (polysaccharides), and lipids are broken down into simple soluble molecules such as amino acids, sugars, long-chain fatty acids (LCFAs),

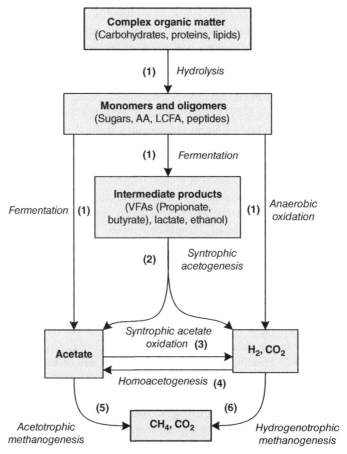

FIG. 17.2   Anaerobic conversion steps: 1. Fermentative bacteria; 2. Acetogenic bacteria; 3. Syntrophic acetate oxidizing bacteria; 4. Homoacetogens; 5. Acetotrophic methanogens; 6. Hydrogenetrophic methanogens. Notes: AA: amino acids; LCFA: long-chain fatty acids; VFAs: volatile fatty acids. Source: Adapted from Khanal, 2008.

Table 17.2 Hydrolytic bacteria with their enzymes

Substrates	Exo-enzymes involved	Hydrolytic bacteria	Products
Polysaccharides (cellulose, starch, and pectin)	Cellulases, amylases, and pectinases	*Clostridium, Bacillus,* and *Lactobacillus*	Simple sugars
Proteins	Protease and peptidases	*Bacteroides, Butyrivibrio, Clostridium, Fusobacterium, Selenomonas,* and *Streptococcus*	Amino acids
Lipids	Lipase	*Mycobacterium,* and *Clostridium*	Fatty acids

and glycerine along with a small amount of acetic acid, H_2, and CO_2 by the action of extracellular enzymes excreted by hydrolytic microorganisms (e.g., *Streptococcus, Clostridium,* etc. Table 17.2). Based on the energy content of the initial complex organic molecule, H_2 and acetic acid account for 5% and 25%, respectively, of total energy, and the remaining 70% of energy is associated with other simple soluble organic molecules. The hydrolytic bacteria can be either facultative anaerobes (surviving under both aerobic and anaerobic conditions) or obligate anaerobes (surviving only in the

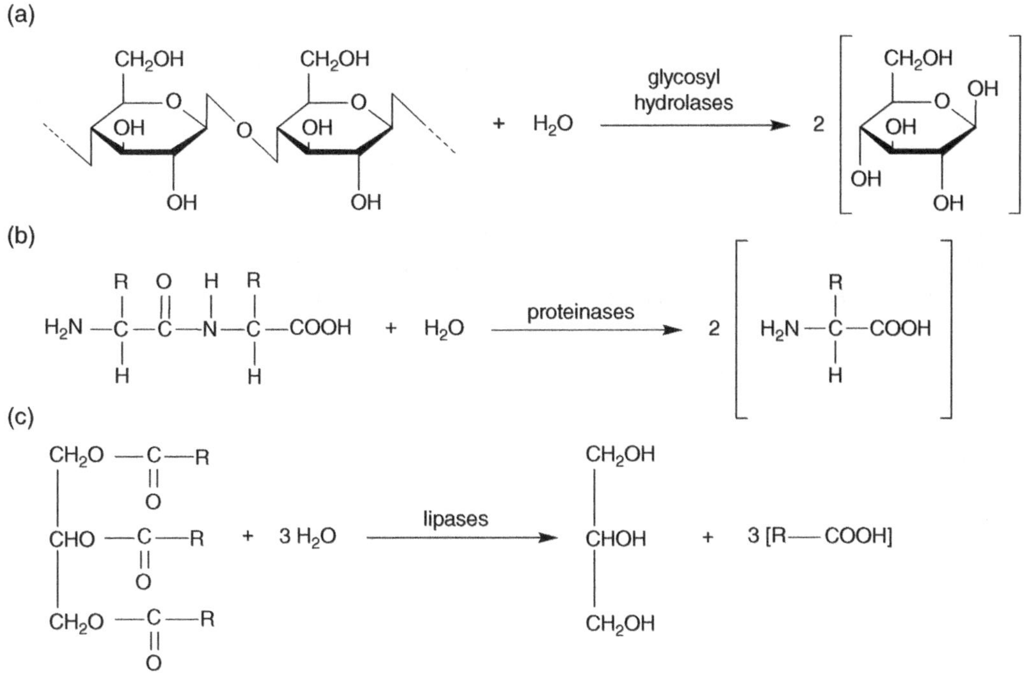

(a)

(b)

(c)

FIG. 17.3 Hydrolysis of (a) cellulose, (b) protein, (c) lipid.

---

absence of free O_2). The representative biochemical reactions for the production of glucose, amino acids, and fatty acids via the hydrolysis of polysaccharides (e.g., cellulose), proteins, and lipids, respectively, are shown in Figure 17.3. Hydrolysis is often a rate-limiting step in the overall AD process, especially for feedstocks containing high lipids, and/or a significant amount of particulate matter and plant materials (e.g., sewage sludge, animal manure, food waste, and lignocellulosic biomass). For such feedstocks, the rate of CH_4 production during AD is proportional to the net rate of particle solubilization. *Hydrolysis* can last for hours to several days, depending on the complexity of feedstocks.

Acidogenesis

The simple soluble molecules such as sugars, amino acids, peptides, and LCFAs generated during hydrolysis are fermented into short-chain fatty acids (SCFAs; commonly known as *volatile fatty acids*, *VFAs*), CO_2, and H_2. The process is known as *acidogenesis*, which is carried out by facultative and obligate anaerobic bacteria such as *Bacteroides*, *Clostridium*, *Butyribacterium*, *Propionibacterium*, *Pseudomonas*, and *Ruminococcus*. The SCFAs produced during acidogenesis primarily consist of acetic, propionic, and butyric acids, with a small amount of valeric, lactic, and succinic acids. A small amount of alcohol, especially ethanol, is also produced during acidogenesis. In terms of the energy distribution of products, H_2 and acetic acid account for 10% and 35%, respectively, and the remaining energy is present in VFAs and alcohol. Acidogenesis occurs rapidly (minutes to days), and may lead to the accumulation of VFAs, which often inhibit the AD process, especially when readily degradable substrates are used.

Acetogenesis

VFAs, along with ethanol produced during *acidogenesis*, are metabolized to acetate and H_2 by H_2-producing acetogenic bacteria, as shown by Equations 17.1, 17.2, and 17.3. These microbes are commonly known as syntrophic acetogens because of their syntrophic relationship with H_2-consuming bacteria. Thermodynamically, a high concentration of H_2 (i.e., high H_2 partial pressure) is unfavorable for acetogenesis to proceed, as is evident from positive Gibb's free energy changes (Equations 17.1, 17.2, and 17.3). In particular, elevated H_2 partial pressure inhibits propionic acid degradation. In an anaerobic system, low H_2 partial pressure ($<10^{-4}$ atm) is achieved via efficient removal of H_2 by H_2-consuming microbes such as hydrogenotrophic methanogens and/or homoacetogens. Thus, there exists a symbiotic (syntrophic) relationship between H_2-producing acetogens and H_2-consuming methanogens/homoacetogens, and the phenomenon is known as *interspecies H_2 transfer*.

$$CH_3CH_2COO^- + 3H_2O \rightarrow CH_3COO^- + HCO_3^- + H^+ + 3H_2 \quad \Delta G^{0'} = +76.1\,kJ/mol$$

Propionate

$$(17.1)$$

$$CH_3CH_2CH_2COO^- + 2H_2O \rightarrow 2CH_3COO^- + H^+ + 2H_2 \qquad \Delta G^{0'} = +48.1\,kJ/mol$$

Butyrate

$$(17.2)$$

$$CH_3CH_2OH + H_2O \rightarrow CH_3COO^- + H^+ + 2H_2 \qquad \qquad \Delta G^{0'} = +9.6\,kJ/mol$$

Ethanol

$$(17.3)$$

The syntrophic acetogens responsible for the production of acetate are all obligate anaerobes. Two species of syntrophic acetogens commonly found in anaerobic digesters are *Syntrophomonas wolfei* and *Syntrophobacteri wolinii* (Ariesyady *et al.*, 2007). Although the reaction time for acetogenesis is short, the growth rate of syntrophic acetogens is low, with a generation time longer than a week.

Homoacetogenenesis

Acetate, an important precursor to CH_4 production, is formed through *homoacetogenesis*. The microbes involved are known as homoacetogens, and are either autotrophs or heterotrophs. Autotrophic homoacetogens use CO_2 and H_2 to produce acetate, where CO_2 acts as the carbon source and electron acceptor, and H_2 as an electron donor (Equation 17.4). Heterotrophic homoacetogens use organic substrates such as formate and methanol as the carbon source to produce acetate.

$$4H_2 + 2HCO_3^- + H^+ \rightarrow CH_3COO^- + 4H_2O \quad \left(\Delta G^{0'} = -104.6\,kJ/mol\right) \qquad (17.4)$$

Interspecies hydrogen transfer

Interspecies H_2 transfer is critical in AD to maintain low H_2 partial pressure in the reactor, which helps to efficiently oxidize intermediate products (e.g., acetate, propionate, butyrate, ethanol, etc.). H_2 produced during acidogenesis and acetogenesis must be consumed rapidly by H_2-consuming microbes such as hydrogenotrophic methanogens, sulfate-reducing bacteria, or homoacetogens to prevent the accumulation of these intermediates in the digester. High-rate bioreactors (i.e., biofilm reactors, UASB) and high biomass inventory provide an excellent opportunity for the proximate growth of diverse microbial communities. Thus, such AD systems are considered to be highly efficient for stable operation.

Energy yield during H_2 utilization by hydrogenotrophs ($\Delta G^{0'}$ = -135.6 kJ/mol) is comparable to that of H_2 utilization by homoacetogens ($\Delta G^{0'}$ = -104.6 kJ/mol). Thus, there is a competition for available H_2 between these two microbial groups (Khanal, 2008). At low H_2 partial pressure (10^{-5}–10^{-4} atm), hydrogenotrophs dominate over homoacetogens, while under low temperature and low H_2 partial pressure (low hydrogen concentration) psychroactive homoacetogens, such as *Acetobacterium bakii*, outcompete hydrogenotrophs. The ability of homoacetogens to use a wide range of substrates and to adapt to low temperatures gives them a competitive advantage over methanogens.

Syntrophic Acetate-Oxidizing Bacteria (SAOB)

Acetate-oxidizing bacteria exist in syntrophic association with hydrogenotrophs. Both methyl and carboxyl groups of acetate are oxidized to CO_2, producing H_2 (Equation 17.5) by SAOB. The overall reaction is thermodynamically favorable, as evident from a negative Gibb's free energy change (Equation 17.6) when SAOB work symbiotically with hydrogenotrophic methanogens.

$$CH_3COO^- + 4H_2O \rightarrow 4H_2 + 2HCO_3^- + H^+ \qquad \Delta G^{0'} = +104.6\,kJ/mol \qquad (17.5)$$

$$4H_2 + HCO_3^- + H^+ \rightarrow CH_4 + 3H_2O \qquad \Delta G^{0'} = -135.6\,kJ/mol \qquad (17.6)$$

At a high concentration of inhibitors, particularly ammonia (mainly released during the degradation of nitrogen-rich compounds such as protein and urea) and VFAs at which the aceticlastic methanogens are inhibited, syntrophic acetate oxidation (SAO) is the main pathway for acetate degradation (Schnurer and Nordberg, 2008; Schnurer *et al.*, 1999). Other factors such as acetate concentration (Hao *et al.*, 2010), HRT (Westerholm, 2012; Shigematsu *et al.*, 2004), methanogenic population (Karakashev *et al.*, 2006), and temperature (Rademacher *et al.*, 2012) also determine the dominance of SAO.

Methanogenesis

Methanogenesis is carried out by a specialized group of microorganisms, classified as *Archaea*. Methanogens are strict anaerobes, and are known to produce CH_4 via aceticlastic and hydrogenotrophic pathways. Methanogens grow very slowly, and are extremely sensitive to environmental conditions such as pH, temperature, and the presence of inhibitory compounds, among others. Therefore, methanogenesis is often a rate-limiting step in AD, especially for readily degradable substrates. The two pathways for methane production are discussed in the following sections.

Acetotrophic or Aceticlastic Methanogens

Acetotrophic or aceticlastic methanogens metabolize acetate directly into CH_4 (Equation 17.7). Acetate is cleaved to methyl and carboxyl groups. The methyl group is reduced to CH_4 via several biochemical reactions, whereas the carboxyl group is oxidized to CO_2. Only two genera of aceticlastic methanogens, *Methanosaeta* and *Methanosarcina*, have been identified. Aceticlastic methanogens belong to the genera *Methanosarcina*, which uses methanol, methylamines, and sometimes H_2/CO_2 besides acetate. *Methanosarcina* is unique in the sense of being both aceticlastic and hydrogenotrophic methanogens. *Methanosarcina* has a typical doubling time of 1–2 days on acetate. The aceticlastic methanogen *Methanosaeta* only uses acetate (Demirel and Scherer, 2008). It has a doubling time of 4–9 days and is a rod-shaped filamentous methanogen, able to grow only at low acetate concentration (7–70 μM). *Methanosarcina* is a coccoid-shaped large aggregate and not only has a higher acetate threshold (0.2–1.2 mM), but also has a higher growth rate and yield coefficient compared to *Methanosaeta* (Conklin *et al.*, 2006; Jetten *et al.*, 1992). This explains the dominance of *Methanosaeta* and *Methanosarcina* in a variety of anaerobic systems under low and high acetate concentrations, respectively (Williams *et al.*, 2013; De Vrieze *et al.*, 2012; Krakat *et al.*, 2010; Hori *et al.*, 2006). The substrate feeding frequency has a significant effect on the dominance of methanogens.

$$CH_3COO^- + H_2O \rightarrow CH_4 + HCO_3^- \quad \Delta G^{0'} = -31\,kJ/mol \tag{17.7}$$

Hydrogenotrophic Methanogens

Hydrogenotrophs reduce CO_2 to CH_4 (Equation 17.6), utilizing H_2 derived from the catabolic reactions of H_2-producing acetogenic bacteria (Equations 17.1, 17.2, and 17.3) and SAOB (Equation 17.5). Hydrogenotrophic methanogens are critical for the stable and efficient operation of anaerobic digesters, as they immediately consume H_2, thereby maintaining the H_2 partial pressure in digesters at an acceptable level to both acetogenic bacteria and SAOB, as discussed earlier. Hydrogenotrophs are distributed in all six phylogenetic orders: *Methanobacteriales*, *Methanococcales*, *Methanomicrobiales*, *Methanopyrales*, *Methanosarcinales*, and *Methanocellales* (Angelidaki *et al.*, 2011). Besides H_2, many hydrogenotrophs can also use formate as their electron donor (Equation 17.8; Khanal, 2008).

$$4HCOO^- + H_2O + H^+ \rightarrow CH_4 + 3HCO_3^- \quad \Delta G^{0'} = -130.4\,kJ/mol \tag{17.8}$$

According to ADM1 (Batstone *et al.*, 2002), two-thirds of total CH_4 are generated from the aceticlastic pathway and one-third of CH_4 is produced through the hydrogenotrophic pathway. This assumption has, however, been contradicted by several recent studies.

17.3 Stoichiometry of Methane Production

The theoretical CH_4 yield, also known as the *stoichiometric CH_4 yield*, of a given substrate can be estimated using the Bushwell equation if the chemical composition of the substrate is known. Organic compounds are composed of the basic elements, C, H, O, N, and S, which are converted into CH_4, CO_2, NH_3, and H_2S during AD, as illustrated by the general biochemical reaction (Equation 17.9):

$$C_aH_bO_cN_dS_e + fH_2O \rightarrow gCH_4 + hCO_2 + iNH_3 + jH_2S \tag{17.9}$$

Conducting a mass balance for each element of the organic compound, we have:

For "S" balance,	$j = e$	(17.10)
For "N" balance,	$i = d$	(17.11)
For "O" balance,	$2h = c + f$	(17.12)
For "H" balance,	$2j + 3i + 4g = b + 2f$	(17.13)
For "C" balance,	$h + g = a$	(17.14)

Solving the mass balances for each element, the variables f, g, h, i, and j can be obtained in terms of a, b, c, d, and e, resulting in:

$$C_aH_bO_cN_dS_e + \left(a - \frac{b}{4} - \frac{c}{2} + \frac{3d}{4} + \frac{e}{2} \right) H_2O \rightarrow$$

$$\left(\frac{a}{2} + \frac{b}{8} - \frac{c}{4} - \frac{3d}{8} - \frac{e}{4} \right) CH_4 + \left(\frac{a}{2} - \frac{b}{8} + \frac{c}{4} + \frac{3d}{8} + \frac{e}{4} \right) CO_2 + dNH_3 + eH_2S \quad (17.15)$$

For glucose ($C_6H_{12}O_6$), the CH_4 yield and CH_4 content can be obtained stoichiometrically by simplifying Equation 17.15:

$$C_6H_{12}O_6 \rightarrow 3CH_4 + 3CO_2 \quad (17.16)$$

The CH_4 yield (m^3 CH_4/kg glucose) can be calculated by combining Equation 17.16 and the ideal gas law (one mole of any gas at STP occupies a volume of 22.4 L):

$$CH_4 \text{ yield at STP} = \frac{(3 \text{ mol } CH_4)(22.4 \text{ L/mol})}{(1 \text{ mol } C_6H_{12}O_6)(180.2 \text{ g/mol})} = 0.373 \text{ L/g} = 0.373 \text{ m}^3/\text{kg}$$

where STP is standard temperature (i.e., 0 °C) and pressure (i.e., 1 atm).

The CH_4 content (%) can be calculated from the molar ratios of CH_4 and $CO_2 + CH_4$:

$$CH_4 \text{ content} = \frac{(3 \text{ mol } CH_4)}{(3 \text{ mol } CH_4) + (3 \text{ mol } CO_2)} = 50\%$$

Table 17.3 shows the calculated theoretical CH_4 yields and contents of common substrates.

Example 17.1 shows how the stoichiometric approach can be applied to calculate the theoretical methane yield based on the known chemical composition of the substrate.

Table 17.3 Theoretical CH_4 yields and contents of common feedstocks

Substrate	Common chemical formula	CH_4 yield at STP (m^3/kg)	CH_4 (% by mass)
Cellulose/starch	$(C_6H_{10}O_5)_n$	0.415	50
Hemicellulose	$(C_5H_8O_4)_n$	0.424	50
Protein (cellular)	$C_5H_7NO_2$	0.496	50
Lipid (triolein)	$C_{57}H_{104}O_6$	1.014	70
Primary sludge	$C_{10}H_{19}O_3N$	0.70	62.5
Secondary sludge	$C_5H_7O_2N$	0.50	50

Example 17.1: Calculate the maximum methane potential of food waste based on the following composition: total solids (TS): 92.6%; volatile solids (VS): 94.8% of TS; fiber (cellulose polymers): 6.5% of VS; starch: 40.8% of VS; lipids: 15.5% of VS; crude protein: 29.5% of VS.

Solution:

1 kg of food waste contains:

TS: 1 kg × 0.926 = 0.926 kg
VS: 0.926 kg × 0.948 = 0.878 kg

Cellulose/starch: 0.878 kg × (0.065 + 0.408) = 0.415 kg
Lipid: 0.878 kg × 0.155 = 0.136 kg
Protein: 0.878 kg × 0.295 = 0.259 kg

Theoretical CH_4 yield for 1 kg food waste

$$(415\,L/kg) \times (0.415\,kg) + (1014\,L/kg) \times (0.136\,kg) + (496\,L/kg) \times (0.259\,kg) = \textbf{439 L}$$

When the chemical composition of substrates is unknown or difficult to determine, the theoretical (maximum) CH_4 yield can be estimated based on the *chemical oxygen demand* (COD), which indicates the total organic content of the substrate. Theoretically, 1 kg COD generates 0.35 m^3 CH_4 at STP, as elucidated in Example 17.2.

Example 17.2: If the chemical oxygen demand (COD) of a substrate is known, the maximum CH_4 yield can be calculated. Calculate the theoretical CH_4 yield of 1 kg COD.

Solution:

Step 1: Calculation of COD equivalent of CH_4

$$CH_4 + 2O_2 \rightarrow CO_2 + 2H_2O$$
16 g   64 g
$$16\,g\,CH_4 \sim 64\,g\,O_2(COD)$$
$$1\,g\,CH_4 \sim 64/16 = 4\,g\,COD$$

Step 2: Conversion of CH_4 mass to equivalent volume
Based on the ideal gas law, 1 mole of any gas at STP occupies a volume of 22.4 L:

$$1\,Mole\,CH_4 \sim 22.4\,L\,CH_4$$
$$16\,g\,CH_4 \sim 22.4\,L\,CH_4$$
$$1\,g\,CH_4 \sim 22.4/16 = 1.4\,L\,CH_4$$

Step 3: CH_4 generation rate per unit of COD
From Steps 1 and 2, we have:

$$1\,g\,CH_4 \sim 4\,g\,COD \sim 1.4\,L\,CH_4$$
$$1\,g\,COD \sim 1.4/4 = 0.35\,L\,CH_4$$
or 1 kg COD $\sim$ 0.35 m^3CH_4

So, complete anaerobic digestion of 1 kg COD produces 0.35 m^3 CH_4 at STP.

The theoretical CH_4 yield often overestimates the actual methane production. The organic matter is not completely metabolized to CH_4 gas during AD, due to either low biodegradability or the generation of other organic byproducts such as VFAs (lost in effluent) and microbial biomass. Serum bottle tests or *biochemical methane potential* (*BMP*) assays are frequently performed in the laboratory to determine the actual CH_4 yield and biodegradability of substrates.

17.4 Important Considerations in Anaerobic Digestion

Although AD is a mature technology, stable operation of the AD process requires special consideration of many factors. Many of those factors primarily revolve around anaerobic microorganisms, especially methanogens, which are considered highly susceptible to changes in environmental and operating conditions. The performance of an AD system is often evaluated based on CH_4 yield/production. Thus, methanogenesis is regarded as a rate-limiting step in the AD process. The high vulnerability and extremely low growth rate of methanogens in an AD system require careful maintenance and monitoring of the environmental and operating conditions. Some of these conditions are discussed in the following sections.

17.4.1 Temperature

Anaerobic processes, like most other biological processes, are strongly dependent on temperature. Although anaerobic microorganisms, especially methanogens, are viable at different temperatures, there exist three optimal temperature ranges for methanogens: psychrophilic, mesophilic, and thermophilic. Thus, methanogens are accordingly classified as psychrophiles, mesophiles, and thermophiles. The anaerobic conversion rates generally increase with temperature up to 60 °C (Pohland, 1992). Studies have shown that anaerobic conversion has its highest activity at 5–15 °C for psychrophiles, at 35–40 °C for mesophiles, and at about 55–60 °C for thermophiles, with decreased activity between these optima, as shown in Figure 17.4 (Lettinga *et al.*, 2001; van Haandel and Lettinga, 1994). As a rule of thumb, the biological activity doubles for every 10 °C increase in temperature within the optimal temperature range.

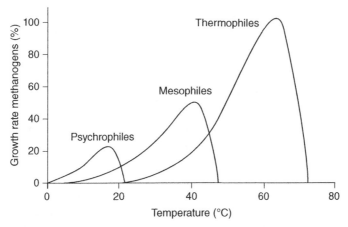

FIG. 17.4   Relative growth rate of psychrophilic, mesophilic, and thermophilic methanogens. Source: Lettinga *et al.*, 2001. Reproduced with permission of Elsevier.

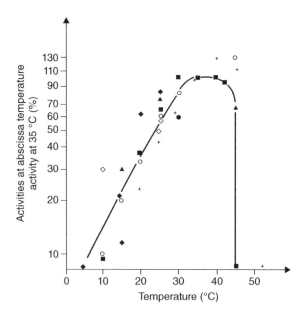

FIG. 17.5 Activities of mesophilic methanogens at different temperatures. Source: Lettinga *et al.*, 2001. Reproduced with permission of Elsevier.

Most of the full-scale anaerobic systems are operated at mesophilic temperature. Figure 17.5 shows the effects of temperature on the activities of mesophilic methanogens. As evident from the graph, methanogens have the highest activity at around 35 °C. It is, however, not uncommon to see a full-scale anaerobic digester operating at a mesophilic temperature of 40–45 °C.

The anaerobic process is recommended to operate for temperature variations not exceeding 0.6–1.2 °C per day (WPCF, 1987), although some studies suggest that the anaerobic process is capable of withstanding larger temperature variations. Methanogenic activity is negatively affected when the temperature is reduced, but it recovers immediately when the temperature returns to an optimum value. It is important to note that temperature has a significant effect on growth kinetics, such as microbial yield, decay rate, half velocity constant, and maximum specific growth rate of anaerobes. Therefore, the effect of temperature on methanogenesis is the result of the collective effect of temperature on the growth kinetics. Temperature also affects many physical phenomena in the digester, such as mixing, substrate diffusion, stratification within the reactor, and gas solubility, among others.

Thermophilic anaerobic digestion

During thermophilic AD, the CH_4-generation rate is nearly constant, independent of temperature at the range of 50–65 °C. The rate is about 25–50% higher than that at mesophilic conditions, depending on substrate type. The HRT of thermophilic digestion can be reduced in comparison to mesophilic conditions due to the more rapid growth of thermophilic and acid-consuming microorganisms. Thermophilic digestion also leads to the destruction of pathogens. Thermophilic digestion, however, possesses a low net biomass yield (about 50% that of mesophilic conditions), thereby resulting in slow start-up and susceptibility to loading variations, substrate changes, or toxicity. Furthermore, the lysis rate of thermophilic microorganisms is relatively high, and as a result they exist only in an exponential growth phase (Zinder and Mah, 1979).

17.4.2 pH and Alkalinity

pH plays a critical role in the successful operation of the AD process and a slight shift in pH from the optimum range adversely affects biogas production. Methanogens are more susceptible to pH change than other microbes in an AD system. Anaerobes can be grouped into two separate pH groups: acidogens and methanogens. The optimum pH is 5.5–6.5 for acidogens and 7.8–8.2 for methanogens. The optimum pH for the combined cultures ranges from 6.8–7.4, with a neutral pH being the ideal. Since methanogenesis is often considered to be the rate-limiting step in AD, it is necessary to maintain the digester pH close to neutral. Acidogens are significantly less sensitive to low or high pH values and acid fermentation prevails over methanogenesis, which may result in souring of the reactor contents (van Haandel and Lettinga, 1994).

Methanogenic activity (i.e., the acetate utilization rate) versus pH is shown in Figure 17.6. This clearly demonstrates that the highest methanogenic activity occurs around a neutral pH range. The drastic drop in methanogenic activity at pH 8.0 and above could be due to a shift of NH_4^+ to the more toxic unionized form, NH_3 (Seagren *et al.*, 1991).

In AD, a pH drop often results from the excessive generation of VFAs and/or CO_2. One of the first options to rectify the problem is to reduce the organic loading rate (OLR) to the point where the accumulated VFAs are allowed to exhaust at a rate faster than their production rate. Once the excess VFAs are exhausted, the pH of the reactor starts to increase to a normal operating range and the methanogens begin to rejuvenate. The OLR can then be increased gradually as the process recovers to full loading capacity. Under extreme circumstances, the decrease in OLR should be coupled with the supplementation of alkaline chemicals for pH adjustment. Since methanogens are vulnerable to abrupt changes in pH from the optimum range, AD system needs sufficient buffering capacity (alkalinity) to mitigate the pH change. When VFAs start to accumulate in the digester, they are

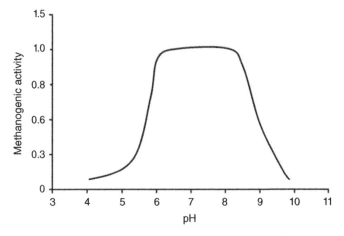

FIG. 17.6   pH dependence of methanogenic activity. Source: Adapted from Khanal, 2008.

neutralized by the alkalinity present in the digester. Thus, alkalinity helps maintain a stable pH in the digester, as shown by Equation 17.17:

$$HCO_3^- + HAc \Leftrightarrow H_2O + CO_2 \uparrow + Ac^- \tag{17.17}$$
$$\underset{\text{(Alkalinity)}}{} \quad \underset{\text{(VFA)}}{}$$

Under unstable conditions, there is an excessive loss of alkalinity coupled with the generation of CO_2, as shown by Equation 17.17. Thus, both the loss of alkalinity and the generation of CO_2 lead to a decrease in digester pH. The relationships among alkalinity, pH, and gas-phase CO_2 are given by Grady *et al.* (1999) and are shown in Equation 17.18:

$$Alkalinity, \text{ as } CaCO_3 = 6.3 \times 10^{-4} \left[\frac{p_{CO_2}}{10^{-pH}} \right] \tag{17.18}$$

where p_{CO_2} is the partial pressure of carbon dioxide in the gas phase in the head space (atm).

Alkalinity supplementation in the digester to maintain an optimum pH is carried out using chemicals such as sodium bicarbonate, sodium carbonate, ammonium hydroxide, and lime. Sodium bicarbonate is preferred because of its high solubility, long-lasting impact, and low toxicity to methanogens. In addition, direct addition of bicarbonate ions results in a direct pH increase, contributing to gas-phase CO_2 consumption. The downside of sodium bicarbonate is cost. Lime is the cheapest option and does not increase the Ca^{2+} toxicity through precipitation of Ca^{2+} as a $CaCO_3$. It can, however, cause serious scaling-up problems in the bioreactor due to precipitation of $CaCO_3$ at the bottom of the reactor. Hydroxide ions from slaked lime react with CO_2 to form alkalinity, as shown in Equation 17.19:

$$Ca(OH)_2 + 2CO_2 \Leftrightarrow Ca^{2+} + 2HCO_3^- \tag{17.19}$$

Alkalinity is affected by the composition of feedstocks, which thereby affects the digester pH. pH generally decreases with the production of VFAs at the beginning of the AD process, but stays within the optimal range of 6.5–7.8 when the VFAs are consumed by the methanogens. The alkalinity and pH of AD may change substantially due to the degradation of substrates. Degradation of protein produces ammonia and results in an increase in alkalinity, while degradation of carbohydrates and fats produces organic acids that consume the alkalinity, thereby lowering the pH.

Importance of co-digestion in alkalinity contribution

Carbohydrate-rich feedstocks (e.g., lignocellulosic biomass, brewery waste, molasses, etc.) do not contribute alkalinity because they lack organic nitrogen. The successful AD of such feedstocks requires supplementation of alkalinity. Thus, co-digestion of carbohydrate-rich feedstocks with nitrogen-rich feedstocks such as animal manure, slaughterhouse wastes, algal cells, etc. can adequately contribute alkalinity, as follows:

$$RCHNH_2COOH + 2H_2O \rightarrow RCOOH + NH_3 + CO_2 + 2H_2$$
$$NH_3 + H_2O + CO_2 \rightarrow NH_4^+ + HCO_3^-$$
$$\underset{\text{(Alkalinity)}}{}$$

Example 17.3: An anaerobic digester treating carbohydrate-rich wastewater encountered a severe souring problem with a VFA/alkalinity ratio of 0.6. The plant operator decided to use lime to rectify the problem. Comment on the various aspects of lime addition.

Solution: The addition of hydrated lime produces alkalinity (calcium bicarbonate). When the point of maximum solubility of calcium bicarbonate is reached, the addition of lime will generate insoluble calcium carbonate precipitate. This will produce no additional alkalinity, but will continue to consume CO_2. If the gas-phase CO_2 is below 10%, pH will increase uncontrollably. The chemical reactions are as follows:

Lime (500 to 1,000 mg/L)

$$Ca(OH)_2 + 2CO_2 \rightarrow Ca(HCO_3)_2$$

Lime (>1,000 mg/L)

$$Ca(OH)_2 + CO_2 \rightarrow CaCO_3 \downarrow + H_2O$$

The other effect of lime addition is the creation of vacuum or negative pressure in the bioreactor due to the dissolution of gas-phase CO_2. This may result in air intrusion and, in severe cases, in the collapse of the digester.

17.4.3 Nutrients

Like all biological processes, an anaerobic process also requires both macronutrients (e.g., nitrogen and phosphorus) and micronutrients (i.e., trace elements) to support the synthesis of new cells. Nutrient requirements for AD vary depending on OLR and feedstock types. There are various methods of determining the macro/micronutrient requirements for AD. Nutrient supplementation can be calculated based on the nutrients required for biomass synthesis by assuming the empirical formula of the microbial cells as $C_5H_7O_2N$ (Speece and McCarty, 1964). The cell mass consists of about 12% nitrogen, which means that about 12 g nitrogen is needed for every 100 g of anaerobic biomass produced. Phosphorus demand accounts for one-seventh to one-fifth of the nitrogen demand. As a rule of thumb, it is assumed that about 10% of the COD removed (i.e., 0.10 kg volatile suspended solids (VSS)/kg COD removed) during the AD process is utilized for biomass synthesis, which can be used to calculate the nitrogen and phosphorus needs. Another approach for calculating the macronutrient requirements is based on the organic strength (i.e., COD) of the substrate (mainly liquid stream). For a highly loaded (0.8–1.2 kg COD/(kg VSS·day)) AD system, the theoretical minimum COD:N:P ratio of 350:7:1 is recommended, whereas for a lightly loaded (<0.5 kg COD/(kg VSS·day)) AD system, the recommended COD:N:P ratio is 1,000:7:1 to calculate the nitrogen and phosphorus needs (Henze and Harremöes, 1983).

For AD of high-solids feedstocks such as lignocellulosic biomass, food wastes, animal manures, OFMSW, and so on, nutrient requirements are often estimated based on the *carbon-to-nitrogen (C/N) ratio*. The recommended C/N ratio for high-solids AD is 20–30, with 25 being the optimal. Co-digestion of high-carbon feedstocks such as fiber-rich biomass with high-nitrogen feedstocks such as animal manures is a good strategy to fulfill the nutrient requirements. Typical C/N ratios of some common feedstocks are shown in Table 17.4.

Table 17.4 Carbon-to-nitrogen (C/N)
ratios for selected feedstocks

Feedstocks	C/N ratio
Kitchen waste	14–16
Dog food	10
Chicken manure	5–15
Swine manure	10–20
Dairy manure	10–15
Corn silage	30–50
Bagasse	110
Corn stover	70–90
Miscanthus	50–90
Switchgrass	70–90
Rice straw	40–60

Nitrogen is most commonly supplemented as urea, aqueous ammonia, or ammonium chloride, and phosphorus is supplemented as phosphoric acid or phosphate salt. In addition to nitrogen and phosphorus, other trace nutrients are also essential for anaerobic microorganisms. Trace metals such as iron (Fe), nickel (Ni), cobalt (Co), molybdenum (Mo), selenium (Se), calcium (Ca), magnesium (Mg), zinc (Zn), copper (Cu), manganese (Mn), tungsten (W), and boron (B) in the mg/L level and Vitamin B_{12} in the μg/L level have been found to enhance CH_4 production (Speece, 1988).

Anaerobic digestion of diverse feedstocks

The versatility of AD is its ability to digest diverse feedstocks ranging from wastewaters, food wastes, sewage sludge, OFMSW, and livestock manures to lignocellulosic biomass. Depending on the feedstocks, the goals of AD could be quite different. For wastewaters, the primary goal is the treatment and possibly generation of bioenergy. The AD of food wastes, sewage sludge, OFMSW, and livestock manures aims at waste stabilization, volume/mass reduction, and bioenergy production. The digestate (the stabilized byproduct) is often land applied as a soil conditioner. In the USA, EPA's AgSTAR Program focuses on AD systems for livestock manures. As of 2014, there are nearly 239 farm-based anaerobic digesters in operation (AgSTAR, 2014). For lignocellulosic biomass such as energy crops, forestry-, and agri-residues, the primary goal is bioenergy generation.

17.4.4 Toxic Materials and Inhibition

The AD process is highly susceptible to toxic substances. The toxic substances are present in either the influent or the byproducts of the metabolic activities of the microorganisms. Heavy metals, halogenated compounds, cyanide, phenol, and so on are examples of the former, while ammonia, sulfide, and long-chain fatty acids belong to the latter group.

Non-ionized sulfide, ammonia, and VFAs are more toxic to anaerobic microbes, especially methanogens, than their ionized forms. The relative distribution of ionized and non-ionized forms is governed by the operating pH, with alkaline pH favoring the more ionized form of these compounds.

Soluble heavy metals are regarded as more critical to failure of an AD system than insoluble forms (Stronach *et al.*, 1986). The generation of sulfide benefits AD by reducing metal toxicity through the formation of insoluble metal sulfides, with the exception of chromium (Cr). Approximately 0.5 mg of sulfide is needed to precipitate 1.0 mg of heavy metal. Heavy metal toxicity follows the following order: Ni > Cu > Pb> Cr > Zn (Hayes and Theis, 1978), with Fe considered more beneficial than detrimental because it mediates sulfide toxicity.

Acclimation to toxic compounds is a key feature to reduce the inhibitory effect on anaerobic microbes. Digesters with a long SRT, high biomass inventory, and biofilm/UASB are more tolerant to toxic compounds.

17.4.5 Total Solids Content

AD can be carried out at very low total solids (TS) concentration (<1.0%) using high-rate anaerobic reactors, such as an anaerobic biofilter, UASB, and expanded/fluidized bed reactor. The anaerobic process for such a low TS content is mainly employed for wastewater treatment coupled with bio-energy production. AD at up to 15% TS content is categorized as liquid/wet digestion, and is carried out in a continuous-stirred tank reactor (CSTR). Examples of such feedstock include animal manure, lignocellulosic biomass, food-processing waste, food waste, sewage sludge, biofuel residues etc. In solid-state digestion, the TS content is >15% and is often around 20–25%. OFMSW is digested at very high TS contents (20–40%) and is commonly known as *dry digestion*. Dry digestion is attractive because the quantity of water generated from the digester is substantially low and, consequently, reduces the digester size significantly.

The volumetric methane production rate during AD often increases with the increase in TS content until a threshold TS content of 15–20% is reached (Figure 17.7). The increase in the volumetric methane production rate is mainly associated with the increase in substrate availability for microbial growth, while the decrease in the volumetric methane production rate is mainly caused by the mass diffusion limitation at high TS levels (Xu *et al.*, 2014).

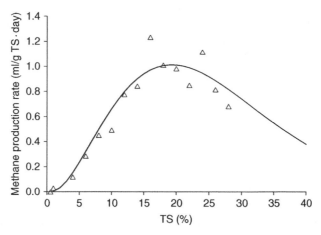

FIG. 17.7   Effect of TS content on volumetric methane production rate. Source: Xu *et al.*, 2014. Reproduced with permission of Elsevier.

17.4.6 Volumetric Organic Loading Rate (VOLR)

The VOLR or simply OLR is often expressed as the mass of organic matter fed per unit reactor volume per unit time. Based on the TS content of the substrate, there are two ways of expressing the OLR. For highly dilute substrates, such as municipal and industrial wastewaters (with TS content <1–2%), OLR is expressed in terms of COD, and is written as kg COD/(m^3·day) (g COD/(L·day)). Many high-rate anaerobic reactors such as UASB, anaerobic filter, and expanded/fluidized bed reactors are employed for liquid waste streams with an OLR of 10–40 kg COD/(m^3·day), and on occasion the OLR can exceed 100 kg COD/(m^3·day) in fluidized bed reactors. For substrates with high TS contents (e.g., energy crops, sewage sludge, food wastes, OFMSW), OLR is expressed in terms of volatile solids (VS), and is written as kg VS/(m^3·day) (g VS/(L·day)). OLR for high-solids AD ranges from 1–5 kg VS/(m^3·day). Some AD systems operate at an OLR as high as 8–9 kg VS/(m^3·day). A high OLR means that more substrate can be digested per unit reactor volume. OLR is one of the most important factors in designing or sizing an anaerobic digester.

Design based on volumetric organic loading rate (VOLR)

$$\text{VOLR} = \frac{S \cdot Q}{V}$$

where VOLR is the volumetric organic loading rate (kg COD/(m^3·day) or kg VS/(m^3·day)), S is biodegradable COD or VS (kg/m^3), Q is wastewater flow rate (m^3/day), and V is bioreactor volume (m^3).

17.4.7 Hydraulic Retention Time (HRT) and Solids Retention Time (SRT)

HRT and SRT are two important design parameters in an AD system. HRT indicates the time for which the substrate remains in the bioreactor in contact with the microbes. The time required to achieve a given degree of digestion depends on the rate of microbial metabolism. Simple soluble substrates are readily degradable, requiring a short HRT, whereas complex substrates (e.g., lignocellulosic biomass) are slowly degradable and need a longer HRT for their digestion. SRT is a measure of the time for which the microbes stay in the bioreactor. Maintaining a long SRT helps maintain a more stable operation, better toxic or shock load tolerance, and a quick recovery from toxicity, among other benefits. The permissible OLR in an AD system is also governed by the SRT.

For slow-growing anaerobic microorganisms, especially methanogens, care must be exercised to prevent their washout from the bioreactor. CSTRs without solid separation and recycling are often prone to failure due to excessive biomass washout unless a long HRT (SRT; note that in CSTR, HRT = SRT) is maintained. A wide variety of high-rate AD reactors are able to maintain an extremely long SRT irrespective of HRT due to biomass immobilization or agglomeration. Such systems operate under a short HRT without any fear of microbial biomass washout, and thus maintain an extremely high SRT/HRT ratio. These bioreactors are mainly suitable for liquid substrate with low suspended solids. For substrates with high solids, such as food wastes, sewage sludge, and energy crops, it is almost impossible to separate HRT from SRT. Thus, such substrates are digested in CSTRs in which the SRT (HRT) is often 20–50 days.

17.4.8 Start-up

Start-up is the initial commissioning period during which the process is brought to a point at where normal performances of the AD can be achieved with continuous substrate feeding (Khanal, 2008). Start-up time is one of the major considerations in AD because of the slow growth rate of anaerobic microorganisms, especially methanogens and their susceptibility to changes in environmental factors, and rate-limiting hydrolysis step for high solids feedstocks. AD systems often need quite a long start-up time. The start-up time could be reduced considerably if the exact microbial culture for the substrate in question is used as a seed. For example, the use of rumen microbes (microbes present in the gut of ruminant animals, e.g., cattle) for fiber-rich feed may greatly reduce the start-up time. A start-up time of 1–3 months is quite common at a mesophilic temperature (37 °C). Periods exceeding 3–6 months may be needed under thermophilic conditions (55 °C) due to the high decay rate of biomass. The start-up time also depends on the initial biomass inventory (i.e., the initial amount of seed added in the reactor). The more seed is used, the shorter the start-up time would be. OLR and environmental factors, such as pH, nutrient availability, temperature, redox conditions, etc., should be maintained within the limits of microbial comfort during start-up.

> **Example 17.4:** A 10,000-head swine farm plans to develop a facility for bioenergy production. Estimate the daily methane production (ft^3/day) and the annual energy production (Btu/yr).
>
> Assume manure production of about 4 kg manure/(animal · day); TS: 10%; VS: 80% of TS; C:N ratio:10; biogas yield: 0.5 m^3/kg VS added; methane content: 60%; energy content of methane: 1,000 Btu/ft^3.

Solution:

Total volatile solids production per day of this swine farm:

$$10,000\,\text{animals} \times 4\,\frac{\text{kg manure}}{\text{animal·day}} \times 10\% \times 0.8\,\frac{\text{kg VS}}{\text{kg TS}} = 3,200\,\text{kg/day}$$

Daily methane production:

$$= 3,200\,\frac{\text{kg}}{\text{day}} \times 0.5\,\frac{m^3}{\text{kg VS}} \times 60\% = 960\,\frac{m^3}{\text{day}} \times 35.32\,\frac{ft^3}{m^3} = 33,907.2\,ft^3/\text{day}$$

The daily methane production is about 33,907 ft^3

Annual energy production:

$$= 33,907.2\,\frac{ft^3}{\text{day}} \times 365\,\frac{\text{day}}{\text{yr}} \times 1,000\,\frac{\text{btu}}{ft^3} = 12,376\,\text{MMBtu/yr}$$

The annual total energy production from the swine manure via anaerobic digestion is about 12,376 MMBtu.

17.5 Anaerobic Digestion Model No. 1 (ADM1)

There have been a wide variety of models that describe the processes involved in AD. However, these models have inherited limitations, as most of them are highly specific to the reactor type, and hence they cannot be widely applied to a variety of processes. In order to overcome this limitation, the International Water Association (IWA) Task Group on Mathematical Modelling of Anaerobic Digestion Processes, consisting of experts from multiple anaerobic process technology disciplines around the world, has developed *ADM1* (Batstone *et al.*, 2002).

ADM1 is a structured model that simulates major processes associated with the conversion of complex organic substrates (e.g., carbohydrates, proteins, fats, and inert solids) into CH_4, CO_2, and inert byproducts, as shown in Figure 17.8. The degradation products are then hydrolyzed to sugars, amino acids, LCFAs, and sugars, respectively. Volatile organic acids (acetate, propionate,

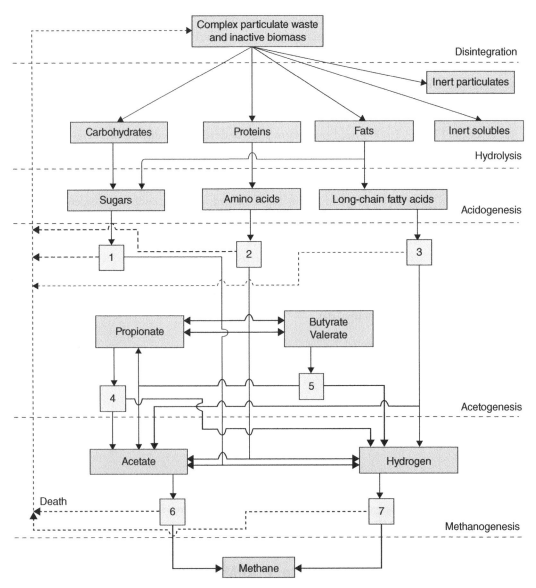

FIG. 17.8   Main pathways of the ADM1: 1. Acidogenesis from sugars; 2. Acidogenesis from amino acids; 3. Acidogenesis from LCFA; 4. Acetogenesis from propionate; 5. Acetogenesis from butyrate and valerate; 6. Aceticlastic methanogenesis; 7. Hydrogenotrophic methanogenesis.

butyrate, and valerate) and H_2 gas are included as intermediate products. Methane is produced either by aceticlastic methanogenic cleavage of acetate or by hydrogenotrophic methanogenic reduction of CO_2 by H_2.

The ADM1 includes a multitude of simultaneous and sequential reactions, and involves complex reaction kinetics. The reactions are broadly divided into two classes: biochemical and physicochemical.

The biochemical reactions involve the uptake of organic substrate and both intra- and extra cellular enzymes are assumed to catalyze them. The biochemical reactions catalyzed by the intracellular enzymes are modeled using Monod-type substrate uptake kinetics, as shown in Equation 17.20:

$$\text{Rate of substrate uptake, } r = k_m \cdot Y \cdot X \cdot \frac{S}{K_s + S} \tag{17.20}$$

where k_m is the specific substrate consumption rate (mg substrate/(mg microbial biomass· hr)), Y is the yield coefficient (mg microbial biomass/mg substrate), X is biomass concentration (mg/L), S is substrate concentration (mg/L), and Ks is the half-saturation constant (mg/L).

pH inhibition to all groups, hydrogen inhibition to acetogenic groups, and free ammonia inhibition to aceticlastic methanogens have been considered in the model. Growth (uptake of substrate) under nitrogen-limited conditions has been regulated using secondary Monod kinetics for inorganic nitrogen. In addition, competitive uptake of butyrate and valerate by the single group that utilizes these two organic acids has been included using an uptake-regulating function.

The biochemical reactions catalyzed by the extracellular enzymes are modeled using first-order kinetics, which is a simplified approach to reflecting the cumulative effect of a multi-step process. Death of biomass is modeled using first-order kinetics, and dead biomass is maintained in the system as composite particulate material for degradation as substrate.

The physicochemical reactions not mediated by microorganisms include ion association/dissociation and gas–liquid transfer of gases. These reactions are very important to correctly predict the pH, concentration of free acids, carbonate alkalinity, and dissolved gas concentration, which directly or indirectly influence the biochemical reactions.

Key features of ADM1 are as follows:

- ADM1 implements conventional nomenclature, measurement units, and model structure in agreement with existing and popular wastewater treatment models such as the activated sludge model (ASM).
- ADM1 consists of 24 state variables and 19 biochemical processes. The biological kinetic rate expressions and coefficients are presented in the form of the Peterson matrix and follow the format of Henze *et al.* (2000). COD balancing is implicit in these equations.
- In order to consider the effect of the physicochemical state (such as pH and gas concentrations) on biochemical reactions, the physicochemical reactions (ion association/dissociation and liquid–gas exchanges) are also included in the model. Although liquid–solid reactions (precipitation and solubilization of ions) are very important in an anaerobic system due to the presence of high levels of cations, ADM1 does not include these due to the complications involved in modeling the precipitation reactions.

An AD system normally consists of a reactor with a liquid volume, and a sealed gas headspace at atmospheric pressure with an exiting gas stream, as shown in Figure 17.9 representing a CSTR with single input and output. A set of mass balance equations for each state component in the liquid and

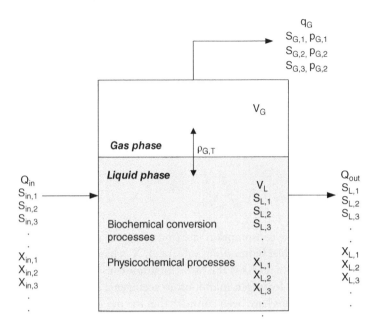

FIG. 17.9   Schematic of typical CSTR.

gas phases is developed and this forms the model of the system. Dynamic equations are used for biochemical conversion as well as gas transfer. An algebraic equation representing the charge balance for the system is included in the model for modeling the pH change.

Aiming to "be as widely applicable as possible for anaerobic process," ADM1 was developed to be a relatively comprehensive model, with 19 biochemical processes involved and the related coefficients provided. After its release, ADM1 has been utilized as the basis for many studies, and different modifications have been made to simplify it and adapt it to specific processes. However, as ADM1 assumes ideal mixing and homogeneous reactor content, it cannot be applied to plug-flow digesters or solid-state digesters, among others. Different digester configurations and their design considerations are discussed in Chapter 18.

References

AgSTAR (2014). US EPA AgSTAR program, http://www.epa.gov/agstar/about-us/index.html, accessed November 2014.

Angelidaki, I., Karakashev, D., Batstone, D.J., Plugge, C.M., and Stams, A.J.M. (2011). Biomethane and its potential. *Methods in Enzymology* 494: 327–51.

Ariesyady, H.D., Ito, T., and Okabe, S. (2007). Functional bacterial and archaeal community structures of major trophic groups in a full-scale anaerobic sludge digester. *Water Research* 41: 1554–68.

Batstone, D., Keller, J., Angelidaki, I., *et al.* (2002). *Anaerobic Digestion Model No. 1 (ADM1).* London: IWA Publishing.

Bushwell, A.M., and Hatfield, W.D. (1938). *Anaerobic Fermentation.* Bulletin No. 32. Urbana, IL: State Water Survey.

Conklin, A., Stensel, D., and Ferguson, J. (2006). Growth kinetics and competition between *Methanosarcina* and *Methanosaeta* in mesophilic anaerobic digestion. *Water Environmental Research* 78(5): 486–96.

Demirel, B., and Scherer, P. (2008). The roles of acetotrophic and hydrogenotrophic methanogens during anaerobic conversion of biomass to methane: A review. *Reviews in Environmental Science and Biotechnology* 7: 173–90.

De Vrieze, J., Hennebel, T., Boon, N., and Verstraete, W. (2012). *Methanosarcina*: The rediscovered methanogen for heavy duty biomethanation. *Bioresource Technology* 112: 1–9.

European Biogas Association (2013). *Biogas Production in Europe: Biogas Report 2013*. Brussels: European Biogas Association.

Grady, C.P.L, Jr., Daigger, G.T., and Lim, H.C. (1999). *Biological Wastewater Treatment*, 2nd edn. New York: Marcel Dekker.

Hao, L.P., Lu, F., He, P.J., Li, L., and Shao, L.M. (2010). Predominant contribution of syntrophic acetate oxidation to thermophilic methane formation at high acetate concentrations. *Environmental Science and Technology* 45(2): 508–13.

Hayes, T.D., and Theis, T.L. (1978). The distribution of heavy metals in anaerobic digestion. *Journal of the Water Pollution Control Federation* 50: 61–72.

Henze, M., and Harremoës, P. (1983). Anaerobic treatment of wastewater in fixed film reactors: A literature review. *Water Science and Technology* 15: 1–101.

Henze, M., Gujer, W., Mino, T., and van Loosdrecht, M. (2000). *Activated Sludge Models ASM1, ASM2, ASM2d and ASM3*. London: IWA Publishing.

Hori, T., Haruta, S., Ueno, Y., Ishii, M., and Igarashi, Y. (2006). Dynamic transition of a methanogenic population in response to the concentration of volatile fatty acids in a thermophilic anaerobic digester. *Applied Environmental Microbiology* 72: 1623–30.

Imhoff, K. (1938). Sedimentation and digestion in Germany. In: L. Pearse (ed.), *Modern Sewage Disposal*. Lancaster, PA: Lancaster Press, p. 47.

Jetten, M., Stams, A., and Zehnder, A. (1992). Methanogenesis from acetate: A comparison of the acetate metabolism in *Methanothrix soehngenii* and *Methanosarcina* spp. *FEMS Microbiological Review* 88: 181–97.

Karakashev, D., Batstone, D.J., Trably, E., and Angelidaki, I. (2006). Acetate oxidation is the dominant methanogenic pathway from acetate in the absence of Methanosaetaceae. *Applied and Environmental Microbiology* 72 (7): 5138–41.

Krakat, N., Schmidt, S., and Scherer, P. (2010). Mesophilic fermentation of renewable biomass: Does hydraulic retention time regulate methanogen diversity. *Applied Environmental Microbiology* 76: 6322–6.

Khanal, S.K. (2008). *Anaerobic Biotechnology for Bioenergy Production: Principles and Applications*. Ames, IA: Wiley-Blackwell.

Lettinga, G., Rebac, S., and Zeeman, G. (2001). Challenges of psychrophilic anaerobic wastewater treatment. *Trends in Biotechnology* 19(9): 363–70.

Pohland, F.G. (1992). Anaerobic treatment: Fundamental concept, application, and new horizons. In: J.F. Malina, Jr., and F.G. Pohland (eds.), *Design of Anaerobic Processes for the Treatment of Industrial and Municipal Wastes*. (eds.). Lancaster, PA: Technomic Publishing, pp. 1–40.

Rademacher, A., Zakrzewski, M., Schlüter, A., *et al.* (2012). Characterization of microbial biofilms in a thermophilic biogas system by high-throughput metagenome sequencing. *FEMS Microbiology and Ecology* 79(3): 785–99.

Schnurer, A., and Nordberg, A. (2008). Ammonia, a selective agent for methane production by syntrophic acetate oxidation at mesophilic temperature. *Water Science and Technology* 57(5): 735–40.

Schnurer, A.G., Zellner, G., and Svensson, B. (1999). Mesophilic syntrophic acetate oxidation during methane formation in biogas reactors. *FEMS Microbiology and Ecology* 29: 249–61.

Seagren, E.A., Levine, A.D., and Dague, R.R. (1991). High pH effects in anaerobic treatment of liquid industrial byproducts. *45th Purdue Industrial Waste Conference Proceedings*. Chelsea, MI: Lewis Publishers, pp. 377–86.

Shigematsu, T., Tang, Y., Kobayashi, T., Kawaguchi, H., Morimura, S., and Kida, K. (2004). Effect of dilution rate on metabolic pathway shift between aceticlastic and nonaceticlastic methanogenesis in chemostat cultivation. *Applied and Environmental Microbiology* 70(7): 4048–52.

Speece, R.E. (1988). Advances in anaerobic biotechnology for industrial wastewater treatment. In: M.F. Torpy (ed.), *Anaerobic Treatment of Industrial Wastewaters*. Park Ridge, NJ: Noyes Data Corporation, pp. 1–6.

Speece, R.E., and McCarty, P.L. (1964). Nutrient requirements and biological solids accumulation in anaerobic digestion. *Advances in Water Pollution Control Research* 2: 305–22.

Stronach, S.M., Rudd, T., and Lester, J.N. (1986). *Anaerobic Digestion Processes in Industrial Wastewater Treatment*. Berlin: Springer.

Surendra, K.C., Takara, D., Hashimoto, A.G., and Khanal, S.K. (2014). Biogas as a sustainable energy source for the developing countries: Opportunities and challenges. *Renewable and Sustainable Energy Reviews* 31: 846–59.

van Haandel, A.C., and Lettinga, G. (1994). *Anaerobic Sewage Treatment: A Practical Guide for Regions with a Hot Climate*. Chichester: John Wiley & Sons Ltd.

Verma, S. (2002). Anaerobic digestion of biodegradable organics in municipal solid wastes. MS thesis, Columbia University, New York.

Westerholm, M. (2012). Biogas production through the syntrophic acetate-oxidising pathway. PhD thesis, Swedish University of Agricultural Sciences, Uppsala.

Williams, J., Williams, H., Dinsdale, R., Guwy, A., and Esteves, S. (2013). Monitoring methanogenic population dynamics in a full-scale anaerobic digester to facilitate operational management. *Bioresource Technology* 140: 234–42.

WPCF (1987) *Anaerobic Sludge Digestion*. Manual of Practice No. 16. Alexandria, VA: Water Pollution Control Federation.

Xu, F., Wang, Z., Tang, L., and Li, Y. (2014). A mass diffusion-based interpretation of the effect of total solids content on solid-state anaerobic digestion of cellulosic biomass. *Bioresource Technology* 167: 178–85.

Zinder, S.H., and Mah, R.A. (1979). Isolation and characterization of a thermophilic strain of Methanosarcina unable to use H_2-CO_2 for methanogenesis. *Applied Environmental Microbiology* 38: 996–1008.

Exercise Problems

17.1. The anaerobic process is a multi-step process mediated by different groups of microorganisms. Why is a symbiotic relation extremely important in an anaerobic process? Explain with examples. What are the strategies to maintain a well-balanced symbiotic condition?

17.2. Why is a long solids retention time important in anaerobic digestion?

17.3. What is the rate-limiting step in anaerobic digestion of high-solids feedstocks?

17.4. Acetoclastic methanogenesis is carried out by *Methanosaeta* and *Methanosarcina*. Explain some of the kinetic behaviors of these microbes.

17.5. Calculate the theoretical methane yield of homopolysaccharides (starch and cellulose), heteropolysaccharides (hemicellulose), protein, and lipid.

17.6. A small noodle factory generates 10 m^3 of wastewater daily with an organic content of 40 g COD/L. What is the maximum methane-generation rate at STP and 35 °C?

17.7. How would you design an anaerobic system? Design an anaerobic digester treating vinasse produced from a 25-MGY sugarcane-to-ethanol plant. Make all valid assumptions.

17.8. An anaerobic filter with an effective volume of 2.0 m^3 is treating wastewater from a molasses-fermentation unit at mesophilic conditions (37 °C). The COD loading rate is 30 kg/m^3-day. The COD removal efficiency is 85%, of which 10% goes to biomass synthesis. Calculate the following:

(a) The maximum methane-generation rate.
(b) The biogas-generation rate in m^3/day (the biogas composition is 60% CH_4, 39% CO_2, and 1% H_2S).
(c) The total energy that could be generated from CH_4 in kWh.
(d) The biomass yield in kg VSS/kg $COD_{removed}$.

17.9. The city of Honolulu generates around 150 metric tons of food waste daily. How much methane can be produced from the food waste? How much electricity can be generated from the gas?

17.10. A brewery produces 0.5 MGD (1,892.5 m^3/day) of carbohydrate-rich wastewater. The wastewater has a total COD of 7,000 mg/L and is deficient in macronutrients. The plant treats the wastewater anaerobically using a UASB reactor. The COD-removal efficiency of the plant is 85% and 10% of the removed COD goes to biomass synthesis. Calculate the daily nutrient (nitrogen and phosphorus) requirement in lb (or kg). What would be the best sources of nitrogen and phosphorus for daily operation?

17.11. The University of Hawaii-Manoa (UHM) campus produces 5 metric tons (wet) of food waste daily. The total solids (TS) content of the food waste is 60% and the volatile solids (VS) (organic content) is 70% of the TS. A bench-scale study showed that food waste could generate as much as 5.1 ft^3 CH$_4$/lb VS at 35 °C. UHM decided to adopt anaerobic digestion to produce bioenergy for heat and steam production. How much methane gas is produced in ft^3/day? The biogas contains 75% methane gas. Calculate the total heating value that the waste could produce in MMBtu.

Biogas Production and Applications

Samir Kumar Khanal and Yebo Li

What is included in this chapter?

This chapter covers various reactor configurations, biogas cleaning and upgrading, and different uses of biogas. Solid-state anaerobic digestion systems, household digesters, and digestate applications are also discussed briefly.

18.1 Introduction

As a mature technology, *anaerobic digestion* (AD) is finding widespread applications ranging from waste (water) treatment and stabilization to bioenergy production. AD is now widely applied for bioenergy production from food waste, sewage sludge, animal manure, crop residues, and energy crops in developed countries, with thousands of commercial-scale plants in operation. Interestingly, small-scale digesters, also known as "household digesters", have been providing much-needed energy for cooking and lighting in millions of homes in developing countries. A detailed discussion of the history, biochemical pathways, and important factors affecting AD performance can be found in Chapter 17. This chapter focuses mainly on the application aspects of AD, including different AD systems, biogas cleaning and upgrading, and various uses of biogas. Furthermore, the chapter also covers household digester and digestate applications.

18.2 Anaerobic Digestion Systems

The digester is considered to be the heart of an AD system, and is where all the biochemical reactions take place. "Bioreactor" is a general term applicable to a reactor in which diverse biochemical reactions occur, whereas the term "digester" is specific to a reactor in which AD is carried out. However, digester and bioreactor are often used interchangeably.

Bioenergy: Principles and Applications, First Edition. Edited by Yebo Li and Samir Kumar Khanal.
© 2017 John Wiley & Sons, Inc. Published 2017 by John Wiley & Sons, Inc.
Companion website: www.wiley.com/go/Li/Bioenergy

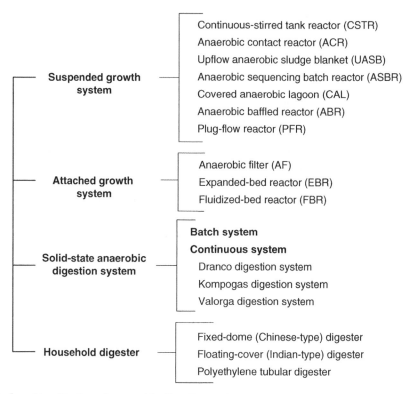

```
                                    ┌─ Continuous-stirred tank reactor (CSTR)
                                    │  Anaerobic contact reactor (ACR)
                                    │  Upflow anaerobic sludge blanket (UASB)
         Suspended growth ─────────┤  Anaerobic sequencing batch reactor (ASBR)
              system                │  Covered anaerobic lagoon (CAL)
                                    │  Anaerobic baffled reactor (ABR)
                                    └─ Plug-flow reactor (PFR)

                                    ┌─ Anaerobic filter (AF)
         Attached growth ──────────┤  Expanded-bed reactor (EBR)
              system                └─ Fluidized-bed reactor (FBR)

                                    ┌─ Batch system
                                    │  Continuous system
         Solid-state anaerobic ────┤     Dranco digestion system
           digestion system         │     Kompogas digestion system
                                    └─    Valorga digestion system

                                    ┌─ Fixed-dome (Chinese-type) digester
         Household digester ───────┤  Floating-cover (Indian-type) digester
                                    └─ Polyethylene tubular digester
```

FIG. 18.1   Classification of anaerobic digestion systems.

One of the important considerations in anaerobic digester design is microbial biomass retention capacity, because anaerobes, especially methanogens, grow slowly. Thus, it is essential to maintain a long solids retention time (SRT), irrespective of hydraulic retention time (HRT). Such decoupling can maintain a significantly high SRT/HRT ratio, which prevents the washout of slow-growing anaerobes, especially methanogens. Some of the approaches of decoupling are biomass immobilization in attached growth systems; granulation and floc formation; biomass recycling; and biomass retention. For dilute substrates (e.g., wastewaters), decoupling is extremely important and is achieved through one of the approaches already discussed. Decoupling is extremely difficult for high-solids substrates, which are often employed for bioenergy production in AD. Such feedstocks are frequently digested in a completely mixed reactor commonly known as continuous-stirred tank reactor (CSTR), in which HRT = SRT.

Based on this discussion, an AD system can be classified into two broad categories, namely, suspended growth and attached growth systems, as shown in Figure 18.1. Anaerobic reactors are also classified based on flow mode such as continuous and batch reactors; based on microbial activities such as low-rate and high-rate reactors; based on total solids (TS) contents such as dry digesters (solid-state anaerobic digestion system) and wet (slurry) digesters; or based on different stages (phases) of digestion such as one-stage, two-stage and acid-phase; based on temperature such as mesophilic and thermophilic digesters, and so on.

18.2.1 Suspended Growth System

The majority of anaerobic reactors currently in operation are suspended growth systems, in which microbes are in suspension in the reactor. Examples of suspended grown anaerobic reactors are presented in Figure 18.1. Some of the important suspended growth reactors are presented in the following sections.

Continuous-Stirred Tank Reactor (CSTR)

CSTR is the most commonly employed reactor configuration in AD. In CSTR, the contents in the reactor are completely mixed by intermittent or continuous stirring. Thus, the concentrations of all constituents are nearly the same throughout the reactor and in the digestate (effluent). CSTRs are easy to build and operate, and are operated at an HRT of 20–50 days. Since the substrate (influent) gets diluted rapidly in the reactor, a CSTR is less sensitive to shock loading or toxicity.

Design consideration: Digester design is typically based on an empirical approach. A fundamental approach can also be employed in sizing a digester, which includes conducting either substrate or microbial balance around the digester. Digester sizing based on microbial mass balance is presented in Khanal (2008). Digester sizing based on substrate balance is discussed here.

The mass balance of substrate around the digester (Figure 18.2) can be written as follows:

(Rate of substrate inflow) + (Rate of substrate degradation) = (Rate of substrate out) + (Rate of substrate accumulation)

$$Q\,C_o + r(V_r) = QC_e + \frac{dC}{dt}(V_r) \tag{18.1}$$

where C_o is the influent substrate concentration (mg/L); C_e is the effluent substrate concentration (mg/L); C is the substrate concentration (mg/L) in the reactor in a given time (t); Q is the substrate flow rate (m^3/sec); V_r is the reactor working volume (m^3); and r is the rate constant for degradation of substrate in the reactor (mg/(L·sec)).

For the steady-state condition, there is no accumulation of substrate in the reactor $\left[\left(\frac{dC}{dt}(V_r) = 0\right)\right]$. Thus, Equation 18.1 can be simplified to:

$$Q\,C_o + r(V_r) = QC_e \tag{18.2}$$

$$V_r = \frac{Q(C_o - C_e)}{-r} \tag{18.3}$$

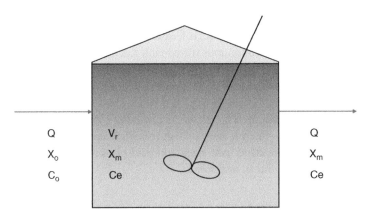

FIG. 18.2 Schematic diagram of a continuous-stirred tank reactor: X_o and X_m represent microbial biomass concentration in the influent and the reactor, respectively.

Plug-Flow Digester

A *plug-flow digester* is a relatively simple and low-cost reactor configuration. The concept was first demonstrated by Jewell's research group at Cornell University in the 1970s for cattle manure digestion (Tilche *et al.*, 1986), and currently it is one of the most preferred configurations for the digestion of dairy manure. Plug-flow digesters consist of a long rectangular concrete tank or polyethylene tube. The tank is sealed with a hard (concrete) or flexible (polypropylene) cover. Plug-flow digesters do not have any mixing device and are mostly operated at mesophilic (35–37 °C) conditions. Temperature control can be achieved through hot-water recirculation using pipes along the concrete tank/polyethylene tube. Figure 18.3 shows a simplified diagram of a plug-flow digester in which substrate movement resembles flow through pipes. Plug-flow digesters are well-suited for high-TS feedstocks, such as cattle manure with a TS content of 10–15%. As feedstock enters the digester, it is decomposed and the digested material is pushed out toward the outlet, resembling the movement of a piston. The HRT (SRT) of a plug-flow digester ranges from 15–30 days. From the kinetic standpoint, plug-flow digesters are more efficient than CSTR on volumetric biogas production.

Design consideration: Unlike in a CSTR, the substrate concentration in the plug-flow digester changes along the direction of flow due to anaerobic degradation. The mixing occurs only in the lateral (radial) direction and there is no mixing in the direction of flow (longitudinal direction). All the elemental volumes of the reacting substrate in the digester remain for the same period of time, and the change in substrate concentration with time is identical for each elemental volume.

The mass balance for substrate flow through the plug-flow digester can be conducted by considering an elemental volume, ΔV, at a distance of X from the inlet, as shown in Figure 18.4. The substrate balance around the elemental volume (ΔV) can be written as:

$$QC + r(\Delta V_r) = Q(C + \Delta C) + \frac{dC}{dt}(\Delta V_r) \qquad (18.4)$$

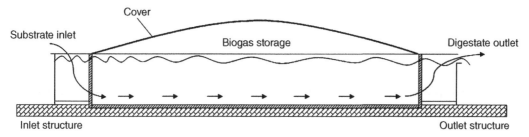

FIG. 18.3   Schematic diagram of a plug-flow digester.

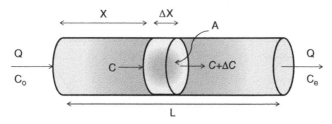

FIG. 18.4   Mass balance of substrate in a plug-flow digester.

where C is the substrate concentration (mg/L) entering the elemental volume at distance X from the inlet; $C + \Delta C$ is the substrate concentration (mg/L) exiting from the elemental volume at a distance of $X + \Delta X$ from the inlet; ΔV is the elemental volume (m³) at a distance X from the inlet = $A \cdot \Delta X$; A is the cross-sectional area of the plug-flow digester under consideration (m²); and ΔX is the small distance through which the element moves in the direction of flow (m). Under the steady-state condition, there is no accumulation of substrate in the digester $\left[\left(\frac{dC}{dt}(\Delta V_r) = 0\right)\right]$. Thus, Equation 18.4 can be simplified to:

$$Q\Delta C = r\left(A \cdot \Delta X\right) \tag{18.5}$$

$$\frac{\Delta C}{\Delta X} = \frac{A}{Q}r \tag{18.6}$$

Taking the limit when $\Delta X \to 0$, Equation 18.6 can be written as:

$$\frac{dC}{dX} = \frac{A}{Q}r \tag{18.7}$$

Equation 18.7 shows that there exists a substrate gradient in the direction of flow in a plug-flow digester. The equation can be rearranged to calculate the volume of a plug-flow digester required to achieve a given degree of digestion:

$$V = \int_{Co}^{Ce} \frac{Q}{r}dC \tag{18.8}$$

Covered Anaerobic Lagoon

Covered anaerobic lagoons are widely adopted as a low-cost option for digesting/stabilizing waste streams with a TS content of 0.5–3%, such as flushed dairy manure, swine manure, or industrial wastewater (Figure 18.5). Lagoons are earthen structures/pits constructed with impermeable liners such as clay or plastic at the bottom and sides to prevent liquid seepage. Although some covered lagoons are designed with a mixing unit, most lagoons are not mixed and are unheated. Lagoons

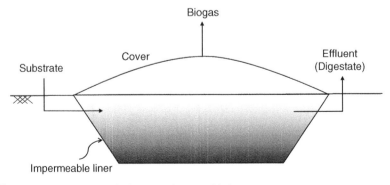

FIG. 18.5 Schematics of a typical covered anaerobic lagoon.

that operate without mixing cannot handle dairy manure containing coarse solids such as hay and silage fibers. Under these situations, stratification occurs within the lagoon, with supernatant on the top, suspended solids as an intermediate layer, and settled solids at the bottom. Covered lagoons have a much longer SRT than the HRT. Since lagoons are operated at ambient temperature, their digestion efficiency and biogas production are strongly dependent on geographical location and climate. Lagoons are operated at an HRT of 3–6 months; thus, lagoons are also considered a waste storage unit.

> **Example 18.1:** The Quasar digester in Columbus, Ohio is fed with high-solids organic substrate. The density of the substrate is 0.85 kg/L. Assume that the average TS content is 10%; VS content is 80% of TS; digester tank capacity is 2.1 million gallons; HRT is 28 days. Calculate the annual substrate input to the digester and the OLR. Evaluate whether the HRT is acceptable based on the recommended OLR of 1–4 kg VS/(m^3· day).
>
> *Solution:* TS (dry matter) input to the digester:
>
> $$2,100,000 \, \text{gal} \times 3.785 \frac{\text{L}}{\text{gal}} \times 0.85 \frac{\text{kg}}{\text{L}} \times 10\% = 675,623 \, \text{kg}$$
>
> Wet biomass input to the digester:
>
> $$\frac{675,623 \, \text{kg}}{10\%} \times \frac{1}{28 \, \text{days}} \times \frac{365 \, \text{days}}{1 \, \text{yr}} = 880,723 \, \text{kg/yr}$$
>
> $$\text{Organic load} = 675,623 \, \text{kg TS} \times 0.8 \frac{\text{kg VS}}{\text{kg TS}} \times \frac{1}{28 \, \text{days}} = 19,304 \, \text{kg/day}$$
>
> $$\text{Thus, the volumetric organic loading rate} = 19,304 \, \frac{\text{kg VS}}{\text{day}} \times \frac{1}{2,100,000 \, \text{gal}} \times \frac{1 \, \text{gal}}{3,785 \, \text{L}}$$
>
> $$\times \frac{1000 \, \text{L}}{\text{m}^3}$$
>
> $$= \textbf{2.43 kg VS/(m}^3\textbf{·day)}$$

The organic loading rate of 2.43 kg VS/(m^3 · day) is within the recommended range and the HRT of 28 days is acceptable.

Upflow Anaerobic Sludge Blanket (UASB) Reactor

The UASB reactor was developed in the 1970s by Gatze Lettinga and co-workers at Wageningen University, The Netherlands. UASB reactors are essentially a suspended growth system in which proper HRT/OLR is maintained in order to facilitate the dense biomass aggregation known as granulation. The size of the granules is about 1–3 mm diameter. Since the granules are large in size and dense, they settle and are retained within the reactor. The volatile suspended solids (VSS) concentration in the reactor may go as high as 50 g/L. Thus, a UASB reactor is able to maintain a very high SRT irrespective of a very short HRT of 4–8 hr. UASB reactors are suitable for soluble carbohydrate-rich industrial wastewaters, which facilitate better granulation. Although originally conceived for industrial wastewater treatment with a low solids content, UASB reactors also provide the

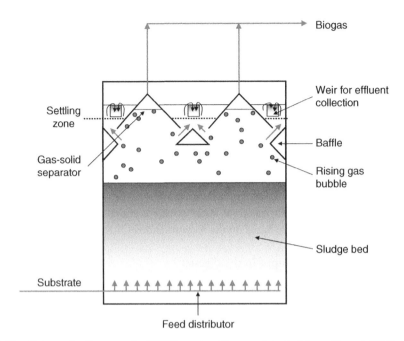

FIG. 18.6   Schematic diagram of a UASB reactor. Source: Adapted from Khanal, 2008.

opportunity for bioenergy recovery from diverse high-strength wastewaters. A schematic diagram of a UASB is shown in Figure 18.6. The working principle and design considerations of UASB reactors can be found in Khanal (2008).

Example 18.2: A 50-million gallon sugarcane-to-ethanol plant plans to build a covered lagoon to treat/store vinasse and also to produce biogas for in-house use. If the amount of vinasse produced is 10 gal per gal of ethanol and the COD is 70 g/L, calculate the maximum CH_4 generation rate in m^3/day at 30 °C. What would be the biogas generation rate at 60% COD removal efficiency when 10% of the removed COD is utilized for biomass synthesis? The mean CH_4 content of biogas is 60%. Assume that the milling season lasts for 7 months.

Solution:

1. Vinasse production rate

$$50 \times 10^6 \frac{\text{gal}}{\text{yr}} \times \left(\frac{3.785\,\text{L}}{1\,\text{gal}} \times \frac{1\text{m}^3}{1,000\,\text{L}} \right) \times 10 \frac{\text{gal}}{\text{gal}} \times \frac{1\,\text{yr}}{365\,\text{day}} = 5,185\,\text{m}^3/\text{day}$$

2. Maximum CH_4 generation rate
The complete degradation of organic matter in the waste could only lead to maximum methane generation and is also regarded as the theoretical methane generation rate.

$$\text{Total COD removed} = 70 \frac{\text{g}}{\text{L}} \times \left(\frac{1\,\text{kg}}{1,000\,\text{g}} \times \frac{1,000\,\text{L}}{1\,\text{m}^3} \right) \times 5,185 \frac{\text{m}^3}{\text{day}} = 362,950\,\text{kg/day}$$

Based on Example 17.1 (Chapter 17), 1 kg COD produces 0.35 m³ CH$_4$ at STP

362,950 kg COD produces 0.35 × 362,950 = 127,033 m³ CH$_4$/day at STP

At 30 °C, CH$_4$ gas generation = $127,033 \times \dfrac{303}{273} = 140,993$ m³/day

The maximum CH$_4$-generation rate = 140,993 m³/day

3. Biogas-generation rate

Not all COD (organic matter) is completely degraded. The fate of COD during anaerobic digestion can be viewed as:

Residual COD (in effluent)
COD converted to CH$_4$ gas
COD diverted to biomass synthesis
COD utilized for sulfate reduction (if sulfate is present)

Total COD removed at 60% efficiency = $362,950 \dfrac{\text{kg}}{\text{day}} \times 0.60 = 217,770$ kg/day

Since 10% of the removed COD is diverted to new biomass synthesis, only 90% of this removed COD is available for methane production.

COD utilized for CH$_4$ generation = $217,770 \times 0.9 = 195,993$ kg/day

At 30 °C, the CH$_4$ generation = $0.35 \times 195,993 \times \dfrac{303}{273} = 76,136$ m³/day

Biogas generation rate $= \dfrac{76,136}{0.60} = \mathbf{126,893}$ **m³/day**

Anaerobic Contact Reactor (ACR)

An ACR is essentially a completely mixed digester coupled with a downstream settling tank. The settled microbial biomass is recycled back to the digester, as shown in Figure 18.7. Hence ACR is able to maintain a high concentration of microbial biomass in the digester. A degassifier allows the removal of biogas bubbles (i.e., CO_2 and CH$_4$) attached to microbial biomass that may otherwise float to the surface. ACR is suitable for the waste streams containing suspended solids (e.g., in a meat packing plant), which allow the microorganisms to attach and form settleable flocs.

Anaerobic Baffled Reactor (ABR)

In an ABR, the substrate flows over and under the baffles (Figure 18.8). The microbial biomass accumulates between the baffles and may in fact form granules with time. In an ABR each chamber acts as a CSTR in series and thus the flow through an ABR resembles the plug flow. The baffles prevent the horizontal movement of biomass in the reactor. Thus, a high concentration of microbes is achieved.

Anaerobic Sequencing Batch Reactor (ASBR)

An ASBR was developed by Dague and co-workers at Iowa State University in the early 1990s (Dague *et al.*, 1992). The ASBR system was developed as a high-rate anaerobic reactor to treat

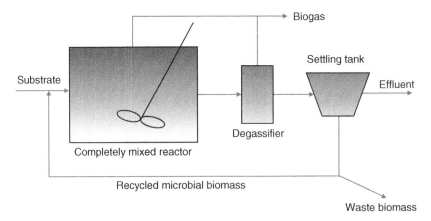

FIG. 18.7   Schematics of anaerobic contact reactor. Source: Adapted from Khanal, 2008.

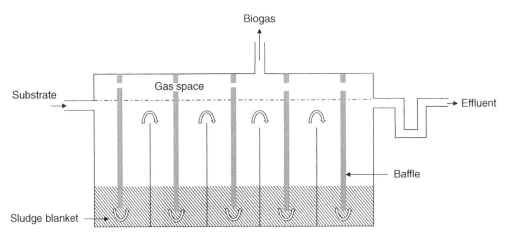

FIG. 18.8   Schematics of an anaerobic baffled reactor.

high-strength and medium solids content feeds (TS: 1–4%). In ASBR, substrate is fed into the reactor during the "Feed" cycle. The organic matter in the substrate is then metabolized by the microbial biomass during the "React" cycle (Figure 18.9). A fraction of the biogas is pumped back into the bottom of the reactor for mixing. During the "Settle" phase, the microbial biomass is allowed to settle in the reactor once the biodegradable organic matter is nearly consumed. The clear supernatant is decanted during the final "Decant" period. Because of the sequential operation, a single reactor can be used as a reaction vessel and as a settling tank, achieving high microbial biomass levels in the reactor independent of HRT. The ASBR retains biomass due to bioflocculation followed by biogranulation, similar to a UASB reactor. The ASBR is ideal for bioenergy production from animal manure and other biowastes with medium TS contents.

18.2.2 Attached Growth System

An *attached growth system*, also known as a biofilm system, is a high-rate anaerobic system in which anaerobic microorganisms, especially methanogens, get attached to an inert media through

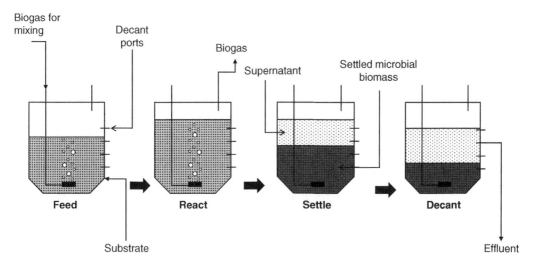

FIG. 18.9   Schematics of anaerobic sequential reactor. Source: Adapted from Khanal, 2008.

self-immobilization. Different inert media or biocarriers such as crushed rock, tiles, activated carbon, shredded tire beads, and plastic of different shapes are used for microbial attachment. These media usually have a very high surface area, which facilitates biofilm formation. Thus, an attached growth anaerobic system maintains a very long SRT independent of HRT. Some examples of attached growth anaerobic systems are discussed here.

Anaerobic Filter (AF)

One of the very first studies using packed-bed reactors, commonly known as an anaerobic filter, was conducted by Young and McCarty (1969). The anaerobic filter was primarily developed to treat soluble organic wastewater. Depending on feeding mode, an anaerobic filter is classified as an *upflow anaerobic filter* (*UAF*; Figure 18.10a), a *downflow anaerobic filter* (*DAF*; Figure 18.10b), or a *multifed anaerobic filter* (*MFAF*; Figure 18.10c).

In a UAF, substrate flows upward through a media bed and the entire bed is submerged. Although a UAF is a fixed-film bioreactor, a significant portion of the biomass remains entrapped within the interstices (voids) between the media. The non-attached microbial biomass forms a bigger floc and eventually takes a granular shape due to the rolling action of rising biogas bubbles. Thus, non-attached biomass contributes significantly to biological activity. The biofilm growth on support media in a UAF is shown in Figure 18.11.

Originally, lava rocks were employed as packing media in anaerobic filters. But due to a very low void volume (i.e., 40–50%), a serious clogging problem occurred. Currently, media are often synthetic plastic or ceramic tiles of different configurations. The void volume of plastic media ranges from 80–95% and provides a high specific surface area, typically $100 \text{ m}^2/\text{m}^3$ or higher, which enhances biofilm growth. Since an anaerobic filter retains a large amount of biomass, a long SRT can be maintained independent of HRT.

A DAF is similar to a UAF except that the microbial biomass is truly attached to the media. Loosely held biomass in a DAF gets washed out of the reactor. The specific surface area of the media plays a more important role in a DAF than in a UAF. Clogging is less of a problem with a DAF, and it

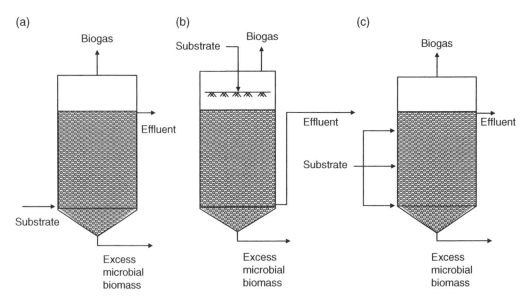

FIG. 18.10 Schematic diagram of anaerobic filters: (a) upflow anaerobic filter, (b) downflow anaerobic filter, and (c) multifed anaerobic filter. Source: Adapted from Khanal, 2008.

FIG. 18.11 Anaerobic biofilm on plastic media. Source: Adapted from Khanal, 2008.

can accommodate feed streams with some suspended solids. Although a DAF has a low biomass inventory, the specific activity of its biomass is relatively high.

In an MFAF, the feed enters the bioreactor through several points along the filter depth. The MFAF maintains a completely mixed regime throughout the reactor, preventing short-circuiting and accumulation of VFAs. In an MFAF substrate is uniformly distributed throughout the reactor, which prevents heavy microbial growth at the bottom of the reactor and minimizes clogging of the bed.

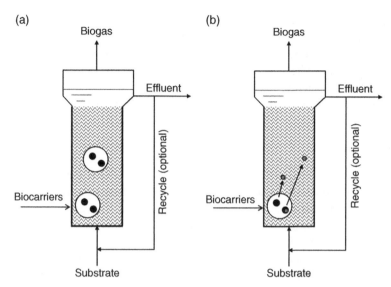

FIG. 18.12   Schematics of (a) an expanded bed reactor and (b) a fluidized bed reactor.

Expanded Bed Reactor (EBR)

An EBR is an attached growth system with some suspended biomass. The microbes get attached on biocarriers such as sand, activated carbon, pulverized polyvinyl chloride, and shredded tire beads. The biocarriers are expanded by the upflow influent velocity and recirculated effluent. In expanded bed reactors, sufficient upflow velocity is maintained to expand the bed by 15–30%. The expanded bed reactor has fewer clogging problems and better substrate diffusion within the biofilm. The bio-carriers are partly supported by fluid flow and partly by contact with adjacent biocarriers, and they tend to remain at the same relative position within the bed (Figure 18.12a).

Fluidized Bed Reactor (FBR)

Although an FBR is similar to an EBR in terms of configuration, an FBR is truly a fixed film reactor, as the suspended microbial biomass is washed out due to the high upflow liquid velocity. The bed expansion is 25–300% of the settled bed volume, which requires a much higher upflow velocity (10–25 m/hr). The biocarriers are supported entirely by the upflow liquid velocity and are therefore able to move freely in the bed (Figure 18.12b). The FBR is free from clogging and short-circuiting problems and results in better substrate diffusion within the biofilm.

18.2.3 Solid-State Anaerobic Digestion System

A *solid-state anaerobic digestion (SS-AD)* system is widely employed for digestion of high-solids feed-stocks, especially OFMSW. An SS-AD system plays an important role in the waste-to-energy field. Such digesters have high volumetric biogas productivity with low parasitic energy requirements for heating and mixing (note: parasitic energy is the energy required for the daily operation of a digester). The maximum OLR can vary from 5–10 kg VS/(m^3·day), with CH$_4$ yield ranging from 200–300 Nm^{3}CH$_4$/kg VS$_{fed}$ depending on the substrate composition. The possible factors affecting the max-imum OLR are substrate composition, microbial concentration, mass transfer rate of substrates to microbes, and accumulation of inhibitory compounds (commonly VFAs and ammonia, among others). The major weaknesses of an SS-AD system include the requirement for a large volume of inoculum, long retention times, and the need for nutrient (macro/micronutrient) supplementation, especially when carbon-rich fiber (e.g., yard waste) is used. SS-AD systems are operated either in batch or continuous mode. They are also classified as "dry" systems, in which the waste is digested as received (20–40% TS content; Vandevivere *et al.*, 2003). Dry systems have widely been employed in France and Germany for biomethane production from mechanically sorted OFMSW.

Batch System

In a batch SS-AD system, the digestate is used as an inoculum and is mixed with fresh materials before loading into the digester. In a percolation system, leachate is collected at the bottom of the digester and is recirculated back into the digester (Figure 18.13). Leachate recirculation enables colonization of microbes throughout the digester. The amount of digestate required for initiating the digestion process can be reduced with leachate recycling. Leachate may also replace the digestate by directly inoculating the fresh materials prior to digestion. Figure 18.13 shows a *garage-type percolation SS-AD system*. It is a rectangular concrete bay with a gas-tight door. A wheel loader feeds the inoculated feedstock into the bay through the door. No mixing is used during digestion.

Continuous System

Continuous SS-AD systems are similar to a plug-flow system in which the incoming feed pushes the digesting materials toward the outlet due to high viscosity. As discussed earlier, a plug-flow system offers technical simplicity due to absence of a mixing unit. Some of the commonly employed commercial SS-AD systems are Dranco (Figure 18.14a), Kompogas (Figure 18.14b), and Valorga (Figure 18.14c) systems (Vandevivere *et al.*, 2003).

The **Dranco digestion system** uses a vertical design. There is no mechanical mixing unit in the digester, but mixing is achieved via recirculation of the digestate drawn from the bottom, mixed with the fresh feedstocks (one part fresh feedstock is mixed with six parts of digestate), and then feeding it

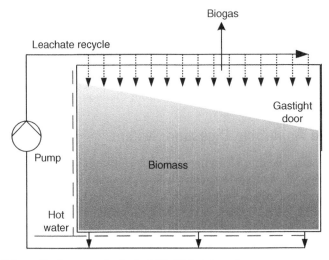

Fig. 18.13   Schematic diagram of a typical SS-AD batch system.

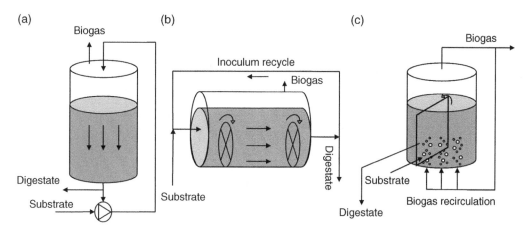

FIG. 18.14   Schematic diagrams of SS-AD systems: (a) Dranco, (b) Kompogas and (c) Valorga. Source: Adapted from Vandevivere *et al.*, 2003.

into the top of the digester (Figure 18.14a). The Dranco system can handle feedstocks with a high TS content of 20–50%.

The **Kompogas digestion system** consists of a cylindrical steel tank placed horizontally where inoculated feedstock moves as a plug flow (Figure 18.14b). The plug flow is assisted by slowly rotating mixers placed inside the digester, which also facilitate homogenization, degassing, and resuspension of heavier particles. In a Kompogas, part of the digestate is recirculated as inoculum and the TS content is adjusted to around 23%. The capacity of a Kompogas is often limited by mechanical constraints.

The **Valorga digestion system** consists of a vertical cylindrical steel tank with a central baffle to move the feedstocks in a circular plug flow (Figure 18.14c). The TS content of the feed is typically around 30%. Pressurized biogas sparging provides mixing and allows for adequate contact between fresh feedstock and the digestate. Only leachate is recirculated to inoculate the fresh feed. Like in the Kompogas process, liquid recirculation is primarily to maintain a TS content of 30% inside the digester. One main issue with biogas injection is the tendency for gas nozzle clogging. Also, for wet (slurry) feedstock (<20% TS), heavy particles settle down in the digester.

18.2.4 Household Digester

Household digesters are commonly employed in rural areas of developing countries, primarily for producing biogas for cooking and lighting. The average volume of household digesters range from 5–7 m^3 and provides about 0.5 m^3 biogas per m^3 of digester volume (Omer and Fadalla, 2003; Akinbami *et al.*, 2001). The daily biogas production varies from 1.0–1.5 m^3. The two types of household digesters most common in developing countries are the Chinese fixed-dome digester, also known as a hydraulic-type digester (Figure 18.15), and the Indian floating-drum digester (Figure 18.16). These digesters are primarily fed with human and animal wastes generated from one household to produce biogas for cooking and lighting. The floating-drum digester is constructed with concrete and steel, whereas the fixed-dome digester is usually built with locally available materials like bricks and stones. There are also modular-type household digesters made of polyethylene (PE), polyvinyl chloride

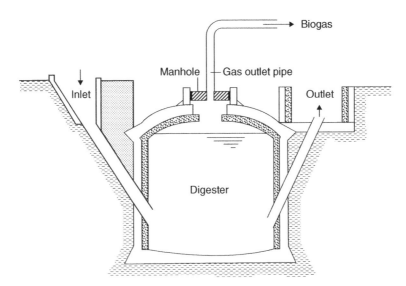

FIG. 18.15   Fixed-dome (Chinese type) digester. Source: Reproduced with permission of Gunnerson and Stuckey, 1986.

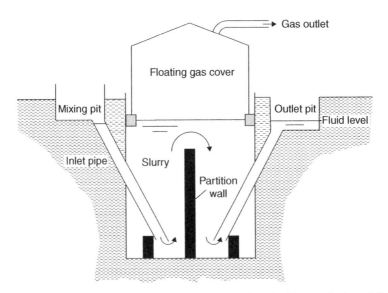

FIG. 18.16   Floating cover (Indian-type) digester. Reproduced with permission of Gunnerson and Stuckey, 1986.

(PVC), or reinforced fiberglass (RBG). Only the cover in a floating-drum digester is above ground, whereas the rest of the parts of both digesters are below ground.

Both types of digesters have a similar working principle. Substrate is fed through the inlet pipe either directly or after mixing in a pit into the digester tank. The biogas is collected above the slurry and exits the tank through a gas pipe connected to the top of the digester. Finally, the digestate leaves the digester through an outlet pipe and is collected in a receiving pit or tank. The digester tank may have either one or two compartments, with an average HRT of 20–30 days. Floating-cover digesters have a steel cover floating on the slurry that moves vertically to accommodate the constant biogas pressure. Higher gas pressures can be achieved by adding a weight on top of the holder. In the case of fixed-dome digesters, the gas is kept roughly at a constant volume while the pressure varies. Both types of digesters lack adequate mixing and are operated without temperature control. Moreover, there is no provision for removing settled inert materials that considerably reduce the effective volume of the digester over time.

A lack of moving parts and simple construction make the digester easy to operate and maintain. However, the cost of installation is still high for rural farmers and skilled craftsmen are required for construction, which are major constraints for the widespread adoption of this technology in developing countries (Surendra *et al.*, 2010).

To reduce the installation cost, as well as simplify the operation and maintenance of digesters, low-cost digesters are being constructed using polyethylene tubular film in developing countries (Solarte, 1995; Sarwatt *et al.*, 1995; Khan, 1996). A low-cost polyethylene tubular digester (Figure 18.17) has no mixing devices and/or heating systems and thus avoids sophisticated monitoring needs. It is fabricated using readily available materials, usually tubular polyethylene for the main tank and polyvinyl chloride (PVC) pipes for the biogas transport. Feedstock passes through the bioreactor from the inlet to the outlet, while biogas is collected by means of a gas pipe connected to the headspace. In order to maintain the higher process temperature and minimize overnight temperature fluctuations in cold mountainous areas, the tubular plastic bag is buried in a trench and the trench is covered with a greenhouse. Design criteria and dimensions for the digester, trench, and greenhouse depend on location; a longer HRT of 60–90 days is generally used when the digester is built at high altitude (Ferrer *et al.*, 2011). Simple design, ease of installation, and little specialized labor demands make the technology affordable for household applications in developing countries such as Colombia, Ethiopia, Tanzania, Vietnam, Cambodia, China, Costa Rica, Bolivia, Peru, Ecuador, Argentina, Chile, and Mexico. The relatively fragile nature of the polyethylene film, however, makes it vulnerable to damage and rupture.

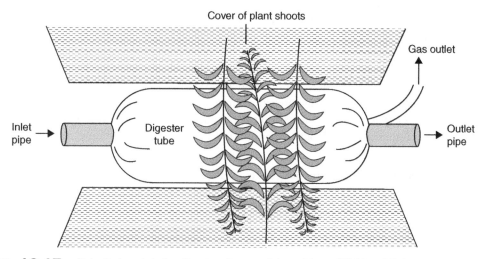

FIG. 18.17   Polyethylene tubular digester. Source: Adapted from Plöchl and Heiermann, 2006.

18.3 Biogas Cleaning and Upgrading

Biogas is primarily composed of CH_4 (50–75%) and CO_2 (25–40%), and small amounts of other compounds such as H_2O, H_2S, NH_3, N_2, O_2, and H_2 (see Table 17.1 in Chapter 17). CH_4 is the major compound with energy value in the biogas. Some of the major impurities in biogas are CO_2, moisture, H_2S, siloxanes, and NH_3. Moisture can be removed in the very early stages of biogas cleaning/upgrading by piping biogas through buried pipes to cool the gas and condense the water. A refrigerated condenser can be used to further reduce the dew point. For biogas cleaning, the focus is primarily on sulfide removal; whereas upgrading requires removal of all types of impurities in biogas. H_2S is one of the highly problematic impurities in biogas due to its corrosive nature. Biogas quality requirements with respect to H_2S for various applications are outlined in Table 18.1.

There are different methods for cleaning/upgrading of biogas. These include physical (e.g., water scrubbing, pressure swing adsorption, and membrane permeation), chemical (e.g., iron sponge and alkali treatment), and biological (e.g., bioscrubber, biofilter, and biotrickling) methods. Discussion of biogas cleaning/upgrading is presented in the following section.

Biogas cleaning and biogas upgrading

Biogas cleaning is the removal of gas components, e.g., H_2O, H_2S, NH_3, etc., from biogas (total concentration <5% by vol.). Biogas cleaning does not target the removal of CO_2. Thus, for electricity generation via CHP and direct burning in boilers, biogas cleaning is sufficient.

Biogas upgrading is the further separation/removal of CO_2 from biogas with a CH_4 content of over 95% (by vol.). Biogas upgrading is required if the biomethane is to be used as a transportation fuel (e.g., compressed natural gas (CNG)) or to be injected into the natural gas grid.

18.3.1 Physical Methods

Water scrubbing or pressure water scrubbing (absorption), also known as gas washing, is the simplest method for removing CO_2 from biogas and is one of the most widely employed methods for biogas upgrading. The process simultaneously removes H_2S and NH_3 from biogas. Biogas is compressed and fed from the bottom of a packed column while clean water is pumped in at the top of the column. The water absorbs the acid gases as it trickles down the packed bed in contact with the biogas. The CO_2/H_2S is removed from the water using an air stripper, before being recycled to the tower. The removal efficiency depends on the pressure, temperature, and pH of the water.

Pressure swing adsorption (PSA) is the second most commonly adopted method for biogas upgrading. PSA removes impurities from biogas under pressure according to their affinities to an adsorbent.

Table 18.1 H_2S limits for different biogas applications

Applications	H_2S limit (ppm)
Direct combustion	1000
Internal combustion engine (ICE) fuel	100
Compressed natural gas (CNG) for transportation fuel	16
Injection into natural gas pipeline grid	4

It works at an ambient temperature and uses adsorbents such as activated carbon, zeolites, or molecular sieves. High pressure is favorable for H_2S separation from biogas. The adsorbents are regenerated once they reach their full capacity by depressurising the bed, which facilitates the release of the adsorbed gas from the adsorbents.

Membrane separation takes advantage of the difference in gas permeability rates to purify biogas. Under a certain pressure, gases with high permeability (small molecular size and low affinity), such as CO_2, O_2, and H_2O, quickly pass through the membrane as permeate, while CH_4 is retained and can be collected.

Cryogenic separation removes biogas impurities using compression and refrigeration, and is based on the principle that gas components liquefy at different temperatures and pressures. This process is conducted at a temperature as low as $-100\,°C$ and a pressure near 40 bar. At these conditions, gas impurities (except N_2), such as H_2S and CO_2, are condensed into liquid while CH_4 remains in a gaseous form. As a result, the gas impurities are separated from the biogas. High-purity dry ice can be produced from this separation process.

18.3.2 Chemical Methods

Chemical methods can be used to remove H_2S and CO_2 from biogas. Sulfide reacts with metal ions (e.g., Fe^{2+}, Fe^{3+}, Cu^{2+}, Zn^{2+}, and Mg^{2+}) to form insoluble metal sulfides. The commonly used metal for sulfide removal is iron based (e.g., ferric chloride ($FeCl_3$), ferrous hydroxide ($Fe(OH)_2$), and ferric oxide (Fe_2O_3)) for economic reason. These reactions are shown in Equations 18.9–18.11. In practice, iron sponges, which are typically made of iron oxide supported on porous media or iron-impregnated wood, are most commonly used for sulfide removal from biogas.

$$Fe(OH)_2 + H_2S \rightarrow FeS + 2H_2O \tag{18.9}$$

$$2FeCl_3 + 3H_2S \rightarrow Fe_2S_3 + 6H^+ + 6Cl^- \tag{18.10}$$

$$2Fe_2O_3 \cdot H_2O + 3H_2S \rightarrow Fe_2S_3 + 4H_2O + heat \tag{18.11}$$

Sulfide-laden biogas is passed through a permeable bed of iron sponge (hydrated ferric oxide). Sulfide is removed in a solid form as iron sulfide (Fe_2S_3). The iron sponge must be kept damp. Once the iron is exhausted, it can be regenerated by aerating the bed under humid conditions, which releases sulfur from iron to form hydrated ferric oxide (Equation 18.12). The reaction is exothermic and generates heat, thus keeping the bed moist is important to prevent fire.

$$2Fe_2S_3 + 3O_2 \rightarrow 2Fe_2O_3 + 6S + heat \tag{18.12}$$

The use of alkali chemicals (e.g., NaOH, $Ca(OH)_2$, and KOH) can also remove acid gases (i.e., H_2S and CO_2) from biogas. These chemicals easily dissolve in water and are used for liquid scrubbing, or can also be packed as granular solids in a fixed bed. Among these alkali chemicals, $Ca(OH)_2$ is the most economic for CO_2 removal. CO_2 reacts with alkali to form bicarbonates, as shown in Equations 18.13–18.15:

$$CO_2 + NaOH \rightarrow NaHCO_3 \tag{18.13}$$

$$CO_2 + KOH \rightarrow KHCO_3 \tag{18.14}$$

$$2CO_2 + Ca(OH)_2 \rightarrow Ca(HCO_3)_2 \tag{18.15}$$

Amine solvent has a much higher absorption selectivity of CO_2 over CH_4; therefore, it can be used to separate CO_2 from biogas. Commonly used solvents are alkanolamines, such as monoethanolamine (MEA), diethanolamine (DEA), or methyldiethanolamine (MDEA). MEA is the most widely used solvent for low-pressure absorption. The reactions during absorption and desorption processes are shown in Equations 18.16 and 18.17:

$$\text{Absorption of } CO_2: 2RNH_2 + H_2O + CO_2 \rightarrow RNH_3{}^+ + HCO_3{}^- \tag{18.16}$$

$$\text{Desorption of } CO_2: RNH_3{}^+ + HCO_3{}^- \rightarrow 2RNH_2 + H_2O + CO_2 \tag{18.17}$$

where R is an organic component (e.g., R is $-(CH_2)_2OH$ for MEA). These reactions are mainly affected by temperature and pressure. Low temperature and high pressure favor absorption, while high temperature and low pressure promote desorption.

18.3.3 Biological Methods

Sulfide can be biologically oxidized to elemental sulfur under aerobic (oxygen-limiting) conditions, as shown by the Equation 18.18:

$$2HS^- + O_2 \rightarrow 2S^0 + 2OH^- \left(\Delta G^{o\prime} = -129.50 \, kJ/mol \, HS^-\right) \tag{18.18}$$

The bacteria involved in sulfide oxidation belong to a group of colorless sulfur bacteria, of which *Thiobacillus* is the best known. *Thiobacillus* is mostly a facultative autotroph, and utilizes reduced inorganic sulfur compounds as an electron donor and CO_2 as a carbon source. A biological sulfide removal process known as the Shell–Thiopaq process (Figure 18.18), developed by Paques Biosystems, The Netherlands, has been adopted commercially.

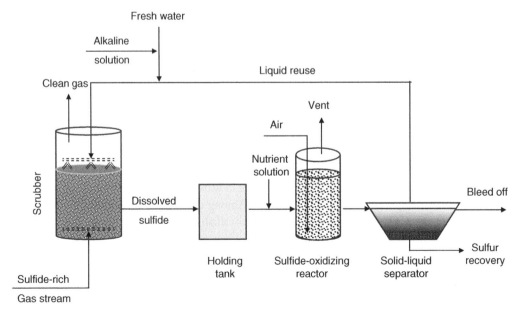

FIG. 18.18   Schematic of a gaseous sulfide-removal process. Source: Adapted from Khanal, 2008.

18.4 Biogas Utilization

In many developing countries, the biogas produced using household digesters is primarily used for cooking and lighting. In developed countries, biogas has been commercially used for generating electricity and heat via a combined heat and power (CHP) unit. A CHP unit is a modified natural gas engine that powers a generator. It is important to note that only 35–40% of the energy in the biogas is converted into electricity and the remaining energy is released as heat. Heat can be recovered from the engine water jacket and the exhaust. Recovery and utilization of heat energy from a CHP unit are critical in the overall economic viability of biogas-to-electricity conversion. Biogas is also either used as a transportation fuel (e.g., CNG) or injected into the natural gas grid after upgrading it to biomethane ($CH_4 > 95\%$). Figure 18.19 shows some of the options for biogas utilization.

How much energy can be produced from biogas?

CH_4 is the only significant energy-rich constituent in biogas. Thus, the energy-production potential of the biogas depends on the concentration of CH_4. The energy content of CH_4 is around 1,000 Btu/ft^3 (35,846 kJ/m^3) at 25 °C and 1 atm. As a rule of thumb, the energy content of biogas is 600 Btu/ft^3 (22,400 kJ/Nm^3) based on a CH_4 content of 60%. The daily gross energy available from a biogas plant can be calculated as follows:

$$E = G \times 22,400\,\text{kJ/m}^3$$

where E is the gross energy generated (kJ/day) and G is the biogas production rate at STP (m^3/day).

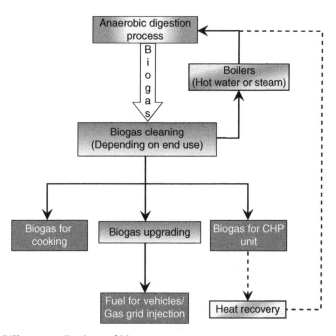

FIG. 18.19   Different applications of biogas.

FIG. 18.20   Pile of dewatered digestate from an AD process with co-digesting animal manure and maize silage. Source: Photo courtesy of Samir Khanal, University of Hawaii.

18.5 Digestate

Residue produced after AD of various organic feedstocks is known as *digestate*. It consists of a mixture of microbial biomass and undigested materials. Digestate often looks like slurry, with a TS content of 2–20%. In some AD plants, digestate is dewatered (as shown in Figure 18.20), especially when it needs to be transported over long distances. Digestate is rich in nutrients (N, P, and K) and trace elements, which get concentrated during the AD process. Digestate is often land applied as organic fertilizer. When liquid slurry is land applied, it is commonly known as *fertirrigation*, which means both fertilization and irrigation.

Application of digestate as a nitrogen source to agricultural land is regulated. Land application of the digestate depends on the N content, location, and crop demand. Since ammonia is the major form of nitrogen in digestate, application of digestate through injection application can help reduce the nitrogen loss compared to surface application. One of the important aspects regarding recycling of digestate is the load of nutrients on farmland. Nitrate leaching or phosphorus overloading can occur due to overapplication of digestate as fertilizer. In Europe, the Nitrate Directive restricts the input of nitrogen on farmland (maximum 170 kg N/(ha·year)), aiming to protect the ground and surface water from nitrate pollution. As a result, the disposal of digestate can be challenging in some regions.

References

Akinbami, J.F.K., Ilori, M.O., Oyebisi, T.O., Akinwumi, I.O., and Adeoti, O. (2001). Biogas energy use in Nigeria: Current status, future prospects and policy implications. *Renewable and Sustainable Energy Reviews* 5(1): 97–112.

Dague, R.R., Habben, C.E., and Pidaparti, S.R. (1992). Initial studies on an anaerobic sequential batch reactor. *Water Science and Technology* 26(9–11): 2429–32.

Ferrer, I., Garfi, M., Uggetti, E., Ferrer-Marti, L., Calderon, A., and Velo, E. (2011). Biogas production in low-cost household digesters at the Peruvian Andes. *Biomass and Bioenergy* 35: 1668–74.

Gunnerson, C.G., and Stuckey, D.C. (1986). *Anaerobic Digestion: Principles and Practices for Biogas Systems.* Technical Paper WTP.49. Washington, DC: World Bank.

Khan, S.R. (1996). *Low-Cost Biodigesters.* Programme for Research on Poverty Alleviation Report. Dhaka: Grameen Trust.

Khanal, S.K. (2008). *Anaerobic Biotechnology for Bioenergy Production: Principles and Applications.* Ames, IA: Wiley-Blackwell.

Omer, A.M., and Fadalla, Y. (2003). Biogas energy technology in Sudan. *Renewable Energy* 28(3): 499–507.

Plöchl, M., and Heiermann, M. (2006). Biogas farming in Central and Northern Europe: A strategy for developing countries. *Agricultural Engineering International,* 8.

Sarwatt, S.V., Lekule, F.P., and Preston, T.R. (1995). Biodigesters as means for introducing appropriate technologies to poor farmers in Tanzania. 2nd International Conference on Increasing Animal Production with Local Resources, Zhanjiang, China.

Solarte, A. (1995). Sustainable livestock systems based on local resources: CIPAV's experiences. 2nd International Conference on Increasing Animal Production with Local Resources, Zhanjiang, China.

Surendra, K.C., Khanal, S.K., Shrestha, P., and Lamsal, B. (2010). Current status of renewable energy in Nepal: Opportunities and challenges. *Renewable and Sustainable Energy Reviews* 15: 4107–17.

Tilche, A., Poli, F.D., Ercoli, L., Tesini, O., Cortellini, L., and Piccinini, S. (1986). An improved plug-flow design for the anaerobic digestion of dairy cattle waste. In: M.M. El-Halwagi (ed.), *Biogas Technology, Transfer and Diffusion.* Amsterdam: Elsevier Applied Science, pp. 259–69.

Vandivivere, P., Baere, P.D., and Verstraete, W. (2003). Types of anaerobic digester for solid wastes. In: J. Mata-Alvarez (ed.), *Biomethanization of the Organic Fraction of Municipal Solid Wastes.* London: IWA Publishing, pp. 111–40.

Yang, L., Ge, X., Wang, C., Yu, F., and Li, Y. (2014). Progress and perspectives in converting biogas to transportation fuels. *Renewable and Sustainable Energy Reviews* 40: 1133–52.

Young, J.C., and McCarty, P.L. (1969). The anaerobic filter for waste treatment. *Journal WPCF,* 41(5): R160–R173.

Exercise Problems

18.1. Explain the importance of decoupling SRT from HRT. What are the different approaches for such decoupling?

18.2. Why is CSTR often the preferred reactor configuration for digestion of high-solids feedstocks? What is the ideal TS level for a plug-flow digester? Name a few feedstocks that are currently being digested using a plug-flow digester.

18.3. A CSTR reactor is shown in Figure 18.21. Calculate HRT, SRT, and the ratio of SRT and HRT.

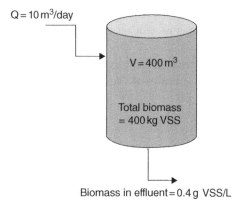

$Q = 10\,m^3/day$

$V = 400\,m^3$

Total biomass = 400 kg VSS

Biomass in effluent = 0.4 g VSS/L

FIG. 18.21   CSTR reactor.

18.4. A UASB-AF hybrid reactor has been employed to treat distillery slops at 35 °C. The waste flow rate is 2,500 m³/day with a mean soluble COD of 9,500 mg/L. Calculate the maximum CH_4 generation rate in m³/day. What would be the biogas-generation rate at 80% COD removal efficiency if 12% of the removed COD is utilized for biomass synthesis? The mean CH_4 content of biogas is 70%. Also calculate the energy-generation potential (kWh).

18.5. A 25-MGY cellulosic ethanol plant plans to employ AD for digesting the stillage following ethanol distillation. The plant generates 10 L stillage per L of ethanol produced. The mean TCOD of stillage is around 150 g/L. How much energy could be generated from the CH_4 gas? Assume that 75% of the stillage is biodegradable and COD removal efficiency is 80%, biomass yield is 0.10 g VSS/gCOD$_{removed}$, and methane yield is 0.30 m³ CH_4/kg COD$_{removed}$.

18.6. An anaerobic digestion system generates 10,000 m³ (353,150 standard ft³) of biogas per day. The produced biogas is to be used in an engine generator for generating electricity for wastewater treatment plant operations. How much electricity can be generated from the biogas?

18.7. An anaerobic digester receiving 7,600 m³/day (2.0 million gal/day) of food-processing wastewater generates 2,500 m³ of biogas per day at 37 °C. If the biogas contains 65% (by volume) CH_4, calculate the organic feeding rate (as COD). Make all valid assumptions.

18.8. A farm anaerobic digester receives 350 m³/day (92,470 gal/day) of cow slurry with 6% TS and specific gravity of 1.04. The biogas generated from the digester is used in an engine generator to generate electricity for the farm operations.

   (a) Determine the size of the engine generator required to use all the biogas effectively.
   (b) If the total electricity demand for the farm operations is 10 MWh per year, what fraction of the total electricity demand can be met by biogas?

   Assume a VS fraction of 75% of TS and a VS reduction of 40%. The biogas yield is 0.25 m³/kg VS$_{removed}$.

18.9. For Problem 18.4, if the wastewater also contains 3.0 g/L of sulfate, how much sulfide will be in the gas phase? Assume that 100% of the sulfate gets reduced during digestion.

18.10. A full-scale anaerobic digester is employed to digest 1,000 m³/day of mixed sludge from a conventional biological waste-treatment plant treating wastewater from an industrial complex. The digester operates at mesophilic temperature (35 °C). The sludge consists of 40% (by weight) thickened secondary sludge (specific gravity 1.01 and TS content of 2%) and 60% primary solids (specific gravity 1.02 and TS content of 3%). The VS contents of primary solids and secondary sludge are 65% and 70%, respectively. Calculate the maximum CH_4 yield for each sludge in m³/day. If the average CH_4 content is 70%, how much biogas could be generated if VS reduction efficiencies are 60% and 50%, respectively, for primary and secondary sludge?

   Make all assumptions as needed. You need to find out the chemical formula for primary and secondary sludge and use Bushwell's equation to calculate the maximum CH_4 yield.

Microbial Fuel Cells

Hongjian Lin, Hong Liu, Jun Zhu, and Venkataramana Gadhamshetty

---

What is included in this chapter?

This chapter covers the basics of microbial fuel cells (MFCs) for the generation of electricity from various substrates. The microbiology, electrochemistry, and engineering considerations of MFC technology are discussed. Pertinent examples and calculations are also included.

19.1 Introduction

Many electrochemical devices convert chemical energy to electrical energy to provide power for electronics. Among the many devices, battery and fuel cell are two common types. A **battery** stores chemical reagents inside the enclosed unit, and gets recharged or replaced once the reagents are eventually exhausted. In contrast, a *fuel cell* is an open system so that once the cell is provided with a fuel (hydrogen, natural gas, methanol, etc.) and an oxidizer (oxygen), it generates direct current (DC) electricity. So the lifetime of a fuel cell is much longer than that of a battery. A battery and a typical chemical fuel cell require expensive chemical reagents or catalysts (e.g., platinum) in their routine operation. A *microbial fuel cell* (MFC) is a device that directly converts chemical energy stored in substrates (organic matter or inorganic electron donors) into electricity by the catalytic activity of microbes (Arends and Verstraete, 2012). An MFC does not require expensive catalysts on the anode; instead, it uses microorganisms to transform the chemical energy in substrates to electricity.

The phenomenon of electricity generation by microorganisms was first discovered by M.C. Potter in 1911, and was utilized to develop MFCs that generated voltages exceeding 35 V in 1931 (Davis and Higson, 2007). The recent development of electrode and catalytic materials as well as the research interest in renewable energy have drawn more attention to MFCs. Since the 1990s, significant research efforts have been made to produce high power and energy density.

Bioenergy: Principles and Applications, First Edition. Edited by Yebo Li and Samir Kumar Khanal.
© 2017 John Wiley & Sons, Inc. Published 2017 by John Wiley & Sons, Inc.
Companion website: www.wiley.com/go/Li/Bioenergy

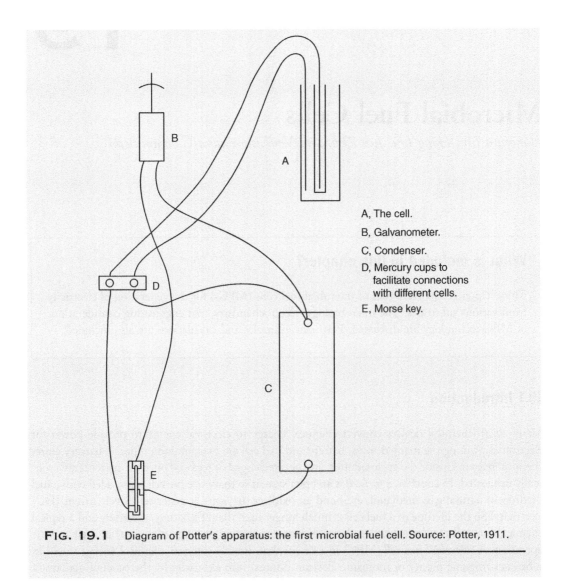

A, The cell.

B, Galvanometer.

C, Condenser.

D, Mercury cups to facilitate connections with different cells.

E, Morse key.

FIG. 19.1   Diagram of Potter's apparatus: the first microbial fuel cell. Source: Potter, 1911.

MFC is an interesting field as it uses microbes to generate electricity under ambient conditions. The fuel in MFCs can be wastewater from the home, food wastes, or organic-rich industrial waste-waters. The electrical current generated by a bench-scale MFC system is too small to run electrical equipment, however. MFCs therefore need to be stacked for practical applications. Although MFCs have the potential to generate electricity at low cost due to the ubiquitous fuels used, there are several technical challenges that need to be addressed in order to improve the viability of MFC technology. The research on electrochemically active bacteria, electrode materials, and cell configurations is evolving rapidly. Meanwhile, the concept of a microbial electrochemical system derived from MFCs for electricity production has been extended to other applications such as the electrochemical treatment of contaminants and microbial electrosynthesis for the production of valuable chemicals. This chapter provides an overview of MFC technology for electricity generation, and the working principles of MFCs with respect to microbiology, electrochemistry, and process engineering.

Example 19.1: Albany County Wastewater Sewer District (ACSD) uses an activated sludge process to treat 22 million gallons of water on a daily basis. The typical chemical oxygen demand (COD) in the incoming wastewater is 160 mg/L. What is the maximum power (MW) that can be generated from COD of incoming wastewater to ACSD? Assume that a gram of COD contains 15 kJ of energy.

Solution: The total power is contained in the wastewater in the form of COD.
The total COD rate entering ACSD is:

Total COD load

$= $ Flow rate of wastewater $\times$ COD concentration

$$= 22 \times 10^6 \frac{\text{gallons}}{\text{day}} \times 160 \frac{\text{mg}}{\text{L}}$$

$$= 22 \times \frac{10^6 \text{gallons} \left(\dfrac{3.785\,\text{L}}{\text{gallon}}\right)}{\text{day} \left(\dfrac{24 \times 60 \times 60\,\text{sec}}{\text{day}}\right)} \times 160 \frac{\text{mgCOD}}{\text{L}} \left(\frac{10^{-3}\text{g}}{\text{mg}}\right)$$

$$= 152.2 \frac{\text{gCOD}}{\text{sec}}$$

The total power is:

$\text{P} = $ Total COD rate $\times$ Energy content of unit COD

$$= 152.2 \frac{\text{gCOD}}{\text{sec}} \times 15 \frac{10^3\text{J}}{\text{gCOD}} = 152.2 \frac{\text{gCOD}}{\text{sec}} \times 15 \frac{10^3\text{J}}{\text{gCOD}}$$

$$= 2.313 \times 10^6 \text{W or } \mathbf{2.3\,MW}$$

19.2 How Does a Microbial Fuel Cell (MFC) Work?

An MFC consists of an anode and a cathode, split by a separator (e.g., a proton exchange membrane), as shown in Figure 19.2. The anaerobic microbes in the anode chamber oxidize the organic matter to release electrons (e^-) and protons (H^+). The protons enter the cathode compartment through a separator. Some of the electrons flow through the external load to the cathode, where they combine with protons and oxygen to form water. The flow of electrons from the anode to the cathode results in electricity production. Graphite of different forms (e.g., rod, felt, foam, etc.) can be used as anode and cathode base materials. The electrochemically active bacteria, *exoelectrogens*, responsible for fuel oxidation (electron release) are inoculated on the anode. The use of a cathode catalyst (e.g., platinum) can catalyze the cathode reaction.

A simplified reaction for the anodic oxidation of glucose into electrons and protons is given in Equation 19.1:

$$C_6H_{12}O_6 + 6H_2O \rightarrow 6CO_2 + 24\,H^+ + 24e^- \tag{19.1}$$

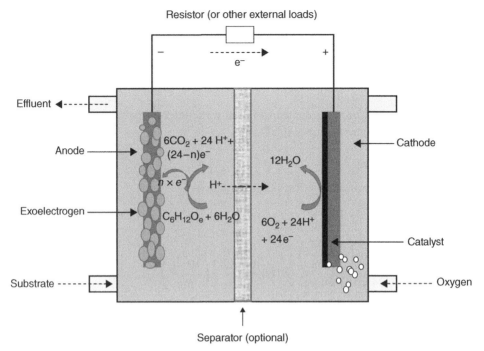

FIG. 19.2 Schematic diagram of a typical microbial fuel cell.

The electrons and protons combine with the oxidizer (oxygen) to produce water on the cathode:

$$6O_2 + 24H^+ + 24e^- \rightarrow 12H_2O \tag{19.2}$$

The overall reaction can be obtained by combining Equations 19.1 and 19.2:

$$C_6H_{12}O_6 + 6O_2 \rightarrow 6CO_2 + 6H_2O \tag{19.3}$$

The electrical power generated can be utilized by external loads (e.g., a LED light) or stored using energy storage devices (e.g., capacitors or other temporary energy storage devices).

A number of electrochemically active bacteria have been identified that are capable of directly transferring electrons to the anode. Some of these microbes include Fe (III) reducing bacteria, e.g., *Shewanella putrefacians, Geobacter sulfurreducens*, and *Rhodoferax ferrireducens*. The use of microorganisms on an anode to generate electricity makes MFCs more sustainable than conventional fuel cells that use precious metal catalysts. Another noticeable advantage of MFCs is that they generate electrical power by using easily available biodegradable organic substrates and environmentally available electron acceptors (e.g., oxygen). MFCs take advantage of using low-cost substrates, such as marine or river sediment and organic wastes, while fuel cells are unable to utilize these substrates to generate electricity (Table 19.1). Although MFCs produce lower power density than batteries and hydrogen fuel cells, future study of MFCs will definitely improve its power density and, more importantly, the feature of MFC technology for power generation directly from wastewater is unique.

Table 19.1 Comparisons between typical microbial fuel cells, batteries, and chemical fuel cells

Device	Typical reaction	Catalyst	Power density (kW/m^3)	Feature
Alkaline battery	Reaction between zinc (anode) and manganese dioxide (cathode)	None	30	Non-rechargeable
Li-ion battery	Reaction between lithium (anode) and lithium cobalt oxide (cathode)	None	90	Rechargeable
Hydrogen fuel cell	Reaction between hydrogen (anode) and oxygen (cathode)	Platinum	140	Clean energy from hydrogen
Microbial fuel cell	Reaction between organic matter (anode) and oxygen (cathode)	Microbes	0.01–2	Extracting energy from waste

Table 19.2 Standard reduction potentials (potential vs. standard hydrogen electrode, SHE) of common electron transport carriers at pH = 7 and T = 298 K

Redox couple	No. of electron transfer	Potential (V)
NADP$^+$/NADPH	2	−0.320
NAD$^+$/NADH	2	−0.315
FAD/FADH$_2$	2	∼0
Ubiquinone ox/red (Q/QH$_2$)	2	+0.045
Cytochrome b ox/red	1	+0.077
Cytochrome c_1 ox/red	1	+0.22
Cytochrome c ox/red	1	+0.235
Cytochrome a ox/red	1	+0.29
Cytochrome a_3 ox/red	1	+0.385

19.3 Electron Transfer Processes

The microorganisms use a respiration process to decompose the carbon source to release electrons that pass through a series of intracellular electron carriers and obtain energy in the form of adenosine triphosphate (ATP). The respiration fulfills both carbon transformation (e.g., through the citric acid cycle) and intracellular electron transfer. Only the electron transfer part is discussed in this section because it directly relates to power generation in MFCs. *Intracellular electron transfer* is carried out with the help of an *electron transport chain* (ETC), which consists of a series of membrane-associated electron carriers with different reduction potentials (Table 19.2). The electron carriers are oriented in the cytoplasmic membrane in such a way that their reduction potentials are increasing at each step of the electron transfer process. Electrons are first transferred to the reduced nicotinamide adenine dinucleotide (NADH), and step by step along the ETC, finally reaching the terminal electron acceptors (TEA) within a cell. Along the ETC, protons accumulate at the outside surface of the cell membrane, forming a proton motive force that drives ATP synthesis or ion transportation and flagella rotation. The intracellular electron transfer is terminated when the electrons are accepted by a TEA. The respiration is termed aerobic respiration when the microbes use oxygen as the TEA or anaerobic respiration when they use non-oxygen compounds such as sulfate and nitrate. Typical bacteria require the TEA to be available in the dissolved form in a cell.

One basic requirement for electricity production in MFCs is that the bacteria use an insoluble and solid anode as the TEA, which cannot diffuse into the bacterial cell membranes. To overcome this challenge, MFCs use a special type of bacteria called exoelectrogens (e.g., *S. putrefacians* and

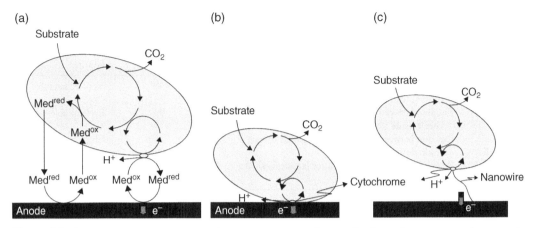

FIG. 19.3   Three common extracellular electron transfer mechanisms between exoelectrogenic bacteria and anodes: (a) redox shuttles, (b) direct contact electron transfer and (c) bacterial nanowires. Source: Schröder, 2007. Reproduced with permission of the Royal Society of Chemistry.

Table 19.3 Standard reduction potentials of endogenous (self-excreted) mediators at pH = 7 and T = 298 K

Endogenous mediators	Redox potential (V)
Phenazine-1-carboxamide	−0.115
Pyocyanine (phenazine)	−0.034
Riboflavin	−0.208
2-Amino-3-carboxy-1,4-naphthoquinone (ACNQ)	−0.071

G. sulfurreducens), which are capable of transferring the electrons from inside the cell to the external anode. The process by which exoelectrogens transfer electrons from the cellular cytoplasm to the insoluble anode is known as the *extracellular electron transfer* (EET) process. The EET process allows for direct harness of electrons from the redox reactions by exoelectrogens in substrate degradation. The exoelectrogens may use three different EET mechanisms to relay the electrons within their cytoplasm to the external anode. Figure 19.3 shows these three types of EET mechanism: mediated electron transfer using electron shuttles; direct electron transfer through physical contact between the cell membrane proteins (e.g., cytochromes) and the anode; and direct electron transfer through bacterial nanowires.

19.3.1 Mediated Electron Transfer

The bacteria use *redox shuttles* (i.e., mediators) to relay the electrons from the cell wall to the anode (Figure 19.3a). These redox shuttles dissolve and diffuse in water (and across the outer membrane), and are electrochemically active. The mediators in oxidized form penetrate the cell membrane, and interact with the reducing agents (e.g., NADH) to accept electrons. The reduced mediators transfer out from the cell membrane and then donate the electron to the anode. The oxidized mediators are then transported back into the cell membrane to continue the next cycle of electron transfer. Some exoelectrogens produce their own redox shuttles to generate electricity in MFCs. Well-known examples of *endogenous* (self-excreted) mediators include pyocyanin, riboflavin, and 2-amino-3-dicarboxy-1,4-naphthoquinone (ACNQ). The typical concentration of the endogenous mediators is around 5–10 μM. Table 19.3 shows the redox potentials of typical endogenous mediators in MFCs.

Table 19.4 Standard reduction potentials of artificial redox shuttles at pH = 7 and T = 298 K

Substance class	Redox shuttles	Potential (V)
Phenazine	Neutral red	−0.32
	Safranine	−0.29
	Phenazine ethosulfate	+0.06
Phenothiazine	New methylene blue	−0.02
	Toluidine blue O	+0.03
	Thionine	+0.06
	Phenothiazinone	+0.13
Phenoxazine	Resorufin	−0.05
	Gallocyanine	+0.02
Quinone	2-hydroxy-1,4-naphthoquinone	−0.14
	Anthraquinone-2,6-disulfonate	−0.18

In the case of non-exoelectrogens, the chemical mediators (*artificial redox mediators*) can be artificially supplemented to generate electricity in MFCs. In some early studies, mediators, such as neutral red, potassium ferricyanide, thionine, or 2-hydroxyl-1, 4-naphthoquinone (HNQ), were employed to facilitate the transport of electrons from inside the microbial cell membrane to the electrode. However, the use of mediators would be impractical due to both the increased cost and the toxicity on microorganisms. Table 19.4 outlines the redox potentials of the mediators from four different classes of chemicals.

Fundamentally, *mediated electron transfer* (MET) can be described in two steps. First, the mediators should be able to physically transport between the cell membrane and the anode. Second, these mediators should be able to participate quickly in redox reactions (i.e., oxidation at the anode and reduction in the bacterial cell). Compared to the redox kinetics, the MET reactions are constrained by the rate-limiting transport step. Therefore, when the mechanism of redox mediators is dominant in MFCs, the maximum current density depends on the rate of shuttle transport by diffusion shown in Equation 19.4 (Torres *et al.*, 2010):

$$j = nF\left(\frac{D_{shuttle}\Delta C_{shuttle}}{\Delta z}\right) \tag{19.4}$$

where $D_{shuttle}$ is the diffusion coefficient of the electron shuttle (m^2/s), Δz is the transport distance between anode and bacteria (m), and $\Delta C_{shuttle}$ is the concentration gradient of the reduced shuttles (mol/m^3).

19.3.2 Direct Electron Transfer (DET)

Direct electron transfer (Figure 19.3b) is the most convenient method for exoelectrogens to produce electricity in MFCs. *S. oneidensis* MR-1 and *G. sulfurreducens* are two model organisms that have been extensively explored for the DET strategy. DET occurs when the cell membrane is in direct contact with the anode. For many years, living organisms have been considered as electrically non-conductive entities. However, the cytochromes in the exoelectrogens have been found to be electrically conductive, thus able to participate in DET mechanisms. Nearly 42 genes encoding *c*-type cytochromes have been discovered in the full *S. oneidensis* genome, and over 100 genes encoding *c*-type cytochromes have been identified for *G. sulfurreducens*, indicating the complexity of the direct contact mechanisms. Commonly studied multi-heme cytochromes include CymA, MtrA,

MtrB, MtrC, and OmcA in *Shewanella*, and OmcB, OmcE, OmcS, OmcT, OmcZ, OmpB, and PpcA in *Geobacter*. Some of these proteins are outer-membrane cytochromes and aggregate to themselves when excreted by bacterial cells. The rate of electron transfer at the bacteria–anode interface can be described by the Butler–Volmer equation (Equation 19.5):

$$j = j_{oA} e^{\frac{nF\alpha_A F \left(E_A \cdot E_A^{o'} \right)}{RT}} \tag{19.5}$$

where j is the electrical current density (A/m^2, j_{oA} is the exchange current density of the anode (A/m^2), α_A is the transfer coefficient of the anodic reaction, E_A is the anode potential (V), and $E_A^{o'}$ is the standard potential at the interface (V).

Under this DET mechanism, bacteria that have direct contact with the electrode contribute to electricity generation in the MFC, meaning that the current density is limited by the amount of monolayer biomass at the anode and the bacterial respiration rate. So the maximum current density can be quantified by Equation 19.6:

$$j = \gamma_s q_{max} X_f L_{fa} \tag{19.6}$$

where γ_s is a dimensionless conversion factor from mass of substrate to coulombs, q_{max} is the maximum specific rate of substrate utilization (mol substrate/(g·hr)), X_f is the concentration of active biomass (g/m^3), and L_{fa} is the electroactive biofilm thickness (m).

> **Example 19.2:** Calculate the upper limit of current density of an MFC fed with sufficient acetate with 20% Coulombic efficiency (see Section 19.4.4 for definition of Coulombic efficiency). Assume that the DET is dictated by the direct contact mechanism. The following parameters are given: $q_{max} = 540$ mg acetate/(g·hr), $X_f = 2.8 \times 10^5$ g/m^3, and $L_{fa} = 1$ μm.
>
> **Solution:** The conversion factor from the mass of substrate to available coulombs, in the case of acetate:
>
> $$\gamma_s = C_E nF / MW_{sub} = 20\% \times 8 \times 96,485 \frac{c}{mol} / 60 \frac{g}{mol} = 2,573 \frac{c}{g}$$
>
> The maximum current density can be obtained by the following equation:
>
> $$j = \gamma_s q_{max} X_f L_{fa}$$
>
> When the transport distance is 1 μm,
>
> $$j = 2573 \frac{C}{g} \times 540 \, mg/(g \cdot hr) \times 2.8 \times 10^5 \, g/m^3 \times 1 \mu m = \mathbf{216 \, mA/m^2}$$

19.3.3 Bacterial Nanowires

Exoelectrogens can synthesize *nanowires* (electrically conductive pili) to promote the DET mechanism for electricity production in MFCs (Figure 19.3c). Both *S. oneidensis* and *G. sulfurreducens* strains can produce long nanowires (nearly 10 μm), which enable them to reach the solid anode at a distance. The nanowires are directly connected to the membrane-bound cytochrome and transfer the electrons

(a)

(b)

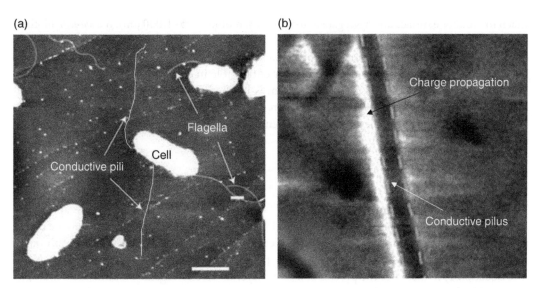

FIG. 19.4   Visualization of charge propagation along pili filaments. (a) image of conductive pili (bacterial nanowires) of *G. sulfurreducens* via atomic-force microscope (AFM); and (b) image showing charge propagation along a pilus, illustrated via electrostatic force microscope (EFM). Source: Malvankar *et al.*, 2014. Reproduced with permission of Nature Publishing Group.

to the outside of the cell wall. This mechanism allows bacteria to transfer electrons efficiently from inside the cell membrane to the anode that may not have direct contact with the outer membranes, and therefore it is not limited by the amount of monolayer biomass, as in the direct contact mechanism. This mechanism is also not limited by the slow molecular diffusion that occurs in mediated electron transfer, because nanowires are attached to the membranes. Due to these reasons, the presence of nanowires improves the EET process and the performance of MFCs.

  G. sulfurreducens is able to produce conductive pili (or bacterial nanowires; Figure 19.4) from protein subunits of PilA, along which there are functional *c*-type cytochrome OmcS. PilA-pili of *G. sulfurreducens* are the first known proteins in biological processes where the metallic-like conductivity dictates. For electrically insulating proteins, electron transport is distinctly realized by electron tunneling that requires the distance between active sites to be approximately within 2–3 nm. In *G. sulfurreducens*, nanowires are essential to form aggregates that are conductive for long-range electron transfer. Conductivity along the length of nanowires of *S. oneidensis* and *G. sulfurreducens* is observed using scanning tunneling microscopy and electrostatic force microscopy.

19.3.4 Long-Range Extracellular Electron Transfer

Individual bacterial cells can form aggregates on a surface. On MFC anode biofilm, the electron transfer may occur between the bacterial cells within the biofilm, and this process is termed *long-range extracellular electron transfer* (LEET). This is a common phenomenon in MFCs using a mixed microbial population (i.e., a mixed culture) – a mixture of genetically diverse bacteria will naturally attach to the anode surface and aggregate to form a thick biofilm (~5–1,000 μm). *Shewanella* biofilm is thinner than *Geobacter* biofilm, and contains a high level of redox proteins of *c*-type outer-membrane cytochromes. For a monolayer of *Shewanella* biofilm, about 10–30% of the anode surface is covered by these cytochromes, indicating their critical role in electron transfer.

A decisive parameter of the conductive biofilm is its conductivity. This parameter is critical in current generation by *Geobacter* biofilm. Assuming that biofilm conductivity is the limiting factor, highly conductive biofilm achieves a higher current density. This relationship can be described by Ohm's law in a modified form (Equation 19.7):

$$j = -\sigma_{biofilm} \frac{E_{OM} - E_{interface}}{\Delta z} \tag{19.7}$$

where $\sigma_{biofilm}$ is biofilm conductivity (Siemens/m), Δz is the distance over which electrons are transported (m), E_{OM} is the potential at which the electrons are released by the microorganism through the outer membrane (OM) (V), and $E_{interface}$ is the potential at the biofilm/anode interface (V). Note that Equation 19.7 describes the process of a single layer of exoelectrogenic biofilm, while multiple layers of biofilm are actually involved in electricity generation in many situations. Therefore, integration of the current density along the distance to the bacteria–electrode interface is needed when the bulk biofilm is exoelectrogenic.

19.4 Electrical Power and Energy Generation

Electricity is generated by an MFC as a result of electron flow from the anode to the cathode. Oxidation occurs at the anode and reduction at the cathode. The electricity can then be harvested by external devices, for example by an energy storage system or by an electric load (Logan, 2008). Explanation for the capability of an MFC for electrical power and energy generation starts with the definitions of a *redox* reaction (redox, a combination of reduction and oxidation), and its *reduction potential*.

19.4.1 Redox Reaction and Electrode Potential

Oxidation is a reaction in which electrons are lost from a chemical species (called a *reductant* or an electron donor), and *reduction* is the reverse process in which electrons are gained by a chemical species (called an *oxidant* or an electron acceptor). Since electrons do not exist alone in solution, a *half-reaction* (e.g., either oxidation or reduction) proceeds with another half-reaction in pairs to fulfill the electron balance in a complete net reaction. For example, a typical air-cathode MFC (see Section 19.5) couples the oxidation of acetate with the reduction of oxygen for electricity generation. The oxidation of acetate at the anode is shown in Equation 19.8, and the reduction of oxygen at the cathode is displayed as a separate half-reaction in Equation 19.9:

$$CH_3COO^- + 4H_2O \rightarrow 2HCO_3^- + 9H^+ + 8e^- \tag{19.8}$$

$$O_2 + 4H^+ + 4e^- \rightarrow 2H_2O \tag{19.9}$$

$$2O_2 + 8H^+ + 8e^- \rightarrow 4H_2O \tag{19.10}$$

The electron balance needs to be maintained so that the number of electrons lost by the reductant is the same as the electrons accepted by the oxidant. In this case, multiplying the half-reaction of oxygen reduction by two yields the same number of transferred electrons (i.e., $8e^-$) as in the other half-reaction, and results in Equation 19.10. The net reaction can be obtained by combining Equations 19.8 and 19.10 and canceling out common terms on both sides:

$$CH_3COO^- + 2O_2 \rightarrow 2HCO_3^- + H^+ \tag{19.11}$$

The chemical constituents that change the oxidation state and appear on each side of a half-reaction are considered a *redox couple*, for example HCO_3^-/CH_3COO^-. By convention, a redox couple is written in such a way that the oxidized form of the species is on the left side of the slash (/), the reduced form on the right side, and the stoichiometry is usually normalized to one free electron transfer. So the respective two redox couples can also be written as $\frac{1}{4}HCO_3^-/\frac{1}{8}CH_3COO^-$, and $\frac{1}{2}O_2/H_2O$.

The *reduction potential*, **E**, is a term indicating the tendency of the oxidized constituent in a redox couple to donate electrons, which is equivalent to the statement that a redox couple with a more positive reduction potential tends to be an oxidant in the net reaction, for example oxygen gas, which has a standard reduction potential of +0.82 V (at pH 7.0). The reduction potential of a reaction depends on the *standard reduction potential* of the redox couple and the concentration of the active species, and can be calculated according to the Nernst equation (Equation 19.12):

$$E = E^o + \frac{RT}{nF} ln \prod \frac{c_{ox}}{c_{red}} \quad \text{or} \quad E = E^o - \frac{RT}{nF} ln \prod \frac{c_{red}}{c_{ox}} \tag{19.12}$$

where E is the reduction potential of a redox couple (V), E^o is standard reduction potential in reference to the standard hydrogen electrode (SHE or NHE) at thermodynamic standard conditions (V), F is the Faraday constant (96,485 C/(mol·electron)), n is the number of electron transfer in the reaction, c_{ox} is the concentration of the species on the oxidized side (mol/L), and c_{red} is the concentration of the species on the reduced side (mol/L). For simplification, the terms *reduction potential* and *electrode potential* are interchangeable in this chapter.

Thermodynamic *standard conditions* refer to conditions in which the partial pressure of any gas involved is 1 atm pressure (1×10^5 Pa), the concentrations of all species (and protons) in aqueous solutions are 1 mol/L, and the temperature is 25 °C (298 K). Since the pH of a biochemical system is generally close to neutral, more commonly $E^{o'}$ at biological standard conditions, instead of E^o, is used. Values of $E^{o'}$ for some common substrates are given in Table 19.5. The biological standard potential of a redox couple can calculated from the Gibbs free energy of a half-reaction ($G^{o'}$, in J/mol), another thermodynamic measurement of a chemical equilibrium, shown in Equation 19.13:

$$E^{o'} = -G^{o'}/(nF) \tag{19.13}$$

Example 19.3: The half-reaction of an MFC anode is based on acetate oxidation, as shown in Equation 19.8. The standard reduction potential of the half-reaction, E^o, is 0.187 V. Calculate its electrode potential (E) at 25 °C at an acetate concentration of 10 mM, bicarbonate concentration of 10 mM, and pH 7.0.

Table 19.5 Standard reduction potential (E°′) of substrates at biological standard conditions (pH = 7)

Substrate	Half-reaction	E°′ Volt vs. SHE
Formate	Formate$^-$ + H$_2$O → HCO$_3^-$ + 2 H$^+$ + 2 e$^-$	−0.408
Acetate	Acetate$^-$ + 4 H$_2$O → 2 HCO$_3^-$ + 9 H$^+$ + 8 e$^-$	−0.280
Propionate	Propionate$^-$ + 7 H$_2$O → 3 HCO$_3^-$ + 16 H$^+$ + 14 e$^-$	−0.281
Butyrate	Butyrate$^-$ + 10 H$_2$O → 4 HCO$_3^-$ + 23 H$^+$ + 20 e$^-$	−0.282
Pyruvate	Pyruvate$^-$ + 6 H$_2$O → 3 HCO$_3^-$ + 12 H$^+$ + 10 e$^-$	−0.356
Oxalacetate	Oxalacetate^{2-} + 7 H$_2$O → 4 HCO$_3^-$ + 12 H$^+$ + 10 e$^-$	−0.230
Lactate	Lactate$^-$ + 6 H$_2$O → 3 HCO$_3^-$ + 14 H$^+$ + 12 e$^-$	−0.328
Malate	Malate^{2-} + 7 H2O → 4 HCO$_3^-$ + 14 H$^+$ + 12 e$^-$	−0.348
β–Hydroxybutyrate	β–Hydroxybutyrate$^-$ + 9 H2O → 4 HCO$_3^-$ + 21 H$^+$ + 18 e$^-$	−0.315
Glucose	Glucose + 12 H$_2$O → 6 HCO$_3^-$ + 30 H$^+$ + 24 e$^-$	−0.414
Methanol	Methanol + 2 H$_2$O → HCO$_3^-$ + 7 H$^+$ + 6 e$^-$	−0.375
Ethanol	Ethanol + 5 H$_2$O → 2 HCO$_3^-$ + 14 H$^+$ + + 12 e$^-$	−0.317
Glycerol	Glycerol + 6 H$_2$O → 3 HCO$_3^-$ + 17 H$^+$ + 14 e$^-$	−0.392
Glycine	Glycine + 4 H$_2$O → 2 HCO$_3^-$ + 8 H$^+$ + NH$_4^+$ + 6 e$^-$	−0.369
Glutamate	Glutamate$^-$ + 11 H$_2$O → 5 HCO$_3^-$ + 21 H$^+$ + NH$_4^+$ + 18 e$^-$	−0.178
Alanine	Alanine + 7 H$_2$O → 3 HCO$_3^-$ + 14 H$^+$ + NH$_4^+$ + 12 e$^-$	−0.318

Note: All the thermodynamic data in this table are calculated based on Thauer *et al.*, 1977.

Solution: According to the Nernst equation

$$E = E^\circ + \frac{RT}{nF} \ln \Pi \frac{c_{ox}}{c_{red}}$$

$$E = E^\circ + \frac{RT}{nF} \ln \frac{[HCO_3^-]^2 [H^+]^9}{[CH_3COO^-]} = 0.187\,V + \frac{8.31 \left(\frac{1}{molK}\right) \times 298K}{8 \times 96,485\,C/mol} \ln \left(\frac{10^{-2 \times 2} 10^{-7 \times 9}}{10^{-2}}\right)$$

$$= 0.187\,V + \left(-0.480\right) V = \mathbf{-0.293\ V}$$

19.4.2 Electromotive Force and Cell Potential

When two redox half-reactions are connected in a closed electric circuit, an *electrochemical cell* is formed. An *anode* is the electrode where the oxidation half-reaction occurs, and a *cathode* is the electrode where the reduction half-reaction occurs. In the electrochemical cell that consists of half-reactions in Equations 19.8 and 19.9, the oxidation of acetate to bicarbonate takes place at the anode, and the reduction of oxygen gas occurs at the cathode. Thus, the redox net reaction can be written as Equation 19.14:

$$CH_3COO^-(aq) | HCO_3^-(aq) \| O_2(g, p = 0.2\,atm) | H_2O$$
$$\text{Anode} \qquad\qquad\qquad \text{Cathode}$$

(19.14)

The potentials of the anode and cathode (represented by E$_a$ and E$_c$, respectively) at equilibrium can be calculated from the Nernst equation (Equation 19.12). The *electromotive force* (E$_{emf}$) is the maximum cell potential, and it can be calculated theoretically by subtracting the anode potential from the cathode potential. For acetate oxidation by oxygen at the given condition, the electromotive can be calculated as:

$$E_a = E_a^o + \frac{RT}{nF} \prod \frac{c_{ox}}{c_{red}} = -0.335\,V; \left[Ac^-\right] = 0.04\,M, \left[HCO_3^-\right] = 10^{-4.5}\,M \tag{19.15}$$

$$E_c = E_c^o + \frac{RT}{nF} \prod \frac{c_{ox}}{c_{red}} = -0.805\,V; Po_2 = 0.2\,atm \tag{19.16}$$

$$Net: E_{emf} = E_c - E_a = 0.805 - (-0.335) = 1.14\,V$$

This electromotive force is positive (1.14 V), so the redox reaction is spontaneous in the given direction. Alternatively, the standard electromotive force of a net reaction ($E_{emf}^{o'}$) can be calculated from the Gibbs free energy of a reaction, ΔG_r^o (J/mol):

$$E_{emf}^{o'} = -\Delta G_r^{o'}/(nF) \tag{19.17}$$

19.4.3 Electrical Power

Electrical power (P) is the product of the electric current and the voltage across a resistor:

$$P = IE_{ext} \tag{19.18}$$

where I is the electrical current (A) and E_{ext} is the external voltage (V) across a resistor, R_{ext} (Ω). The current is calculated as follows:

$$I = E_{ext}/R_{ext} \tag{19.19}$$

The capacity of an MFC for electrical power generation is commonly expressed as the *power density*, which in general is the electrical power normalized to the surface area of electrode, A (m^2 or cm^2). When the cell volume is a sensitive design parameter, the densities may be normalized by the volume of either the entire reactor or the anode/cathode chamber (m^3 or L).

A *polarization curve* is a plot of the external voltage versus the current, and is one of the most common tools to illustrate the performance of MFCs. The *power curve*, a plot of the external power against the current, can also be used for defining an MFC's power output.

> **Example 19.4:** Table 19.6 shows polarization data for two-compartment MFCs (anode surface area = 4.9 cm^2) using two different electron acceptors: ferricyanide and dissolved oxygen. Plot the power curves in each case.
>
> **Solution:** Calculate the current using Ohm's law, $I = E_{ext}/R_{ext}$. So the power $P = IE_{ext}$. Plot the current (mA) on the x-axis and power (mW) on the y-axis (Figure 19.5):

Table 19.6 Polarization data for two-compartment MFCs

Resistance [Ω]	Voltage [V]	
	Ferricyanide	Oxygen
0	0.7	0.45
100	0.19	0.082
330	0.43	0.14
1,000	0.53	0.21
2,170	0.58	0.26
4,700	0.61	0.30
10,000	0.63	0.34
15,000	0.65	0.36
22,000	0.67	0.42
56,000	0.68	0.41
100,000	0.70	0.42
120,000	0.70	0.42
330,000	0.70	0.42
470,000	0.70	0.43

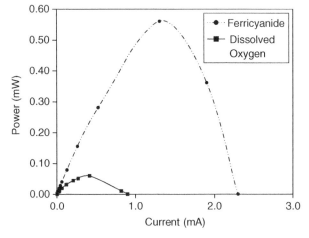

FIG. 19.5   Power curves.

The maximum power (P_{max}), maximum current (I_{max}), and internal resistance (R_{int}) can further be derived from these curves. Polarization curves typically follow a linear relationship when the current is in the normal range (neither too high nor too low).

Example 19.5: A polarization curve (Figure 19.6) is obtained from a microbial fuel cell with a total volume of 0.1 L and both electrode surface areas are 100 cm². Calculate the maximum power and current as well as internal resistance from the polarization curve. For the maximum power, also give the normalized results based on the cathode surface and reactor volume.

Solution: Regression analysis of the polarization curve results in a relationship between voltage and current as follows:

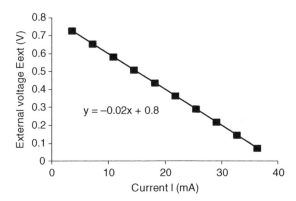

FIG. 19.6   Polarization curve.

$$E_{ext} = 0.8 - 20I$$

Assuming $E_{ext} = 0$, the maximum current can be calculated as 40 mA.

From $E_{int} = E_{OCV} - E_{ext}$ and $E_{int} = IR_{int}$, it can be obtained that $E_{ext} = E_{OCV} - IR_{int}$. The internal resistance, therefore, is the opposite of the slope of the polarization curve, $20\,\Omega$.

$$P = IE_{ext} = I \times (0.8 - 20I)$$

Solving this quadratic equation gives the result that when $\mathbf{I = 20\,mA}$ (at $\mathbf{R_{ext} = 20\Omega}$), the power reaches its maximum value of **8 mW**.

The normalized maximum power is as follows:

$$P_{area} = \frac{P_{max}}{A} = \frac{8\,mW}{0.01\,m^2} = \mathbf{0.8\,W/m^2}$$

$$P_{vol} = \frac{P_{max}}{V} = \frac{8\,mW}{0.1\,m^3} = \mathbf{0.08\,W/m^3}$$

19.4.4 Coulombic and Energy Efficiency

The substrate and chemical energy extraction process in MFCs is more diverse and stratified than that in a typical electrochemical cell, because biological growth and metabolism processes consume part of the substrate and energy available in the substrate. These performances can be evaluated by *Coulombic efficiency* (C_E) and *energy efficiency* (η), respectively, quantifying the ratio of harvested electrons (charges) to electrical energy from substrate oxidation (Logan, 2008). These two parameters can be calculated using chemical oxygen demand (COD) to represent the amount of organic matter in the wastewater:

$$C_E = \frac{\text{Charge recovered in MFCs}}{\text{Charge released from substrates}} = \frac{\int_0^t Idt/F}{-b_{O2}V\Delta COD/M_{O2}} \qquad (19.20)$$

$$\eta = \frac{\text{Energy recovered in external loads}}{\text{Energy released from substrates}} = \frac{\int_0^t E_{ext}Idt}{\Delta G_r V\Delta COD/M_{O2}} \qquad (19.21)$$

where ΔCOD is the change of COD concentration (g/L), b_{O2} is the number of moles of electrons accepted in the reduction of 1 mole of oxygen gas (4 mol/L), M_{O2} is the molar mass of oxygen gas (32 g/mol), and ΔG_r is the Gibbs free energy of the oxidation reaction of the organic substrates (J/mol). For a stable external voltage (E_{ext}), the relationship between Coulombic and energy efficiencies can be deduced, shown in Equation 19.22:

$$\eta = \frac{E_{ext}}{E_{emf}}C_e \qquad (19.22)$$

Coulombic efficiency (C_E) quantifies the ratio of electrons that are harvested at an electrical circuit to the electrons that are released from the oxidation of substrate. *Coulombic loss* is mainly originated from the microbial degradation and utilization of substrates, either by attached or suspended biomass, but irrelevant to the anode reaction. Energy efficiency (η) quantifies the ratio of electrical energy that is harvested in the electrical circuit to the chemical energy that is released in substrate oxidation. *Energy loss* in MFCs can be categorized into two major types: biological loss and polarization loss. Biological loss is associated with the growth and metabolism of both attached and suspended microbial cells, whereas polarization loss can be further categorized into activation overpotential, ohmic loss, and mass transfer overpotential. Activation overpotential indicates the energy loss required for transferring electrons from/to the electrode, which depends on the activation energy of the redox reaction. Ohmic loss is caused by the resistance of the medium and proton exchange membranes. Mass transfer overpotential is a result of the diffusion of substrate from the bulk to the electrode interface, which can be large under high current density conditions. These energy losses (biological and polarization losses) typically limit the net recovered energy to lower than 6 MJ/kg COD from organic substrates in most experimental studies, despite organic substrates containing chemical energy between 15 and 20 MJ/kg COD.

> **Example 19.6:** An air-cathode microbial fuel cell with a graphite brush anode is used to treat domestic wastewater and generate electricity. The MFC has a liquid volume of 0.05 L and an anode surface area of 0.25 m². The external voltage output reaches a steady state 2 hr after wastewater addition, stabilized at 0.52 V across a 500 Ω resistor. The voltage starts declining after another 8 hr operation. The COD concentrations at 2 hr and 10 hr after wastewater addition are 800 mg/L and 10 mg/L, respectively. What is the Coulombic efficiency and energy efficiency of the MFC during this time? For calculating the energy efficiency, an electromotive force of 1.05 V is assumed.
>
> **Solution:** The current of the MFC is $I = E_{ext}/R_{ext} = 0.52\,V/500\,\Omega = 1.04\,mA$.
> According to Equation 19.20, the Coulombic efficiency:

$$C_E = \frac{\int_0^t Idt/F}{-b_{O_2}V\Delta COD/M_{O_2}}$$

$$= \frac{1.04\,mA \times (10-2)hr/\left(96485\dfrac{1}{mol}\right)}{-4 \times 0.05\,L(10-800)\dfrac{mg}{L}/\left(32\dfrac{g}{mol}\right)} = \mathbf{6.3\%}$$

According to Equation 19.22, the energy efficiency:

$$\eta = \frac{E_{ext}}{E_{emf}}C_E$$

$$= \frac{0.52\,V}{1.05\,V}C_E = \mathbf{3.1\%}$$

19.5 Design and Operation of an MFC

The Coulombic efficiency, external voltage, and power outputs from MFCs are dependent on the MFC configuration, electrode and catalyst materials, substrate characteristics, and operating conditions. Current studies typically demonstrate electrical power production of up to 4 W/m^2 (2 kW/ m^3). This section summarizes common practices in choosing appropriate materials, configurations, and substrates for MFCs. It is worthwhile mentioning that a set of electrochemical analysis techniques is universally exploited to optimize the electrochemical performance of MFC materials and configurations. Typical techniques include the polarization curve, cyclic voltammetry, and electrochemical impedance spectroscopy. Students are encouraged to read other references for basic information and experimental skills in relation to these methods (Barsoukov and Macdonald, 2005; Holze, 2009).

19.5.1 MFC Configurations

MFC electrodes should be closely situated so as to decrease the resistance (to the diffusion of ions) between the electrodes, thereby improving the energy efficiency and the power outputs. If additional functions are to be met, such as COD and contaminant removal, special design considerations are needed to address these requirements appropriately. MFC configurations (Figure 19.7) are quite diverse due to many choices in chamber shape, flow channel, electrode material, separator, reactant, and operating mode. The MFC configuration in early studies was primarily a two-chambered, H-type configuration (Figure 19.7a), with chambers connected by a salt bridge or an ion exchange membrane. Rectangular reactors (Figure 19.7b) were later adopted to increase the size of the ion exchange membrane and to reduce the electrode spacing, thereby reducing the internal resistance. A tubular and upflow configuration (Figure 19.7c) is aimed at achieving the dual goals of effective electricity generation and wastewater treatment, in which methanogens form granules and are retained to increase the solids-retention time for better COD degradation. The oxygen is primarily

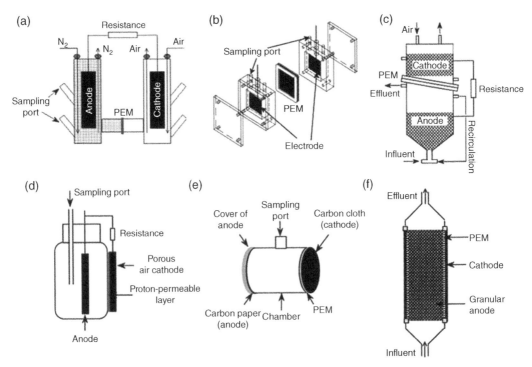

FIG. 19.7 Schematics of MFC configurations. Two-chamber configurations: (a) H-type, (b) rectangular shape, (c) upflow configuration; and single-chamber configurations: (d) with an ion exchange membrane coated air-cathode, (e) an air-cathode with electrodes placed on the opposite side, and (f) a tubular MFC with an outer cathode and an inner anode. Source: Du *et al.*, 2007. Reproduced with permission of Elsevier.

provided by aeration, and in some studies soluble oxidants, for example ferricyanide, are added in the cathode chamber. These configurations have an ion exchange membrane (e.g., Nafion) inserted between the anode and cathode chambers, so they are termed *two-chamber* configurations.

Another type of configuration is a *single-chamber MFC* (Figure 19.7d, e, and f). It usually adopts a *membrane electrode assembly* (MEA), in which an ion exchange membrane is directly bound to the cathode as a complete assembly. The MEA allows a substantial decrease of electrode space between the anode and the cathode without short-circuiting. The *separator electrode assembly* has a similar structure to an MEA, but it uses separators, for instance cloth and glass fiber instead of membranes. The replacement of membrane by a separator reduces the membrane overpotential. Another MEA design, a tubular and upflow configuration (Figure 19.7f), is further designed to increase the ratio of electrode surface area to reactor volume, in which the vessel is made of cathode and filled with granular graphite as anode. In single-chamber configurations, oxygen is provided by passive diffusion through the cathode, so these types of cathode are termed *air-cathodes* (Liu and Logan, 2004).

A single small MFC, whatever configuration it adopts, usually generates a limited voltage, power, and energy; therefore, some strategies are necessary to augment these outputs to meet real application requirements. Power output can be magnified by up-scaled MFCs that have a larger electrode surface area and reactor volume; multiple MFCs can also be stacked together to increase the power output. When an intermittent mode of power supply satisfies the application requirements, the power can be boosted by storing electrical energy and then releasing it in a short period controlled by a power management system (Donovan *et al.*, 2011).

19.5.2 Separators

An *ion exchange membrane* is a common *separator*, referring to either a cation exchange membrane (CEM) or an anion exchange membrane (AEM), which allows the transfer of cation or anion across the membrane. A proton exchange membrane (PEM) is a particular CEM that allows protons to pass between two sides. An ion exchange membrane is also able to prevent O_2 diffusion to the anode chamber. The diffusion of oxygen into an anode chamber negatively affects exoelectrogenesis and encourages aerobic metabolism. Finally, with an ion exchange membrane in place, the electrode space between anode and cathode can be very small but without short-circuiting. In the absence of the membrane, the cathode and the catalyst are exposed to chemical foulants and other biological metabolites, so the air-cathode can be fouled shortly after operation. Additional problems such as fouled catalysts, reduced diffusion of oxygen, undesirable side reactions, or substrate oxidation on the cathode may occur, which reduce the overall performance of MFCs as a result.

The ion exchange membrane is not an essential component in some MFCs, because an aqueous and conductive environment in MFCs can easily allow charge transfers to take place. There are some examples where a membrane is not required, for example in air-cathode MFCs where oxygen is gradually provided by passive diffusion, or in benthic MFCs where the electrode is naturally separated. On the contrary, the inclusion of a membrane brings about two additional types of overpotential that may lower overall power generation in MFCs: the ohmic overpotential due to its resistance and the overpotential due to the ion concentration splitting across a membrane. The latter resembles the *membrane potential* of a biological membrane. In the simplest case, this overpotential is a result of pH imbalance built up on the two sides, so its value can be calculated from the proton concentration or pH difference (Equation 19.23):

$$\eta = \frac{RT}{nF} \ln \frac{[H^+]_{anode}}{[H^+]_{cathode}} \text{ or } \eta = 0.058(pH_{cathode} - pH_{anode}) \tag{19.23}$$

Example 19.7: The operation of a two-chamber MFC generates a pH difference between the two sides of its proton exchange membrane; that is, a pH of 6.8 at the anode and 7.3 at the cathode. Calculate the membrane overpotential that is created by the pH gradient across the proton exchange membrane at room temperature.

Solution: This overpotential is the result of pH splitting across a membrane, so it can be calculated according to Equation 19.23:

$$\eta = 0.058 \times (pH_{cathode} - pH_{anode}) = 0.058 \times (7.3 - 6.8) = \textbf{0.029 V}$$

Some other materials can be alternatives to ion exchange membranes as separators. Filtration membranes (micro- and ultrafiltration membranes) and some cloths (J-cloth and polyester wipe cloth) can prevent the electrode from short-circuiting while maintaining a small distance between electrodes. Materials costs are also greatly reduced. However, a PEM or other ion exchange membranes may still be used in certain MFCs when fluid isolation is necessary. For example, when the toxic ferricyanide is used as the electron acceptor, inclusion of a membrane will allow charge transfer to proceed, but keep the toxic reactant away from the anode biofilm. When oxygen is mechanically aerated to the cathode medium in an MFC, a membrane helps to eliminate the oxygen gas diffusion into anode chamber and to improve the Coulombic efficiency.

19.5.3 Anode Materials and Catalysts

Microorganisms are catalysts for organic matter oxidation at the anode, and their species and mechanisms were discussed earlier. This section focuses on the anode material. The anode provides a surface for microbial attachment and an interface for electron transfer between microbes. The choice of MFC anodes should favor a low material cost, high conductivity, chemical inertness, microbial compatibility, and high specific surface area. Graphite and non-corrosive metals such as carbon cloth, carbon paper, carbon mesh, graphite felt, graphite granules, graphite brushes, graphite rod, stainless steel mesh, plate, and felt (Figure 19.8) are frequently employed as anode materials (Wei *et al.*, 2011). Graphite-fiber brushes have a high specific surface area, and have been found to be a promising anode material. An intertwined carbon nanotube increases the anode surface area with better microbial attachment attributes, and thus lowers the charge transfer resistance at the interface. Besides the direct application of these conventional materials, structural and compositional modification of anode by nanostructured materials (e.g., carbon nanotubes), conductive polymers (e.g., polyaniline and polypyrrole), and ammonia treatment may also improve anode performance.

19.5.4 Cathode Materials and Catalysts

Cathode provides an interface for a catalytic reduction reaction. In most cases, the reduction reaction is the *oxygen reduction reaction* (ORR). Inefficient cathode ORR generally constitutes the major limitation on the power generation of MFCs. Thermodynamically, side reactions, such as the formation of hydrogen peroxide instead of water, decrease the cathode potential; and kinetically, the catalytic effect of ORR may be slow in some cases. An ideal cathode should have a high specific surface area,

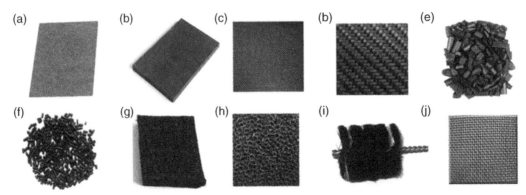

FIG. 19.8 Photographs of common electrode materials used for MFC: (a) carbon paper; (b) graphite plate; (c) carbon cloth; (d) carbon mesh; (e) granular graphite; (f) granular activated carbon; (g) carbon felt; (h) reticulated vitrified carbon; (i) carbon brush; and (j) stainless steel mesh. Source: Wei et al., 2011. Reproduced with permission of Elsevier.

good corrosion resistance, and cost-effectiveness. Carbon-based cathodes, especially wet-proofed carbon cloth, are among the most extensively studied materials.

The rate of ORR is primarily dictated by the catalysts. That is to say, the reaction rate can be driven by reducing the activation overpotential via the use of suitable catalysts. Platinum (Pt), applied at a load between 0.1 and 0.5 mg/cm^2, is the most commonly used catalyst for ORR in MFCs. The problem with Pt is that it is a noble metal of high cost, and is vulnerable to a gradual poisoning process. Some non-precious metals can be alternatives to the Pt catalyst, such as Fe, Co, and Mn-based macrocyclic complexes combined with tetra-methoxy-phenyl-porphyrin (TMPP) or phthalocyanine (Pc). Besides these materials, activated carbon coated on the cathode by polytetrafluoroethylene (PTFE) or other binders is an effective ORR catalyst. Certain species of microorganisms catalyze ORR (biocathode), but their performance is usually poor compared to chemical catalysts.

19.5.5 Substrates

Substrates provide either reducing power, or a carbon source, or both for microbial growth, metabolism, and electricity generation. The appropriateness of a substrate for MFCs depends not only on its biodegradability, but also on the ability for electron transfer by the specific microorganism to a solid anode surface. However, substrate utilization for electricity generation can be facilitated by the symbiosis of microbial consortia consisting of hydrolytic and fermentative bacteria, and exoelectrogens. In some cases, chemoautotrophic bacteria (e.g., *Paracoccus denitrificans* and *Paracoccus pantotrophus*) utilize sulfide as an electron donor for electricity generation in MFCs. Ammonium oxidation by exoelectrogens at the anode can also contribute to electricity generation.

The substrates can cover a wide range of compounds, for example volatile fatty acids, glucose, glycerol, lactate, and complex organic substrates such as starch and cellulose. Acetate is a common intermediate of microbial metabolisms, and is commonly used in MFC studies. Wastewaters suitable for MFC contain readily biodegradable organic matter and nutrients necessary for microbial growth, and have a near neutral pH, good buffering capacity, and conductivity. Domestic wastewater, some industrial wastewater (biofuel, textile, and tannery), agro-industrial wastewater, the liquid fraction of animal waste, anaerobic digestate, and food-processing wastewater (winery, olive mill, and potato processing) are promising candidates for electricity generation in MFCs. Unlike pure substrates, wastewater streams are more complex, and may contain recalcitrant or even inhibitory components to microbial electrochemical systems. So the maximum power density and Coulombic efficiency of MFCs fed with wastewater are consistently lower than those fed with acetate solution. Another category of MFC substrate is sulfide- or organic matter-containing sediment, such as river and marine sediments, and organic-rich soil. Due to the ubiquity of these sediments, they can be good substrates for MFC applications as well.

References

Arends, J., and Verstraete, W. (2012). 100 years of microbial electricity production: Three concepts for the future. *Microbial Biotechnology* 5(3): 333–46.

Barsoukov, E., and Macdonald, J.R. (2005). *Impedance Spectroscopy: Theory, Experiment, and Applications*. Chichester: John Wiley & Sons Ltd.

Davis, F., and Higson, S.P. (2007). Biofuel cells: Recent advances and applications. *Biosensors and Bioelectronics* 22 (7): 1224–35.

Donovan, C., Dewan, A., Peng, H., Heo, D., and Beyenal, H. (2011). Power management system for a 2.5 W remote sensor powered by a sediment microbial fuel cell. *Journal of Power Sources* 196(3): 1171–7.

Du, Z., Li, H., and Gu, T. (2007). A state of the art review on microbial fuel cells: A promising technology for wastewater treatment and bioenergy. *Biotechnology Advances* 25(5): 464–82.

Holze, R. (2009). *Experimental Electrochemistry*. Chichester: John Wiley & Sons Ltd.

Liu, H., and Logan, B.E. (2004). Electricity generation using an air-cathode single chamber microbial fuel cell in the presence and absence of a proton exchange membrane. *Environmental Science and Technology* 38(14): 4040–46.

Logan, B.E. (2008). *Microbial Fuel Cells*. Hoboken, NJ: John Wiley & Sons, Inc.

Malvankar, N.S., Yalcin, S.E., Tuominen, M.T., and Lovley, D.R. (2014). Visualization of charge propagation along individual pili proteins using ambient electrostatic force microscopy. *Nature Nanotechnology* 9(12), 1012–17.

Potter, M. C. (1911). Electrical Effects Accompanying the Decomposition of Organic Compounds. *Proceedings of the Royal Society of London. Series B, Containing Papers of a Biological Character*, 84(571), 260–276.

Rabaey, K. (2010). *Bioelectrochemical Systems: From Extracellular Electron Transfer to Biotechnological Application.* London: IWA Publishing.

Schröder, U. (2007). Anodic electron transfer mechanisms in microbial fuel cells and their energy efficiency. *Physical Chemistry Chemical Physics* 9(21): 2619–29.

Thauer, R.K., Jungermann, K., and Decker, K. (1977). Energy conservation in chemotrophic anaerobic bacteria. *Bacteriological Reviews* 41(1): 100.

Torres, C.I., Marcus, A.K., Lee, H.S., Parameswaran, P., Krajmalnik-Brown, R., and Rittmann, B.E. (2010). A kinetic perspective on extracellular electron transfer by anode-respiring bacteria. *FEMS Microbiology Reviews* 34(1): 3–17.

Wei, J., Liang, P., and Huang, X. (2011). Recent progress in electrodes for microbial fuel cells. *Bioresource Rechnology* 102(20): 9335–44.

Exercise Problems

19.1. What are the two features (catalyst and fuel) that distinguish microbial fuel cells (MFCs) from conventional electrochemical fuel cells?

19.2. Consider the following reactions that may occur in the cytoplasm of bacteria at pH = 7 and T = 298 K.

$$2NAD^+ + 2H^+ + 2e^- \rightarrow 2NADH; E^{0'} = -0.32V$$

$$2H^+ + 2e^- \rightarrow H_2; E^{0'} = -0.414V$$

Estimate the hydrogen partial pressure (kPa) at which the NADH pathway (NAD$^+$/ NADH) becomes feasible for transferring electrons to a proton reduction reaction (H$^+$/H$_2$) for hydrogen production in bacteria.

19.3. Consider the following half-cell reaction for the oxidation of lactate to acetate:

$$CH_3COO^- + HCO_3^- + 5H^+ + 4e^- \rightarrow CH_3CH(OH)COO^- + 4H_2O; E^{0'} = 0.09V$$

Use the Nernst equation to calculate the reduction potential (lactate oxidation to acetate) using the conditions of HCO$_3^-$ = 5 mM, CH$_3$COO$^-$ = 5 mM, CH$_3$CH(OH)COO$^-$ = 100 mM, pH = 7, and T = 298 K.

(a) Plot the anode potential (lactate oxidation to acetate) as a function of lactate concentration. Show the plot ranging from 100–500 mM at an increment of 100 mM.

(b) Plot the anode potential (lactate oxidation to acetate) as a function of pH. Show the plot ranging from 5–9 at an increment of 1 unit.

19.4. Table 19.7 shows the polarization data for a two-compartment MFC operated at different values of external load. Use Ohm's law to fill the blanks in the table. Obtain values of current density and power density by normalizing current and power with an anode surface area of 7 cm^2 and a reactor volume of 0.05 L.

Table 19.7 Polarization data for a two-compartment MFC

External resistor Ω	External voltage V	Current A	External power W	Current density A/m^2	Power density W/m^2	Power density W/m^3
1000	0.63					
470	0.6					
100	0.43					
68	0.34					
47	0.27					
33	0.209					
15	0.106					
6.8	0.053					
3.3	0.027					

19.5. A small air-cathode microbial fuel cell has a reactor volume of 30 mL. The specific surface area of electrode to volume is $25 \text{ m}^2/\text{m}^3$. Table 19.8 lists points in a power curve of this MFC.

(a) What is the internal resistance of the MFC?
(b) What is the maximum power that this MFC can provide? Normalize the result to the reactor volume.

Table 19.8 Points in a power curve of an MFC

I	Pext	I	Pext
A/m^2	W/m^3	A/m^2	W/m^3
0.25	5	5	43
1	15	6	39
1.8	27	8	32
2.1	31	9	25
3	38	10	16

19.6. In a two-chamber microbial fuel cell study, glucose is used as the anode substrate and potassium hexacyanoferrate $(K_3Fe(CN)_6)$ as the oxidant. Glucose is fermented to acetate, which is then consumed as reactants for electricity generation catalyzed by electroactive bacteria. The oxidant loses electrons by the half-reaction $Fe(CN)_6^{3-} + e^- \rightarrow Fe(CN)_6^{4-}$, and this redox couple has a standard reduction potential of 0.361 V (E^o). A microbial fuel cell of this type is fed with 20 mM glucose solution, and 30% of the glucose almost instantaneously goes to fermentation via acetogenesis:

$$C_6H_{12}O_6 + 2H_2O \rightarrow 2CH_3COO^- + 2HCO_3^- + 4H_2$$

Calculate the anode potential and electromotive force of the microbial fuel cell when 50% of the produced acetate remains in the solution. Assume pH = 6.5, $[HCO_3^-]$ = 5 mM, and $[Fe(CN)_6^{3-}]$ = $[Fe(CN)_6^{3-}]$ = 50 mM. The reduction potential (E°) of the redox couple HCO_3^-/acetate is 0.187 V:

$$2HCO_3^- + 9H^+ + 8e^- \rightarrow CH_3COO^- + 4H_2O$$

19.7. Extracellular electron transfer from exoelectrogenic bacteria and anodes serves a critical role in MFC electricity generation. What are the three common extracellular electron transfer mechanisms?

19.8. In a microbial fuel cell, riboflavin shuttle is the dominant extracellular electron transfer mechanism, and shuttle transport is found to be the rate-limiting step. Given the following parameters, calculate the maximum current density of the MFC at a shuttle transport distance of $\Delta z = 0.2\,\mu m$: $D_{shuttle} = 6.7 \times 10^{-10}\,m^2/s$, $\Delta C_{shuttle} = 0.1\,\mu M$, and $n = 2$.

19.9. The extracellular electron transfer of a microbial fuel cell fed with acetate is dictated by the direct contact mechanism. The reactor has a Coulombic efficiency of 30%. Calculate the upper limit of current density given the following parameters: $q_{max} = 400\,mg$ acetate/(g dry wt.·hr), $X_f = 2.5 \times 10^5$ g dry wt./m^3, and $L_{fa} = 2\,\mu m$.

19.10. An air-cathode microbial fuel cell with a graphite brush anode is used to treat domestic wastewater and generate electricity. The MFC has a liquid volume of 0.1 L and an anode surface area of 0.5 m^2. The external voltage output reaches a steady state 2 hr after wastewater addition, stabilized at 0.43 V across a 100 Ω resistor. The voltage starts declining after another 5 hr operation. The COD concentrations at 2 hr and 7 hr after wastewater addition are 1,000 mg/L and 10 mg/L, respectively. What are the Coulombic efficiency and energy efficiency of the MFC during this time? For calculating the energy efficiency, an electromotive force of 1.05 V is assumed.

SECTION IV
Thermal Conversion Technologies

Combustion for Heat and Power

Sushil Adhikari, Avanti Kulkarni, and Nourredine Abdoulmoumine

---

What is included in this chapter?

This chapter focuses on the fundamentals of biomass combustion, biomass properties and issues related to biomass combustion, different combustion cycles for power generation, environmental impacts, and some case studies of biomass combustion in large-scale applications for heat production.

20.1 Introduction

Early humans' discovery of the ability to control fire was a pivotal point in our civilization. Fire provided protection against predators, enabled cooking, and facilitated human expansion by allowing settlement in colder and harsher environments. Over the course of history, fire and combustion have been inextricably linked by human civilizations attempting to understand their mysteries, often through heavenly explanations. The Greeks rationalized combustion through the phlogiston theory. From the ancient Greek word *phlogos*, meaning "flame" or "blaze," this theory viewed combustion as a process that released "phlogiston," a substance similar to fire and contained in combustible materials. This theory, however, was erroneous and ultimately disproved. As the limitations of the phlogiston theory became apparent, Sir Francis Bacon (1620), Otto von Guericke (1650), and Robert Hooke (1665) all noted the importance of air in the combustion process. However, it was Antoine Lavoisier, a French chemist, who first understood the true nature of combustion and suggested that it is a chemical reaction that requires oxygen.

Bioenergy: Principles and Applications, First Edition. Edited by Yebo Li and Samir Kumar Khanal.
© 2017 John Wiley & Sons, Inc. Published 2017 by John Wiley & Sons, Inc.
Companion website: www.wiley.com/go/Li/Bioenergy

FIG. 20.1   Antoine Lavoisier (1743–94). Source: https://en.wikipedia.org/wiki/Conservation_of_mass#/media/File:Antoine_laurent_lavoisier.jpg. Public domain.

Antoine Lavoisier (Figure 20.1) was a French chemist who is credited for providing the first proper explanation of the combustion reaction. In 1772, Lavoisier suggested that the weight increase upon burning of phosphorous and sulfur was due to the combination of these elements with air providing the first evidence of oxygen and its central role in combustion. Lavoisier's experimental approach to chemistry and subsequent work paved the road for modern chemistry and shed light on one of the most important reactions in chemistry.

Despite the fact that much of the recent focus is on liquid biofuels production, combustion is the most widely used thermochemical conversion process for heat and power generation from biomass. In many developing countries, over 75% of the primary energy consumed is derived from the combustion of biomass.

Burning of biomass in open fires and simple stoves is a common example of combustion around the world for space heating and cooking purposes. However, these methods are less desirable from an indoor air pollution perspective, especially due to their potential to release carbon monoxide (CO) and poly-aromatic hydrocarbons (PAHs) as a result of incomplete combustion. The focus in this chapter is on the combustion of biorenewable resources such as biomass for heat and power production. Biomass is either directly combusted to generate steam or converted into flue gas, which is then used in steam or gas turbines, respectively, for producing power.

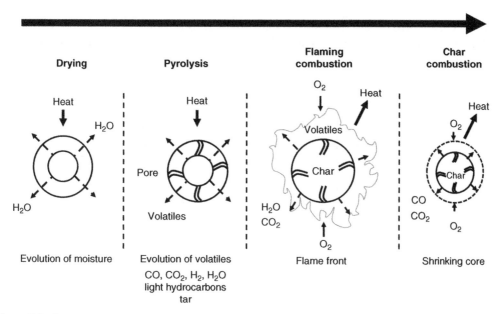

FIG. 20.2 Progression of combustion for a particle of biomass.

20.2 Fundamentals of Biomass Combustion

20.2.1 Biomass Combustion Phases

Combustion of biomass takes place at temperatures of 700–1,000 °C. Biomass combustion can be divided into four phases: *drying, pyrolysis or devolatilization, volatile combustion, and char combustion.* *Heating* and *drying* occur when biomass is first introduced into the combustion chamber and rapidly heated from its ambient temperature to the onset temperature of pyrolysis, approximately between 200 and 225 °C (Tillman *et al.*, 1981; Basu, 2010).

In the *pyrolysis (devolatization) phase*, biomass is devolatilized into condensable and non-condensable gases, resulting in a rapid mass loss at 200–400 °C (Brown, 2011). In this phase, the decomposition mechanisms of biomass constituents (cellulose, hemicelluloses, and lignin) vary and occur at different temperature ranges: cellulose and hemicelluloses are decomposed through the cleavage of the glycosidic bonds between 200 and 350 °C, while lignin is decomposed at 350–500 °C (Tillman *et al.*, 1981). Factors affecting devolatilization include heating rate, residence time, initial and final temperature, particle size of the biomass, types of biomass, and pressure. Volatiles and char produced during the pyrolysis phase further undergo oxidation. Oxygen and volatiles react in a diffusion flame to produce CO_2 and H_2O. This step is called *flaming (volatile) combustion* and produces heat. After volatiles have been oxidized, *char combustion* occurs, produces CO and CO_2, and releases heat in the process, as illustrated in Figure 20.2.

Pyrolysis profile of different types of biomass

Figure 20.3 shows the mass loss curve of pine, sweetgum, and switchgrass in nitrogen atmospheres at a heating rate of 10 °C/min. Degradation of lignocellulosic biomass under inert conditions can be broken down into three distinct stages. The first stage is dehydration

(<150 °C), followed by fast pyrolysis (150–400 °C) and, lastly, slow pyrolysis (>400 °C). The initial stage of weight loss in the curve can be attributed to the removal of the moisture in the biomass.

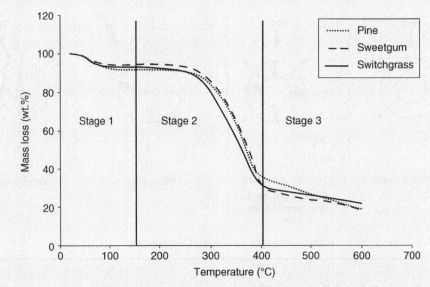

FIG. 20.3   Pyrolysis profile of different types of biomass.

20.2.2 Biomass Combustion Reaction and Stoichiometry

Overall, biomass combustion conditions can be represented by Equation 20.1:

$$C_\alpha H_\beta O_\gamma S_\theta N_\kappa + m(O_2 + 3.76N_2) \rightarrow aCO_2 + bH_2O + cSO_2 + dN_2 \tag{20.1}$$

Here, we have assumed that all the nitrogen in biomass was converted in nitrogen gas for the simplicity of calculation. Usually, nitrogen in the biomass would form NOx. If NOx is included in the product, the elemental balance would be different than what is presented herein. In Equation 20.1, the coefficient for nitrogen (N_2), 3.76, represents the number of moles of nitrogen in air for every mole of oxygen. The Greek coefficients α, β, γ, θ, and κ are known from the elemental composition of biomass, as determined from the ultimate analysis. The coefficients a, b, c, d, and m are unknown and can be determined by balancing the reaction. Based on Equation 20.1, the following set of algebraic expressions can be deduced for the unknown coefficients:

C balance: $\alpha = a$ $\quad\quad\quad\quad\quad \Rightarrow a = \alpha$

H balance: $\beta = 2b$ $\quad\quad\quad\quad \Rightarrow b = \beta/2$

S balance: $\theta = c$ $\quad\quad\quad\quad\quad \Rightarrow c = \theta$

O balance: $\gamma + 2m = 2a + b + 2c$ $\quad \Rightarrow m = (2a + b + 2c - \gamma)/2$

$\quad\quad\quad\quad\quad\quad\quad\quad\quad\quad\quad \Rightarrow m = \alpha + \beta/4 + \theta - \gamma/2$

N balance: $\kappa + (2m \times 3.76) = 2d \Rightarrow d = 3.76m + \kappa/2$

Thus, the general reaction can be written as shown in Equation 20.2:

$$C_\alpha H_\beta O_\gamma S_\theta N_\kappa + \left(\alpha + \frac{\beta}{4} + \theta - \frac{\gamma}{2} \right) (O_2 + 3.76N_2)$$

$$\rightarrow \alpha CO_2 + \left(\frac{\beta}{2} \right) H_2O + \theta SO_2 + \left(3.76 + \frac{\kappa}{2} \right) N_2 \tag{20.2}$$

In most biomass, nitrogen and sulfur contents are small and their products can be ignored. Accordingly, $m = (\alpha + \beta/4 - \gamma/2)$ and $d = 3.76\,m$.

> **Example 20.1:** The chemical elemental composition of southern pine is $C = 50.89\%$, $H = 6.11\%$, $O = 41.83\%$, $S = 0.06\%$, $N = 0.48\%$, and ash $= 0.63\%$ (dry wt. basis). Determine coefficients a, b, c, and d and balance the general biomass combustion equation (based on 100 g biomass).
>
> **Solution:**
>
> $m_c = 0.5089 \times 100\,g = 50.89\,g \quad n_c = 50.89\,g/12\,(g/mol) = 4.24\,mol$
>
> $\alpha = 4.24$
>
> Using a similar approach, $\beta = 6.11$, $\gamma = 2.62$, $\theta = 0.002$, and $\kappa = 0.03$
>
> $C_{4.24}H_{6.11}O_{2.62}S_{0.002}N_{0.03} + 4.46(O_2 + 3.76N_2)$
> $\rightarrow 4.24CO_2 + 3.06H_2O + 0.002SO_2 + 16.80N_2$
>
> or if coefficients are normalized with respect to C
>
> $C_1H_{1.44}O_{0.62}S_{0.0004}N_{0.008} + 1.05(O_2 + 3.76N_2)$
> $\rightarrow CO_2 + 0.72H_2O + 0.0004SO_2 + 3.96N_2$

It is useful to present the quantity of air required for complete combustion in terms of a ratio based on unit mass of biomass. The theoretical ratio (stoichiometric ratio) of mass of air to mass of biomass $(A/B)_{stoichiometric}$ can be calculated by Equation 20.3 using the coefficients of the balanced Equation 20.1:

$$\left(\frac{A}{B} \right)_{stoichiometric} = \frac{m_{air}}{m_{biomass}} = \frac{\Sigma(n_iM_i)_{air}}{\Sigma(n_iM_i)_{biomass}} = \frac{m(M_{O_2} + 3.76M_{N_2})}{\alpha M_C + \beta M_H + \lambda M_O + \theta M_S + \kappa M_N} \tag{20.3}$$

where M_i is the molecular weight of constituents i.

Another ratio that is often encountered is known as the *equivalence ratio*, λ, and is a measure of the actual oxygen supplied to the combustion process to that required stoichiometrically, as outlined in Equation 20.4:

$$\lambda = \frac{(A/B)_{actual}}{(A/B)_{stoichiometric}} \tag{20.4}$$

where $(A/B)_{actual}$ and $(A/B)_{stoichiometric}$ are the actual and stoichiometric air to biomass ratios, respectively.

The air/biomass equivalence ratio is used to define combustion regimes. In a biomass-rich regime, insufficient air is available for complete combustion and $\lambda < 1$. When excess air is supplied, $\lambda > 1$ and the regime is deemed biomass lean. If the stoichiometric amount of oxygen required for complete combustion is supplied, there is no excess air and $\lambda = 1$. The term "e," expressed in percentage, is used to determine the excess air in the combustion process and is defined as in Equation 20.5 (Annamalai and Puri, 2007):

$$e = 100(\lambda - 1) \tag{20.5}$$

Example 20.2: For the same biomass as in Example 20.1, find the air flow rate required to achieve an air/biomass equivalence ratio (λ) of 1.25, if the biomass feeding rate is 1 kg/sec.

Solution:

1. **Find the stoichiometric air to fuel ratio**
   For m = 1.05, α = 1, β = 1.44, and θ = 0.0004

$$\left(\frac{Air}{biomass}\right)_{stoichiometric} = \frac{1.05[32 + (3.76 \times 28)]}{(1 \times 12) + (1.44 \times 1) + (0.62 \times 16) + (0.0004 \times 32) + (0.008 \times 14)}$$

$$= 6.17 \frac{kg\,air}{kg\,biomass}$$

2. **Find the actual air to fuel ratio $(A/B)_{actual}$ required and the air flow rate**

$$ER = 1.25 \rightarrow (A/F)_{actual} = 1.25 \times 6.17 = 7.71 \text{ kg air/kg biomass}$$

$$\dot{m}_{air} = \frac{7.71\,kg\,air}{kg\,biomass} \times \frac{1\,kg\,biomass}{sec} = 7.71 \frac{\mathbf{kg\,air}}{\mathbf{sec}}$$

The combustion reaction always occurs in an excess oxygen environment (up to 50% of stoichiometric amount) to ensure that there is complete combustion of biomass. However, combustion seldom yields only the products in Equation 20.1, even in the presence of excess oxygen. In addition to CO_2, H_2O, SO_2, and N_2, other compounds such as CO and NO_x are formed. The general combustion reaction can be rewritten as in Equation 20.6 to account for CO, NO_X, and unreacted oxygen:

$$\begin{aligned} C_\alpha H_\beta O\gamma S_\theta N_\kappa + m(O_2 + 3.76N_2) \\ \rightarrow aCO_2 + bH_2O + cSO_2 + dN_2 + eCO + fNO + gN_2O + hNO_2 + iO_2 \end{aligned} \tag{20.6}$$

where, e, f, g, h, and i are stoichiometric coefficients for the additional gaseous products formed in real combustion flue gas.

Example 20.3: 1 kg of biomass (C = 50.18, H = 6.06, O = 40.43, S = 0.03, N = 0.6, and ash = 2.7% [dry wt. basis]) is burned with 20% excess air. Determine the amount and volumetric concentration of carbon dioxide, water, sulfur, nitrogen, and oxygen in the products. Assume that all the S and N in the fuel are converted into SO_2 and NO_2.

Solution: First, use the balanced Equation 20.1 and determine the stoichiometric moles of air required for combustion:

$$C_1H_{1.45}O_{0.60}S_{0.00022}N_{0.010} + 1.07(O_2 + 3.76N_2)$$
$$\rightarrow 1CO_2 + 0.725H_2O + 0.00022SO_2 + 4.03N_2$$

If 20% excess air is used, 120% air is used or 1.2 times the stoichiometric amount:

$$C_1H_{1.45}O_{0.60}S_{0.00022}N_{0.010} + 1.28(O_2 + 3.76N_2)$$
$$\rightarrow CO_2 + 0.725H_2O + 0.00022SO_2 + 4.83N_2 + 0.010NO_2 + 0.21O_2$$

From the balanced Equation 20.1, the molar concentration of all products can be determined according to the following equation:

$$y_i = 100 \times \frac{n_i}{n_{tot}}$$

n_{CO2} = 1 mole, n_{H2O} = 0.725 mole, n_{SO2} = 0.00022 mole, n_{N2} = 4.83 mole, n_{NO2} = 0.010 mole, n_{O2} = 0.21 mole, and n_{tot} = 6.78 moles.

$y_{CO_2} = 100 \times \dfrac{1}{6.78} = 14.75\%$; Similarly, y_{H2O} = 10.68%, y_{N_2} = 71.25%, y_{O_2} = 3.16%, y_{NO_2} = 0.15% and y_{SO_2} = 33.07 ppmv.

20.3 Biomass Properties and Preprocessing

20.3.1 Biomass Properties

In general, lignocellulosic biomass is composed of mainly carbon (43–51% wt. basis), hydrogen (4–6%), and oxygen (34–48%), with small amounts of other elements such as N, S, and Cl. The concentration of N, S, and Cl varies with biomass type. Biomass also contains inorganic elements such as Ca and Mg, which increase the ash melting temperature, while K and P have the opposite effect. In general, these constituents are at a much lower concentration in woody biomass compared to grasses and agricultural residues, thereby resulting in a higher ash melting temperature. For example, some grasses have an ash softening temperature as low as 800 °C due to ash composition (34% K_2O in ash), while wood (oak) has an ash melting temperature of 1360 °C (9.9% K_2O in ash). Furthermore, elements such as Cl, S, K, and Na in biomass play major roles in corrosion. Volatile matter in biomass varies between 70% and 86% (wt. basis). The ash content of biomass varies from 0.5% to as high as 35% (wt. basis) depending on the type of biomass, maturity, harvesting, and processing methods.

During combustion, nitrogen bound in biomass results in the formation of NO_X together with NH_3 and HCN. For an optimized process, NH_3 and HCN would be converted into oxides of

Fig. 20.4 Common shape of biomass feedstock for thermochemical conversion technologies. Source: Photo courtesy of Center for Bioenergy and Bioproducts, Auburn University.

nitrogen (NO_X). Emission of SO_2 is not significant for woody biomass combustion, but may be significant for grasses such as miscanthus, switchgrass, and straw, depending on the sulfur content of the feedstock. Oxides of sulfur (SO_X) are the result of complete combustion. However, a significant portion remains in the residual ash in the form of alkali sulfate ($NaSO_4$, K_2SO_4, etc.) and H_2S. Cl bound in biomass is primarily released as salts ($NaCl$, KCl, etc.), with the remainder released as HCl.

20.3.2 Biomass Preprocessing

The most common biomass preparation steps prior to combustion are size reduction and drying. Biomass feedstocks are available in different shapes and sizes. Typically, size reduction includes chunking, chipping and grinding. The extent of biomass preprocessing depends on the type and quality of biomass, types of combustor and biomass handling system. Sometimes biomass is preprocessed to a very uniform feedstock such as pellets to minimize feedstock variability and increase energy density. As shown in Figure 20.4, biomass feedstocks with varying geometries and sizes can be produced. As size is critical for feeding, the acceptable size distribution of biomass feedstock is decided in the early stages of process design to fit the feeding mechanism.

As biomass feedstocks are available at widely varying moisture contents, drying is often required to improve combustion efficiency. In addition, long-term storage of wet biomass results in dry matter loss and microbial degradation due to fungal growth.

Adiabatic flame temperature

Combustion produces heat and raises the temperature of both the reactor and the exit reaction products. The theoretical maximum temperature of the reactor and these products is called the *adiabatic reaction temperature* or *adiabatic flame temperature*. It represents the combustion process temperature if first, the reaction is carried out under adiabatic conditions (no heat loss); and second, there are no other effects such as electrical effects and work. In reality, the process temperature is always less than the adiabatic flame temperature.

20.4 Biomass Furnaces

There are various furnace designs in use today for biomass combustion. The focus here will be primarily on two main types of furnaces: fixed-bed and fluidized-bed furnaces.

20.4.1 Fixed-Bed Furnaces

Fixed-bed furnaces are classified based on the way the biomass is fed onto the grate as *overfeed, underfeed*, and *crossfeed stokers* (Miller, 2011), where stoker refers to a continuous fuel feed system. The fuel feeding is most commonly achieved using screw conveyors, hydraulic feeders, and mechanical or pneumatic spreaders. In overfeed stokers, biomass is fed onto the grate from the top and forms a bed with different layers that moves down as biomass is decomposed, with the produced flue gas exiting at the top. With underfeed stokers, biomass is fed from the bottom along with air while the produced flue gas exits at the top. For crossfeed stokers, biomass moves horizontally while air is supplied from the bottom and flue gas exits at the top. While the classification of fixed-bed furnaces based on the biomass feeding pattern is more convenient, commercial furnaces often exhibit several biomass/air flow patterns along with other important design features. Therefore, it is worth discussing a few well-known patterns used in commercial systems. The combustion systems commonly used for biomass combustion are traveling-, vibrating-, inclined-, and rotating-grate furnaces and underfeed furnaces.

In the *traveling-grate furnace*, biomass is fed onto a conveyor grate and transported through the combustion chamber with minimal disturbance of the biomass bed. Traveling-grate systems offer a uniform combustion condition for wood chips and pellets, minimal dust production as the bed is not mixed, and easy maintenance. On the other hand, these systems require a long residence time and high excess air (Van Loo and Koppejan, 2008). A *vibrating-grate furnace* consists of an inclined grate placed on springs, where biomass transport down the grate is induced by vibrations and gravity. Vibrating-grate systems are well suited for biomass with slagging and sintering tendencies (e.g., straw, waste wood), but result in high fly-ash emission due to bed vibrations. The grates in *inclined-grate* or *horizontal moving grate furnaces* are made of fixed and movable grate bars. The movable bars are alternatively moved horizontally, forward, and backward, causing transport of the biomass and a localized mixing of the bed.

Moving-grate furnaces can handle many kinds of wet or dry biomass feedstocks (bark, wood chips, and sawdust). Illustrations of traveling- and vibrating-grate furnaces are in Figures 20.5 and 20.6.

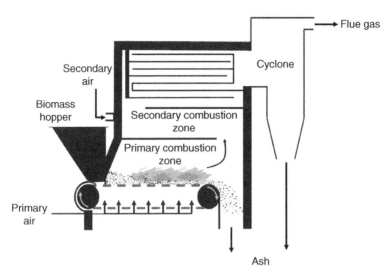

FIG. 20.5   Traveling-grate furnace. Source: Adapted from Van Loo and Koppejan, 2008.

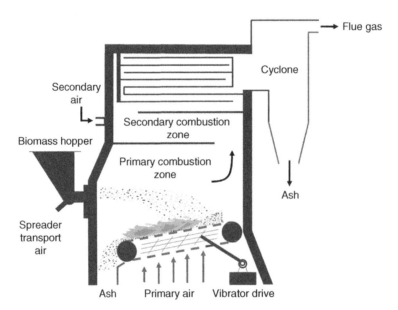

FIG. 20.6   Vibrating-grate furnace with spreader feeding systems. Source: Adapted from Van Loo and Koppejan, 2008.

20.4.2 Fluidized-Bed Furnaces

Unlike the earlier *fixed-bed furnaces, fluidized-bed furnaces* do not have a grate on which the biomass can burn. As shown in Figure 20.7, these furnaces consist of a hot inert material in the reactor, maintained in a suspended or "fluidized" state using air. As biomass enters the reactor, particles mix with the fluidized hot inert material and begin burning. The fluidization results in good mixing, uniform temperature in the bed, lower excess air, and therefore higher combustion efficiency (Van Loo and

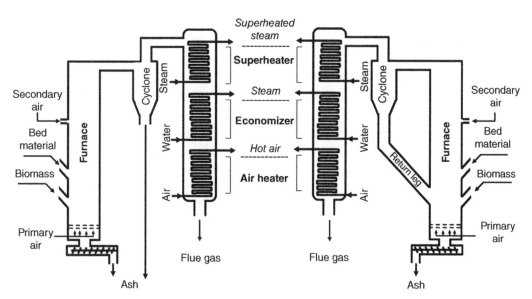

FIG. 20.7   Fluidized bed furnaces: bubbling fluidized-bed furnace (left) and circulating fluidized-bed furnace (right). Note: Depending on the design demands, the heat exchanger configuration (superheater, economizer, and air heater) may not be needed as depicted in the figure. Source: Adapted from Van Loo and Koppejan, 2008; Basu, 2006.

Koppejan, 2008). Size reduction of biomass is often necessary in these systems to ensure complete decomposition of biomass as well as proper fluidization and feeding. Knowledge of ash melting and sintering behavior of biomass is equally important to avoid agglomeration, which adversely affects fluidization (Van Loo and Koppejan, 2008).

Fluidized-bed furnaces are suitable for various biomass mixtures, and are typically attractive for medium to large-scale applications (>20 MW$_{th}$). Relative to fixed-bed furnaces, fly-ash and dust entrainment is quite significant in fluidized-bed systems, leading to a higher cost for particulate removal. Fluidized-bed systems are further classified as *bubbling fluidized-bed* (BFB) and *circulating fluidized-bed* (CFB) systems. In BFB, the air velocity is maintained low (1.0–2.5 m/sec) in order to minimize entrainment, whereas in circulating fluidized-bed combustors, a high fluidization velocity (5–10 m/sec) is deliberately set to entrain bed materials, which are separated from flue gas and circulated back into the combustor.

20.5 Power Generation

Biomass combustion can be used to generate power using either internal combustion (IC) engines or external combustion engines (Stirling engine, steam turbines). For medium and large-scale power generation (>500 kWe), biomass can be directly combusted to produce steam that can be used in steam turbines. Alternatively, biomass can be converted via gasification into producer gas (see Chapter 21), which can fuel internal combustion engines for small-scale applications or gas turbines for larger-scale applications.

Power cycles

A power cycle is a cycle that takes heat and uses it to perform work on the surroundings. The relevant power cycles for power generation from biomass combustion are the Rankine, Stirling, and Brayton cycles.

In a power plant, when the generator shaft is turned, electricity is produced. The generator shaft is turned either by a steam turbine or a gas turbine. In a steam turbine, the steam generated in the boiler expands to turn the turbine blades, resulting in shaft rotation. Similarly, in a gas turbine, the flue gas from combustion, which is at high temperature and pressure, expands over the turbine blades, causing shaft rotation. Steam turbine plants are based on the *Rankine cycle*, whereas gas turbine plants are based on the Brayton cycle. The following section covers Rankine and Brayton cycles and introduces the Carnot cycle.

20.5.1 Carnot Cycle

The *Carnot cycle* is the simplest form of power cycle. While impractical, it sets the maximum attainable efficiency of any cycle or heat engine operating between a higher-temperature (T_H) reservoir with heat energy Q_H and a lower-temperature (T_L) sink with heat energy Q_L. Based on the first law of thermodynamics, the Carnot cycle efficiency is defined as:

$$\eta_{Carnot} = \frac{Q_H - Q_L}{Q_H} = 1 - \frac{Q_L}{Q_H} \tag{20.7}$$

The heat transfer ratio is approximated by the ratio of its equivalent absolute temperature, thus reducing Equation 20.7 to:

$$\eta_{Carnot} = 1 - \frac{T_L}{T_H} \tag{20.8}$$

Figure 20.8 shows the Carnot cycle and its T-S diagram, respectively. The first step represents the isothermal heat addition of working fluid (Steps 1–2), followed by isentropic expansion (Steps 2–3),

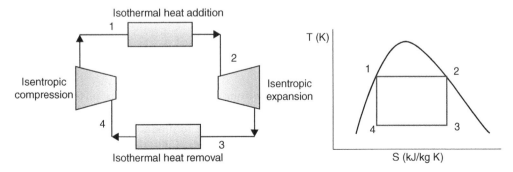

FIG. 20.8   Schematic of Carnot cycle (left) and T-S diagram (right).

isothermal heat removal (Steps 3–4), and isentropic compression (Steps 4–1). All these steps are assumed to be completely reversible. This is an ideal cycle and no such cycle exists in reality, as there are heat and friction losses associated with every process.

> **Example 20.4:** A Carnot engine receives 750 kJ of heat from a heater at 700 °C. It rejects heat to a heat sink at 25 °C. Calculate (a) the Carnot cycle efficiency; and (b) the amount of heat rejected at the lower temperature.
>
> ***Solution:*** Given: $T_H = 700+273 = 973$ K, $T_L = 25+273 = 298$ K, $Q_H = 750$ kJ
> **(a) The Carnot cycle efficiency is calculated as:**
>
> $$\eta_{Carnot} = 1 - \frac{T_L}{T_H}$$
>
> $$\eta_{Carnot} = 1 - \frac{298}{973} = 0.693 = \mathbf{69.3\%}$$
>
> The maximum efficiency that a heat engine operating between 700 °C and 25 °C can obtain is a Carnot cycle efficiency of 69.3%.
> **(b) The heat rejected to sink is calculated as:**
>
> $$Q_L = (Q_H) \times (T_L/T_H) = 750 \times (298/973) = \mathbf{229.70\,kJ}$$

20.5.2 Rankine Cycle

The *Rankine cycle* is a power cycle applicable to steam turbine applications. The processes in this cycle are assumed to be irreversible. The schematic of the Rankine cycle and its T-S diagram are presented in Figure 20.9. An ideal Rankine cycle consists of a pump, boiler, turbine, and condenser. A pump is used to feed water into the boiler, where water absorbs heat from the combustion of biomass to produce steam. This steam expands in a turbine producing shaft work, which is used to generate electricity. The steam is then collected in a condenser, where it is condensed back to water and fed into the pump.

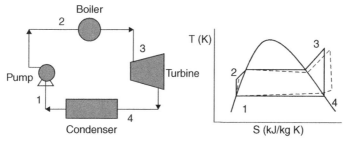

FIG. 20.9   Schematic of Rankine cycle (left) and T-S diagram (right).

The energy analysis of the Rankine cycle can be performed to calculate the efficiency by considering the following four steps:

1. Isentropic compression: $W_{pump, in} = h_2 - h_1$
2. Isobaric heat addition: $Q_{in} = h_3 - h_2$
3. Isentropic expansion: $W_{turbine, out} = h_4 - h_3$
4. Heat removal at constant pressure: $Q_{out} = h_4 - h_1$

Consequently, the efficiency can be expressed as in Equation 20.9:

$$\eta_{th, rankine} = \frac{W_{net}}{q_{in}} = 1 - \frac{q_{out}}{q_{in}} = 1 - \frac{(h_4 - h_1)}{(h_3 - h_2)} \tag{20.9}$$

Heat rate

In the USA, the efficiency of a power plant is measured in terms of heat rate. Heat rate is defined as the amount of energy required in Btu to generate 1 kWh of electricity. Since 1 kWh = 3,412 Btu, without taking into consideration the losses associated with the shaft power, the thermal efficiency is given as:

$$\eta_{th} = \frac{3412 \left(\dfrac{Btu}{kWh} \right)}{Heat\ rate \left(\dfrac{Btu}{kWh} \right)}$$

The actual power cycle deviates slightly from the ideal Rankine cycle, due to friction and heat losses and the pressure drop associated with the processes. This deviation from the ideal Rankine cycle is represented in the T-S diagram in Figure 20.9 with dotted lines. To compensate for these heat losses, additional energy is consumed in the boiler to maintain the temperature of the steam, thus reducing the energy efficiency of the process. Three major steps are taken to increase the efficiency of the actual Rankine cycle: lowering the condenser pressure, superheating the steam, and increasing the boiler pressure.

Example 20.5: A combustion turbine has a thermal efficiency of 35% and a generator has an efficiency of 95%. What is the heat rate?

Solution: Given: $\eta_{turbine} = 35\%$ and $\eta_{generator} = 95\%$

Thermal efficiency of the system $\eta_{th} = \eta_{turbine} \times \eta_{gen} = 0.35 \times 0.95 = 0.3325$

$$Heat\ rate = \frac{3412 \left(\dfrac{Btu}{kWh} \right)}{\eta_{th}}$$

$$Heat\ rate = \frac{3412}{0.3325} = \mathbf{10261\ Btu/kWh}$$

20.5.3 The Air-Standard Brayton Cycle

Airplane engines are based on gas turbines. In such engines, combustion flue gases expand on the turbine blades, producing the thrust required for an airplane to fly. The *Brayton cycle* is often used as a close representation of a gas turbine. Unlike the steam turbine, it does not require a boiler to produce steam, because turbines use exhaust gases from the combustion chamber as working fluid. In this cycle, air enters the compressor, where it is compressed and fed into the combustion chamber. Combustion takes place in the chamber, producing exhaust gases that are expanded in the turbine to generate power. The air-standard Brayton cycle is the simplest model of a gas turbine, and its schematic and T-S diagram are presented in Figure 20.10.

To calculate the efficiency of the air-standard Brayton cycle, it is convenient first to calculate the energy supplied and energy removed in the cycle. The energy supplied to the turbine (q_{in}) can be calculated as the difference between the enthalpy of gas exiting (h_3) and entering (h_2) the combustion chamber. Similarly, we can define the energy removed as the difference between the enthalpy of gas exiting (h_4) and entering (h_3) the heat exchanger. Enthalpy of gas can also be written as a function of the specific capacity (c_p) of the heat and temperature of the gas. Using these relations we get:

$$q_{in} = h_3 - h_2 = c_p(T_3 - T_2) \tag{20.10}$$

$$q_{out} = h_4 - h_1 = c_p(T_4 - T_1) \tag{20.11}$$

Thus, thermal efficiency can be defined as a ratio of net work done (W_{net}) to the total energy supplied, as illustrated by Equations 20.12 and 20.13:

$$\eta_{th} = \frac{W_{net}}{q_{in}} = \frac{(q_{in} - q_{out})}{q_{in}} \tag{20.12}$$

$$\eta_{th} = 1 - \frac{q_{out}}{q_{in}} = 1 - \frac{(T_4 - T_1)}{(T_3 - T_2)} \tag{20.13}$$

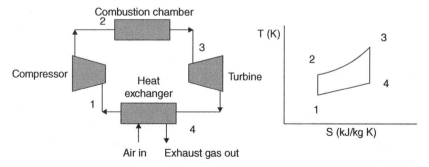

FIG. 20.10   Schematic of air-standard Brayton cycle (left) and T-S diagram (right).

Example 20.6: The turbine inlet and outlet temperatures for a gas turbine are 1,000 K and 700 K, respectively. The specific heat C_p = 1.148 kJ/(kg·K). Calculate the work extracted by the turbine.

Solution: Given: T_{in} = 1,000 K, T_{out} = 700 K, C_p = 1.148 kJ/(kg·K).

$$W_{turbine} = C_p(T_{in} - T_{out}) = 1.148\,(1,000 - 700) = \textbf{344.4 kJ/kg}$$

20.5.4 Combined Gas Turbine and Steam Turbine Power Cycles

When vapor and gas cycles are used to obtain power from fuel, the exit temperature of the gases in the gas cycle is very high. Regeneration is an excellent option to extract energy and to increase the efficiency of the gas cycle. The exhaust gases in the vapor cycle are used to produce steam, which is then expanded in another steam turbine. This process is called a combined gas and steam turbine power cycle, or combined gas–vapor power cycle. In a combined gas and steam turbine scheme, biomass can be converted into synthesis gas, which can then be used as natural gas in a gas turbine, and the hot exit gas can be used to generate steam and then used in a steam turbine. This significantly increases the efficiency of a power plant. Figure 20.11 shows the schematics of an integrated biomass gasification combined cycle (IGCC).

20.6 Biomass Co-firing with Coal

Coal and natural gas are currently the two major feedstocks employed for power generation in the USA. In 2012, 37.5% of the total electricity generated in the USA was obtained from coal (EIA, 2012). This was owed to a relatively cheap supply from the large reserves. With the advent of fracking, there has been substantial increase in use of natural gas (NG) for electricity production. In 2012, around 30.4% of total electricity generated in the USA came from natural gas. However, there are several environmental issues associated with burning fossil fuels like coal and NG. Coal contains high amounts of sulfur and nitrogen. Consequently, combustion of coal results in the emission of higher amounts of SO_X and NO_X, resulting in environmental problems such as acid rain and ozone depletion. NG, on the other hand, burns cleaner than coal and has lower SOx and NOx emissions. However, combustion of NG and coal also results in the release of greenhouse gases (GHGs) such as CO_2 and CH_4 (Mann and Spath, 2001). In order to curb SOx, NO_x, and GHG emissions from power plants, biomass can be co-fed with coal. In contrast, combustion of biomass feedstock results in lower emissions of NOx, SOx, and other GHGs. Therefore, biomass co-combustion with coal leads to a reduction in NO_x and SO_x emissions (Van Loo and Koppejan, 2008).

Co-firing is defined as the simultaneous burning of biomass and coal. It is cheaper to co-fire biomass with coal than to build stand-alone plants for biomass combustion. A number of types of biomass, such as sawdust, wood chips, straw, energy crops, and manure blend, can be co-fired with coal. Several power companies have been testing co-firing biomass with coal either using a separate biomass injector or co-milling with coal. The amount of biomass that can be used in co-milling is about 3–5% of total energy input, but can be increased up to 10% if a separate injector is used. There are some challenges associated with co-firing biomass with coal. First,

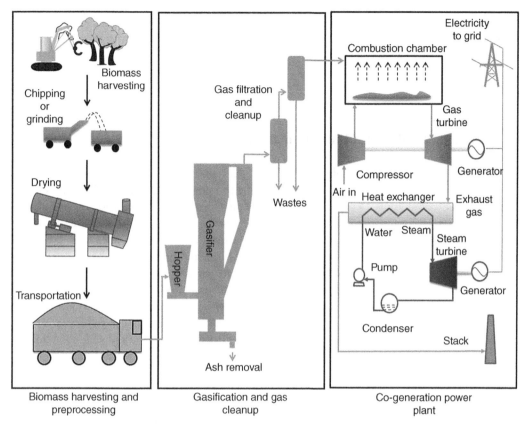

FIG. 20.11   Schematics of an integrated biomass gasification combined cycle (IGCC).

the feed-handling system in existing coal power plants must be adapted; an endeavor that is still expensive, particularly since maintaining a constant supply of biomass throughout the year is a major challenge in some parts of the world. Second, the impact of metals that are present in biomass on pollutant-control equipment is not well studied. Third, the composition of ash, a valuable byproduct, is likely to change with co-firing with biomass. Fly-ash, captured in power plants, is used as a raw material in building material/construction. Change in its composition can result in changing the composition of building material, thus requiring new regulations for the use of such building materials.

100 MWe biomass combustion power plant

Southern Company (Atlanta, Georgia) has built a 100 MW_e wood-fueled power plant, the largest such plant in the USA, near Sacul in Nacogdoches County, Texas. The plant began operation in June 2012 and currently runs entirely on woody biomass. This facility features state-of-the-art condensing steam turbine generators and a sophisticated fuel-handling system. Feedstock is obtained from recycled wood waste from local timber operations, manufacturing facilities, and municipalities. It is estimated that 1 million metric tons of biomass is consumed annually using a bubbling fluidized-bed boiler in this facility. The plant covers 165 acres (667,000 m^2) and biomass feedstock is received from a 75-mile (120 km) radius.

20.7 Environmental Impact and Emissions of Biomass Combustion

The emissions from all combustion devices consist of gases, particulates, and the residual ash. These can be broadly classified as first, species produced due to poor mixing – smoke, PAHs, CO, and volatile organic compounds (VOC); second, pollutants formed in the reaction – NO, NO$_2$, N$_2$O, SO$_x$, H$_2$S, HCl, etc.; and third, stable species emitted – potassium salts and other inorganic aerosols. Table 20.1 summarizes typical emissions for various combustion applications. Pollutants are formed mainly from the combustion reactions of N, S, Cl, and K as well as other trace elements contained in the volatiles and char. Carbon monoxide, PAHs, and soot, together with characteristic smoke compounds of biomass combustion such as levoglucosan, are released due to poor air–fuel mixing, low temperature, and shorter residence time during the combustion. Thus, the atmospheric emissions can contain tar aerosols and soot along with fine char particles and metal-based aerosols such as KCl. The nitrogen compounds are partially released with the volatiles, while some form a C–N matrix in the char, which is then released during the char combustion stage, forming NOx, HCN, and HNCO. Sulfur is released as SO$_2$ during both volatile and char combustion. KCl, KOH, and other metal-containing compounds together with the sulfur compounds form a range of gas phase species, which can be released as aerosols, but also deposited in combustion chambers. The CO produced due to incomplete combustion is the biggest concern for the flue gas, as it is detrimental to both health and environment. This also results in poor combustion efficiency.

Combustion plants use both primary and secondary measures to reduce emissions. Primary emission-reduction measures consist of modifying the combustion process to avoid or limit the formation of harmful emissions, whereas secondary measures deal with the abatement of such emissions after combustion. As such, primary measures are employed in the combustion chamber, while secondary measures are applied downstream. In practice, several strategies are employed to reduce emissions. Table 20.2 shows common primary and secondary emission-control measures.

Table 20.1 Typical emissions for wood combustion applications

Combustion applications	NOx (mg/MJ)	Particulates (mg/MJ)	Tar (mg/MJ)	CO (mg/MJ)	Unburned hydrocarbons (mg/MJ)	VOC (mg/MJ)	PAH (mg/MJ)
Fluidized-bed boilers	170	2	N/A	0	1	N/A	4
Grate furnaces	111	122	N/A	1,846	67	N/A	4,040
Wood boilers	101	N/A	499	4,975	1,330	N/A	30
Modern wood stoves	58	98	66	1,730	200	N/A	26
Traditional wood stoves	29	1,921	1,842	6,956	1,750	671	3,445
Fireplaces	N/A	6,053	4,211	6,716	N/A	520	105

N/A: Not available
Source: Adapted from Van Loo and Koppejan, 2008.

Table 20.2 Primary and secondary measures for biomass combustion emission control

Measure	Approach	Goal	Impact on emission
Primary			
Change in fuel composition	Washing and/or leaching	Reduction of fuel-bound S, N, K, Na, Cl	Reduction of SO_x, NO_x, HCl, KCl, and NaCl
Change in fuel moisture content	Reduction of moisture content	Reduction of incomplete combustion	Reduction of CO
Lime or limestone injection	Injection of CaO or $CaCO_3$ in chamber	SO_2 capture	Reduction of SO_2
Secondary			
Particle separation	Cyclone	Particle and fly-ash removal from flue gas	Reduction of particulate matter
	Bag filter		
	Electrostatic precipitator (ESP)		Reduction of SO_2, NO_2, and HCl (scrubber)
	Scrubber		Reduction of tars
	Settling chamber		(scrubber and ESP)
Selective catalytic reduction of NO_x	Catalytic reactor with platinum, titanium, and vanadium oxide catalysts	Conversion of NO_x to N_2 and H_2O in the presence of ammonia or urea	Significant reduction of NOx in exhaust gas
Selective non-catalytic reduction	High-temperature reduction of NOx	Conversion of NO_x to N_2 and H_2O in the presence of ammonia or urea	60–90% reduction of NOx
Wet limestone scrubbing	Flue-gas scrubbing in water and finely ground limestone	SO_2 absorption in slurry and reaction with limestone ($CaCO_3$) to produce $CaSO_3$ and CO_2	SO_2, NO_2, and HCl reduction

References

Annamalai, K., and I.K. Puri (2007). *Combustion Science and Engineering.* Boca Raton, FL: CRC Press.

Basu, P. (2006). *Combustion and Gasification in Fluidized Beds.* Boca Raton, FL: CRC Press.

Basu, P. (2010). *Biomass Gasification and Pyrolysis: Practical Design and Theory.* Burlington, MA: Academic Press.

Brown, R.C. (ed.) (2011). *Thermochemical Processing of Biomass: Conversion into Fuels, Chemicals and Power.* Chichester: John Wiley & Sons Ltd.

EIA (2012). *U.S. Energy in Brief.* Washington, DC: U.S. Energy Information Administration, http://www.eia.gov/energy_in_brief/article/role_coal_us.cfm, accessed November 2013.

Mann, M., and Spath, P. (2001). A life assessment of biomass co-firing in a coal fired power plant. *Clean Products and Processes* 3(2): 81–91.

Miller, B.G. (2011). *Clean Coal Engineering Technology.* Burlington, MA: Butterworth- Heinemann.

Tillman, D.A., Rossi, A.J., and Kitto, W.D. (1981). *Wood Combustion: Principles, Process and Economics.* New York: Academic Press.

Van Loo, S., and Koppejan, J. (2008). *The Handbook of Biomass Combustion and Co-firing.* London: Earthscan.

Exercise Problems

20.1. Discuss how the biorefinery concept applies to the combustion of lignocellulosic biomass.

20.2. Highlight and summarize the major differences between the two types of combustors.

20.3. Propane (C_3H_8) is combusted in air. Write a balanced equation for complete combustion with 30% excess air and determine the volumetric concentration of products.

20.4. The elemental analysis of a switchgrass sample is as follows (% dry wt. basis): C = 48.26%, H = 5.98%, O = 42.5%, S = 0.10%, N = 0.62%, and ash = 2.71%. Determine coefficients a, b, c, and d, and balance the general biomass combustion equation. Also, find the air flow rate required to achieve an air/fuel equivalence ratio (λ) of 1.50 if the biomass feeding rate is 60 kg/min.

20.5. Pine, pretreated via torrefaction, has the following composition (dry wt. basis): C = 56.70%, H = 5.49%, O = 36.28%, S = 0.09%, N = 0.74%, and ash = 0.98%. Determine coefficients a, b, c, and d, and balance the general biomass combustion equation. Also, find the air flow rate required to achieve an air/fuel equivalence ratio (λ) of 1.3 if the biomass feeding rate is 1 kg/s.

20.6. 1 kg of torrefied biomass (C = 56.70%, H = 5.49%, O = 36.28%, S = 0.09%, N = 0.74%, and ash = 0.98% [dry wt. basis]) is burned with 30% excess air. Determine the amount and volumetric concentration of carbon dioxide, water, sulfur, nitrogen, and oxygen in the products. Assume that all the S and N in the fuel are converted to SO_2 and NO_2.

20.7. Estimate the fuel-to-air ratio of fuel gas with an energy content of 5.0 MJ/kg if the combustor and compressor exit temperatures are 1,500 K and 700 K, respectively. The specific heat of air is 1.004 kJ/(kg·K).

20.8. A Carnot engine receives 1,000 kJ of heat from a heater at 900 °C. It rejects heat to a heat sink at 25 °C. Calculate (a) the Carnot cycle efficiency; and (b) the amount of heat rejected at the lower temperature.

20.9. A gas turbine's inlet and outlet temperatures are 1,200 K and 800 K, respectively. If the specific heat is C_p = 1.148 kJ/(kg·K), calculate the work extracted by the turbine.

20.10. A combustion turbine utilizing natural gas (46 MJ/kg) as a fuel has a combustor exit temperature of 1,400 K and a compressor exit temperature of 600 K. Estimate the fuel-to-air ratio. What would the fuel-to-air ratio be if syngas with 5.5 MJ/kg is being used? The specific heat of air is 1.004 kJ/(kg·K).

CHAPTER 1

Gasification

Sushil Adhikari and Nourredine Abdoulmoumine

What is included in this chapter?

This chapter covers the gasification of various lignocellulosic biomass feedstocks for synthesis gas production. The fundamentals of gasification, types of gasifier, biomass preprocessing, mass and energy balances, and gas cleanup are discussed, with relevant examples. Various applications of synthesis gas are also discussed.

21.1 Introduction

Gasification is a thermal conversion process of carbonaceous material at elevated temperatures (700–1,400 °C) in the presence of limited oxygen to produce a gaseous product known as *producer gas* (CO, H_2, CH_4, CO_2, H_2O) and a solid residue or *char*. Much like combustion, gasification has been around for more than 200 years (Breault, 2010). One of the first practical applications of gasification was demonstrated by William Murdoch in 1792 when he used gasification-derived town gas to light his house and office in Redruth, England. Town gas was subsequently used to light London, Boston, Baltimore, and Berlin, among other cities in Europe and North America. Later, in the 20th century, industrial processes employed gasification as a source of synthesis gas (or syngas in brief), which is primarily a mixture of CO and H_2. One such noteworthy process is the Fischer–Tropsch (FT) process, developed in the 1920s by Franz Fischer and Hans Tropsch. This demonstrated a method to combine carbon monoxide and hydrogen to produce hydrocarbon fuel such as gasoline and diesel. Today, gasification is predominantly used for power generation. There is, however, a renewed interest in chemicals and fuel production from renewable feedstocks as an alternative to fossil fuels.

Bioenergy: Principles and Applications, First Edition. Edited by Yebo Li and Samir Kumar Khanal.
© 2017 John Wiley & Sons, Inc. Published 2017 by John Wiley & Sons, Inc.
Companion website: www.wiley.com/go/Li/Bioenergy

Two German scientists, Franz Fischer and Hans Tropsch, developed a process to produce fuel and chemical from syngas (CO and H_2). The Fischer–Tropsch process, named after its inventors, was used during and after World War II at an industrial scale.

21.2 Fundamentals of Gasification

One important distinguishing characteristic in thermochemical conversion processes (pyrolysis, combustion, and gasification) is the amount of oxygen supplied (Figure 21.1). For example, supplying excess air results in combustion (Chapter 20), while conversion in the absence of air or oxygen results in pyrolysis (Chapter 22).

21.2.1 Gasifying Agents

In gasification, a limited amount of oxygen is supplied to facilitate the conversion of feedstocks into producer gas. Oxygen is supplied as a *gasifying or oxidizing agent*, which can be pure oxygen, air, steam, or mixture of oxygen and steam or air and steam. When using air as a gasifying agent, inert nitrogen (N_2) is present in the producer gas.

In Chapter 20, we introduced the term equivalence ratio, a ratio of the actual amount of air supplied during the conversion process to that required for complete combustion. The same concept is also applied in gasification and is commonly known as ER (equivalence ratio), instead of λ as discussed in Chapter 20. In gasification, ER varies between 0 and 1, where 0 indicates a pyrolysis mode and 1 or higher a combustion mode. In practice it varies between 0.20 and 0.35 for biomass gasification. When steam is used as a gasifying agent, the steam to biomass (S/B) or steam to carbon (S/C) ratios are used to express the quantity of steam supplied relative to the biomass or carbon fed in:

$$S/B = \frac{\dot{m}_s}{\dot{m}_b} \tag{21.1}$$

$$S/C = \frac{\dot{m}_s}{\dot{m}_c} \tag{21.2}$$

where $\dot{m}_s$, $\dot{m}_b$, and $\dot{m}_c$ are the mass flow rates (dry basis) of steam, biomass, and carbon, respectively.

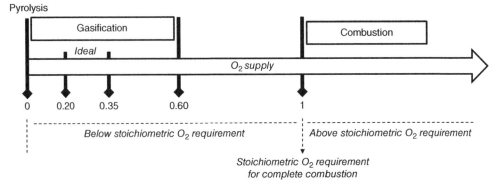

FIG. 21.1 Oxygen supply relative to stoichiometric requirement for complete combustion in thermochemical conversion.

Example 21.1: A pine biomass sample is gasified with a mixture of air and steam. The mass flow rates of dry air and steam into the gasifier are 1.54 kg/s and 1.1 kg/s, respectively. If the feeding rate of pine is 1 kg/s (on a dry basis), the $(A/B)_{stoichiometric}$ (see Chapter 20) is 6.17 kg dry air/kg dry biomass, and the carbon content of pine is 50.89% (dry basis), determine ER and S/C.

Solution:

1. **Determine ER**

$$(A/B)_{actual} = \frac{1.54 \text{ kg dry air/s}}{1 \text{kg dry biomass/s}} = 1.54 \text{ kg dry air/kg dry biomass}$$

$$ER = \frac{(A/B)_{actual}}{(A/B)_{stoichiometric}} = \frac{1.54 \text{ kg dry air/kg dry biomass}}{6.17 \text{ kg dry air/kg dry biomass}} = \mathbf{0.25}$$

2. **Determine S/C**

$$\frac{S}{C} = \frac{\dot{m}_s}{\dot{m}_c} = \frac{1.1 \text{ kg/s}}{0.5089 \times 1 \text{kg dry biomass/s}} = \mathbf{2.16}$$

21.2.2 Gasification Reactions

Biomass gasification is a highly complex chemical process and is divided into four major steps, as depicted in Figure 21.2.

When biomass is first introduced into the reactor, it undergoes rapid heating, causing evaporation of water and drying of biomass (**Step I**). This process occurs below 150 °C. It is followed by a

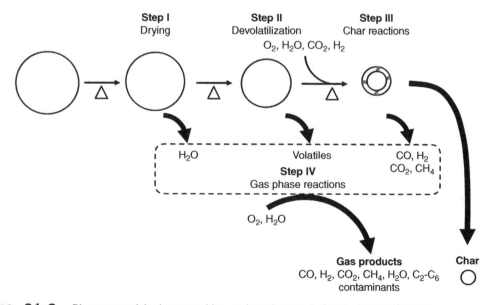

FIG. 21.2   Biomass particle decomposition and product evolution during gasification.

devolatilization or pyrolysis process (**Step II**), which leads to the removal of volatile compounds between 200 °C and 400 °C, resulting in a rapid weight loss and an increase in porosity in the biomass particles. After devolatilization, the carbon-rich char undergoes various reactions with reactive gases such O_2, H_2O, CO_2, and H_2, either from the gasifying agents or from the products of all major gasification steps. This is the char reaction stage (**Step III**) and is governed by the following reactions:

$$C + O_2 \quad \rightarrow \quad CO_2 \tag{21.3}$$

$$C + 0.5O_2 \quad \rightarrow \quad CO \tag{21.4}$$

$$C + H_2O \quad \leftrightarrow \quad CO + H_2 \quad \text{(Water–gas reaction)} \tag{21.5}$$

$$C + CO_2 \quad \leftrightarrow \quad 2CO \quad \text{(Boudouard reaction)} \tag{21.6}$$

$$C + 2H_2 \quad \leftrightarrow \quad CH_4 \tag{21.7}$$

In **Step IV**, the gas phase CO, H_2, CO_2, CH_4, and H_2O can react with each other as well as with the gasifying agents, as depicted by Equations 21.8–21.11.

These four steps occur simultaneously during gasification to produce the final products in the producer gas: CO, H_2, CO_2, CH_4, H_2O, N_2 (when air is used), C_2–C_6 hydrocarbons, and a small amount of undesirable byproducts, known as contaminants (see Section 21.6). The final proportions of constituents in producer gas depend on the relative amount of gasifying agent present (ER and S/C), reaction temperature and pressure, gasifier design, and feedstock types, among others.

$$CO + 0.5O_2 \quad \rightarrow \quad CO_2 \tag{21.8}$$

$$H_2 + 0.5O_2 \quad \rightarrow \quad H_2O \tag{21.9}$$

$$CO + H_2O \quad \leftrightarrow \quad CO_2 + H_2 \quad \text{(Water–gas shift reaction)} \tag{21.10}$$

$$CO + 3H_2 \quad \leftrightarrow \quad CH_4 + H_2O \quad \text{(Methanation reaction)} \tag{21.11}$$

21.3 Gasifiers

A *gasifier* is a reactor where gasification is carried out. There are several types of gasifier, classified as fixed-bed, moving-bed, fluidized-bed, and entrained-flow depending on the reactor design and the feedstock and gas-flow patterns. In Table 21.1, the product gas composition of woody biomass using air as a gasifying agent is presented for various gasifiers.

21.3.1 Moving-Bed Gasifiers

Moving-bed gasifiers are further divided into co-current (downdraft) and counter-current (updraft) gasifiers, as illustrated in Figure 21.3. In co-current or downdraft gasifiers, the feedstock and syngas flow **co-currently** from top to bottom. In counter-current gasifiers, feedstock and syngas flow **counter-currently** as feedstock is fed from the top and syngas is collected from the top. Moving-bed gasifiers are suitable for small to medium-scale applications (10 kW–10 MW; Basu, 2010).

Table 21.1 Gas product composition of various gasifiers with air or oxygen as the gasifying agent

Gasifier types	Updraft	Downdraft	Bubbling fluidized-bed	Circulating fluidized-bed	Entrained-flow
Feedstock types	Pine sawdust	Wood chips	Wood pellet	Sawdust	Sawdust
ER	0.10–0.50	0.32–0.37	0.19–0.32	0.24–0.45	0.29–0.50
Temperature, °C	750–950	800–1000	750–800	728–805	1,000–1,400
Gas composition (% by vol)					
CO	8–32	15	12–16	10–20	40–50
H_2	7–18	16	14–17	5–7	33–48
CO_2	10–22	15.3	16–17	14–18	10–20
CH_4	4–6	2	4–5	1–4	0–9
N_2 and others	36–65	50	46–52	55–65	–

Source: Data from Liu *et al.*, 2012; Son *et al.*, 2011; Kim *et al.*, 2013; Li *et al.*, 2004.

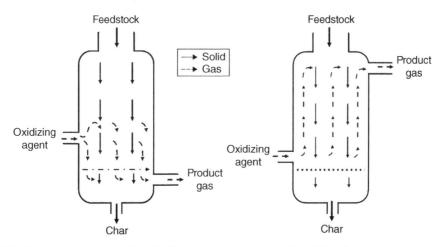

FIG. 21.3   Moving-bed gasifiers: (left) co-current or downdraft and (right) counter-current or updraft.

The product gas flow direction has significant implications on product gas quality. For example, downdraft yields "cleaner" producer gas because it passes through the hottest zone in the reactor, resulting in further thermal cracking. Downdraft gasifiers are preferred for small-scale power generation due to the low tar content in the producer gas. The problem with moving-bed gasifiers is their inability to maintain uniform radial temperatures, which results in local slag, bridging, and clinkering problems. Lack of a uniform radial temperature is one of the reasons why this kind of gasifier is difficult to scale up, rendering it of limited application.

21.3.2 Fluidized-Bed Gasifiers

Fluidized-bed gasifiers provide a higher throughput than fixed-bed gasifiers. Fluidization enhances mass and heat transfers, and can be operated at a higher pressure, making it an ideal choice for biomass gasification. In fluidized-bed gasifiers, feedstock is fluidized with bed material such as sand, silica, or sometimes catalysts along with a gasifying agent.

Fluidized-bed gasifiers are further classified as bubbling fluidized-bed and circulating fluidized-bed (Figure 21.4). Bubbling fluidized-bed gasifiers (BFBGs) are simpler in design and thus preferred in small to medium-scale applications (a few kW to 25 MW), whereas circulating fluidized-bed gasifiers (CFBGs) are emerging as the choice for medium to large-scale applications (25–60 MW). Examples of commercial pilot-scale biomass gasifiers are shown in Figure 21.5.

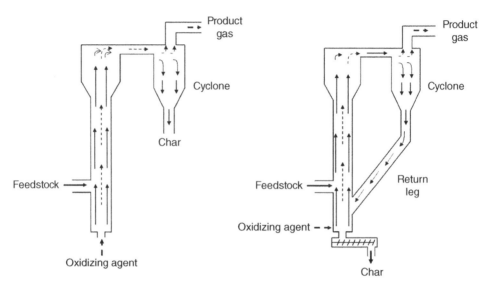

FIG. 21.4   Bubbling fluidized-bed (left) and circulating fluidized-bed (right) gasifiers.

FIG. 21.5   Examples of commercial pilot-scale biomass gasifiers installed at Auburn University: 28 kg/hr downdraft built by Community Power Corporation (left) and 45 kg/hr pressurized bubbling fluidized-bed built by Gas Technology Institute (right). Source: Photo courtesy of Sushil Adhikari, Auburn University.

21.3.3 Entrained-Flow Gasifiers

Entrained-flow gasifiers are commonly divided into top-fed and side-fed, as shown Figure 21.6. In top-fed entrained-flow gasifiers, biomass and oxidizing agent are introduced from the top of the reactor and solid and gas products flow co-currently downward. In side-fed entrained-flow gasifiers, biomass and oxidizing agent are fed in at the bottom, from the side, and product gas is removed from the top. Entrained-flow gasifiers are generally employed in very large applications (over 50 MW) due to their high throughput (Basu, 2010). Several designs of entrained-flow gasifiers have been in use over the years and several companies, including Shell, General Electric, and ConocoPhillips, have their own proprietary designs (Figure 21.7). This type of gasifier is primarily used for coal gasification and operates at fairly high temperatures (1,350–1,650 °C), which results in relatively clean product gas with high conversion efficiency of carbon. Ash from this type of gasifier is typically removed in a molten stage and is known as "slag."

Factors affecting gas composition

In Table 21.1, the product gas composition of woody biomass using air as an oxidizing agent is presented for various gasifiers. In addition to the type of gasifier, several factors, such as biomass, oxidizing agents, reactor hydrodynamics, temperature, and pressure, affect biomass gasification. These factors often have synergistic interactions that affect product gas composition. There is wide variability among biomass feedstocks and gasifier designs at bench, pilot, and industrial scales. Consequently, isolating the effect of the main factors from their interaction in gasification is a very complex task that requires computational modeling. While there are many computational modeling tools, Aspen and ANSYS Fluent are commonly used in gasification modeling.

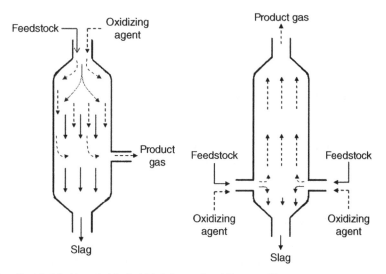

FIG. 21.6   Top-fed (left) and side-fed (right) entrained-flow gasifiers.

(a)   (b)   (c)

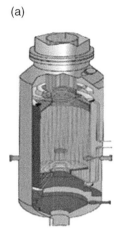   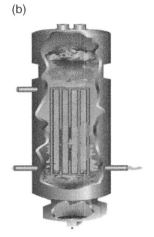   

Shell Coal Gasification Process (SCGP)

The SCGP is built around a side-fed entrained-flow gasifier that uses a dry, pulverized coal feed. SCGP operates at high temperature (1,500–1,600 °C) and pressure (24–40 bar). Steam and oxygen-enriched air (>95% O_2) are used as oxidizing agents. The SCGP is well established commercially, with over 50 Shell gasifiers in operation. worldwide.

GE gasifier

The GE process revolves around a top-fed entrained-flow gasifier using coal slurry (~65% water on a wt. basis) and O_2-enriched air. It is operated at high temperature (1,250–1,450 °C) and pressure (up to 80 bar). This technology has achieved great commercial success, with 65 existing plants worldwide, including a 250 MW plant in Florida.

Conoco Phillips E-Gas gasifier

The E-Gas gasifier is the heart of a gasification technology owned by CB&I. It is a side-fed entrained-flow gasifier and uses coal slurry and oxygen-enriched air. It can be operated at high temperature (1,425 °C) and pressure (28 bar). A notable commercial success of this system is the 260 MW Wabash River IGCC Plant near West Terre Haute, IN.

FIG. 21.7 Commercial examples: (a) Shell, (b) GE, and (c) E-Gas gasifiers. Source: Bell *et al.*, 2010; photo source: U.S. Department of Energy – Gasifipedia (www.netl.doe.gov). Public domain.

21.4 Feedstock Preparation and Characterization

Feedstocks, especially biomass, come with different physical and chemical properties, and different sizes, which affect the gasification efficacy. In most cases, minimal feedstock preparation is required depending on the type of gasifier and the applications of product gas. Biomass feedstock used for gasification needs to be characterized for designing the process and conducting energy and mass balance.

21.4.1 Feedstock Preparation

The most common feedstock preparation steps prior to gasification are size reduction and drying. Size reduction is generally required to enhance the reactivity of the feedstocks, especially biomass, and in some cases to ensure proper feeding. Typically, particle size is determined by the type of gasifier and the biomass feeding system. Although smaller particles have a higher reactivity, the benefits should outweigh the cost of grinding biomass to finer particles. Of all gasifiers, fluidized-bed gasifiers have the most stringent size requirement. Higher moisture content in biomass is undesirable as it reduces the efficiency of the gasifier. Every kg of moisture requires a minimum of 2,260 kJ (2,142 Btu) of energy to remove the moisture. Since fresh biomass has a high moisture content of around 50% (on a wet wt. basis), it is often necessary to dry the feedstocks prior to gasification. Gasifiers have different moisture content tolerance: fluidized-bed gasifiers require much drier feedstock (~10% on a wet wt. basis), whereas counter-current fixed-bed gasifiers can handle higher moisture content (up to 50% on a wet wt. basis). Other preparation steps such as densification and torrefaction are also performed prior to gasification to obtain more uniform feedstock.

21.4.2 Feedstock Characterization

Feedstock properties are important factors that affect gasification. *Proximate analysis* provides information on moisture and ash content, volatile matters, and fixed carbon. A high moisture content also leads to higher biological activity during storage of feedstock, especially biomass, as well as loss in gasification efficiency. The volatile matter content, which is the measure of the organic content of biomass, is important because it provides an indication of biomass reactivity during gasification and is related to the amount of tar formation. Tar formation is a serious challenge in biomass gasification. Most biomass contains alkali and alkaline earth metals that remain as inorganic solid residues in char and as metal vapors in producer gas. High biomass ash content, which is an indication of high inorganic content, may result in equipment damage due to corrosion and is likely to adversely affect gasification by ash melting in the gasifier and catalyst deactivation in downstream processes. Table 21.2 shows the properties of different types of biomass feedstocks.

The composition in terms of the basic chemical elements is equally important to perform basic gasification calculations. The elemental composition, determined by *ultimate (elemental) analysis*, provides the weight percentage of C, H, N, S, O, and other minor elements such as Cl. Besides the proximate and ultimate analyses, it is equally important to know the energy content (heating value) of the biomass for the energy balance calculation. A more detailed discussion on the heating value of common fuels/energy resources is presented in Chapter 2. In the following sections, information obtained from these analyses will be employed to perform mass and energy balances.

Table 21.2 Proximate, ultimate, and heating value of selected biomass feedstocks

	Switchgrass	Hybrid poplar	Pine chips	Sugar cane bagasse	Corn stover
Proximate analysis (% dry wt. basis)					
Fixed carbon	14.34	12.49	18.01	11.95	11.40
Volatile matter	76.69	84.81	81.71	85.61	71.80
Ash	8.97	2.70	0.28	2.44	16.70
Ultimate analysis (% dry wt. basis)					
Carbon	46.68	50.18	49.33	48.64	46.23
Hydrogen	5.82	6.06	5.03	5.87	5.34
Nitrogen	0.77	0.60	0.53	0.16	0.85
Oxygen	37.38	40.43	44.70	42.82	35.85
Sulfur	0.19	0.02	0.13	0.04	0.06
Chlorine	0.19	0.01	0.003	0.03	0.27
HHV, MJ/kg	18.06	19.02	19.40	18.99	17.23

Source: Data from Jenkins *et al.*, 1998; Gautam *et al.*, 2011; Carpenter *et al.*, 2010.

21.5 Gasification Mass and Energy Balance

21.5.1 Mass Balance

For brevity, the discussion only focuses on carbon balance, which is sufficient to demonstrate a good mass balance, since carbon is the primary constituent of biomass. The general steps to determine carbon conversion to producer gas are outlined in Example 21.2.

> **Example 21.2:** For the biomass in Example 21.1, perform a carbon balance if the exit syngas has the following composition (mole percent on dry basis): $y_{CO} = 25\%$, $y_{CO2} = 8\%$, $y_{CH4} = 5\%$, $y_{H2} = 16\%$, and $y_{N2} = 46\%$.

Solution:

1. **Determine the carbon output**

   Assuming nitrogen as an inert gas, and air contains 77% of N_2 by mass%.

   Mass flow rate of N_2 into gasifier $= 0.77 \times 1.54 \, kg/s = 1.186 \, kg \, N_2/s$

   Molar flow rate of N_2 into gasifier $= 1.186 \, kg \, N_2/s/28 \, kg/kmol = 0.042 \, kmol \, N_2/s$

   Nitrogen balance

   Molar flow rate of N_2 in = Molar flow rate of N_2 out

   $$= y_{N2} \times \dot{n}_{producer \, gas}$$

   $$\dot{n}_{producer \, gas} = 0.042 \, kmol \, N_2/s \, / \, 0.46 = 0.09 \, kmol \, gas/s$$

   Carbon balance

   $$Carbon_{out} = \dot{n}_{syngas} \times \left(y_{CO} + y_{CO2} + y_{CH4}\right)$$

   $$= 0.09 \, kmol/s(0.25 + 0.08 + 0.05) = 0.034 \, kmol \, C/s$$

   $$Carbon_{in} = [(50.89/100) \times 1 kg/s] \, / \, 12 \, kg/kmol = 0.042 \, kmol \, C/s$$

2. **Determine carbon conversion to producer gas**

   $$Carbon \, conversion = 100 \, x \, [Carbon_{out}/Carbon_{in}] = 100 \times [0.034 \, / \, 0.042] = 81\%$$

In practice, it is often desirable to maximize carbon conversion to CO, CO_2, and CH_4 and minimize carbon loss in char and other gas or liquid phase compounds. While Example 21.2 focuses on carbon conversion to CO, CO_2, and CH_4, other carbon-containing compounds can also be included in the carbon conversion calculation if their concentrations are known. The same procedure can also be applied to other elements. An overall carbon balance includes carbon in all product streams including producer gas, char, and any liquid condensate that might contain carbonaceous compounds.

21.5.2 Energy Balance

In addition to the mass balance, it is equally important to conduct an energy balance to assess process efficiencies. Three efficiencies, namely cold gas efficiency, hot gas efficiency, and net efficiency, are commonly determined through the energy balance. The cold gas efficiency (η_{cold}), evaluates the gasification performance after the product gas has been cooled to ambient temperature. Cold gas efficiency represents the fraction of the biomass energy in the product gas and is expressed as:

$$\eta_{cold} = 100 \times \left[\frac{\dot{V}_g Q_g}{\dot{m}_b Q_b} \right] \tag{21.12}$$

where $\dot{V}_g$ is the volumetric flow rate (m³/sec) of product gas, Q_g the volumetric energy content (MJ/m³) of product gas, $\dot{m}_b$ the mass flow rate (kg/sec) of biomass, and Q_b the biomass energy content (MJ/kg). The product gas volumetric energy content (Q_g) can be calculated using the individual constituents' heat of combustion, as presented in Table A21.1 in the Appendix to this chapter. It is important to be consistent while using heating values in the numerator and denominator.

$$Q_g = \sum y_i \Delta H^0_{c_i} \tag{21.13}$$

In Equation 21.13, y_i is the volume fraction of individual gas constituents and $\Delta H^0_{c_i}$ is the heat of combustion at 25 °C and 1 atm pressure.

Example 21.3: What is the cold gas efficiency of the biomass discussed in Example 21.2 if the higher heating value of biomass is 18.50 MJ/kg and the syngas flow rate is 1.71 m³/s?

Solution:

1. **Find the volumetric energy content of the syngas**
   According to Table A21.1 in the Appendix to this chapter, only CO, CH₄, and H₂ have heat of combustion.

$$Q_g = \left(\frac{0.25\,m^3 CO}{m^3 syngas} \times \frac{12.63\,MJ}{m^3 CO} \right) + \left(\frac{0.05\,m^3 CH4}{m^3 syngas} \times \frac{39.82\,MJ}{m^3 CH4} \right)$$
$$+ \left(\frac{0.16\,m^3 H_2}{m^3 syngas} \times \frac{12.74\,MJ}{m^3 H2} \right) = 7.19\,MJ/m^3$$

2. **Determine the cold gas efficiency of the process**

$$\eta_{cold} = 100 \times \frac{1.71\,m^3/\sec}{1\,kg/\sec} \times \frac{7.19\,MJ/m^3}{18.50\,MJ/kg} = 100 \times \frac{12.27\,MJ/\sec}{18.50\,MJ/\sec} = 66.5\%$$

In some cases, it is more appropriate to evaluate gasifier performance based on the hot gas efficiency, which accounts for the sensible heat in the product gas at the exit temperature.

$$\eta_{hot} = 100 \times \left[\frac{\dot{V}_g Q_g + H_{sensible}}{\dot{m}_b Q_b}\right] \qquad (21.14)$$

$$H_{sensible} = \dot{n} \sum y_i C_{p_i} \left(T_{gas} - T_R\right) \qquad (21.15)$$

In Equation 21.15, $\dot{n}$ (kmol gas/s) is the producer gas molar flow rate, C_{p_i} (kJ/kmol.K) is the individual specific heat capacity of the producer gas constituents, T_{gas} (K) is the temperature of gas products exiting the gasifier, and T_R (K) is the reference temperature, usually taken as 298.15 K (25 °C).

Example 21.4: In Example 21.3, product gas is sent directly to a furnace for burning. If the exit temperature of the gasifier is 850 °C and the ambient air temperature is 25 °C, what is the hot gas efficiency of the process at the gasifier exit temperature?

Solution:

1. Find the heat capacity for each constituent

$T_{gas} = 850\,°C + 273.15 = 1,123.15\,K \quad T_R = 25\,°C + 273.15 = 298.15\,K$

Using the equations in Table A21.2 (Appendix), we can find Cp for each constituent.

$Cp_{co} = 27.62 + 0.005(1,123.15) = 33.24\,kJ/(kmol \cdot K)$

Using the same approach, $Cp_{CO2} = 56.73\,kJ/(kmol \cdot K)$, $Cp_{CH4} = 76.26\,kJ/(kmol \cdot K)$, $Cp_{H2} = 31.53\,kJ/(kmol \cdot K)$, and $Cp_{N2} = 31.93\,kJ/(kmol \cdot K)$

2. Find the enthalpy for each constituent

$$H_{co} = \frac{0.09\,kmol\;syngas}{s} \frac{0.25\,kmol\;CO}{Kmol\;gas} \frac{33.24\,kJ}{kmol\,CO\,K} \frac{(1123 - 298)K}{}$$

$H_{CO} = 617.01\,kJ/s$ and, using a similar approach, $H_{CO2} = 336.97\,kJ/s$,

$H_{CH4} = 283.11\,kJ/s$, $H_{H2} = 374.57\,kJ/s$, and $H_{N2} = 1,090.57\,kJ/s$

$H_{sensible} = 617.01\,kJ/s + 336.97\,kJ/s + 283.11\,kJ/s + 374.57\,kJ/s + 1,090.57\,kJ/s$

$\qquad = 2,702.23\,kJ/s = 2.70\,MJ/s$

$$\eta_{hot} = 100 \times \frac{12.27\,MJ/s + 2.70\,MJ/\,sec}{18.5\,MJ/\,sec} = 80.9\%$$

Finally, the net efficiency can be determined by considering the net energy input. The net input is the energy of biomass plus the heat in the oxidizing agent after accounting for the energy recovered through burning the tars and steam condensation.

Table 21.3 Problems of contaminants and approaches to their removal

Contaminants	Problems	Removal approach
Particulates	Emission Fouling Corrosion	Filtration with cyclones, candle and baghouse filters, or electrostatic precipitators
Tar	Fouling Catalyst deactivation Wastewater pollution	Wet scrubbing in water, organic solvents, or vegetable oil Reaction with mineral-based (e.g., dolomite) or nickel-based catalysts
N contaminants **(NH_3, HCN, etc.)**	NOx emissions Catalyst deactivation	Wet scrubbing in water or dilute acid solution Catalytic decomposition with Ru-based catalysts
S contaminants **(H_2S, COS, etc.)**	Corrosion Catalyst deactivation	Chemical adsorption in amine-based solution (e.g., MEA) Physical adsorption in solutions of ethylene glycol alkyl ethers (e.g., Selexol) or methanol (Rectisol) Chemical reaction with solid metal oxides (e.g., ZnO or CuO)
Hydrogen halides **(HCl, HF, etc.)**	Emission Fouling Corrosion	Wet scrubbing in adequate solution (e.g., water or caustic water) Absorption on solid materials (e.g., $CaCO_3$)
Trace metals **(Na, K, P, etc.)**	Emission Corrosion Catalyst deactivation	Cooling of gas to induce condensation Gas filtration through a bed (e.g., activated carbon bed)

$$\eta_{net} = 100 \times \left[\frac{\text{Net energy output from syngas}}{\text{Net energy input}} \right] \tag{21.16}$$

21.6 Gas Cleanup

Biomass-derived product gas contains a significant amount of undesirable components such as fine particulates, tars, sulfur- and nitrogen-containing compounds, hydrogen halides, and trace metals. The undesirable components are collectively known as *contaminants*. The process of removing contaminants from product gas is called *gas cleanup*. Table 21.3 summarizes the problems of contaminants and approaches to their removal.

21.7 Applications of Biomass Gasification

Biomass-derived syngas has numerous applications, as illustrated in Figure 21.8. However, for all these applications, the concentrations of contaminants in the syngas must adhere to strict limits in order to achieve the best performance. The product gas from gasification has two major applications: heat and power production; and fuels and chemicals synthesis.

Syngas can be used for heat and power generation through internal combustion engine, gas or steam turbines, combined with heat recovery for improved efficiency and revenue generation. When power generation is coupled with heat recovery, the process is referred to as *combined heat and power (CHP) or cogeneration*. Power production using gasification product gas is also feasible with fuel cells.

Furthermore, cleaned and conditioned syngas can be used in liquid transportation fuel (gasoline or diesel) synthesis through the well-known Fischer–Tropsch (FT) process. In this process, H_2 and CO react to form hydrocarbon of various chain lengths depending on the process conditions:

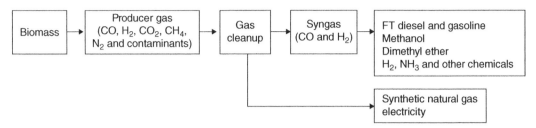

FIG. 21.8   Potential applications of gasification-derived producer gas after adequate cleanup. Source: Adapted from Huber *et al.*, 2006.

$$\left(\frac{n}{2} + m\right)H_2 + mCO \rightarrow C_mH_n + mH_2O \qquad (21.17)$$

Depending on the type of catalyst employed, the FT process can yield gasoline or diesel. In addition, methanol synthesis from syngas can also be achieved through CO or CO_2 hydrogenation over a copper-zinc oxide catalyst, according to Equations 21.18 and 21.19. Gasoline can then further be produced by reacting methanol over a zeolites catalyst.

$$2H_2 + CO \rightarrow CH_3OH \qquad (21.18)$$

$$3H_2 + CO_2 \rightarrow CH_3OH + H_2O \qquad (21.19)$$

Currently, no commercial plant produces liquid transportation fuel from biomass-derived syngas in the USA. However, there have been several pilot-scale and demonstration facilities built in the USA in recent years.

Biomass to fuel: Pilot-scale demonstration of an advanced biorefinery in the USA

In 2010, a group of thermochemical technology developers and feedstock suppliers were awarded a 4-year grant by the US Department of Energy to develop a pilot-scale demonstration of a biomass-to-gasoline biorefinery. The project aimed to investigate the entire production cycle, from biomass harvesting and supply to fuel distribution to end-user. The pilot-scale design is intended to serve as a basis for future commercial plant of "*drop-in*" green gasoline. Drop-in liquid fuels are fuels produced from renewable sources that can be used without major changes in existing infrastructures or engine designs. In order to address all aspects of production from biomass supply to drop-in gasoline distribution, four technology providers and a feedstock supplier entered into partnership: Haldor-Topsoe, GTI, Andritz Carbona, and Conoco Phillips provided expertise in process design, gas cleanup, catalytic conversion of gas to liquid fuel, fuel characterization, handling, and marketing; and UPM-Kymmene provided expertise in biomass logistics (harvesting, gathering, handling, and transporting of wood). In early 2013, the pilot successfully demonstrated the production of 20 barrels per day of gasoline from biomass.

References

Basu, P. (2010). *Biomass Gasification and Pyrolysis: Practical Design and Theory.* Cambridge, MA: Academic Press.

Bell, D.A., Towler, B.F., and Fan, M. (2010). *Coal Gasification and Its Applications.* Burlington, MA: Elsevier.

Breault, R.W. (2010). Gasification processes old and new: A basic review of the major technologies. *Energies* 3: 216–40.

Carpenter, D.L., Bain, R.L., Davis, R.E., *et al.* (2010). Pilot-scale gasification of corn stover, switchgrass, wheat straw, and wood: 1. Parametric study and comparison with literature. *Industrial and Engineering Chemistry Research* 49 (4): 1859–71.

Gautam, G., Brodbeck, C., Adhikari, S., Bhavnani, S., Fasina, O., and Taylor, S. (2011). Gasification of wood chips, agricultural residues, and waste in a commercial downdraft gasifier. *Transactions of the ASABE* 54(5): 1801–7.

Huber, G.W., Iborra, S., and Corma, A. (2006). Synthesis of transportation fuels from biomass: Chemistry, catalysts, and engineering. *Chemical Reviews* 106(9): 4044–98.

Jenkins, B.M., Baxter, L.L., Miles, T.R., Jr., and Miles, T.R. (1998). Combustion properties of biomass. *Fuel Processing Technology* 54(1–3): 17–46.

Kim, Y.D., Yang, C.W., Kim, B.J., *et al.* (2013). Air-blown gasification of woody biomass in a bubbling fluidized bed gasifier. *Applied Energy* 112: 414–20.

Li, X.T., Grace, J.R., Lim, C.J., Watkinson, A.P., Chen, H.P., and Kim, J.R. (2004). Biomass gasification in a circulating fluidized bed. *Biomass and Bioenergy* 26(2): 171–93.

Liu, H., Hu, J., Wang, H., and Li, J. (2012). Experimental studies of biomass gasification with air. *Journal of Natural Gas Chemistry* 21(4): 374–80.

Press, N.A. (1983). *Another Fuel for Motor Transport.* Washington, DC: National Academy Press.

Son, Y.-I., Yoon, S.J., Kim, Y.K., and Lee, J.G. (2011). Gasification and power generation characteristics of woody biomass utilizing a downdraft gasifier. *Biomass and Bioenergy* 35(10): 4215–20.

Exercise Problems

21.1. Discuss how biomass can be used to produce gasoline and diesel through gasification.

21.2. Explain proximate and ultimate analyses of a biomass and discuss their utility in gasification.

21.3. In your own words, summarize the differences between various gasifiers.

21.4. What are the three major differences between updraft and downdraft gasifiers?

21.5. Write the general combustion reaction for sweetgum with the following elemental composition: C 49.76%, H 5.96%, N 0.36%, and O 42.53% on a dry wt. basis. Ash content is 1.39% (dry wt. basis).

21.6. Determine the air to biomass stoichiometric ratio, $(A/B)_{stoichiometric,}$ for sweetgum in Problem 21.5.

21.7. Using the same fuel as in Problem 21.6, determine the mass flow rate of air and steam in order to obtain ER = 0.30 and S/C = 1.5 if the feeding rate of sweetgum is 1.67 kg/s (dry wt. basis).

21.8. Determine the heating value of producer gas (MJ/m^3) with the following composition: CO 22.8%, CO_2 8.9%, CH_4 2.7%, and H_2 20.1%. The balance in the gas is nitrogen and the gas composition was measured at ambient temperature (25 °C).

21.9. The mass flow rate of wood chips into a gasifier is 26.8 kg/hr and they have the following elemental composition: C 46.26%, H 5.62%, N 0.13%, and O 47.66%. Determine the carbon conversion and cold gas efficiency if the producer gas flow rate out of the gasifier is 65 Nm^3/hr (normal cubic meter per hour) and it has following composition: CO 22.8%, CO_2 8.9%, CH_4 2.7%, and H_2 20.1%. The balance in the producer gas is nitrogen and the gas composition was measured at ambient temperature (25 °C). The heating value of the biomass is 18.5 MJ/kg. Assume other information if required.

21.10. Assuming that air was used as a gasifying agent in Problem 21.9, determine the equivalence ratio (ER).

Appendix

Table A21.1 Heat of combustion of syngas components at T = 298 K and P = 1 atm

Producer gas components	Volumetric heat of combustion (MJ/Nm^3)
CO	12.63
H$_2$	12.74
CH$_4$	39.82
C$_2$H$_2$	58.06
C$_2$H$_4$	63.41
C$_2$H$_6$	70.29
CH$_3$H$_6$	93.57
CH$_3$H$_8$	101.24
NH$_3$	13.07
H$_2$S	25.10

Source: Basu, 2010.

Table A21.2 Specific heat of select producer gas constituents as function of producer gas exit temperature (T) in Kelvin

Gas	Specific heat function, kJ/kmol.K	Temp range of validity (K)
CO	$27.62 + 0.005T$	273–2,500
CO$_2$	$43.28 + 0.0114T - 818363T^{-2}$	273–1,200
CH$_4$	$22.35 + 0.048T$	273–1,200
H$_2$	$27.71 + 0.0034T$	273–2,500
N$_2$	$27.21 + 0.0042T$	300–5,000
H$_2$O(g)	$34.4 + 0.000628T + 0.0000052T^2$	300–2,500
H$_2$S	$30.139 + 0.015T$	300–600

Source: Basu, 2010.

Pyrolysis

Manuel Garcia-Perez

What is included in this chapter?

This chapter covers slow and fast pyrolysis for the production of charcoal (bio-char) and bio-oils from lignocellulosic feedstocks. The stoichiometry and kinetics of the most important pyrolytic reactions, as well as a bio-oil refining strategy for the production of fuels, are examined with pertinent examples.

22.1 Introduction

The word "pyrolysis" is derived from *pyro*, the Greek word for "fire," and the word *lysis*, for "separating." Pyrolysis is a thermochemical process that occurs in the absence of an oxidizing agent (air/oxygen) at temperatures between 350 °C and 600 °C. This process is used for the production of charcoal or liquid products (also known as tar or bio-oil). The development of the pyrolysis process is one of the most important advances in the history of industrial chemistry. Charcoal first produced with primitive pyrolysis technologies, also known as carbonization or slow pyrolysis, is considered to be the first synthetic material created by humankind. It was originally used for melting tin in the manufacture of bronze tools. Charcoal was also critical for the expression of the human spirit, as seen from the magnificent drawing in the Grotte Chauvent, over 38,000 years old (Antal and Grønli, 2003).

The recovery of liquid products (tars) from charcoal production pits was first mastered by the Macedonians. Then in 1658, a liquid known as "pyroligneous acid" was identified as being similar to the acid contained in vinegar (acetic acid). Acetic acid was later isolated from pyrolysis liquid by Russian pharmacist Johann Tobias Lowitz in 1870. Methyl alcohol (also known as wood alcohol) was another key chemical isolated from pyrolysis liquids by French chemists Jean-Baptiste Dumas and Eugène-Melchior Peligot in 1835. In 1856 the demand for

---

Bioenergy: Principles and Applications, First Edition. Edited by Yebo Li and Samir Kumar Khanal.
© 2017 John Wiley & Sons, Inc. Published 2017 by John Wiley & Sons, Inc.
Companion website: www.wiley.com/go/Li/Bioenergy

methyl alcohol increased due to the production of aniline purple (the first synthetic organic dye). The identification, purification, and demand for these important chemicals were the origin of the wood distillation industry, which flourished between 1850 and the 1920s (see Figure 22.1).

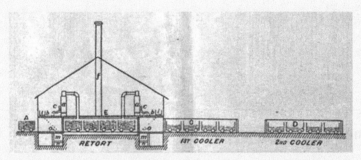

FIG. 22.1   A wood distillation plant. Source: Veitch, 1907. Public domain.

The decline of the wood-distillation industry at the beginning of the 20th century was the result of competition with the growing petroleum and coal gasification industries, with their cheap and abundant feedstock, and capacity to produce methanol and acetic acid at lower cost. From 1920 until the 1960s, the scope of pyrolysis shrunk to small ovens without liquid recovery, focusing on the production of charcoal for metallurgical applications and for cooking (backyard home barbecues). However, the economic support for energy-related projects due to the oil crisis in 1970 and the high yields of oil (close to 70% by mass) achieved when small lignocellulosic particles were pyrolyzed in fluidized-bed reactors promoted spectacular growth in pyrolysis research and development projects in the 1980s and 1990s. Developments in the high-yield production of bio-oil and identification of numerous potential end-products, including transportation fuel and bioplastics, have contributed to significant interest in *bio-oil refinery* concepts. Bio-oil refining is also progressing rapidly and the first bio-oil refinery broke ground in Hawaii, USA in 2011.

According to FAOSTAT (http://faostat.fao.org/), in 2012 the world production of charcoal was over 50 million metric tons. It is estimated that nearly 220 million metric tons of biomass are used worldwide for charcoal production annually. In 2012 Brazil produced 6.8 million metric tons of charcoal, making it by far the largest charcoal producer in the world. Other important charcoal-producing countries are Nigeria (4.1 million metric tons/year), Ethiopia (3.9 million metric tons/year), Democratic Republic of Congo (2.1 million metric tons/year), Mozambique (1.0 million metric tons/year), India (2.8 million metric tons/year), China (1.7 million metric tons/year), and the USA (0.8 million metric tons/year). While the main exporter of bio-char is Paraguay (0.2 million metric tons/year), the main importers are Germany (0.2 million metric tons/year) and China (0.2 million metric tons/year). Billions of people in developing nations use charcoal for cooking. Despite its advantages over biomass for cooking, large-scale production of charcoal in developing nations is causing serious environmental problems due to deforestation and gaseous emissions from kilns. Nonetheless, charcoal is likely to remain the fuel of choice in many developing countries as long as feedstock supply and demand from impoverished people both exist.

22.2 Slow vs. Fast Pyrolysis

The pyrolysis process can be controlled to maximize the production of either charcoal or liquid products, and as such is classified generally as slow or fast pyrolysis.

22.2.1 Slow Pyrolysis

Slow pyrolysis (also known as carbonization) is a process in which relatively large biomass particles (chips, logs) are heated in the near absence of air/oxygen for the production of charcoal. Biomass is pyrolyzed at slow heating rates (5–10 °C/min), resulting in reduced liquid and enhanced char production. Carbonization reactors are classified by Emrich (1985) into three types: kilns, retorts, and converters. Kilns are used in traditional charcoal making, solely to produce charcoal, and are typically autothermal processes. The term retort refers to a reactor that has the ability to pyrolyze pile-wood or wood logs over 30 cm long and over 18 cm in diameter. Converters produce charcoal from the carbonization of chips or pelletized wood. The main carbonization kilns are earth kilns, pits, Brazilian beehive brick kilns, Argentine beehive brick kiln, Adam retorts, and Missouri kilns. The most common retorts and converters used today for charcoal production are wagon retort, Reichert, French SIFIC Process, Lambiotte, Lurgi, Herreshof multiple-hearth furnace, and rotary drums (Garcia-Perez *et al.*, 2011). Figure 22.2 shows schematics of an earth kiln and a Brazilian kiln.

Despite the growing interest in charcoal production, many of the technologies currently available do not recover the energy contained in the volatiles. Recovery of the heat contained in the volatiles (Figure 22.3) is critical to enhance the economic viability of slow pyrolysis plants and to reduce their negative environmental impact. Typical product yields for slow and fast pyrolysis are presented in Table 22.1.

22.2.2 Fast Pyrolysis

Fast pyrolysis is a process in which very small biomass particles (less than 2 mm diameter) are rapidly heated to between 450 °C and 600 °C in the absence of *air/oxygen* to produce high yields of bio-oil (60–75% by wt.). Bio-oil production requires a very low vapor residence time, typically 2 seconds or less, to minimize secondary reactions. Researchers have shown that maximum liquid yields are obtained with high heating rates at temperatures around 500 °C. The liquids are typically recovered in a condensation step (Figure 22.3).

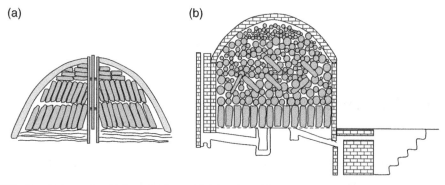

(a)   (b)

FIG. 22.2   Schematic of (a) earth kiln and (b) Brazilian kiln.

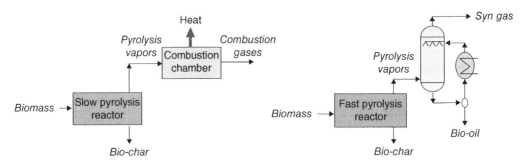

FIG. 22.3 Slow and fast pyrolysis processes.

Table 22.1 Summary of processing conditions and yield of products typically obtained in slow or fast pyrolysis technologies

Pyrolysis technology	Conditions	Liquid yield (% w/w)	Char yield (% w/w)	Gas yield (% w/w)
Slow	Slow heating rates Large particles Long residence time of vapors	30–45	25–35	25–35
Fast	High heating rates Small particles Short residence time of vapors	60–75	12–20	13–20

Common pyrolysis reactors used for fast pyrolysis include ablative reactors, fluidized-bed, circulating fluidized-bed, and vacuum reactors (Figure 22.4; Vanderbosch and Prins, 2010).

In *ablative pyrolysis* reactors, wood is pressed against a heated surface and rapidly moved. The wood melts at the heated surface and leaves an oil film behind, which evaporates. The main advantage of this process is that it can use larger particles of wood and that the reactor is quite compact.

In the *fluidized-bed* and *circulating fluid-bed* reactors, pyrolysis heat is transferred by a mixture of convection and conduction. The main drawback of these processes is that a substantial amount of carrier gas is needed for fluidization or transport.

In *vacuum pyrolysis*, the heating rates are typically lower than in ablative or fluidized-bed reactors, but the rapid removal of vapors from the reactor by a vacuum pump allows for high oil yields to be obtained.

The main hurdle for the commercialization of fast pyrolysis technologies is the lack of refineries capable of converting the bio-oils into biofuels and high-value chemicals.

22.3 Pyrolysis Reactions and Mechanisms

22.3.1 Pyrolysis Reactions

Lignocellulosic feedstocks are typically made up of three major bio-polymers (cellulose, hemicellulose, and lignin) and materials extractable with organic solvents. Biomass is typically composed of fibrous material and when ground it results in particles that are cylindrical in shape. When a biomass

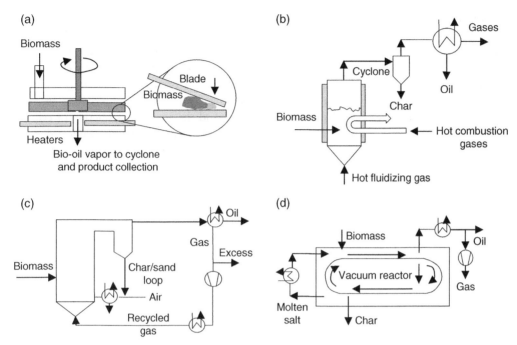

FIG. 22.4   Scheme of fast pyrolysis reactors: (a) ablative; (b) fluidized-bed; (c) circulating-bed; and (d) vacuum pyrolysis reactor. Source: Adapted from Vanderbosch and Prins, 2010.

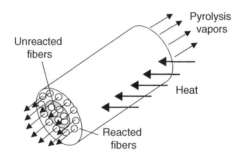

FIG. 22.5   Scheme of a reacting biomass particle.

particle is exposed to a hot environment (over 200 °C), pyrolysis reactions take place as the temperature front progresses toward the core of the particle. The species formed in these initial reactions may not be volatile enough to evaporate (escape the surface) and might further react (secondary thermal decomposition reactions) to form smaller volatile products, or undergo condensation/polymerization reactions to form charcoal. The released volatile species may undergo further reactions within the pores of the particles (intra-particle secondary reactions), either homogeneously in the gas phase or heterogeneously by reaction with the partially converted solid biomass or char. Due to the orientation of the biomass cell structure, most of the vapors will diffuse in the axial direction while the heat is transferred mostly in the radial direction (Figure 22.5). When very small particles (diameter below 1 mm) are used, the cell structure is almost completely destroyed, exposing most cell walls to the reaction environment, and significantly reducing heat transfer and diffusion limitations.

Thermochemical reactions are typically classified as primary thermochemical reactions that occur in the solid phase; or secondary thermochemical reactions that occur inside the particles (intraparticle); after the vapors leave the particle and encounter another particle (interparticle); or in the path of the vapor out of the pyrolysis reactor (homogeneous secondary reactions in vapor phase). At slow heating rates (10 °C/min), hemicellulose and the amorphous fraction of cellulose typically break down at temperatures between 200 °C and 350 °C, and crystalline cellulose degrades between 300 °C and 400 °C. Lignin pyrolysis happens in a much broader range, between 200 °C and 600 °C.

The yields of charcoal and volatiles produced by pyrolysis reactions are dramatically affected by the presence of alkalines. In materials with a low fraction of alkalines, the yield of charcoal can be estimated as the additive contribution of the yield of charcoal produced by each of the biopolymers forming the biomass. In the range of heating rates between 2 °C and 50 °C/min, the charcoal yields of cellulose typically vary between 3% and 8% (by wt.), hemicellulose between 21% and 25% (by wt.), and lignin between 26% and 39% (by wt.), depending on the type of lignin.

> **Example 22.1:** Estimate the charcoal yield of sugar cane bagasse (washed with hot water for the removal of ash) carbonized at 10 °C/min if it contains 23% hemicelluloses, 39% cellulose, 28% lignin, and 10% moisture (dry wt.). Suppose that there is no interaction between the fractions.
>
> ***Solution:***
>
> Contribution of hemicelluloses to charcoal production: 23% × 23% = 5.29%
>
> Contribution of cellulose to charcoal production: 39% × 5% = 1.95%
>
> Contribution of lignin to charcoal production: 28% × 30% = 8.40%
>
> Contribution of water to charcoal production: 10% × 0 = 0.00%
>
> Total charcoal yield of bagasse = **15.64%**

22.3.2 Reaction Mechanisms

Establishing reaction conditions that accurately represent those in the process of interest is the major challenge in the study of pyrolysis reactions. Separating the effect of secondary and primary reactions is extremely challenging. To study the primary reactions it is necessary first to uniformly heat all the particles; second to know the temperature of the solids; and third instantaneously to quench and quantitatively recover volatile products of primary reactions. Thermogravimetric analyzers are typically used to study reactions at relatively slow heating rates. At higher heating rates, drop tube furnaces, wire mesh reactors, and fluidized bed reactors are more commonly used to study fast pyrolysis reactions (Kandiyoti *et al.*, 2006).

As a result of these experimental studies, several pseudo-models have been proposed to describe the pyrolysis of cellulose, hemicelluloses, and lignin. The Broido–Shafizadeh model (Figure 22.6) is the most commonly used to describe the pyrolysis of cellulose. This model postulates that the cellulose is converted into a more active form and that this is a rate-limiting step, followed by the formation of either char or volatile compounds. The values of "a" and "b" in the model depend on the heating rate used. The Shafizadeh model shown in Figure 22.6 uses the following values: a = 0.35; and b = (1 − a) = 0.65.

The relationship between the activation energy and the reaction rate constants (k_1, k_2, k_3) is typically calculated by the Arrhenius equation ($k_i = A_i \cdot e^{(-E_i/RT)}$). Table 22.2 lists the values of pre-exponential factor and activation energy reported in the literature (Di Blasi, 1998) for the Broido–Shafizadeh model.

Knowing the reaction scheme, it is possible to estimate the evolution of all species in the solid phase by solving the system of ordinary differential equations derived from the reaction scheme:

$$dA/dt = -k_1 \cdot A \tag{22.1}$$

$$dB/dt = k_1 \cdot A - (k_2 + k_3) \cdot B \tag{22.2}$$

$$dC/dt = k_2 \cdot B \tag{22.3}$$

$$dD/dt = a \cdot k_3 \cdot B \tag{22.4}$$

$$dE/dt = b \cdot k_3 \cdot B \tag{22.5}$$

where a + b = 1 (mass is conserved). The units of concentration are typically expressed in mass fraction because they can be easily measured by existing characterization schemes.

The Koufopanos mechanism (Figure 22.7) is also an important reaction scheme (Di Blassi, 1998). The kinetic parameters used to describe the thermal behavior of cellulose, hemicelluloses, and lignin with this scheme are reported in Table 22.3.

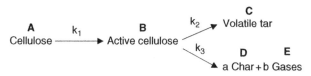

FIG. 22.6   Shafizadeh model for cellulose pyrolysis.

Table 22.2 Pre-exponential factor (A_i) and activation energies (E_i) to calculate the reaction rate constants for the Broido–Shafizadeh model

	A_i (1/min)	E_i (kJ/mol)
Reaction 1 k_1	1.7×10^{21}	242.8
Reaction 2 k_2	1.9×10^{16}	197.9
Reaction 3 k_3	7.9×10^{11}	153.1

Source: Data from Di Blasi, 1998.

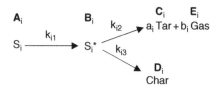

FIG. 22.7   Koufopanos mechanism, where i =1 (hemicellulose); i =2 (cellulose); and i =3 (lignin).

Table 22.3 Pre-exponential factor (A_i) and activation energies (E_i) to calculate the reaction rate constants for the Koufopanos mechanism

	A_i (1/sec)	E_i (kJ/mol)
Hemicellulose (xylan)*		
Reaction 1, k_{11}	3.3×10^6	72.4
Reaction 2, k_{12}	1.1×10^{14}	174.1
Reaction 3, k_{13}	2.5×10^{13}	172.0
Cellulose*		
Reaction 1, k_{21}	2.2×10^{14}	167.5
Reaction 2, k_{22}	9.5×10^{15}	216.6
Reaction 3, k_{23}	3.1×10^{13}	196.0
Lignin*		
Reaction 1, k_{31}	3.3×10^{12}	147.7
Reaction 2, k_{32}	8.6×10^8	137.1
Reaction 3, k_{33}	4.4×10^7	122.0

Note: *For very fast heating rates use $a_1 = 0.68$, $b_1 = (1-a_1) = 0.32$ for hemicellulose (xylan); $a_2 = 0.96$, $b_2 = (1-a_2) = 0.04$ for cellulose; and $a_3 = 0.91$, $b_3 = (1-a_3) = 0.09$ for lignin (estimated from data reported by Hoekstra *et al.*, 2012).
Source: Data from Di Blassi, 1998.

Example 22.2:

(a) Using Euler's method, calculate the time evolution of the yield of tars, active cellulose, and gases as a function of reaction time if a very small biomass particle (assume a kinetically controlled reaction regime) is introduced in a fluidized bed reactor at 680 K (use the Broido–Shafizadeh model).

(b) Repeat the calculations but now considering that the fluidized bed reactor operates at 623 K. How do your results change?

Solution: The partial differential equations can be solved by Euler's method:

$$\mathbf{A}_{(t+\Delta t)} = \mathbf{A}_t + (d\mathbf{A}/dt)_t \cdot \Delta t \tag{22.6}$$

The solutions to (a) and (b) have been obtained in Excel and are presented in Figure 22.8.

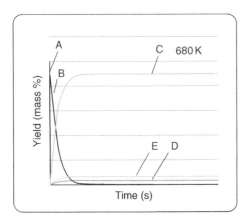
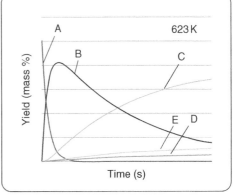

FIG. 22.8   (a) Evolution of cellulose; (b) active cellulose; (c) volatile tar; (d) char; and (e) gases.

22.4 Single-Particle Models

The estimation of the reaction time at which a biomass particle achieves a target conversion is critical for the design of pyrolysis reactors. If the particles are very small (generally below 1 mm) and if the external heat transfer coefficient is very high (this is the case with fluidized-bed reactors), the conversion time is likely to be controlled by the reaction kinetics rather than heat transfer. This is typically the case for fast pyrolysis reactors. In situations where large biomass particles are used, typical of slow pyrolysis, internal heat transfer rates may control the outcome of the pyrolysis process. Given that the localized distribution of temperature controls the rate of primary reactions, the description of the heat transfer rate is critical to describing the overall rate of pyrolysis. Single-particle models have been developed to calculate the conversion rates and to identify whether a controlling mechanism exists during pyrolysis. Simple one-dimensional (1-D) models are commonly used for particles with a length to diameter ratio below 3. More complex two-dimensional models, needed for situations in which the axial heat transfer cannot be disregarded, are important in reactor design, but are beyond the scope of this textbook.

Bamford's model is the simplest 1-D particle model proposed (Equation 22.7; Pyle and Zaror, 1984). It is a simple energy balance inside the biomass particle that considers a constant effective thermal conductivity. Note that T is T(r,t).

$$\frac{\partial(C_p\rho T)}{\partial t} = K_{eff}\left(\left(\frac{\partial^2 T}{\partial r^2}\right) + \frac{b-1}{r}\frac{\partial T}{\partial r}\right) - q\frac{\partial r}{\partial t} \tag{22.7}$$

The Bamford model considers that a first-order reaction for the loss of dry mass is given as:

$$-\frac{\partial \rho}{\partial t} = Ae^{-\frac{E}{RT}}(\rho - \rho_\infty) \tag{22.8}$$

where T is temperature (°C), t is time (sec), k_{eff} is effective conductivity (W/m · °C), ρ is density (kg/m³), ρ_∞ is density at final conversion (kg/m³), q is heat of reaction (kJ/kg), b is geometry factor (slab = 1, cylinder = 2, sphere = 3), c_p is specific heat capacity (J/kg · °C), and R is the ideal gas constant (R = 8.314 J/(mol· K)).

The heat of the reaction depends on the amount of charcoal produced. Typically when the material is heated slowly (slow pyrolysis), large quantities of charcoal (25–35% by wt.) are produced and the overall reaction is exothermic. Fast pyrolysis reactions that result in small quantities of charcoal (12–20% by wt.) are endothermic. So the heat of (slow and fast) pyrolysis reactions is typically between -20 kJ/kg and 453 kJ/kg (Auber, 2009). Note that the negative sign indicates that heat is released (i.e., exothermic reaction).

Bamford's model assumes that the heat transferred to the solid from the surrounding gas is a combination of convection and radiation and that the temperature of the gas remains constant (Pyle and Zaror, 1984).

Pyle and Zaror (1984) used the Bamford model to propose three dimensionless numbers (the Biot number and two pyrolysis numbers) to identify the controlling mechanism of a pyrolysis process:

$$B_i = h\frac{r}{k} \tag{22.9}$$

Table 22.4 Criteria for identifying the controlling mechanism during pyrolysis

Controlling mechanism	Bi	Py	Py'
Internal heat transfer	>50	<10^{-3}	<<1
External heat transfer	<1	>1	>1
Kinetics	<1	>10	>10
Non-controlling conditions		All the other combinations	

Source: Data from Pyle and Zaror, 1984.

$$P_y = \frac{k}{k' \, \rho \, c_p \, r^2} \tag{22.10}$$

$$P'_y = \frac{h}{k' \, \rho \, c_p r} = 2.34 \tag{22.11}$$

where h is the heat transfer coefficient (W/(m^2·K), r is the particle radius (m), k is the thermal conductivity (W/(m·K)), k' is the apparent reaction rate constant (k' = A · exp (-E/RT)) (1/sec), ρ is the bulk density (kg/m^3), and c$_p$ is the specific heat capacity (J/(kg· K)).

The criteria proposed to identify the controlling mechanism are listed in Table 22.4. Use of these numbers suggests or indicates to the operator what controllable factors (h, r, T, ρ) the operator might need to adjust to improve the rate of disappearance. For example, decreasing r, decreases Bi, and increases Py and Py', thereby shifting the controlling mechanism to external heat transfer and/or kinetics.

Example 22.3: Identify the controlling pyrolysis mechanism for a biomass particle with a radius of 0.2 cm that is introduced into a reactor at 680 K. The biomass particle-surrounding gas external heat transfer coefficient for the reactor studied is 0.004 J/(sec·cm^2·K). The main physical properties of the biomass particle are thermal conductivity (k) = 0.001 J/(sec·cm·K), specific heat (C$_p$) = 1 J/(g· K), activation energy (E) = 6.6 • 10^4 J/mol, and pre-exponential factor (A) = 2,000 1/sec.

Solution:

Reaction rate constant: $k' = A \bullet e^{(-E/RT)} = 0.017$ 1/sec

Biot number: $B_i = h\dfrac{r}{k} = 0.8$

Py number: $P_y = \dfrac{k}{k' \, \rho \, c_p \, r^2} = 2.94$

Py' number: $P'_y = \dfrac{h}{k' \, \rho \, c_p r} = 2.34$

According to the criteria listed in Table 22.4, the overall rate of pyrolysis is controlled by the external heat transfer rate. Under these conditions the temperature gradients inside the biomass particles will be negligible compared with the temperature gradient between the biomass surface and the surrounding gas.

22.5 Bio-Oil

Pyrolysis oil, also known as bio-oil, is a black or dark red-brown liquid, derived from plant material via fast pyrolysis. It has a kinematic viscosity that varies from 25–1,000 cSt (centiStoke; 1 cSt = 10^{-6}

Table 22.5 Bio-oil liquid range of physicochemical properties

Properties	Typical range of values
Solid (char) content (wt. %)	0.2–1.0
Specific heat capacity (J/(g·K))	2.6–3.8 (25–70 °C)
Conradson carbon residue content (wt. %)	14–23
Ash content (wt. %)	0.01–0.20
C, H, O, and N contents (wt. % dry basis)	C: 54–58.0
	H: 5.5–7.0
	N: 0–0.2
	O: 35–40
Alkali metals content (Na + K) (ppm)	5–500
Pour point (°C)	–12–36
High heating value (MJ/kg)	16–19
Flash point (°C)	40–110

Source: Czernik and Bridgwater, 2004; Bridgwater and Peacocke, 2000.

Table 22.6 Potential chemicals from bio-oils

Chemicals	Note
Acetic acid	World production 7 million metric tons/year, potential price $0.60/kg
Adhesives	Phenol substitute for the production of adhesives used in the production of wood panels (plywood, medium-density fiberboard [MDF], particle board, and oriented-strand board [OSB])
Antioxidants	Lignin-derived compounds with excellent antioxidant properties
Biocarbon electrodes	Production of electrodes, calcinations at 1,000 °C, and graphitization at 2,700 °C
Slow-release fertilizer	Amides, imines, and mannich-reaction products, produced from the reaction of bio-oil functional groups (carbonyl, carboxyl, hydroxyl, phenolic, and methoxyl) with ammonia, urea, and other amino compounds
Levoglucosan	By using demineralized cellulose, high yields of levoglucosan (up to 46% by wt.) and levoglucosanone (up to 24% by wt.) can be produced
Impermiabilizer	Black residue of tar distillation commercialized to impermiabilize ships
Surfactants	More than 10 commercial grades are used for ore flotation
Wood preservatives	Bio-oils can act as insecticides and fungicides due to the presence of some terpenoid and phenolic compounds

m^2/sec) at 40 °C. The water content typically varies from 15–50% (by wt.). Bio-oil is miscible with polar solvents (methanol, acetone), but it is almost totally immiscible with petroleum-derived fuels. It has a density of approximately 1,200 kg/m^3, and a heating value of 16–19 GJ/metric ton (approximately 55% of the heating value of diesel on a volumetric basis and 45% on a wt. basis). Bio-oil is acidic (pH ~2.3). Table 22.5 summarizes some of the most common physicochemical properties of bio-oil.

Bio-oil can be stored, pumped, and transported like petroleum products. Although bio-oil has been successfully tested in engines, turbines, and boilers, and has been upgraded to high-quality hydrocarbon fuels, it cannot be considered a commercial commodity due to lack of standards to ensure consistent performance.

In the production of high-value chemicals in addition to transportation and heating, bio-fuels are critical to creating economically viable biomass pyrolysis conversion pathways. Table 22.6 summarizes the potential chemical products and transportation fuels that can be obtained from bio-oils.

22.6 Charcoal

Charcoal is a combustible solid (with a high heating value, around 30 MJ/kg) that can be burned to generate energy in most systems that are currently burning coal. The sulfur content of charcoal is low and hence industrial combustion of charcoal generally does not require technology for removing SO_x from emissions to meet US EPA limits. The ash content of charcoal depends substantially on the feedstock used. Some biomass feedstocks, such as corn stover and rice husk, contain high levels of silica (Si), which is concentrated in the biochar after pyrolysis. Combustion of high Si-containing charcoal will cause scaling in the wall of the combustion chamber and decrease the useful life of these chambers.

Low-ash-containing charcoal can be used in metallurgy, and as feedstock for activated carbon production. Activated carbon has many applications, such as an adsorbent to remove odorants/contaminants from air streams, and both organic and inorganic contaminants from wastewater streams.

An emerging new use of charcoal is as soil amendment. The charcoal used for carbon sequestration and for soil amendment is typically known as biochar. The return of biochar, a co-product of pyrolysis, to the soil from which the biomass was harvested has been proposed as a means to enhance soil quality and thereby provide an opportunity for additional revenue generation from the low-value residues. Furthermore, many of the nutrients (except nitrogen, N) in biomass are recovered with the biochar product, offering opportunities for nutrient recycling. Conversion of biomass carbon, C, to biochar (slow pyrolysis) leads to the sequestration of about 50% of the initial C in a stable form, as compared to approximately 3% retained after burning, and <10–20% after 5–10 years from biological decomposition. Therefore, biochar production is being considered as a means to produce more stable soil C than can be achieved by burning or direct land application of biomass. The term biochar has become common in place of charcoal, especially in cases where environmental applications are concerned.

22.7 Bio-oil Refining

Integrating pyrolysis technologies into biorefinery concepts that fully utilize all bio-oil fractions is one of the biggest challenges facing the thermochemical conversion industry. Petroleum refineries cannot refine bio-oils, and some researchers have expressed concerns that using crude bio-oils could create corrosion issues, intensify coke formation, and accelerate the deactivation of expensive catalysts in hydrotreatment units (Elliott, 2007). The negative bio-oil properties largely result from the high proportions of reactive oxygenated compounds present in the bio-oil (mostly aldehydes and acids).

> **Example 23.4:** Determine the carbon and hydrogen efficiency of a fast pyrolysis unit operated at 500 °C that converts 64% of a dry pine wood into a crude bio-oil containing 17% water. Further, 15% of the biomass was converted into charcoal with an elemental composition of (dry wt. basis): C 80%, H 3.5%, N 0.3%, and O 16.2%. The elemental composition of the bio-oil obtained (dry wt. basis) was C 52.2%, H 7.5%, N 0.1%, and O 40.2%. All the gas produced is combusted to produce the heat needed to run the pyrolysis process.
>
> Consider the elemental composition of the biomass processed as C 50.3%, H 6.3%, N 0.1%, and O 43.3% (dry wt. basis).

Solution:

1. Carbon efficiency

Carbon in: 0.503 kg C/kg of biomass

Carbon out: as bio-oil $0.64 \times (1-0.17) \times (0.522) = 0.277$ kg/kg of biomass

as charcoal $(0.15 \times 0.8) = 0.12$ kg/kg of biomass

Carbon efficiency: Carbon out in products/Carbon in = $(0.277 + 0.12)/0.503$
= **0.789 ~79%**

2. Hydrogen efficiency

Hydrogen in: 0.063 kg H/kg biomass

Hydrogen out: as bio-oil $0.64 \times (0.17 \times 2/18 + (1-0.17) \times 0.075) = 0.052$ kg H/kg biomass

as charcoal $0.15 \times 0.035 = 0.005$ kg H/kg biomass

Hydrogen efficiency: Hydrogen out/Hydrogen in = $(0.052+0.005)/0.063$
= 0.904 ~**90%**

Elliott (2007) developed a two-stage hydrotreatment process to upgrade bio-oils that dramatically reduces early catalyst coking. In the first stage of the treatment, bio-oil is allowed to react under mild conditions (200–250 °C) and at a hydrogen pressure ranging from 10–15 MPa with a hydrotreatment catalyst. During this process a large portion of the oxygen contained in bio-oil is combined with hydrogen and is driven off as water, together with the water already present in the oil. Once the mild hydrotreatment step is completed, the stabilized bio-oil is then treated by a catalytic hydrocracking stage to produce the final hydrocarbon mixture (Figure 22.9).

The typical second-stage hydrocracking conditions are at temperatures between 300 °C and 400°C and at hydrogen pressures of 10–15 Mpa, with a hydrocracking catalyst comparable to those used in petroleum refineries to hydrocrack vacuum residual oil. Using this bio-oil refinery approach,

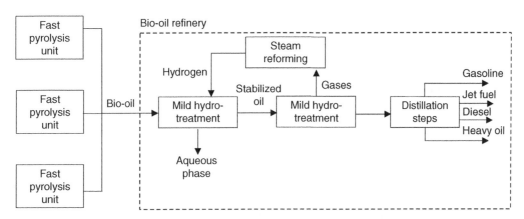

FIG. 22.9 Pyrolysis unit integrated with a two-step hydrotreatment unit. Source: Adapted from Jones *et al.*, 2009.

a total 34% of the original biomass can be converted into hydrocarbon fuels (green gasoline and green diesel) with a thermal efficiency close to 61% (BTU content of the fuel vs. BTU of all the inputs; Elliott, 2010). Although bio-oil refining is currently limited to laboratories and pilot plant studies, rapid developments in this area are occurring. Hydrogen consumption is typically 30–50 g/kg of bio-oil processed (Holmgren *et al.* 2008). The main products are light ends (100–150 g/kg of bio-oil), naphtha (210–300 g/kg of bio-oil), diesel (80–210 g/kg of bio-oil), and water-CO_2 (510–520 g/kg of oil processed; Holmgren *et al.* 2008).

Example 22.5: Calculate the carbon and hydrogen efficiencies of a pyrolysis/bio-oil hydrotreatment system (Figure 22.9) in which the bio-oil produced is subjected to a two-step hydrotreatment system. 64% (by wt.) of the original biomass is converted into bio-oil. Consider that 10% (by wt.) of the oil is converted to lights (consider the lights as methane), 25% (by wt.) is converted into gasoline, and 15% (by wt.) is converted into diesel. The rest is converted into water, CO_2, and heavy fractions. For these calculations, consider the naphtha as C_6H_6 and the diesel fraction as $C_{13}H_{28}$. 0.05 kg of hydrogen is need to hydrotreat 1 kg of bio-oil. 2 kg of methane and 4.5 kg of steam (H_2O) are needed to produce 1 kg of hydrogen in the steam reformer. Consider that half of the lights are used as fuel in the steam reforming and the other half is steam reformed to produce hydrogen. 15% (by wt.) of the biomass is converted into biochar. The elemental composition of the biomass processed and the biochar produced can be found in Example 22.4.

Solution:

Overall mass balance
Base: Biomass 1 kg
Bio-oil produced: 0.64 kg/kg biomass
Light gases (CH_4): 0.10 × 0.64 = 0.064 kg/kg biomass (0.032 kg/kg biomass used for the production of hydrogen 0.032 kg/kg is used for the heat production)

Gasoline (C_6H_6): 0.25 × 0.64 = 0.16 kg/kg biomass
Diesel ($C_{13}H_{28}$): 0.15 × 0.64 = 0.096 kg/kg biomass
Hydrogen needed for the hydrotreatment step: 0.05 × 0.64 = 0.032 kg/kg biomass
Methane needed for hydrogen production: 2 × 0.032 = 0.064 kg/kg biomass
Steam needed for hydrogen production: 4.5 × 0.032 = 0.144 kg/kg biomass
Extra methane needed: 0.064 − 0.032 = 0.032 kg/kg biomass

Carbon efficiency

Carbon in:	as biomass	= 0.503 kg C/kg of biomass
	as natural gas	= 0.032 × (12/16) = 0.024 kg C/kg biomass
	total	= 0.527 kg C/kg biomass
Carbon out:	as gasoline	= 0.16 × (72/78) = 0.148 kg C/kg of biomass
	as diesel	= 0.096 × (156/184) = 0.081 kg C/kg of biomass
	as charcoal	= 0.15 × 0.8 = 0.12 kg C/kg of biomass
	total	= 0.148 + 0.081 + 0.12 = 0.349 kg C/kg biomass
Carbon efficiency:	Carbon out in products/Carbon in = 0.349/0.527 = 0.6622 ~**66%**	

Hydrogen efficiency

Hydrogen in: as biomass   = 0.063 kg H/kg biomass

as natural gas   = 0.032 × (4/16) = 0.008 kg H/kg biomass

as steam   = 0.144 × (2/18) = 0.016 kg H/kg biomass

total   = 0.087 kg H/kg biomass

Hydrogen out: as gasoline   = 0.16 × (6/78) = 0.0123 kg H/kg of biomass

as diesel   = 0.096 × (28/184) = 0.0146 kg H/kg of biomass

as charcoal   = 0.15 × 0.035 = 0.00525 kg H/kg of biomass

total   = 0.0123 + 0.0146 + 0.00525 = 0.0321 kg H/kg biomass

Hydrogen in: Hydrogen out   = 0.0321/0.087 = 0.369 ∼**37%**

in products/
Hydrogen in

References

Antal, M.J., and Grønli, M. (2003). The art, science and technology of charcoal production. *Industrial and Engineering Chemistry Research* 42: 1619–40.

Auber, M. (2009). Effect catalytique de certains inorganiques sur la selectivite des reactions de pyrolyse rapide de biomass et de leurs constituants. PhD thesis, Institut National Polytechnique de Lorraine.

Bridgwater, A.V., and Peacocke, G.V.C. (2000). Fast pyrolysis processes for biomass. *Renewable Energy Reviews* 4(1): 1–73.

Czernik, S., and Bridgwater, A.V. (2004). Overview of applications of biomass fast pyrolysis oil. *Energy and Fuels* 18: 590–98.

Di Blasi, C. (1998). Comparison of semi-global mechanisms for primary pyrolysis of lignocellulosic fuels. *Journal of Analytical and Applied Pyrolysis* 47(1): 43–64.

Elliott, D.C. (2007). Historical developments in hydro-processing bio-oils. *Energy and Fuels* 21: 1792–1815.

Elliott, D. (2010). Advancement of bio-oil utilization for refinery feedstock. Washington Bioenergy Research Symposium, November 8. http://www.pacificbiomass.org/documents/Elliott%20(C1).pdf, accessed April 2016.

Emrich, W. (1985). *Handbook of Biochar Making: The Traditional and Industrial Methods*. Dordrecht: D. Reidel.

Garcia-Perez, M., Lewis, T., and Kruger, C.E. (2011). *Methods for Producing Biochar and Advanced Biofuels in Washington State. Part 1: Literature Review of Pyrolysis Reactors*. First Project Report. Pullman, WA: Department of Biological Systems Engineering and the Center for Sustaining Agriculture and Natural Resources, Washington State University. https://fortress.wa.gov/ecy/publications/publications/1107017.pdf, accessed April 2016.

Hoekstra, E., Van Swaaij, W.P.M., Kersten, S.R.A., and Hogendoorn, K.J.A. (2012). Fast pyrolysis in a novel wire-mesh reactor: Decomposition of pine wood and model compounds. *Chemical Engineering Journal* 187: 172–84.

Holmgren, J., Marinangeli, R., Nair, P., Elliott, D., and Bain, R. Consider up-grading pyrolysis oils into renewable fuels. Special report. *Hydrocarbon Processing*. September: 95–103.

Jones, S.B., Holladay, J.E., Valkenburg, C., et al. (2009). *Production of Gasoline and Diesel from Biomass via Fast Pyrolysis, Hydrotreating and Hydrocracking: A Design Case*. PNNL-18284 Rev. 1. DE-AC05-76RL01830. Washington, DC: U.S. Department of Energy.

Kandiyotl, R., Herod, A., and Bartle, K. (2006). *Solid Fuels and Heavy Hydrocarbon Liquids: Thermal Characterization and Analysis*. Amsterdam: Elsevier.

Pyle, D.L., and Zaror, C.A. (1984). Heat transfer and kinetics in the low temperature pyrolysis of solids. *Chemical Engineering Science* 36(1): 147–58.

Vanderbosch, R.H., and Prins, W. (2010). Fast pyrolysis technology development. *Biofuels, Bioproducts and Biorefining* 2010: 178–208.

Veitch, F.P. (1907). *Chemical Methods for Utilizing Wood*. Washington, DC: US Department of Agriculture.

Withrow, J.R. (1915). The chemical engineering of the hardwood distillation industry. *Industrial and Engineering Chemistry Research* 7(II): 912.

Exercise Problems

22.1. Discuss how the biorefinery concept can be applied to converting lignocellulosic materials into gasoline via pyrolysis.

22.2. Discuss the main differences between fast and slow pyrolysis technologies.

22.3. Calculate the yield of char from a biomass that contains (dry wt. basis) 40% lignin, 30% cellulose, and 30% hemicelluloses in a slow pyrolysis reactor.

22.4. What can be done to improve the carbon efficiency in the production of drop-in biofuels via fast pyrolysis?

22.5. What are the main issues with the direct utilization of bio-oil as a fuel?

22.6. Use the Koufopanos mechanism (Figure 22.7) to estimate the distribution of products as a function of reaction time at 300 °C and 500 °C for a material that contains (dry wt. basis) 50% cellulose, 23% hemicelluloses, and 27% lignin. The kinetic parameters to be used can be found in Table 22.3. For all the polymers consider a = 0.7 and b = 0.3.

22.7. Identify the controlling pyrolysis mechanism for a biomass particle with a radius of 0.04 cm that is introduced into a pyrolysis reactor with a biomass–gas external heat transfer coefficient of 0.01 J/(sec·cm^2·K). The reactor is operated at 773 K. Use the physical properties listed in Problem 22.3.

22.8. Estimate the time for which the particle in Problem 22.7 needs to stay inside the reactor to achieve 95% conversion.

22.9. Propose and discuss two strategies to increase C efficiency in bio-oil hydrotreatment units.

22.10. Propose and discuss two alternatives to increase H efficiency during bio-oil hydrotreatment.

SECTION V
Biobased Refinery

Sugar-Based Biorefinery

Samir Kumar Khanal and Saoharit Nitayavardhana

What is included in this chapter?

This chapter covers sugar-based biorefinery using various sugar-based feedstocks. The process description of ethanol production, stoichiometry, and generation of different co-products as well as mass balance are discussed. The production of important high-value biochemicals and biopolymers using monomeric sugar is also considered.

23.1 Introduction

Sugar-based feedstocks represent first-generation biofuels, and biorefineries using these feedstocks were probably one of the earliest to be put to practical use. In recent years, sugar-based feedstocks have been successfully used for biofuel production at a commercial scale. For example, Brazil has established a sugarcane-based ethanol biorefinery. The major sugar-based feedstocks for biofuel production, especially ethanol, are sugarcane, sweet sorghum, and sugar beet. The sugar-rich juice obtained from these feedstocks is primarily sucrose ($C_{12}H_{22}O_{11}$), which is readily fermentable to ethanol by yeast, specifically *Saccharomyces cerevisiae*. It is important to point out that sugar is primarily produced as an edible staple, with 70% of the world's sugar produced from sugarcane grown mainly in tropical regions, while the remaining 30% is derived from sugar beet grown in industrialized, temperate regions. The European Union, Brazil, and India are the top three sugar-based ethanol producers, whereas the USA is the largest producer of bioethanol derived from corn starch (see Chapter 24).

In this chapter, we discuss how different sugar-based feedstocks are processed into biofuels, and what co-products are formed during these processes. A better understanding of biorefinery concepts for sugar-based biofuels will help students to understand the various processes involved in biofuel production, proper management of waste produced during the conversion process, and quantification of various co-products. The chapter also briefly covers diverse high-value biochemicals and biopolymers that could be produced from monomeric sugars.

Bioenergy: Principles and Applications, First Edition. Edited by Yebo Li and Samir Kumar Khanal.
© 2017 John Wiley & Sons, Inc. Published 2017 by John Wiley & Sons, Inc.
Companion website: www.wiley.com/go/Li/Bioenergy

23.2 Stoichiometry

The biochemical reactions involved in ethanol production from sucrose are given by Equations 23.1 and 23.2:

$$C_{12}H_{22}O_{11} + H_2O \xrightarrow[\text{yeast}]{\text{sucrase}} 2\,C_6H_{12}O_6 \qquad (23.1)$$

Sucrose	Glucose/Fructose
342	180 (2)
100 g	105.26 g

$$2C_6H_{12}O_6 \longrightarrow 4C_2H_5OH + 4CO_2 \qquad (23.2)$$

Glucose/Fructose	Ethanol	Carbon dioxide
180 (2)	46 (4)	44 (4)
105.26 g	53.80 g	51.46 g
or 100 g	51.11 g	48.89 g

Based on stoichiometry, 1 mole of sucrose produces 4 moles of ethanol (53.80 g ethanol per 100 g sucrose) and 1 mole of six-carbon sugars (e.g., glucose and fructose) produces 2 moles of ethanol (51.11 g ethanol per 100 g of six-carbon sugar).

Stoichiometrically, 1 metric ton of sucrose produces:

$(1,000 \text{ kg/metric ton}) \times (53.80 \text{ kg ethanol}/100 \text{ kg sucrose}) \times (1/0.789 \text{ L ethanol/kg}$

$\text{ethanol}) = 682 \text{ L ethanol (or } 180 \text{ gal ethanol)}$

Under actual conditions, a yield of 588 L ethanol per metric ton of sucrose (155 gal of ethanol per metric ton of sucrose) can be expected from sugarcane, and molasses, a byproduct of the sugarcane-to-sugar production process, can produce up to 270 L ethanol per metric ton of molasses (71 gallons ethanol per metric ton of molasses). Note that sugarcane molasses contains about 49.2% total sugars; that is, 35% sucrose and about 14% reducing sugars (glucose and fructose).

The other major product of ethanol fermentation is carbon dioxide. Based on stoichiometry, 4 moles of CO_2 are produced for every mole of sucrose fermented to ethanol.

Stoichiometrically, fermentation of 1 metric ton of sucrose produces:

$= (1,000 \text{ kg/metric ton}) \times (51.46 \text{ kg } CO_2/100 \text{ kg sucrose})$

$= 515 \text{ kg } CO_2 (\text{or } 1,133 \text{ lb } CO_2)$

23.3 Sugarcane Ethanol

23.3.1 Ethanol Production Process

Sugarcane (*Saccharum* spp.) stalks are mechanically pressed to extract juice (Figures 23.1 and 23.2). The juice is primarily composed of water and sucrose, a 12-carbon sugar. The sucrose concentration in sugarcane mixed juice ranges from 10–15%. Yeast, *S. cerevisiae*, ferments sucrose into ethanol directly. Ethanol is then recovered by distillation. Molasses is a byproduct remaining after the recovery of sucrose from sugar-rich juice in the sugar production process. Molasses contains around 49.2% total sugars. The sugarcane-to-sugar processing plant generates around 11.4 L of molasses as a byproduct per 45.5 kg of raw sugar produced (3.0 gal molasses per 100 lb of raw sugar produced). Typically, the sugar concentration in molasses is about 20–30 degree Brix (°Bx). (Note: °Bx is a measurement of dissolved sugar in aqueous phase by mass; for example, 15 °Bx is equal to 15 g sugar per 100 g solution). For ethanol fermentation, the substrate can be up to 30 °Bx, but generally it is diluted to 15 °Bx.

FIG. 23.1 Sugarcane mill with imbibition water added for sugar extraction and bagasse production from Hawaiian Commercial & Sugar Co, Maui, HI. Source: Photo courtesy of Samir Khanal, University of Hawaii.

FIG. 23.2 Sugarcane mixed juice from Hawaiian Commercial & Sugar Co, Maui, HI. Source: Photo courtesy of Samir Khanal, University of Hawaii.

The enzymes present in the yeast, namely *sucrase* or *invertase*, first convert disaccharides, such as sucrose ($C_{12}H_{22}O_{11}$), into simpler carbohydrates (i.e., monosaccharides such as glucose and fructose). These monosaccharides are then converted into ethanol and carbon dioxide by the enzyme zymase excreted by yeast, *S. cerevisiae*.

- **Bagasse:** A fibrous residue that remains after sugarcane/sweet sorghum juice extraction. The fresh bagasse contains around 40–50% moisture. Typically bagasse is used as a fuel in the boiler for the production of steam and electricity for in-plant use. Excess electricity produced may be supplied to the grid.

- **Vinasse:** A liquid fermentation by-product remaining after ethanol recovery. Around 8–15 L vinasse per liter of ethanol is produced. Although in Brazil it is generally disposed off by land application known as fertirrigation, vinasse is a nutrient-rich liquid stream, which can be anaerobically digested to produce biogas or other high-value products.

23.3.2 Sugarcane-to-Ethanol Biorefinery

Figure 23.3 is a schematic diagram of a sugarcane-to-ethanol biorefinery. Imbibition water is added during the juice extraction phase to facilitate the extraction process. A typical water addition rate would be about 325 kg of water per metric ton of cane for optimal extraction. The extracted juice is clarified before fermentation, which removes the non-sugar materials in juice through heat, and the addition of lime, which allows these materials to settle at the bottom of the clarifier. Plants that produce edible sugar generate molasses as a co-product, which can be fermented into ethanol similar to the fermentation of sugarcane juice to ethanol. If molasses is used for ethanol fermentation, dilution may be necessary for efficient fermentation.

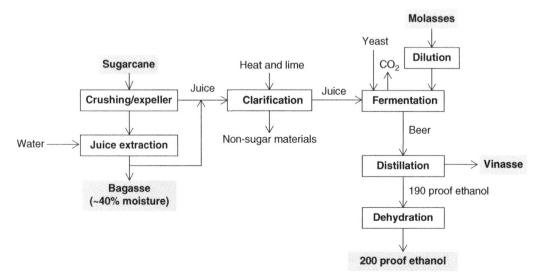

FIG. 23.3   Schematic diagram of the ethanol production process from sugarcane/molasses.

Brazil has successfully established sugarcane-based ethanol biorefineries. The low production cost of sugarcane-based ethanol in Brazil is hard to compete against by other countries, even where the climate is favorable for sugarcane production. A typical sugarcane-to-ethanol plant in Brazil uses 2 million metric tons of sugarcane annually and produces 200 million liters of ethanol per year (52.8 million gallon per year; Godemberg and Guardabassi, 2010). It is important to point out that typical Brazilian mills produce both crystalline sugar (edible) and ethanol, and produce more of either product depending on which pays more. The sugarcane-to-ethanol plant generally runs on a seasonal basis (typically, from April to November). The plantation size for such a biorefinery is on average 30,000 hectares (74,000 acres), and the driving distance to the biorefinery is limited to 70 km (43.5 miles) to minimize the degradation of harvested canes (UNICA, 2013).

Each metric ton of harvested sugarcane yields:

- 740 kg juice (135 kg sucrose and 605 kg water)
- 260 kg moist bagasse (130–140 kg dry bagasse)
- 70 L ethanol
- 560–1,050 L vinasse (total solids (TS) content of 2.2–2.5%)
- 55–60 kg CO_2

In 2008/2009, annual global sugarcane production was around 560 million metric tons, with annual ethanol production from sugarcane of 27.5 billion liters (7.27 billion gallons), primarily in Brazil. Although sugarcane is grown in Hawaii, Florida, Louisiana, and Texas, currently no commercial sugarcane-to-ethanol plant exists in the USA. Economics certainly has been the major factor, as corn-based ethanol can be produced at nearly half the price of sugarcane-based ethanol in the USA due to differences in both feedstock production and processing costs. Also, the traditional conversion of cane juice into crystalline sugar for food consumption is a competitive use for sucrose derived from sugarcane. Similarly, molasses-based ethanol appears to be more cost competitive with corn-based ethanol in the USA. The estimated production costs of ethanol from different sugar-based feedstocks are summarized in Table 23.1. For comparison, the table also shows the production cost of sugarcane-based ethanol in Brazil.

In terms of capital costs, it is estimated that a 20 MGY (million gallon per year) sugarcane-based ethanol biorefinery would cost in the range of $2.10–2.20 per gallon of annual capacity in the continental USA.

Table 23.1 Summary of estimated ethanol production cost (US$/gallon of ethanol)

Cost items	US sugarcane US$/gal	US molasses US$/gal	US corn dry milling[a] US$/gal	Brazil sugarcane US$/gal
Feedstock	1.48	0.91	0.53	0.30
Processing	0.92	0.36	0.52	0.51
Total	2.40	1.27	1.05	0.81

Note: [a]The process generates a high-value co-product known as distiller's dry grains with solubles (DDGS), which is sold as an animal feed at a price of $250–275 per metric ton.
Source: Data from USDA, 2006.

Example 23.1: A Hawaiian sugar company plans to build a molasses-based ethanol plant in Maui. The plant generates 100 gal molasses/hr. Assume that the specific gravity of molasses is 1.4 at 85 °Bx, and the density of ethanol is 789 kg/m³. How much ethanol can be generated per year?

Solution:

1. **Find the mass of molasses**

$$(100 \, \text{gal} \times 3.785 \, \text{L/gal}) \times (1.4 \times 1000 \, \text{kg/m}^3) \times 10^{-3} \text{m}^3/\text{L} = 530 \, \text{kg/hr}$$
$$= \mathbf{4.64 \times 10^6 kg/yr} \, (\textbf{assuming 365 operation days per year})$$

Since the total sugar content of molasses is around 49% (35% sucrose and 14% reducing sugars), total sucrose produced is 1.62×10^6 kg/yr ($4.64 \times 10^6 \times 0.35$) and total reducing sugar is 649,600 kg/yr ($4.64 \times 10^6 \times 0.14$).

2. **Calculate the ethanol produced from sucrose (Equations 23.1 and 23.2)**

$$\left(1.62 \times 10^6 \, \text{kg sucrose/yr}\right) \times (53.8 \, \text{kg ethanol}/100 \, \text{kg sucrose})$$
$$\times (1 \, \text{L ethanol}/0.798 \, \text{kg ethanol})$$
$$= \mathbf{1.10 \times 10^6 L \, ethanol/yr} \, (\textbf{290,620 gal ethanol/yr})$$

3. **Calculate the ethanol produced from reducing sugars (Equation 23.2)**

Note: Reducing sugars in molasses are referred to glucose and fructose, which are six-carbon monosaccharides.

$$= (649,600 \, \text{kg sugar/yr}) \times (51.11 \, \text{kg ethanol}/100 \, \text{kg sugar})$$
$$\times (1 \, \text{L ethanol}/0.798 \, \text{kg ethanol}) = \mathbf{420,799 \, L \, ethanol/yr} \, (\textbf{111,175 gal ethanol/yr})$$

Thus, total ethanol produced is $(1.1 \times 10^6 + 420,799)$ L/yr = $\mathbf{1.52 \times 10^6 \, L/yr}$ (**401,796 gal/yr**).

23.4 Sweet Sorghum Ethanol

Sweet sorghum (*Sorghum bicolor* (L.) Moench) is similar to sugarcane in many aspects, including having a stalk with sugar-rich juice. Sweet sorghum is an alternative sugar-rich crop for ethanol production. Its short growing season, high yield, high carbon assimilation rate, high water efficiency, and lower nutrient requirement are some of the merits of sweet sorghum. Researchers at Oklahoma State University are exploring farm-level juice extraction and fermentation to develop a farm-scale biorefinery. Of the total sugar content in the sweet sorghum juice, 53–60% is sucrose, 28–33% is glucose, and 7–19% is fructose. The sugar concentration in the juice depends on maturity and the average concentration is around 15 °Bx.

The sweet sorghum-to-ethanol-production process is very similar to the sugarcane-to-ethanol process as described earlier (Figure 23.3). Similar to sugarcane-ethanol production, the three major residues, including bagasse, vinasse, and carbon dioxide, are co-generated as byproducts.

> # 1 metric ton of sweet sorghum stalk yields:
>
> - 793 kg juice (113 kg sucrose, 665 kg water, and 16 kg other)
> - 307 kg moist bagasse
> - 69 L ethanol
> - 560–1,050 L vinasse (TS content of 2–3%)
> - 55–60 kg CO_2
>
> Note: 100 kg water is added during juice extraction.

23.5 Sugar Beet Ethanol

Unlike sugarcane or sweet sorghum, sugar beet (*Beta vulgaris* L.) is a root crop and is commonly grown in temperate climates. Sugar beet processing for juice extraction involves several steps. The roots are thoroughly cleaned to remove soil/dirt, and then are sliced into long, thin strips, known as cossettes. Slicing increases the surface area and facilitates the extraction of sugar. Sucrose is then extracted using hot water (50–80 °C or 122–176 °F) through a diffusion process in which the cossettes are passed through continuous diffusers. The diffusers are tanks in which the cossettes and water move in counter-current directions. The tanks may be oriented vertically, horizontally, or at an angle. The sugar solution exiting from the other end is known as juice and contains between 16% and 18% sucrose.

The residual cossettes are separated using a screen, which contains a significant amount of water (~95%) as well as some usable sugar. Screw presses are employed to remove the excess water from the residual cossettes. The squeezed water becomes part of the juice and the remaining pressed cossettes, known as beet pulp, contain around 70–75% water. In general, beet pulp is mainly composed of cellulose (20%), hemicellulose (25%), and pectin (25%), together with small amounts of protein, lignin, and ash, and is commonly used for animal feed applications.

Similarly to other sugar-based feedstocks, the juice becomes a substrate for yeast fermentation. In ethanol production, the fermentation followed by distillation results in a liquid waste stream, vinasse. Although sugar beet vinasse has generally similar characteristics to other vinasse from sugar-based feedstocks, it has a higher protein content from betaine and glutamic acid.

Like sugarcane-based ethanol, there is no sugar beet–based ethanol biorefinery currently in operation in the USA. The primary reason is the high production cost. The capital costs are similar to sugarcane-to-ethanol biorefineries, as discussed earlier. In 2008, the European Union produced nearly 2.7 billion liters (0.71 billion gallons) of ethanol, primarily from sugar beet.

Example 23.2: Calculate the amount of ethanol produced per metric ton of sugar beet.

Solution:

1. **Calculate the amount of sugar (sucrose) produced from 1 metric ton (1,000 kg) of sugar beets**

   Since the sugar recovered from sugar beets is 15.5%, the amount of sugar produced per metric ton of sugar beets is:

   $$= (1,000 \text{ kg}) \times (15.5 \text{ kg sucrose}/100 \text{ kg sugar beet}) = \textbf{155 kg}$$

2. Calculate the total ethanol produced based on stoichiometry (Equations 23.1 and 23.2)

= 155 kg sucrose × (53.8 kg ethanol/100 kg sucrose) × (1 L ethanol/0.789 kg ethanol)

= 106 L ethanol (28 gal ethanol)

Considering an efficiency of fermentation of 86.5%, similar to the cane sugar-to-ethanol process, the ethanol yield would be around 106 × 0.865 = **92 L/metric ton of sugar beets (22 gal per ton)**.

23.6 Biochemicals and Biopolymers

Sugar is a critical substrate for producing virtually all forms of bioenergy (ethanol, butanol, biodiesel, biomethane, biohydrogen, etc.) and several important biobased products (e.g., biochemicals and polymers) via a biochemical platform. The previous sections discussed how sugars derived from sugarcane, sweet sorghum, and sugar beet have been commercially employed for biofuel, especially bioethanol production. Biorefinery is much broader than just producing bioenergy. As a starting chapter on biorefinery, a discussion of the biochemical conversion of monomeric sugars to various high-value platform chemicals and polymers is presented here (see Figure 23.4). Several representative platform chemicals, including lactic acid, succinic acid, 1,3-propanediol, and 3-hydroxypropionic acid, are introduced briefly. The same concept is also applicable to starch-based biorefinery (Chapter 24) and lignocellulosic biomass-based biorefinery (Chapter 25), which yield monomeric sugars.

23.6.1 Lactic Acid

The global demand for lactic acid is estimated to be about 130,000–150,000 metric tons per year (Li and Cui, 2010). Biobased production of lactic acid has been commercialized by a number of companies, such as Archer Daniels Midland (ADM) and Natureworks (a joint venture of Cargill and Teijin) in the USA, and Jiangxi Musashino Bio-chem (JMB) in China. ADM produces lactic acid for polylactic acid and ethyl lactate. Natureworks offers various biopolymers with a capacity of 140,000 metric tons per year at competitive cost and performance to petroleum-based materials (http://www.natureworksllc.com). JMB produces L-lactic acid from grain starch with a capacity of 5,000 metric tons per year.

Lactic acid is a monomer for producing *poly[lactic acid]*, *PLA*), commonly known as bioplastic or biodegradable plastic. Lactic acid has two optical isomers, L-(+)-lactic acid and D-(–)-lactic acid. Homopolymers that are made up of one type of monomer unit usually form regular structures with a crystalline phase. A mixture of different types of monomer will result in unfavorable amorphous materials. Therefore, high-purity L-(+)- or D-(–)-lactic acid is required in order to obtain PLA with desirable properties. Chemical synthesis usually results in a mixture of the two isomers (racemic DL-lactic acid), which are difficult to separate due to their very similar properties. Microbial fermentation is capable of selectively producing pure L-(+)- or D-(–)-lactic acid by using suitable microorganisms. Recovery and purification of lactic acid from the fermentation medium is still a challenge, but it is technically feasible and much easier than separating the two isomers. Various microorganisms, such as *Carnobacterium, Enterococcus, Lactobacillus, Lactococcus, Leuconostoc, Oenococcus, Pediococcus,*

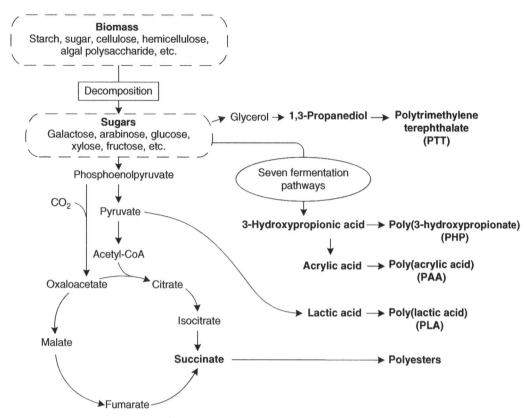

FIG. 23.4   Biorefinery of biomass feedstocks for producing platform chemicals. Source: Adapted from Jang *et al.*, 2012.

Streptococcus, Tetragenococcus, Vagococcus, and *Weissella,* can ferment monomeric sugars to produce lactic acid with pyruvate as the direct precursor (Jang *et al.*, 2012). Moreover, engineered *E. coli* and *Sporolactobacillus* have also been developed to produce D-lactate with productivities of 3.5 g/(L·hr) and 3.8 g/(L·hr), and yields of 0.86 g/g and 0.98 g/g of glucose, respectively (Wang *et al.*, 2011; Zhu *et al.*, 2007). Advances in metabolic engineering have made it possible to reorient the metabolic pathway in host cells for producing either L-(+)-lactic acid or D-(−)-lactic acid (Wee *et al.*, 2006).

23.6.2 Succinic Acid

Succinic acid is an essential platform chemical for food, agricultural, and pharmaceutical products (Zeikus *et al.*, 1999). Succinate is an intermediate in the tricarboxylic acid (TCA) cycle which is a key part of aerobic respiration in cells. In the TCA cycle, succinate is further converted into fumarate by succinate dehydrogenase. Therefore, aerobic accumulation of succinate can be achieved by inactivating the gene encoding succinate dehydrogenase. The primary pathway for anaerobic production of succinate relies on the carboxylation of phosphoenolpyruvate to oxaloacetate, which makes it possible to couple succinic acid production with CO_2 fixation (Cheng *et al.*, 2012). In the past three decades, metabolically engineered strains have been developed for increasing succinic acid production and minimizing byproducts formation. *E. coli* is one of the most extensively studied strains for succinate production because of its rapid cell growth, simple culture medium, and the available genetic tools (Cheng *et al.*, 2012). Furthermore, metabolic engineering of *Corynebacterium glutamicum* has also been studied for either anaerobic or aerobic production of succinic acid. For example, an engineered *C. glutamicum* strain can anaerobically produce 134 g/L of succinic acid with a yield of 1.1 g/g glucose and a productivity of 2.5 g/(L·hr) with glucose, formate, and carbon dioxide as carbon sources (Litsanov *et al.*, 2012). The potential market demand for succinic acid has been increasing, because it is a precursor of many important chemicals and solvents such as 1,4-butanediol and tetrahydrofuran, and biodegradable polymers like polybutyrate succinate (PBS; Cheng *et al.*, 2012). Production of succinic acid from glucose has been commercialized by several companies, such as Myriant (http://www.myriant.com) and BioAmber (http://www.bio-amber.com; Erickson *et al.*, 2012).

23.6.3 1,3-Propanediol

1,3-Propanediol is mainly used as a building block in the production of polyesters, such as polytrimethylene terephthalate (PTT). 1,3-Propanediol can be produced by microbes, such as *Clostridium*, *Enterobacter*, *Klebsiella*, and *Lactobacillus*, using glycerol as the sole carbon source (Oh *et al.*, 2012; Wilkens *et al.*, 2012; Gungormusler *et al.*, 2010). There are no microorganisms in nature that are able to directly convert sugars to 1,3-propanediol. However, conversion of glucose to glycerol is a natural pathway that is found in yeast, *S. cerevisiae*. Dupont and Genencor International have developed an engineered *E. coli* that is able to convert sugars to glycerol, and glycerol to 1,3-propanediol (Emptage *et al.*, 2009). Concentrations as high as 135 g/L of 1,3-propanediol have been reported using glucose as a carbon course with a productivity of 3.5 g/(L·hr) and a yield of 0.51 g/g. Recently, several bio-based 1,3-propanediol systems (including Zemea and Susterra propanediol) have been commercialized by DuPont Tate and Lyle BioProducts (http://www.duponttateandlyle.com).

23.6.4 3-Hydroxypropionic Acid

3-Hydroxypropionic acid is an important compound for producing poly(3-hydroxypropionate), a polyhydroxyalkanoate (PHA) that has excellent material characteristics and exhibits a large variety of applications. 3-hydroxypropionic acid is also a precursor for producing acrylic acid, which has wide applications in coatings, personal care, packaging, building, and construction. Although no known natural microorganisms can produce 3-hydroxypropionic acid as a major metabolic end-product, synthetic pathways have been proposed to produce 3-hydroxypropionic acid from glucose (Jiang *et al.*, 2009). Dow and OPX Biotechnologies are collaborating on the production of biobased *acrylic acid* from glucose with 3-hydroxypropionic acid as the intermediate. Cargill has also been working on a project for 3-hydroxypropionic acid production from glucose in partnership with BASF and Novozymes (Matsakas *et al.*, 2014). Recently, BASF, Cargill and Novozymes announced that they have achieved pilot-scale biobased acrylic acid production with 3-hydroxypropionic acid as the precursor (http://www.novozymes.com).

References

Cheng, K.-K., Zhao, X.-B., Zeng, J., and Zhang, J.-A. (2012). Biotechnological production of succinic acid: Current state and perspectives. *Biofuels, Bioproducts and Biorefining* 6: 302–18.

Emptage, M., Haynie, S.L., Laffend, L.A., Pucci, J.P., and Whited, G. (2009). Process for the biological production of 1, 3-propanediol with high titer. Google Patents.

Erickson, B., Nelson, J.E., and Winters, P. (2012). Perspective on opportunities in industrial biotechnology in renewable chemicals. *Biotechnology Journal* 7: 176–85.

Godemberg, J., and Guardabassi, P. (2010). The potential for first-generation ethanol production from sugarcane. *Biofuels, Bioproducts and Biorefinery* 4: 17–24.

Gungormusler, M., Gonen, C., Ozdemir, G., and Azbar, N. (2010). 1, 3-Propanediol production potential of *Clostridium saccharobutylicum* NRRL B-643. *New Biotechnology* 27: 782–8.

Jang, Y.-S., Kim, B., Shin, J.H., *et al.* (2012). Bio-based production of C2–C6 platform chemicals. *Biotechnology and Bioengineering* 109: 2437–59.

Jiang, X., Meng, X., and Xian, M. (2009). Biosynthetic pathways for 3-hydroxypropionic acid production. *Applied Microbiology and Biotechnology* 82: 995–1003.

Li, Y., and Cui, F. (2010). Microbial lactic acid production from renewable resources. In: O. Singh and S. Harvey (eds), *Sustainable Biotechnology*. Dordrecht: Springer, pp. 211–28.

Litsanov, B., Brocker, M., and Bott, M. (2012). Toward homosuccinate fermentation: Metabolic engineering of Corynebacterium glutamicum for anaerobic production of succinate from glucose and formate. *Applied and Environmental Microbiology* 78: 3325–37.

Matsakas, L., Topakas, E., and Christakopoulos, P. (2014). New trends in microbial production of 3-hydroxypropionic acid. *Current Biochemical Engineering* 1: 141–54.

Oh, B.R., Seo, J.W., Heo, S.Y., *et al.* (2012). Fermentation strategies for 1,3-propanediol production from glycerol using a genetically engineered Klebsiella pneumoniae strain to eliminate by-product formation. *Bioprocess and Biosystems Engineering* 35: 159–65.

UNICA (2013). *Final Report of 2012/2013 Harvest Season-South-center of Brazil*. Sugarcane Harvest Reports. São Paulo: União da Indústria de Cana-de-Açúcar. http://www.unicadata.com.br/listagem.php?idMn=83, accessed October 2013.

USDA (2006). The economic feasibility of ethanol production from sugar in the United States. Washington, DC: U. S. Department of Agriculture. http://www.usda.gov/oce/reports/energy/EthanolSugarFeasibilityReport3.pdf, accessed June 2009.

Wang, L., Zhao, B., Li, F., *et al.* (2011). Highly efficient production of D-lactate by *Sporolactobacillus* sp. CASD with simultaneous enzymatic hydrolysis of peanut meal. *Applied Microbiology and Biotechnology* 89: 1009–17.

Wee, Y.J., Kim, J.N., and Ryu, H.W. (2006). Biotechnological production of lactic acid and its recent applications. *Food Technology and Biotechnology* 44: 163–72.

Wilkens, E., Ringel, A.K., Hortig, D., Willke, T., and Vorlop, K.D. (2012). High-level production of 1,3-propanediol from crude glycerol by *Clostridium butyricum* AKR102a. *Applied Microbiology and Biotechnology* 93: 1057–63.

Zeikus, J.G., Jain, M.K., and Elankovan, P. (1999). Biotechnology of succinic acid production and markets for derived industrial products. *Applied Microbiology and Biotechnology* 51: 545–52.

Zhu, Y., Eiteman, M.A., DeWitt, K., and Altman, E. (2007). Homolactate fermentation by metabolically engineered *Escherichia coli* strains. *Applied and Environmental Microbiology* 73: 456–64.

Exercise Problems

23.1. Discuss how a biorefinery concept can be applied to converting sugarcane to ethanol.

23.2. How is the sugar-rich juice extracted from sugar beets? Discuss using a schematic diagram.

23.3. A sugarcane-based ethanol biorefinery processes 100 metric tons of sugarcane daily. Assume that the density of ethanol is 0.789 kg/L. Compare the heating value of glucose and ethanol.

23.4. In Problem 23.3, the sugarcane is used to produce food-grade sugar. How much molasses can be produced? Calculate the amount of ethanol produced from molasses. If bagasse is used for producing electricity, how much electricity can be generated from the bagasse annually?

23.5. Develop a complete mass balance for Problem 23.2.

23.6. Develop a complete mass balance for a sugar beet–ethanol biorefinery.

23.7. A 5 million gallon sugar beet molasses-to-ethanol plant is to be built in Minnesota. Calculate the amount of sugar beet required (metric tons).

23.8. Calculate the amount of CO_2 produced in Problem 23.7.

23.9. Compare the theoretical ethanol yield on a per metric ton basis of three major sugar-based feedstocks.

23.10. Compare the maximum ethanol yield per hectare of land for three major sugar-based feedstocks based on currently available best yield data.

Starch-Based Biorefinery

Samir Kumar Khanal and Saoharit Nitayavardhana

What is included in this chapter?

This chapter covers different starch-based biorefineries with a focus on process for ethanol production, stoichiometry, generation of different biobased and co-products, and mass balance.

24.1 Introduction

Starch-based ethanol is a typical example of a first-generation biofuel, and several countries have already successfully developed commercial starch-based biorefineries for the production of biofuel and biobased products. The USA, for example, generates nearly all of its ethanol domestically from corn to meet its annual renewable fuel goals and regulations. Similarly, in Thailand, cassava-based ethanol biorefineries have been successfully established.

In general, starch-based feedstocks are converted into ethanol using two types of processes: wet milling and dry milling. Both approaches have been designed and refined over the years to produce ethanol while simultaneously generating various (marketable) co-products. The wet milling process in fact represents a true biorefinery, which produces diverse co-products in addition to ethanol, utilizing each and every component of the feedstock. The oil embargo on the USA during the early 1970s was arguably the starting point for much of the first-generation biofuel technology, which has resulted in the significant growth of corn ethanol industries in the USA. As of January 2014, the USA had nearly 210 installed corn/sorghum-based biorefineries with a total annual capacity of 14.9 billion gallons (56.4 billion liters) and an actual ethanol production of 13.3 billion gallons (50.3 billion liters) in 2012. In recent years, environmental concerns coupled with unstable petroleum prices have led to increased starch-based ethanol production globally.

Bioenergy: Principles and Applications, First Edition. Edited by Yebo Li and Samir Kumar Khanal.
© 2017 John Wiley & Sons, Inc. Published 2017 by John Wiley & Sons, Inc.
Companion website: www.wiley.com/go/Li/Bioenergy

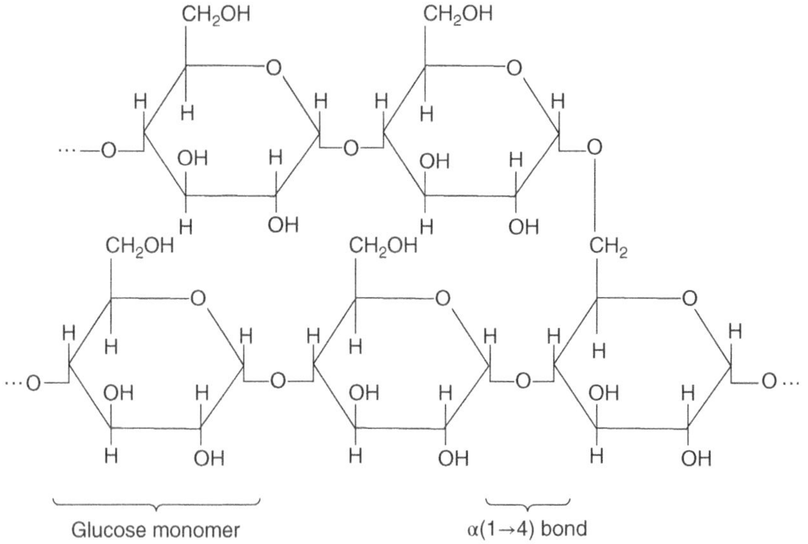

FIG. 24.1   The structure of starch.

As mentioned previously, starch-based (first-generation) feedstocks such as corn, potato (sweet potato), grain sorghum, and cassava are used for ethanol production. Unlike sugar-based (first-generation) feedstocks (i.e., sugar beet and sugarcane juice), starch-based feedstocks cannot be directly fermented to ethanol by traditional yeast such as *Saccharomyces cerevisiae*. Starch hydrolysis (liquefaction and saccharification) is an essential step in the upstream processing of all starch-based biorefineries to release soluble sugars, namely glucose, for downstream fermentation (Figure 24.1 shows the structure of a starch molecule). In traditional starch hydrolysis, the starch-based feedstocks are first cooked at a high temperature (95–100 °C) for starch gelatinization. Starch gelatinization is a process that breaks down the intermolecular bonds of starch molecules in the presence of water and heat. The first enzyme, alpha-amylase, is added which breaks down the long-chain starch molecules to low molecular weight carbohydrate, dextrins. This first step is called liquefaction. Since alpha-amylase is a heat-tolerant enzyme, both gelatinization and liquefaction steps are integrated in full-scale plants. The resulting freed-up starch molecules, dextrins, are then converted to glucose by a process known as saccharification with the addition of another enzyme known as glucoamylase. Saccharification occurs at a lower temperature (55 °C) and can be carried out prior to fermentation or often simultaneously to fermentation, commonly known as simultaneous-saccharification and fermentation (SSF). Currently, with genetic engineering techniques, a newly developed commercial enzyme, *STARGEN*^{TM}, allows starch hydrolysis to occur without the cooking step.

To fully understand and appreciate the merits and demerits of the first-generation starch-based biorefineries, it is necessary to take a closer look at the various steps involved in biofuel production, the co-generation of high-value co-products, and the proper management of waste streams. It is also important to note that starch can be enzymatically hydrolyzed to glucose, which is an ideal substrate to produce diverse platform biochemicals and biopolymers; these were discussed in Chapter 23, so will not be considered here.

24.2 Stoichiometry of Starch to Ethanol

Stoichiometrically, conversion of starch-based feedstocks into ethanol can be illustrated as shown in Equation 24.1:

$$(C_6H_{10}O_5)_n + H_2O \xrightarrow{\overset{\text{amylolytic}}{\underset{}{\text{enzymes}}}} n\,C_6H_{12}O_6 \xrightarrow{\text{yeast}} 2n\,CH_3CH_2OH + 2n\,CO_2 \quad (24.1)$$

Starch	Water	Glucose	Ethanol	Carbon dioxide
162 (n)		180 (n)	46 (2n)	44 (2n)
100 g		111.11 g	56.79 g	54.32 g

Based on Equation 24.1, 100 g of starch is hydrolyzed into 111.11 g of glucose, and 1 mole of glucose is converted into 2 moles of ethanol. The theoretical yield of ethanol is 56.79 g per 100 g starch.

Example 24.1: The moisture content of corn and cassava chips is assumed to be around 15% and 14%, respectively. The starch content (dry wt.) of corn and cassava is 73% and 79%, respectively. How much ethanol can be produced from 1 bushel of dry corn and 1 metric ton of dry cassava chips, respectively?

Solution:

1. Corn to ethanol
   Amount of starch in 1 bushel of dry corn:

$$= \frac{56\,\text{lbs}}{1\,\text{bushel}} \times (1-15\%) \times \frac{0.45\,\text{kg}}{1\,\text{lb}} \times \frac{73\,\text{kg starch}}{100\,\text{kg corn}} = 15.6\,\text{kg starch}$$

   Ethanol production from 1 bushel of corn:

$$= 15.6\,\text{kg starch} \times \frac{0.5679\,\text{kg ethanol}}{1\,\text{kg starch}} \times \frac{1\,\text{L ethanol}}{0.789\,\text{kg ethanol}} \times \frac{1\,\text{gal}}{3.785\,\text{L}} = \mathbf{3.0\,gal}$$

   However, in reality, the ethanol yield is typically around 2.7–2.8 gallons per bushel of corn in the dry-grind process.

2. Cassava to ethanol
   Starch amount in 1 metric ton of dry cassava chips:

$$= \frac{1{,}000\,\text{kg}}{1\,\text{metric ton}} \times (1-14\%) \times \frac{79\,\text{kg starch}}{100\,\text{kg cassava}} = 679.4\,\text{kg starch}$$

   Ethanol production from 1 metric ton of cassava chips:

$$= 679.4\,\text{kg starch} \times \frac{0.5679\,\text{kg ethanol}}{1\,\text{kg starch}} \times \frac{1\,\text{L ethanol}}{0.789\,\text{kg ethanol}} = \mathbf{489\,L\,(129\,gal)}$$

   The actual ethanol yield, however, is around 337 L ethanol/metric ton cassava chips (14% moisture content), which is about 387 L/dry metric ton cassava chips.

24.2.1 Corn-Based Ethanol Biorefinery

The dominant feedstock for ethanol in the USA is corn (*Zea mays* L.), which accounts for over 95% of total ethanol production, but grain sorghum (*Sorghum bicolor*) has also been used in facilities in Nebraska and Kansas. Corn-based ethanol in the USA is produced by either dry-grind milling or wet-milling processes. The former, however, accounts for the majority of ethanol production (>80%).

A typical **dry-grind** corn ethanol process is illustrated in Figure 24.2. The whole corn (kernel) is ground in a hammer mill or roller mill, and then mixed with water to form a mash. The mash is adjusted to pH 6.0 and cooked in a jet cooker at 95–100 °C (203–212 °F) for 15–20 min. A small amount of the alpha-amylase is added during jet cooking to assist in liquefaction. Additional alpha-amylase is added during secondary liquefaction, which occurs for 90 min at 95 °C (203 °F). The cooked mash is then cooled to 55–60 °C (131–140 °F), adjusted to pH 4.5, and mixed with the enzyme glucoamylase to convert the starch to fermentable sugars, a process known as saccharification. This saccharified mash is then fermented to ethanol using yeast, *S. cerevisiae*. In most plants, saccharification and fermentation occur simultaneously, known as simultaneous-saccharification and fermentation (SSF), to minimize the inhibition of enzymes by the product (sugar). Fermentation is typically conducted at a pH of 4.8–5.0 and a temperature of 30–37 °C (86–99 °F) for 48–72 hr.

The fermented mash, often referred to as beer, is distilled to produce 95% ethanol by volume (also known as 190 proof). Further dehydration of ethanol is achieved by molecular sieves, which preferentially retain the water while allowing the ethanol to pass. The product stream is 99.5% ethanol (200 proof). Fermentation residues, left at the bottom of the distillation tanks, are referred to as whole stillage and are centrifuged to obtain wet cake. The wet cake is passed through a series of dryers to obtain *distiller's dry grains* (DDG). Thin stillage, the liquid fraction from centrifugation, is partially dehydrated by evaporative processes to obtain a syrup. The syrup is blended with DDG to form distiller's dried grains with soluble (DDGS; Figure 24.3). Since the syrup needs to be dried further, it is added before or during drying. In two-stage drying, ideally syrup is added just in the second stage of drying. The remainder of the thin stillage (approximately 50%) is then recycled as process water (not

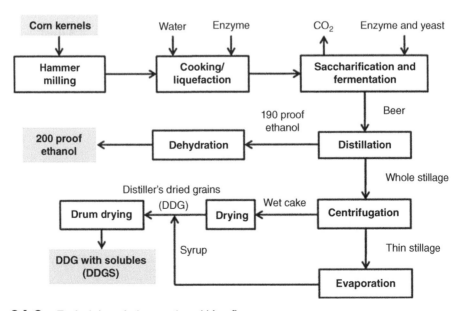

FIG. 24.2   Typical dry-grind corn ethanol biorefinery.

FIG. 24.3   A DDGS pile in a dry-grind corn ethanol plant in Iowa. Source: Photo courtesy of Samir Khanal, University of Hawaii.

shown in Figure 24.2). Dry-grind plants produce ethanol as a major product and DDGS as a byproduct, which is sold as animal feed. The carbon dioxide from the fermentation process can be captured for use in carbonating beverages and manufacturing dry ice. There is also an attempt to use the captured CO_2 in algae cultivation. A typical dry-grind ethanol plant has a production capacity of 50 million gallons per year (MGY) (189.3 million L/year). Due to the low capital cost and other logistical considerations, the dry-grind process is the most common in corn ethanol plants in the USA.

Each bushel (56 lb or 25.4 kg) of corn yields:

- 2.7–2.8 gal (10.2–10.6 L) of ethanol
- 17 lb (7.7 kg) (based on 90% dry wt.) of DDGS
- 1.85–2.2 gal (7.0–8.3 L) of thin stillage
- 17 lb (7.7 kg) (dry wt.) of CO_2

ICM Kansas based ICM is well known for designing and building dry-grind ethanol plants in the USA. ICM's equipment and proprietary technology are used in ethanol plants currently in operation throughout North America that are estimated to produce 6.7 billion gallons of ethanol annually. Further details can be found at http://www.icminc.com/.

Recent **modifications of the dry-grind process** include processing at reduced temperatures and pre-fractionation for the recovery of germ and fiber prior to fermentation of the corn starch portion. This process has undergone extensive investigation by researchers at the University of Illinois. The reduced temperature or non-cooking process uses a special enzyme that works effectively at low temperatures. $STARGEN^{TM}$ is such an enzyme, developed by DuPont. $STARGEN^{TM}$ contains *Aspergillus kawachi* alpha-amylase expressed in *Trichoderma reesei*, and a glucoamylase from *Aspergillus niger* that functions synergistically to hydrolyze starch into glucose. There are several full-scale modified dry-grind plants currently in operation in the USA. Since there is no cooking step involved in the process, the thin stillage is not directly recycled upstream. It is sterilized by boiling to eliminate any possibility of bacterial contamination. Modified dry milling is a relatively new development that incorporates some aspects of both wet- and dry-milling technologies.

Example 24.2: A full-scale 50 MGY dry-grind corn ethanol plant produces 2.8 gal of ethanol/bushel of corn. Calculate the ethanol production efficiency of this ethanol plant based on the stoichiometrical ethanol yield. Assume that the moisture content of corn is 13% and the corn starch content is 73% (dry wt.). Also calculate the amount of DDGS produced annually, if 17 lb (based on 90% dry wt.) of DDGS is co-generated with every 2.8 gal of ethanol produced.

Solution:

1. **Amount of glucose produced from 1 metric ton (1,000 kg) of corn**

   The amount of glucose produced per metric ton of corn can be calculated using Equation 24.1:

   $$= \frac{1,000\,\text{kg corn}}{1\,\text{metric ton corn}} \times (1-13\%) \times \frac{73\,\text{kg starch}}{100\,\text{kg corn}} \times \frac{111.11\,\text{kg glucose}}{100\,\text{kg starch}}$$

   $$= 705.7\,\textbf{kg glucose}$$

2. **Stoichiometric amount of ethanol produced per ton of corn**

   Based on stoichiometry, 1 mole of glucose produces 2 moles of ethanol:

   $$= \frac{705.7\,\text{kg glucose}}{1\,\text{metric ton corn}} \times \frac{56.79\,\text{kg ethanol}}{111.11\,\text{kg glucose}} \times \frac{1\,\text{L ethanol}}{0.789\,\text{kg ethanol}} \times \frac{1\,\text{gal}}{3.785\,\text{L}}$$

   $$= 120.8\,\textbf{gal}$$

3. **Ethanol production efficiency from corn with the dry-grind process**

   Full-scale dry-grind corn ethanol plants produce 2.8 gal ethanol/bushel of corn. Thus, the actual ethanol yield from one ton of corn would be around:

   $$= \frac{\left(\dfrac{2.8\,\text{gal}}{\text{bushel}} \times \dfrac{2205\,\text{lb}}{\text{metric ton}}\right)}{56\dfrac{\text{lb}}{\text{bushel}}}$$

   $$= 110\,\textbf{gal}$$

4. **Efficiency of the dry-grind process**

   $$= \frac{110}{120.8} \times 100\% = \textbf{91\%}$$

5. **Amount of DDGS produced in 50 MGY annually in metric tons**

   $$= \frac{50 \times 10^{6}\,\text{gal ethanol}}{\text{year}} \times \frac{17\,\text{lb DDGS}}{2.8\,\text{gal ethanol}} \times \frac{1\,\text{metric ton}}{2205\,\text{lb}}$$

   $$= \textbf{137,674 metric ton (based on 90\% dry wt.)}$$

Note: The significant amount of DDGS co-generated in the dry-grind corn ethanol process is one of the major revenue sources for ethanol plants, and it is sold at a current market price of $250–270 per metric ton. With nearly 14 billion gallons of ethanol production per year, the amount of DDGS generated is over 30 million tons (English). Mexico, China, Canada, Vietnam, and Korea are the top five countries importing DDGS from the USA.

South Dakota–based POET (formerly known as Broin) is responsible for developing the turnkey project, design, engineering, construction, and management of their non-cooking ethanol plants (http://www.poetenergy.com/).

In the **wet-milling process**, prior to fermentation the corn kernel is separated into three major components: outer layer (known as bran or hull); germ (contains mainly oil); and endosperm (gluten and starch). A multitude of products such as starch, high-fructose corn syrup, germ, gluten feed, and corn gluten meal are produced by wet-milling plants. Thus, wet-milling facilities perhaps best exemplify a true biorefinery. The starch and germ are further processed into ethanol and corn oil, respectively. The process is named *wet milling* because each step involves significant water use for the separation and recovery of different products from the corn kernels.

The wet-milling process is illustrated in Figure 24.4. The corn kernels are screened and cleaned to remove stones, dirt, and other impurities. The cleaned kernels are then soaked in large tanks (steeps)

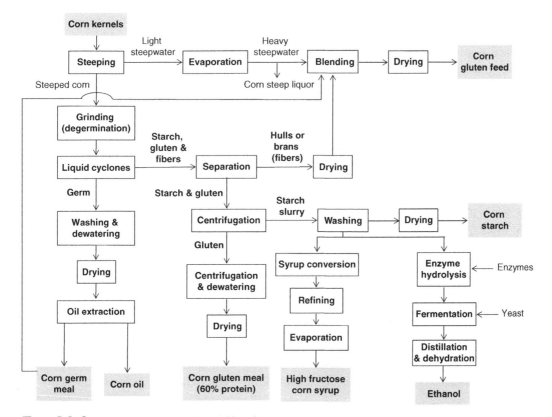

FIG. 24.4   Wet-milling corn ethanol biorefinery.

containing a dilute sulfuric acid solution (0.1–0.2%) at about 52 °C (125 °F) for 28–48 hr. The process is known as steeping (similar to making tea) and helps to soften the corn kernels for the easy separation of starch, gluten, germ, and other components. The moisture uptake by the corn kernels also helps to soften them.

The water separated from the steeped corn, known as light steepwater, contains corn solubles with a high protein content. The light steepwater is passed through evaporators to recover the protein and other nutrients. The resulting concentrated steepwater is now known as heavy steepwater or corn steep liquor, and is blended with fibrous residues (barn or hull) and corn germ meal to produce corn gluten meal. The corn steep liquor is sometime sold as a nutrient broth for fermentation.

Steeped corn is passed through degerminating mills (grinding step) that dislodge the germ as well as nearly half of the starch and gluten from the kernels. The pulpy slurry is then fed to liquid cyclones to separate germs from the mixture of starch, gluten, and fiber. The germ is primarily the oil-rich component of the corn kernel, and is easily separated. Following washing, dewatering, and drying (to <5% moisture content), the germ is processed to extract oil. The corn oil residue, known as corn germ meal, can also be mixed with fibrous fraction and becomes part of the corn gluten feed. The slurry exiting from the degerminating mills contains starch, gluten, and fibers, and is subjected to a series of washing and grinding operations to separate the starch and gluten from the fibers (hulls). The hulls are dried and blended with corn germ meal to produce corn gluten feed.

Low-density gluten is separated from the starch by centrifugation in two steps using a high-speed centrifuge. The resulting high-quality gluten (60–70% protein) and 1.0–1.5% solids are then further centrifuged, dewatered, and dried. Both gluten meal (~60% protein, 2.5% fat, and 1% fiber) and gluten feed (~21% protein, 2.5% fat, and 8% fiber) are widely used as animal feed. The starch fraction is washed to remove the residual gluten and solubles. Purified starch can be used for ethanol fermentation (similar to the dry-grind process), processed into high-fructose corn syrup, or dried to produce starch powder.

The merits of wet-milling plants are better operational flexibility; higher-value co-products generation; continuous operation; and lower equipment fouling and contamination. The primary demerits of wet-milling plants are higher capital costs, and slightly lower ethanol yield than the dry-grind process. In addition, the process also generates a large volume of dilute wastewater.

Each bushel (56 lb or 25.4 kg) of corn yields:

- 31.5–32 lb (14.3–14.5 kg) of starch
- 13.0–12.5 lb (7.7 kg) corn gluten feed
- 2.5–3.0 lb (8.3 kg) corn gluten meal
- 1.6–2.0 lb corn oil
- 2.5–2.6 gal (9.5–9.8 L) ethanol (if all starch is used for ethanol production)
- 17 lb (7.7 kg) (dry wt.) CO_2

Dry-grind and wet-milling processes are compared in Table 24.1.

Archer Daniels Midland (ADM), based in Decator, IL, has adopted wet-milling processes in its plants in Iowa and Illinois for producing ethanol and other co-products (http://www.adm.com).

Table 24.1 Comparison of dry-grind and wet-milling processes

	Dry-grind process	Wet-milling process
Cost	Low capital cost; but high operation and maintenance cost	High capital cost; but low operation and maintenance cost
Capacity	~50 MGY	>100 MGY
Products	Two major products:	Multiple products:
	• Ethanol • Distiller's dried grains with solubles (DDGS)	• High-fructose corn syrup/corn starch/ethanol • Corn oil • Corn gluten meal • Corn gluten feed

24.2.2 Corn-to-Ethanol Plants and Sorghum-to-Ethanol Plants

The USA is the number one ethanol-producing country in the world and over 95% of the total ethanol comes from corn starch using the dry-grind milling and wet-milling processes. The former, however, accounts for the majority of ethanol production (>80%). As of January 2014, the USA had nearly 210 corn/sorghum-based biorefineries with a total production capacity of 14.9 billion gallons per year. Based on 2008 corn production data, the USA used 3,026 million bushels of corn for ethanol production from an available total of 14,362 million bushels ; this means that about 21% of the total corn produced was diverted for ethanol production. In 2012, the USA used 4,550 million bushels of corn out of the total of 10,780 million bushels for ethanol production. This was equivalent to about 42% of the corn grown. The demand for corn has thus grown steadily due to a significant increase in the number of new ethanol plants.

Corn ethanol is sold at competitive prices in the continental USA. The feedstock cost has shown significant variations, ranging from $1.90/bushel in 2005 to $2.00–3.50/bushel in 2006. In 2007 and 2008 the feedstock price crossed $4.00/bushel, and in 2011 and 2012 the corn price exceeded $6.00/bushel. The capital cost for a corn-based ethanol plant is around $1.50/gallon of annual capacity based on a 20-MGY plant.

Although ethanol in the USA is commercially produced from corn in the corn-belt region (the Midwest), grain sorghum (*Sorghum bicolor*) has also been used interchangeably with corn in facilities in Nebraska and Kansas. The dry-grind process is especially adaptable to grain sorghum. In the sorghum belt of the USA, primarily in Kansas and Texas where the weather is not favorable for corn production, grain sorghum ethanol biorefinery could be adopted. The USA is one of the world's largest grain sorghum producers. However, grain sorghum for biofuel production is underutilized, and its major use is for animal feeds. Importantly, in 2012 the US Environmental Protection Agency (EPA) has approved grain sorghum as a feedstock for biofuel production to meet the Renewable Fuels Standard (RFS), which requires that 36 billion gallons of biofuel be produced and blended for the US fuel supply by 2022. Recently, US biofuel production, especially for ethanol, has therefore been moving toward grain sorghum. The starch content of grain sorghum is around 50–75% (dry wt. basis) and the ethanol yield from sorghum grain is comparable to that from corn.

24.2.3 Cassava-Based Ethanol Biorefinery

Cassava (*Manihot esculena*) is a starch-rich feedstock (70–85% starch on a dry wt. basis and 28–35% on a wet wt. basis) and is primarily grown in a tropical climate. It grows well on otherwise infertile land with minimal input of chemicals, such as fertilizers, herbicides, and insecticides, making cassava

Table 24.2 Characteristics of cassava chips (wet wt. basis)

	Percentage	
Composition	Rupert, 1979	Sui Heng Lee[+]
Moisture	10–14	12.0
Starch	70–82	69.7
Total ash	1.8–3.0	–
Crude fiber	2.1–5.5	3.4
Sand/silica	–	2.4
Others	–	12.5

Note: [+]Data on commercial cassava chips provided by Sui Heng Lee Co. Ltd., Bangkok, Thailand.

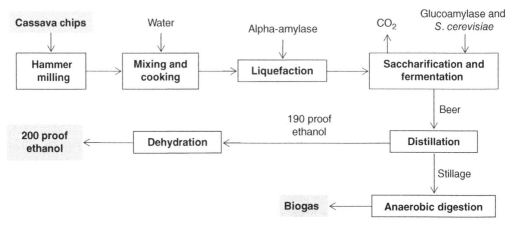

FIG. 24.5 Cassava-based ethanol biorefinery.

one of the cheapest agro-based feedstocks. Thailand is a leading nation in full-scale cassava-based ethanol plants. Sapthip Co., for example, was producing 30 million liters per year (7.93 MGY) in Lop Buri, Thailand in 2009.

Cassava chips are the most common form of cassava feedstock for ethanol production. Although fresh tubers can also be used directly during the harvesting season, drying is often needed for off-season use due to a high water content of around 60–70% (note that drying prevents premature degradation). To prepare cassava chips, the fresh cassava tubers are cut into small pieces without washing and are dried to obtain a moisture content of less than 14% for further storage. Sun drying is the common practice for cassava chip factories in Thailand and other tropical countries. After the chips are dried, they may be passed through the separator to remove dirt and sand. Although uncommon, pretreatment of cassava feedstock like washing and peeling is done for improving the ethanol fermentation efficiency.

The drying process requires a minimum of 2–3 days, and about 2.0–2.5 kg (4.4–5.5 lb) of fresh cassava tuber is required for 1 kg (2.2 lb) of cassava chip production. The important characteristics of cassava chips are shown in Table 24.2. In actual cassava-based ethanol plants, approximately 3 kg (6.6 lb) of cassava chips (14% moisture content) is required to produce 1 L ethanol (25 lb cassava chips/gal ethanol).

The cassava ethanol biorefinery also adopts the same model as the dry-grind corn ethanol biorefinery illustrated in Figure 24.2. Dry-grind cassava ethanol production is presented in Figure 24.5.

After distillation, the liquid left over is called stillage (up to 20 gal/gal of ethanol produced) and contains organic matter from the root fiber, which can be further converted into biogas through anaerobic digestion. The characteristics of cassava ethanol stillage vary depending on the processes involved in ethanol production and the location. In general, it is dark brown in color, high in organic and inorganic matter (chemical oxygen demand [COD] of 100–150 g/L and biological oxygen demand [BOD] of 35-50 g/L), and relatively low pH (~4.0). Stillage also has a high level of sulfate (2–7 g/L) resulting from the sulfuric acid used in pH control during the ethanol fermentation process.

Ethanol from cassava can also be produced by the wet-milling process. Similarly to wet-milling corn ethanol production, the cassava chips are soaked in a dilute acid solution to separate the starch from other components. The fibers are recovered through a series of processes. The fibrous residue obtained after starch extraction is called cassava bagasse and contains 40–60% starch and 15–50% fiber, with small quantities of proteins and lipids. The composition of cassava bagasse differs depending on the origin as well as the processing of cassava.

A full-scale cassava-based ethanol plant produces around 0.33–0.4 L ethanol/kg of cassava (330–400 L ethanol/metric ton of cassava). Cost analyses are based on studies conducted in Asian countries. For example in China, the total production cost, excluding raw materials, was reported to be around $0.27/L ($1.01/gal). In Thailand, the total production cost was reported to be around $0.64–0.67/L ($2.42–2.54/gal). The feedstock cost is expected to remain stable at $38.5/dry metric ton (or $0.33/gallon ethanol).

- **Cassava bagasse** is the fibrous residue that remains after starch extraction from cassava tubers. Typically bagasse is used for the production of heat/steam and electricity for in-plant use.

- **Stillage** is the liquid fermentation byproduct following ethanol recovery. Around 16–20 gallons of stillage are generated for every gallon of ethanol produced. Cassava stillage is not as nutritious as corn stillage and has limited application as animal feed. Anaerobic digestion to produce energy-rich biogas and land application could be possible options for co-product utilization and simultaneous bioremediation.

Example 24.3: A wet-milling cassava ethanol facility uses fresh cassava tubers as a feedstock. The cassava tubers contain 30% starch (wet wt.). After the feedstock passes through the starch extraction process, cassava bagasse (85% moisture content) is generated as a solid residue containing 40% starch (dry wt.). The data from the literature provide that 275 metric tons of cassava roots are processed per day and the generation of cassava bagasse is 280 metric tons per day. Calculate how much ethanol fuel can be produced per day (assuming that the starch to ethanol conversion efficiency is 90%). Also calculate how much cassava root would be needed to produce 1 L ethanol.

Solution:

1. Amount of starch in cassava roots

$$= 275 \text{ metric tons fresh root} \times \frac{30 \text{ metric tons starch}}{100 \text{ metric tons fresh root}} = 82.5 \text{ metric tons starch}$$

2. Amount of starch in cassava bagasse

$$= 280 \text{ metric tons bagasse} \times (1 - 85\%) \times \frac{40 \text{ metric tons starch}}{100 \text{ metric tons dry bagasse}} = 16.8 \text{ metric tons}$$

3. Amount of starch for ethanol production

$$= 82.5 - 16.8 = 65.7 \text{ metric tons}$$

4. Amount of ethanol produced from 65.7 metric tons of starch

The amount of ethanol production from starch can be calculated using Equation 24.1:

$$= 90\% \times \frac{0.5679 \text{ kg ethanol}}{1 \text{ kg starch}} \times \frac{1 \text{ L ethanol}}{0.789 \text{ kg ethanol}} \times 65.7 \text{ metric tons starch} \times \frac{1,000 \text{ kg}}{1 \text{ metric ton}}$$

$$= \mathbf{42,560 \, L \, ethanol \, (11,244 \, gallons)}$$

The amount of ethanol produced from the wet-milling process from 275 metric tons of cassava tubers is 42,560 L per day.

5. Amount of cassava roots used for the production of 1 L ethanol

$$= \frac{275 \text{ metric tons cassava root}}{42,560 \text{ L ethanol}} \times \frac{1,000 \text{ kg}}{1 \text{ metric ton}} = \mathbf{6.5 \, kg \, cassava \, root}$$

6.5 kg of fresh cassava roots are needed to produce 1 L of ethanol by the wet-milling process (or 54 lb of cassava roots are needed to produce 1 gallon of ethanol).

24.3 Integrated Farm-Scale Biorefinery

The current model of a 50-MGY plant for a dry-grind mill cannot be applied in areas where feedstock availability and markets for co-products are limited. An integrated small-scale biorefinery of 0.5–1.0 MGY (1.890–3.785 million L per year) capacity could be one option in rural areas as a close-looped system. Figure 24.6 is a typical example of an integrated farm-scale corn ethanol biorefinery. The same model could also be applied for other starch-based feedstocks.

These biorefineries operate in a similar way to a large-scale dry-grind mill, but with fewer unit processes/operations. In addition, the liquid stream (part of the stillage) is anaerobically digested to produce biogas for in-plant energy generation. The effluent is then land applied. Such biorefining must be integrated with livestock production (dairy/beef cattle or swine) so that the co-product can be fed wet directly to animals without further processing. The manure produced by the livestock can be anaerobically digested to generate methane gas, and the digested slurry can be land applied as biofertilizer to supplement nutrients (N and P).

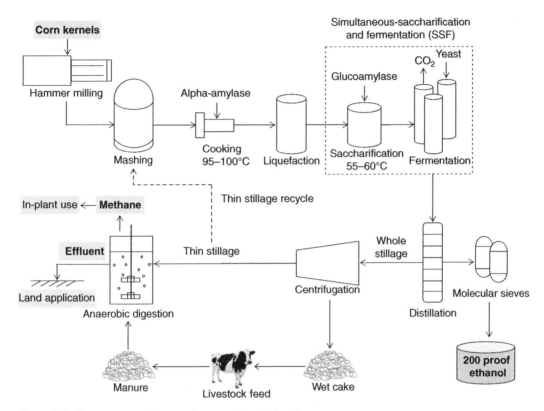

FIG. 24.6   Integrated farm-scale corn ethanol biorefinery.

References

Bangkok Post (2010). Ethanol makers seek more state help. *Bangkok Post January* 5: B10.

Dahlberg, J. (2007). An overview of US sorghum starch and ethanol production. In: K. Rausch, V. Singh, and M. Tumbleson (eds.), *Proceedings of the Fifth International Starch Technology Conference: Energy Issues*, Urbana, IL: University of Illinois at Urbana-Champaign, pp. 50–55.

Dai, D., Hu, Z., Pu, G., Li, H., and Wang, C. (2006). Energy efficiency and potential of cassava fuel ethanol in Guangxi region in China. *Energy Conversion and Management* 47: 1686–99.

de Menezes, T.J.B. (1989). The treatment and utilization of alcohol stillage. In: D.L. Wise (ed.), *International Biosystems, Vol.* III. Boca Raton, FL: CRC Press, pp. 1–14.

Elander, R.T., and Putsche, V.L. (1996). Ethanol from corn: Technology and economics. In: C.E. Wyman (ed.), *Handbook on Bioethanol: Production and Utilization*. Bristol, PA: Taylor & Francis, pp. 329–49.

Erickson, G.E., Bremer, V.R., Klopfenstein, T.J., Stalker, A., and Rasby, R. (2007). Feeding of corn milling co-products to beef cattle. In: *Utilization of Corn Milling Co-products in the Beef Industry*, 2nd edn. Lincoln, NE: Nebraska Corn Board, pp. 3–27.

Foust, T.D., Ibsen, K.N., Dyton, D.C., Hess, J.R., and Kenney, K.E. (2008). The biorefinery. In: M. Himmel (ed.), *Biomass Recalcitrance: Deconstructing the Plant Cell Wall for Bioenergy*. Oxford: Blackwell, pp. 7–137.

IEA Bioenergy (2008). *From 1st to 2nd Generation Biofuel Technologies: An Overview of Current Technologies and RD & D Activities*. Paris: International Energy Agency.

McAloon, A.J. (2007). Ethanol as an economic competitor to gasoline. In: K. Rausch, V. Singh, and M. Tumbleson (eds.), *Proceedings of the Fifth International Starch Technology Conference: Energy Issues*. Urbana, IL: University of Illinois at Urbana-Champaign, pp. 30–33.

Pandey, A., Soccol, C.R., Nigam, P., Soccol, V.T., Vandenberghe, L.P.S., and Mohan, R. (2000). Biotechnological potential of agro-industrial residue II: Cassava bagasse. *Bioresource Technology* 74: 81–7.

Renewable Fuels Association (2009). *Ethanol Industry Outlook*. Washington, DC: Renewable Fuels Association. http://www.ethanolrfa.org/wp-content/uploads/2015/09/RFA_Outlook_2009.pdf, accessed April 2016.

Rupert, B. (1979). *Cassava Drying*. Cali, Colombia: CIAT. http://www.epa.gov/ttn/chief/ap42/ch09/final/c9s09-7.pdf, accessed April 2016.

Sriroth, K., and Piyachomkwan, K. (2005). *Ethanol Production from Cassava Chip*. National Bangkok: Research Council of Thailand.

USDA (2006). *The Economic Feasibility of Ethanol Production from Sugar in the United States*. Washington, DC: U.S. Department of Agriculture. http://www.usda.gov/oce/reports/energy/EthanolSugarFeasibilityReport3.pdf, accessed June 2009.

Exercise Problems

24.1. What are the major starch-based feedstocks for biofuel production?

24.2. What are the major differences in processing sugar and starch-based feedstocks to ethanol?

24.3. Discuss the biorefinery concept as applied to a dry-grind corn ethanol plant.

24.4. What are the major differences between dry-grind and wet-milling processes?

24.5. What are the major benefits of a farm-scale biorefinery?

24.6. A corn starch-based ethanol plant processes 10,000 bushels of corn daily. Assume that the density of ethanol is 0.789 kg/L. Compare the heating value of corn and ethanol. Assume that corn has a moisture content of 14.5%.

24.7. A cassava-based biorefinery is set up for bioethanol production. The plant processes 500 metric ton of chips daily. The composition of cassava chips is 70–82% starch, 10–14% moisture, 2.1–5.5% crude fiber, 1.8–3% ash, and less than 1.5% protein, on a dry wt. basis. How much sugar can be produced daily? Calculate the amount of ethanol produced annually. Assume that the density of ethanol is 0.789 kg/L. If the actual ethanol yield is 0.4 L/kg cassava chips, what would be the total ethanol produced annually?

24.8. Develop a complete mass balance for Problem 24.7.

24.9. The USA is currently producing nearly 14 billion gallons of ethanol primarily from corn using the dry-grind process. Calculate the amount of corn needed to produce the ethanol. How much DDGS and CO_2 would be produced annually? Also calculate the land area required to cultivate the corn assuming an average corn yield of 170 bushels/acre.

24.10. Compare the theoretical ethanol yield per metric dry ton of three major starch-based feedstocks.

Lignocellulose-Based Biorefinery

Scott C. Geleynse, Michael Paice, and Xiao Zhang

What is included in this chapter?

This chapter covers lignocellulose-based biorefinery for the conversion of biomass constituents into fuels, chemicals, and power. Cell structure, stoichiometry, and energy content of plant biomass, and value-added products from lignin and hemicellulose, are introduced. Finally, industrial biorefinery of woody biomass is also discussed.

25.1 Introduction

The term lignocellulosic biomass refers to all plant feedstocks that are made up of three main (structural) components: cellulose, hemicellulose, and lignin. In contrast to sugar-based feedstocks (e.g., sugarcane, sugar beet, and sweet sorghum) and starch-based feedstocks (e.g., corn, cassava, and sweet potato), the use of lignocellulosic biomass for renewable energy and chemical production provides a sustainable approach that does not compete with the food and feed industries. Conversion of biomass to ethanol, butanol, and other advanced biofuels, and other biobased chemicals/materials that replace petroleum feedstock, is the major emphasis of the biorefinery concept.

Initially, the rationale for developing biorefineries centered on the conversion of lignocellulosic biomass to ethanol, either by means of a stand-alone facility or as part of an existing forest industry such as sawmills or pulp and paper mills. Currently, several stand-alone biorefineries use crop residues such as corn stover as a starting feedstock. However, the scale of these operations is much smaller than for petroleum refineries due to the limitations inherent in collecting and transporting the feedstocks, and the relative novelty of bioprocessing (compared to mature petroleum industries). Forestry-based biorefineries have the advantage that the feedstock is often denser and therefore easier to transport to a central location for processing. However, lignocellulose-based biorefineries remain in the early stages of commercialization. It is difficult to generate profit in alternative transportation

---

Bioenergy: Principles and Applications, First Edition. Edited by Yebo Li and Samir Kumar Khanal.
© 2017 John Wiley & Sons, Inc. Published 2017 by John Wiley & Sons, Inc.
Companion website: www.wiley.com/go/Li/Bioenergy

fuels when the technology requires government subsidies to compete with petroleum-based fuels, as is the case for ethanol and butanol today.

This chapter covers the constituents of lignocellulsoic biomass, with a focus on the conversion of each constituent of biomass into biofuel, biochemical, and biopower. This chapter also discusses an example of the industrial biorefinery of woody biomass.

25.2 Cell Structure of Lignocellulosic Feedstocks

Chapter 6 introduces plant biochemistry and the basic structure of plant biomass. Here, the cellular structure of lignocellulosic feedstocks is briefly introduced, with a particular emphasis on the utilization of specific components for bioenergy and biobased chemicals production in biorefineries. Plant tissue is constructed mainly of three types of cells: sclerenchymas, collenchymas, and parenchymas. Fibers are elongated and slender cells that function primarily to support the structure of the plant. Common fiber cells include tracheids, fiber tracheids, and libriform fibers, which are made of thick cell-well sclerenchymas cells. Fiber cells occupy a significant fraction of the mass and volume of the plant (Fengel and Wegener, 1984). Collenchymas are elongated cells with an irregularly thick cell wall, often found in the plant epidermis. Parenchyma cells are short, compact cells with stubby ends and thin cell walls. A sizable quantity of parenchyma cells is also present in many types of monocot grass biomass. Cellulose, hemicellulose, and lignin are the major components of fiber cells. In contrast to sclerenchymas cells, which are considered to be "dead cells," collenchymas and parenchyma cells are living tissue cells that are unlignified.

From a morphological viewpoint, lignocellulosic biomass can be represented at several structural levels: fiber, fibril, and molecular (Figure 25.1). The fiber level (Figure 25.1a) presents the highest magnitude of an intact fiber cell. The morphological properties of fibers such as length, width, cell wall thickness, lumen diameters, etc. are revealed at the fiber level. A typical fiber consists of primary (P) and secondary (S) cell-wall layers. The primary wall (P) is the first layer formed in a fiber cell and is followed by the secondary wall, which is further divided into three layers (S_1, S_2, and S_3) due to different fibril orientations, indicated by the darks lines in Figure 25.1a. The fibril orientation in the P layer is random. The middle lamella (M) is a thin layer that separates individual fiber cells and is primarily composed of lignin. Each of the S and P layers is composed of numerous fibrils that encompass the macro-fibril, micro-fibril, and elementary fibril. At the fibril level, the interactions between different cell-wall chemical components, cellulose, hemicellulose, and lignin (macro-fibril level), as well as the arrangement of amorphous and crystalline cellulose can be revealed (micro-fiber/elementary fibril level; Figure 25.1b). The molecular level describes molecular interaction and chemical bonding between the monomeric constituents (e.g., glucose, xylose, and phenolic compounds) of cell-wall components, as shown in Figure 25.1c. It is clear that plant cells are ingeniously constructed to fulfill at least two primary functions: transportation of nutrients throughout the plant; and structural integrity. However, the structure of the cell wall makes lignocellulosic biomass infamously recalcitrant (i.e., resistant) to biological and chemical processing for biofuels and biochemicals production.

25.3 Stoichiometry and Energy Content

Both cellulose and hemicellulose are polysaccharides with backbones consisting of repeating units of anhydrous sugars bound by glycosidic linkages. Cellulose is a linear polymer constructed of anhydroglucopryranoses linked by $\beta(1\text{-}4)$-glycosidic linkages. By contrast, hemicelluloses are branched polymers composed of shorter chains with various monosaccharides. Based on the number of carbon

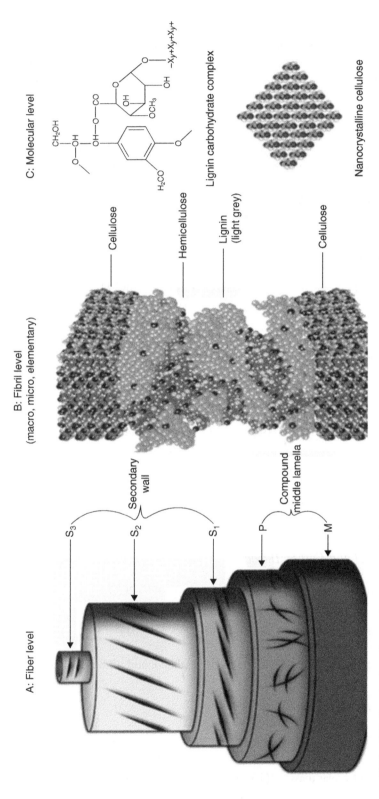

C: Molecular level

Lignin carbohydrate complex

Nanocrystalline cellulose

B: Fibril level
(macro, micro, elementary)

Cellulose

Hemicellulose

Lignin
(light grey)

Cellulose

A: Fiber level

Secondary
wall

S_3

S_2

S_1

Compound
middle lamella

P

M

FIG. 25.1 Intricate interactions among cellulose, hemicellulose, and lignin at different cell-wall structural levels. Source: Ju et al., 2013. Reproduced with permission of Elsevier.

atoms, these monosaccharides are grouped into either hexoses (glucose, galactose, and mannose) or pentoses (arabinose and xylose). Besides monosaccharides, sugar acids such as glucuronic acid and galacturonic acid are also found in hemicelluloses.

25.3.1 Stoichiometry

The stoichiometry for converting polysaccharides to their monomeric constituents is presented here using cellulose and xylan as examples. Xylan is a polysachharide consisting of xylose, and is a primary hemicellulose component of most types of biomass. Because the polysaccharides are anhydrous, the addition of water (known as a hydrolysis reaction) will hydrolyze the glycosidic bonds of the polymer and release its monosaccharide building blocks.

Cellulose into Glucose

Chemical equation :	$(C_6H_{10}O_5)_n + nH_2O \rightarrow nC_6H_{12}O_6$		
Molar mass :	162.13	18.02	180.16
Equation in terms of molar masses :	$n(162.14) + n(18.02) = n(180.16)$		

(25.1)

100 g cellulose can theoretically yield up to 111.11 g glucose.

Xylan into Xylose

Chemical equation :	$(C_5H_8O_4)_n + nH_2O \rightarrow nC_5H_{10}O_5$		
Molar mass :	132.11	18.02	150.13
Equation in terms of molar masse :	$n(132.11) + n(18.02) = n(150.13)$		

(25.2)

100 g xylan can theoretically yield up to 113.64 g xylose.
The stoichiometry to convert glucose and xylose to ethanol and carbon dioxide is also given.

Glucose into Ethanol and Carbon Dioxide

Chemical equation :	$C_6H_{12}O_6 \rightarrow 2C_2H_5OH + 2CO_2$		
Molar mass :	180.16	46.07	44.01
Equation in terms of molar mass :	$180.16 = 2(46.07) + 2(44.01)$		

(25.3)

100 g glucose can theoretically yield up to 51.14 g ethanol and 48.86 g carbon dioxide.

Xylose into Ethanol and Carbon Dioxide

Chemical equation :	$3C_5H_{10}O_5 \rightarrow 5C_2H_5OH + 5CO_2$		
Molar mass :	150.13	46.07	44.01
Equation in terms of molar mass :	$3(150.13) = 5(46.07) + 5(44.01)$		

(25.4)

100 g xylose can theoretically yield up to 51.14 g of ethanol and 48.86 g carbon dioxide (same as glucose).

Table 25.1 Energy content of major biomass components

Biomass component	Higher heating value (kJ/g)
Softwood lignin	26.4
Hardwood lignin	24.7
Cellulose	17.4
Hemicellulose	16.7

Source: Adapted from Domalski, 1987.

Table 25.2 Enthalpy of combustion of monosaccharides, ethanol and butanol

Chemicals	Enthalpy of combustion	
	kJ/mol	kJ/g
Ethanol	–1,370	–29.74
Butanol	–2,670	–36.02
Hexose	–2,805	–15.57
Pentose	–2,350	–15.65

25.3.2 Energy Content

The total energy content of biomass, biomass components, chemicals, and fuels can be expressed as the amount of energy released when the substance undergoes complete combustion. For biomass and biomass components, this is generally represented by the higher heating value (HHV) and varies considerably depending on the source of the material (Table 25.1). The lower heating value (LHV) is also sometimes provided, which represents the same value but assuming that the water vapor formed in the reaction does not condense and remains as a vapor (see Chapter 2). For a pure molecular component (like glucose or ethanol), the standard enthalpy of combustion is typically used. HHV is reported as a positive value but an enthalpy change, while representing the same property is expressed as a negative value (indicating that heat is released from the reaction). It is prudent to double-check your calculations to ensure that your HHV and enthalpy values have appropriate plus/negative signs. Some approximate values for the HHV of biomass constituents and the enthalpies of combustion for important chemicals are shown in Tables 25.1 and 25.2. More information on enthalpies is presented in Chapter 3.

> **Example 25.1:** Using the process demonstrated in Section 25.3.1, determine the theoretical yield of ethanol from a kilogram of pure cellulose.

Solution:

Cellulose has a molecular weight of 162.14
Glucose has a molecular weight of 180.16
Ethanol has a molecular weight of 46.07

So for one kilogram of cellulose, we can produce:

$$1 \text{ kg} \times (180.16/162.14) = 1.111 \text{ kg of glucose}$$

Recall that for each three moles of glucose converted, we can produce 5 moles of ethanol. From the 1.111 kg of glucose, we can produce:

$$1.111 \text{ kg} \times (46.07/180.16) \times (5/3) = 0.474 \text{ kg ethanol}$$

Example 25.2: Estimate the amount of energy available in 1 kg of softwood composed of 30% lignin, 25% hemicellulose, and 45% cellulose.

Solution:

From Table 25.1:

> HHV of softwood lignin = 26.4 kJ/g
> HHV of hemicellulose = 16.7 kJ/g
> HHV cellulose = 17.4 kJ/g

> (Note: 1 kJ/g = 1 MJ/kg)

> Energy in lignin = 0.3 kg × 26.4 MJ/kg = 7.92 MJ
> Energy in hemicellulose = 0.25 kg × 16.7 MJ/kg = 4.175 MJ
> Energy in cellulose = 0.45 kg × 17.4 MJ/kg = 7.83 MJ

> **Total energy = 19.93 MJ/kg**

25.4 Lignocellulosic Biomass Conversion to Fuel

A generic lignocellulose-based biorefinery process consists of at least four steps: biomass pretreatment; enzyme hydrolysis (saccharification); fermentation or catalytic conversion of sugars to fuel and chemicals; and products recovery/separation. Unlike sugar- and starch-based biomass (sugar beet, starch, etc.) where monosaccharides are readily obtainable from the feedstock, pretreatment of lignocellulosic biomass to disrupt cell-wall integrity and to expose structural carbohydrates is critical for the success of the entire biochemical conversion process. A review of the variety of biomass pretreatment methods can be found in Chapter 12. The resulting pretreated solid substrate/fiber is subsequently subjected to a saccharification step to hydrolyze carbohydrate polymers or polysaccharides (i.e., cellulose and hemicellulose) into monosaccharides. This is achieved by a consortium of enzymes, primarily consisting of cellulases and hemicellulases (Chapter 13). The monosaccharides are then converted to fuel and/or fuel precursors by either fermentation (Chapters 14 and 15) or catalytic processes. Separation and purification of fuel products form the final key step.

It is increasingly recognized that fuel, regardless of its type, is a low-value product. Using lignocellulosic feedstocks to produce fuel as the sole product poses a great economic challenge. The economic feasibility is the major bottleneck hindering the commercial implementation of a large-scale lignocellulosic biomass-to-biofuel conversion process. Identifying value-added co-products along with the production of biofuel provides a key solution to overcoming this economic barrier. Therefore, this chapter focuses on discussing the current status in utilizing biomass components for co-product development in a lignocellulose-based biorefinery.

Example 25.3:
1. Calculate the volume of ethanol and mass of residual biomass produced from a process using 100 kg of hardwood biomass of given composition: 50% cellulose, 25% lignin, 25% hemicellulose (dry wt. basis).
2. Determine the amount of energy available in the ethanol fuel and the residual biomass.

Assume 10% degradation of cellulose during pretreatment; 95% conversion of cellulose from hydrolysis; 20% degradation of hemicellulose during pretreatment; complete conversion of non-degraded hemicellulose; hemicellulose is composed of only xylose; 100% ethanol fermentation from glucose and xylose. The density of ethanol is 789 kg/m^3.

Solution:

1. Determination of ethanol yield
Ethanol yield from cellulose

> Theoretical yield of glucose is 1.11 g/g cellulose
> Theoretical yield of ethanol is 0.51 g/g glucose
> Pretreatment yield is 90%, hydrolysis yield is 95% for cellulose

> Actual yield = 100 kg × 0.50 × 1.11 × 0.51 × 0.90 × 0.95
>   = (24.2 kg × 1,000 g/kg)/789 g/L = 30.7 L

Ethanol yield from hemicellulose

> Theoretical yield of xylose is 1.14 g/g xylan
> Theoretical yield of ethanol is 0.51 g/g xylose
> With 20% degradation, but full recovery, 80% of hemicellulose is converted

> Actual yield = 100 kg × 0.25 × 1.14 × 0.51 × 0.80 × 1.00
>   = (11.6 kg × 1,000 g/kg)/789 g/L = 14.7 L

> Total volume of ethanol produced = 30.7 + 14.7 = **45.4 L (12.0 gal)**

Residual biomass

> Lignin: 100 kg × 0.25 = 25 kg
> Cellulose: 100 kg × 0.50 × (1-0.10) × (1-0.95) = 2.25 kg
> **Total residual biomass = 27.25 kg**

2. Determination of energy content
Using the higher heating values from Tables 25.1 and 25.2, the energy contents of each fraction can be calculated.

Energy available from ethanol = 35.8 kg × 29.74 kJ/g × 1,000 g/kg = 1,065 MJ
Remaining cellulose = 2.25 kg × 1,000 g/kg × 17.4 kJ/g = 39.15 MJ
Remaining lignin = 25 kg × 1,000 g/kg × 26.4 kJ/g = 660 MJ
Energy available in unconverted biomass = 699.15 MJ

25.5 Co-Products from Lignocellulose-Based Biorefinery

In a biochemical-based biomass-to-biofuel conversion process, lignin is typically generated as a waste stream. Also, as mentioned earlier, hemicellulose is not well utilized for biofuel production. Therefore, there has been a significant effort toward identifying new valorization pathways for biomass-derived lignin and hemicellulose. A brief summary of potential value-added products that can be derived from lignin and hemicellulose is discussed below.

25.5.1 Products from Lignin

Lignin is a ubiquitous component in almost all plant biomass (Fengel and Wegener, 1984; Sjöström,1993; Davin *et al.*, 2008). Large quantities of lignin are produced annually as a waste product from chemical pulping processes (i.e., kraft pulping, sulfite pulping). Most of the lignin is burned as a fuel in the boiler, providing energy for in-plant use. It is anticipated that the future lignocellulose-based biorefinery will generate a significant amount of lignin. A report from the US Department of Energy estimates that 225 million US short tons of lignin will be generated from processing of 750 million US short tons of biomass feedstock from biofuel production processes (Holladay *et al.*, 2007). The heating value of lignin is higher than the other two major components of lignocellulosic biomass, namely, cellulose and hemicellulose. Thus, utilizing lignin for heat and power generation is a sensible approach. Combustion, gasification, pyrolysis, or hydroliquefaction has been applied to convert lignin to power, liquid fuel, and/or syngas products. Combusting pulping black liquor to recover heat and power is widely employed in the pulp industry. However, the value of lignin realized through heat and power generation only reflects the price of fossil fuel.

Developing high-value marketable products from lignin is much more attractive and has been a major industrial and scientific endeavor since the early 20th century (Pye, 2006). The majority of commercial lignin products are macromolecules, predominantly in the form of lignosulfonates. Lignosulfonates are produced from sulfite pulping, which applies a mixture of sulfurous acid (H_2SO_3) and bisulfite ions (HSO_3^-) to derivatize and dissolve lignin. The lignosulfonates have a wide ranging molecular weight distribution, from a few hundred to over 100,000 dalton. The percentage breakdown of commercial lignosulfonate applications (in 2008) is shown in Figure 25.2.

Lignin is the largest source of renewable material with an aromatic skeleton. Depolymerizing macromolecular lignin to low molecular weight phenolic compounds (LMWPC) offers an attractive pathway for producing a wide range of high-value chemicals (Whiting, 2001; Zhang *et al.*, 2011; Ragauskas *et al.*, 2014; Holladay *et al.*, 2007; Boudet, 2007). Figure 25.3 shows several options

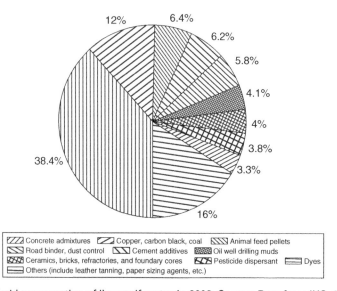

FIG. 25.2   World consumption of lignosulfonates in 2008. Source: Data from IHS, 2009.

for the utilization of lignin. Potential technologies and pathways to produce valuable chemicals from lignin are covered in several recent reviews (Ragauskas *et al.*, 2014; Ma *et al.*, 2014) and are beyond the scope of this textbook.

25.5.2 Products from Hemicellulose

As mentioned before, hemicellulose is generally more susceptible to degradation during pretreatment processes. One effective approach to maximize the recovery of hemicellulose from biomass is to apply a pre-extraction stage prior to pretreatment. A variety of hemicellulose pre-extraction or prehydrolysis methods have been investigated (Al-Dajani and Tschimer, 2008; Sattler *et al.*, 2008; Springer and Harris, 1982; Yoon *et al.*, 2008). Depending on the pre-extraction process, hemicellulose can be recovered as either monosaccharides (e.g., xylose, mannose, arabinose) or as oligosaccharides (e.g., xylooligosaccharides, mannanoligosaccharides; Zhang *et al.*, 2011). A summary of hemicellulose-derived products is shown in Figure 25.4.

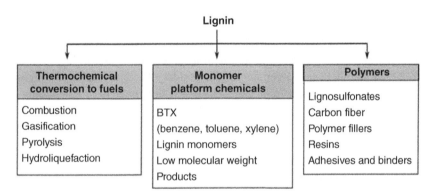

FIG. 25.3   Different applications/products derived from lignin.

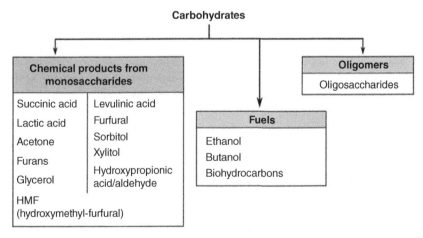

FIG. 25.4   Products derived from hemicellulose.

One existing commercial product from biomass-derived xylose is furfural. Furfural can be further converted to furan derivatives, which are important precursors for manufacturing a number of industrial chemicals and materials (Sain *et al.*, 1982; Zeitsch, 2000; Chheda et al., 2007; Tong *et al.*, 2010). Corncobs and bagasse are the primary biomass feedstocks for industrial furfural production. China is currently the world's largest furfural producer and consumer. Industrial furfural production processes employ inorganic acid, primarily sulfuric acid, to hydrolyze xylan present in dried biomass, and then to dehydrate xylose to furfural. The xylose-to-furfural conversion yield is typically less than 50% of the theoretical yield using these processes (Win, 2005; Hoydonckx *et al.*, 2007), while much lower production yield is reported from the Chinese manufacturing process (Marinova *et al.*, 2010).

Xylitol is another commercial product manufactured predominantly from hardwood and corncobs. Although Danisco is still the world largest xylitol producer, xylitol production has been shifting to China in the last decade due to the lower cost of production (D'Amico, 2009). Prehydrolysis of lignocellulosic biomass by SO_2 produces a high yield of xylose that can be readily converted to furfural and xylitol. The potential economic benefit from producing furfural and xylitol in an existing Canadian dissolving pulp mill has been evaluated (Marinova *et al.*, 2010). This pulp mill implemented hemicellulose prehydrolysis in 2008. Based on a dissolving pulp production of 518 metric tons per day, CAD $12 million and CAD $39 million per year can be generated from the production of furfural and xylitol, respectively, in addition to the pulp production.

Water and alkaline pre-hydrolysis can yield a high concentration of oligosaccharides. Water-soluble oligosaccharides are typically a mixture with a degree of polymerization of 2–7. Biomass-derived xylooligosaccharides and mannanoligosaccharides are non-digestible oligosaccharides (NDOs). These compounds have potential beneficial effects on human health, including chemically stable food-processing treatments, non-cariogenicity, low calorific value, and the ability to stimulate the growth of beneficial bacteria in the colon, mainly the *bifidobacteria* species, and are thus recognized as prebiotics. Non-digestible oligosaccharides are currently supplemented to beverages, milk products, probiotic yogurts, and synbiotic products (Aachary and Prapulla, 2011). Industrial applications have rapidly increased in the last few years, especially in prebiotic formulations and synbiotic products.

25.6 Industrial Lignocellulose-Based Biorefinery

Commercial implementation of a lignocellulose-based biorefinery process faces many techno-economic bottlenecks. While the concept of a true lignocellulose-based biorefinery remains in its infancy, the industrial production of bioethanol and biochemicals from woody biomass has been practiced for several decades. One recent example in North America is the operation of the Tembec chemical plant located at Temiscaming, Quebec, Canada. As illustrated in Figure 25.5, wood chips are treated by ammonium sulfite pulping liquor at a temperature of around 170 °C. This pulping stage removes a significant amount of lignin and most of the hemicelluloses, which are dissolved in spent sulfite liquor (SSL). The cellulose fibers are further processed through bleaching to yield pure cellulose as a feedstock for cellulose acetate and rayon production. The SSL is concentrated to a predetermined solid content. A portion of the concentrated SSL is subjected to yeast fermentation to produce ethanol, while the rest is fed to a boiler and burnt for energy recovery. The lignosulfonate (LS) remains in the SSL as residue after fermentation. The plant produces about 165,000 metric tons per year of pure cellulose as dissolving pulp, which is primarily used for textiles, pharmaceuticals, food additives, nursing pads, and diapers. It has the capacity to produce 170,000 metric tons of

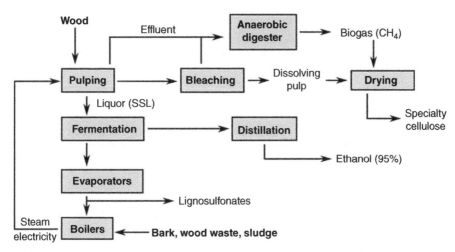

FIG. 25.5   Woody biomass-based biorefinery producing cellulose, ethanol, and lignosulfonate.

lignosulfonates and 18 million liters of ethanol per year. The plant started the sale of lignosulfonate as a commercial product in 1984 and ethanol in 1991. The sugar-free LS is sold for concrete admixture and other applications. This example clearly demonstrates that lignocellulose-based biorefinery can be commercially viable and successful.

Example 25.4: After using pretreatment and enzymatic hydrolysis to remove a majority of the carbohydrates in a hardwood feedstock, a process generates a solid composed of 85% lignin and 15% cellulose. A biorefinery using this process generates 10 metric tons of this residue and is considering two scenarios for its use:

1. Burning the residue for power.
2. Extracting the lignin and selling it as a product, burning the remaining cellulose.

For both options, determine the mass of lignin produced and the energy released through combustion per day.

Solution:

1. Burning the residue for power
The stream contains 8.5 metric tons/day (8,500 kg/day) of hardwood lignin and 1.5 tons/day (1,500 kg/day) of cellulose.
   Using the HHVs from Table 25.1, we have:

Energy from lignin = 24.7 MJ/kg × 8500 kg/day = 210 GJ/day
Energy from cellulose = 17.4 MJ/kg × 1500 kg/day = 26 GJ/day
Total energy released = **236 GJ/day**

2. Extracting the lignin and selling it as a product, burning the remaining cellulose
As calculated earlier, **8,500 kg/day** of lignin is extracted and **26 GJ/day** of energy is produced from cellulose.

Example 25.5: The biorefinery in Example 25.4 wants to produce electrical power from the biomass it is burning; it can do so at 40% efficiency. Calculate the amount of power generated by the biorefinery under both scenarios, assuming a 24 hr/day operation.

Solution:

1. Burning the residue for power

At 40% efficiency, only 40% of the energy available in the biomass (based on HHV) is converted into electricity:

$$214.2\,GJ/day \times 40\% = 214.2\,GJ/day$$

There are 86,400 sec in a day, so this plant produces (214.2 GJ/day)/(86,400 sec/day) =2,480 W

2. Extracting the lignin and selling it as a product, burning the remaining cellulose

$$23.7\,GJ/day \times 40\% = 9.48\,GJ/day$$

This plant produces (9.48 GJ/day)/(86,400 sec/day) = **110 W**

References

Aachary, A.A., and Prapulla, S.G. (2011). Xylooligosaccharides (XOS) as an emerging prebiotic: Microbial synthesis, utilization, structural characterization, bioactive properties, and applications. *Comprehensive Reviews in Food Science and Food Safety* 10(1): 2–16.

Al-Dajani, W.W., and Tschirner, U.W. (2008). Pre-extraction of hemicelluloses and subsequent kraft pulping. Part I: Alkaline extraction. *Tappi Journal* 7: 3–8.

Alonso, D.M., Bond, J.Q., and Dumesic, J.A. (2010). Catalytic conversion of biomass to biofuels. *Green Chemistry* 12(9): 1493–513.

Alvarez-Vasco, C., and Zhang, X. (2013). Alkaline hydrogen peroxide pretreatment of softwood: Hemicellulose degradation pathways. *Bioresource Technology* 150: 321–7.

Boudet, A.M. (2007). Evolution and current status of research in phenolic compounds. *Phytochemistry* 68(22–24): 2722–35.

Chandra, R.P., Bura, R., Mabee, W.E., Berlin, A., Pan, X. and Saddler, J.N. (2007). Substrate pretreatment: The key to effective enzymatic hydrolysis of lignocellulosics? *Biofuels* 108: 67–93.

Chheda, J.N., Roman-Leshkov, Y., and Dumesic, J.A. (2007). Production of 5-hydroxymethylfurfural and furfural by dehydration of biomass-derived mono- and poly-saccharides. *Green Chemistry* 9(4): 342–50.

D'Amico, E. (2009). Danisco issues profit warning: Curtails xylitol production. *Chemical Week* March 9.

Davin, L.B., Patten, A.M., Jourdes, M., and Lewis, N.G. (2008). Lignins: A 21st century challenge. In M.E. Himmel (ed.), *Biomass Recalcitrance Deconstructing the Plant Cell Wall for Bioenergy*. Oxford, Blackwell, pp. 213–305.

Domalski, E.S., and Milne, T.A. (eds.) (1987). *Thermodynamic Data for Biomass Materials and Waste Components*. ASME Research Committee on Industrial and Municipal Wastes. New York: American Society of Mechanical Engineers.

Fengel, D., and Wegener, G. (1984). Wood: Chemistry, Ultrastructure, Reactions. New York: W. de Gruyter.

Gregg, D., and Saddler, J.N. (1996a). A techno-economic assessment of the pretreatment and fractionation steps of a biomass-to-ethanol process. *Applied Biochemistry and Biotechnology* 57–58: 711–27.

Gregg, D.J., and Saddler, J.N. (1996b). Factors affecting cellulose hydrolysis and the potential of enzyme recycle to enhance the efficiency of an integrated wood to ethanol process. *Biotechnology and Bioengineering* 51(4): 375–83.

Helle, S.S., R.A. Petretta, R.A., and Duff, S.J.B. (2007) Fortifying spent sulfite pulping liquor with hydrolyzed reject knots. *Enzyme and Microbial Technology* 41(1–2): 44–50.

Himmel, M. (2008). *Biomass Recalcitrance: Deconstructing the Plant Cell Wall for Bioenergy*. Oxford: Blackwell.

Himmel, M.E., Ding, S.Y., Johnson, D.K., *et al.* (2007). Biomass recalcitrance: Engineering plants and enzymes for biofuels production. *Science* 315(5813): 804–7.

Holladay, J.E., Bozell, J.J., White, J.F., and Johnson, D. (2007). *Top Value-Added Chemicals from Biomass. Vol. II: Results of Screening for Potential Candidates from Biorefinery Lignin*. Washington, DC: U.S. Department of Energy.

Hoydonckx, H.E., Van Rhijn, W.M., Van Rhijn, W., De Vos, D.E., and Jacobs, P.A. (2007). Furfural and derivatives. In: *Ullmann's Encyclopedia of Industrial Chemistry*. Weinheim: Wiley-VCH.

Huang, H.J., Ramaswamy, S., Tschirner, U.W., and Ramarao, B.V. (2008). A review of separation technologies in current and future biorefineries. *Separation and Purification Technology* 62(1): 1–21.

IHS (2009). *Chemical Economics Handbook: Lignosulfonates*. Englewood, CO: IHS.

Ju, X., Grego, C., and Zhang, X. (2013). Specific effects of fiber size and fiber swelling on biomass substrate surface area and enzymatic digestibility. *Bioresource Technology* 144: 232–9.

Klein-Marcuschamer, D., Oleskowicz-Popiel, P., Simmons, B.A., and Blanch, H.W. (2012). The challenge of enzyme cost in the production of lignocellulosic biofuels. *Biotechnology Bioengineering* 109: 1083–7.

Ma, R., Xu, Y., and Zhang, X. (2014). Catalytic oxidation of biorefinery lignin to value-added chemicals to support sustainable biofuel production. *ChemSusChem* 8: 24–51. doi:10.1002/cssc.201402503

Marinova, M., Mateos-Espejel, E., and Paris, J. (2010). From kraft mill to forest biorefinery: An energy and water perspective. II. Case study. *Cellulose Chemistry and Technology* 44(1–3): 21–6.

Pye, K.E. (2006). Industrial lignin production and applications. In: B. Kamm, P.R. Gruber, and M. Kamm (eds.), *Biorefineries: Industrial Processes and Products*. Weinheim: Wiley-VCH, pp. 165–200.

Ragauskas, A.J., Beckham, G.T., Biddy, M.J., *et al.* (2014). Lignin valorization: Improving lignin processing in the biorefinery. *Science* 344(6185): 1246843.

Sain, B., Chaudhuri, A., Borgohain, J.N., Baruah, B.P., and Ghose, J.L. (1982). Furfural and furfural-based industrial chemicals. *Journal of Scientific and Industrial Research* 41(7): 431–8.

Sattler, C., Labbe, N., Harper, D., Elder, T., and Rials, T. (2008). Effects of hot water extraction on physical and chemical characteristics of oriented strand board (OSB) wood flakes. *CLEAN–Soil, Air, Water* 36(8): 674–81.

Sjöström, E. (1993). *Wood Chemistry: Fundamentals and Applications*. San Diego, CA: Academic Press.

Springer, E.L., and Harris, J.F. (1982). Prehydrolysis of aspen wood with water and with dilute aqueous sulfuric acid. *Svensk Papperstidning* 85(15): R152–R154.

Tong, X., Ma, Y., and Li, Y. (2010). Biomass into chemicals: Conversion of sugars to furan derivatives by catalytic processes. *Applied Catalysis A: General* 385(1–2): 1–13.

Werpy, T., Holladay, J., and White, J. (2004). *Top Value Added Chemicals from Biomass: I. Results of Screening for Potential Candidates from Sugars and Synthesis Gas*. Richland, WA: Pacific Northwest National Laboratory.

Whiting, D.A. (2001). Natural phenolic compounds 1900–2000: A bird's eye view of a century's chemistry. *Natural Product Reports* 18(6): 583–606.

Win, D.T. (2005). Furfural-gold from garbage. *AU Journal of Technology* 8(4): 185–90.

Wyman, C.E. (2007). What is (and is not) vital to advancing cellulosic ethanol. *Trends in Biotechnology* 25(4): 153–7.

Wyman, C.E.E. (2013). *Aqueous Pretreatment of Plant Biomass for Biological and Chemical Conversion to Fuels and Chemicals*. Chichester: John Wiley & Sons.

Yoon, S.H., Macewan, K., and Van Heiningen, A. (2008). Hot-water pre-extraction from loblolly pine (*Pinus taeda*) in an integrated forest products biorefinery. *Tappi Journal* 7(6): 27–32.

Zeitsch, K.J. (2000). *The Chemistry and Technology of Furfural and Its Many By-products*. Amsterdam: Elsevier.

Zhang, Y.H.P., and Lynd, L.R. (2004). Toward an aggregated understanding of enzymatic hydrolysis of cellulose: Noncomplexed cellulase systems. *Biotechnology and Bioengineering* 88(7): 797–824.

Zhang, X., Tu, M.B., and Paice, M.G. (2011). Routes to potential bioproducts from lignocellulosic biomass lignin and hemicelluloses. *Bioenergy Research* 4(4): 246–57.

Exercise Problems

25.1. Describe a lignocellulose-based biorefinery. Why is it important to adopt a biorefinery approach for lignocellulosic feedstocks?

25.2. What are the main chemical constituents of lignocellulosic biomass fiber cell wall?

25.3. Describe the structure of wood fiber cell wall.

25.4. Draw a flowchart showing the major bioproducts that can be divided from hemicellulose presented in lignocellulosic feedstocks in addition to biofuels.

25.5. Draw a flowchart showing the major bioproducts that can be divided from lignin presented in lignocellulosic feedstocks in addition to biofuels.

25.6. Discuss bioproducts other than pulp and paper that can be potentially produced from a paper mill.

25.7. A biorefinery uses a dilute acid pretreatment process that results in the complete depolymerization and solubilization of hemicellulose and 15% cellulose solubilization. During the pretreatment process, 30% of the solubilized hemicellulose and all of the solubilized cellulose are degraded into furans. The rest of the solubilized hemicellulose is converted into monosaccharides. Enzymatic hydrolysis converts 95% of the solid cellulose. On average, this biorefinery processes 20 metric tons per day of a softwood feedstock that contains 20% lignin, 20% hemicellulose, and 60% cellulose, assuming that all hemicellulose is xylan.

  (a) How many metric tons of monosaccharides (hexoses and pentoses) does this biorefinery generate in a year?

  (b) If this refinery uses these sugars to produce ethanol at 90% conversion/recovery, how many gallons will it produce per day?

25.8. A biorefinery process yields 80% conversion of cellulose and 90% conversion of hemicellulose into sugars, which are all converted into ethanol. The remaining biomass (lignin and unconverted cellulose/hemicellulose) are burned for power. Use a mass balance to determine the yield in gallons of ethanol and MJ of heat available per metric ton of biomass feedstock processed. (Assume that all hemicellulose is xylan.)

  (a) Softwood with 19% lignin, 17% hemicellulose, and 64% cellulose.

  (b) Hardwood with 30% lignin, 15% hemicellulose, and 55% cellulose.

25.9. The biorefinery in Problem 25.7 would like to change to a new pretreatment process resulting in higher conversions of cellulose, but degrading some hemicellulose, resulting in a 95% conversion of cellulose and 70% conversion of hemicellulose into glucose and xylose, respectively. With a softwood feedstock that is 20% lignin, 20% hemicellulose, and 60% cellulose, does this new process generate more ethanol than before?

25.10. Using a feedstock of pure cellulose, calculate the energy available per kilogram of cellulose if it is converted to ethanol at the maximum theoretical yield. Compare this to the energy output if the cellulose is instead simply burned.

Lipid-Based Biorefinery

B. Brian He, J. H. Van Gerpen, Matthew J. Morra, and Armando G. McDonald

---

What is included in this chapter?

In this chapter, lipid-based biorefinery is discussed with particular attention to conversion of vegetable oils, animal fats, and waste cooking oil to biodiesel. The chemical composition of major oils and fats, seed oil extraction from oil seeds, the stoichiometry of biodiesel production, and byproducts from seed oil extraction and biodiesel production processes are also covered. Additionally, bioproducts from the oilseed-to-biodiesel process for industrial applications are also introduced.

26.1 Introduction

Lipids are the structural building blocks of living cell membranes, and provide a means for energy storage in plants, a number of microbes, and animals. Generally, the term lipid includes oils, fats, waxes, etc. in the forms of acylglycerides (i.e., monoglycerides, diglycerides, and triglycerides), phospholipids, terpenes, etc. In the broadest sense, lipid also includes fatty acids. Triglycerides, which are the primary chemical components of fats and oils, are fatty acid esters of glycerol. Oils are liquid at ambient temperatures and are mainly derived from plants. The triglycerides in the oils contain fatty acid esters of largely unsaturated carbon chains, whereas fats are solid at ambient temperatures, and primarily refer to lipids from animal sources, which are triglycerides of mainly longer and more saturated carbon chains. Phospholipids, also known as gums, are a key component of cell membranes. Chemically different from triglycerides, phospholipids contain a diglyceride and a phosphate group. Phospholipids form double-layered membranes or bilayers, with the water-soluble phosphate ends outward and the fatty acid "tails" inward. Unlike triglycerides, waxes are the esters of long-chain fatty acids with long-chain monohydric alcohols. Terpenes are hydrocarbons produced by many plants such as conifer, lemon, etc., and are characterized by their pleasant smell. Among the various types of lipids, triglycerides are of particular interest for their applications in biodiesel production.

Bioenergy: Principles and Applications, First Edition. Edited by Yebo Li and Samir Kumar Khanal.
© 2017 John Wiley & Sons, Inc. Published 2017 by John Wiley & Sons, Inc.
Companion website: www.wiley.com/go/Li/Bioenergy

Rudolf Diesel

Vegetable oil as a fuel for engines was tested more than 100 years ago by Rudolf Diesel (Figure 26.1), when he used peanut oil to power his newly invented diesel engine in the 1910s. Diesel's famous quote shows his great vision in biofuel application: "The use of vegetable oils for engine fuels may seem insignificant today, but such oils may become, in the course of time, as important as petroleum and the coal tar products of the present time."

FIG. 26.1   Rudolf Diesel (1858–1913), the inventor of the diesel engines. Source: https://es.wikipedia.org/wiki/Rudolf_Diesel#/media/File:Rudolf_Diesel.jpg. Public domain.

Biodiesel is a renewable, clean-burning diesel-replacement fuel. It reduces harmful emissions of air pollutants, including particulate matter, hydrocarbons, and greenhouse gases. Biodiesel from renewable feedstocks also greatly reduces the net carbon dioxide emissions to the atmosphere. Biodiesel can be used in existing diesel engines without modifications. Currently, biodiesel from plant oils and/or animal fats has been designated an *advanced biofuel*, according to the definition established by the US Environmental Protection Agency for the Renewable Fuel Standard. In the past two decades, biodiesel production from plant oils and animal fats has emerged as a new industry. In 2013, biodiesel production in the USA was approximately 6.4 billion liters (1.7 billion gallons), which was seven times higher than in 2003. As the world leader in biodiesel production, the European Union (EU) consumed more than 10.2 billion liters (2.7 billion gallons) of biodiesel in 2013, and still shows an increasing trend. Brazil, another country that advocates biofuel utilization, produced more than 2.8 billion liters (740 million gallons) of biodiesel in 2013. Soybean oil is the major feedstock for biodiesel production in the USA and Brazil, and rapeseed and/or canola oil is the dominant feedstock in Europe, while countries including Malaysia, Indonesia, and Colombia use palm oil for biodiesel production.

Besides the biodiesel produced from plant oils and animal fats, other products from oils and fats, and specialty chemicals from seed meals and crude glycerol, can be important products of an integrated lipid-based biorefinery. They are not only important industrial products, but also generate value-added products to offset part of the biodiesel production costs.

This chapter introduces the lipid-based biorefinery with a focus on biodiesel production from various feedstocks. Biodiesel production from plant oils, animal fats, and waste cooking oils is discussed in greater detail. Other value-added bioproducts in a lipid-based biorefinery are introduced. The fundamentals of lipid properties and their applications for industrial uses, especially for biodiesel production, are also covered.

26.2 Lipid-Based Feedstocks

Lipid-based feedstocks include mainly plant oils (including seed oils such as soybean oil, rapeseed oil, palm oil, and coconut oil), animal fats (such as beef tallow, lard, and chicken fat), microalgal oil, and waste cooking oil and fats (known as yellow grease). A brief review of the types of feedstock used for biodiesel production is included in this chapter. A detailed discussion of various oilseed feedstocks is provided in Chapter 9, whereas algal feedstock is included in Chapter 11.

26.2.1 Plant Oils

There are two types of oilseed that have industrial importance in the USA and the EU, namely, soybean and rapeseed. The latter is commonly referred to as canola in North America. Soybean seeds contain 18–20% oil (wt. basis), which is not a high percentage when compared to other oilseeds. However, due to the large quantity produced (approximately 86 million metric tons or 189 billion lb in 2013), soybean oil dominates the US biodiesel industry with about a 50% market share. Soybean is an annual, leguminous plant that adapts to a wide range of climates and environments. Aside from the USA, the other major soybean producers are Brazil, Argentina, Australia, China, and India.

Rapeseed is the common name for the *Brassica* genus of oilseed crops. Canola is a trade name for a specific cultivar of rapeseed that was developed in Canada for its very low glucosinolates (a group of sulfur-containing chemicals, the hydrolyzed products of which are potentially toxic) and low erucic acid content. Traditional industrial uses of rapeseed oil include lubricants, hydraulic fluids, and plastics, but now the low glucosinolates and low erucic acid varieties are the dominant feedstock for biodiesel production in Europe. Mustard oils (including yellow, white, and/or oriental mustards) possess similar chemical properties to rapeseed, and are also suitable for biodiesel production.

Tropical plant oils from palm and coconut are also used for producing biodiesel in areas where they are grown, such as in Malaysia, Indonesia, Colombia, and Thailand. Biodiesel produced from *palm and coconut oils*, however, tends to have a higher cloud point and higher gelling point, which limit its use in cold environments. Other types of plant oils, including safflower, sunflower, and flaxseed, can also be used for biodiesel production. However, their limited supplies and high-value uses in other applications make them less attractive for commercial biodiesel production. New high oil-yield plant seeds, such as camelina and jatropha, are also being actively studied for their biodiesel production potential.

Microalgal oil is considered a promising feedstock for biodiesel production because, unlike other plant oils, it can potentially be produced in an "industrialized" setting using bioreactors once the biological limitations are overcome. A detailed discussion of algal cultivation is presented in Chapter 11.

26.2.2 Animal Fats

Similar to plant oils, *animal fats* are also composed of triglycerides, which typically have a higher degree of saturation in their chemical bonds. As a result, animal fats are usually solid or semi-solid at room temperature. Due to their low prices, animal fats are affordable feedstocks for biodiesel production, especially for small producers. However, animal fats already have their own established markets, and their availability for biodiesel production is limited.

26.2.3 Waste Cooking Oils

Waste cooking oils, also called yellow grease, are used vegetable oils or mixtures of used vegetable oils and animal fats. Waste cooking oils contain a significant amount of impurities, including water and food particles. These impurities are of concern because they cause processing and fuel quality problems. Pretreatments, such as filtering and dewatering, are required before waste cooking oils can be used for biodiesel production. Waste cooking oils typically have a high free fatty acid content because of the degradation that occurs during cooking. These free fatty acids tend to react with basic catalysts to form soap, which causes emulsification during biodiesel production. Therefore, special processing is needed. Biodiesel produced from animal fats and grease has better oxidative stability than plant oil–based biodiesel due to the high level of saturation in the chemical bonds of the fats and greases. On the other hand, the cold-flow properties of biodiesel from waste cooking oils are usually worse than those from plant oils.

26.3 Chemical Properties of Lipids

26.3.1 Chemical Composition of Lipids

Chemically, oils and fats are composed of the same constituents, triglycerides and free fatty acids. Triglycerides are the fatty acid esters of glycerol. Depending on the source of the lipids, the chemical structures of the triglycerides are quite different, which lead to different fuel properties once they are converted to biodiesel.

As shown in Figure 26.2a, *triglycerides* consist of a three-carbon glycerol backbone with three long-chain fatty acids attached, designated by R_1, R_2, and R_3. Naturally formed fatty acid chains have different lengths, although 16 and 18 carbons are the most common, and have different numbers of carbon–carbon double bonds. The fatty acids are often denoted in short form by CXX:Y, where XX is the number of carbon atoms in the chain and Y is the number of double bonds. For example, C18:1 is the short notation for oleic acid, $HOOC\text{-}C_{17}H_{33}$. The alkyl group attached to the oleic structure is the unsaturated $\text{-}C_{17}H_{33}$. Figure 26.2b shows the expanded form of a triglyceride that is composed of three separate fatty acids: stearic acid (top), oleic acid (middle), and linoleic acid (bottom). Each source of lipid is distinguished by a specific combination of fatty acids, with some lipids having longer chains or more double bonds than others. Table 26.1 lists the representative compositions of fatty acids found in the most common plant oils, animal fats, and microalgal oils.

26.3.2 Average Molecular Weight of Triglycerides

Triglycerides are usually mixtures of fatty acid esters of glycerol. The molecular weight of a specific triglyceride depends on the three fatty acids from which the triglyceride is made (Figure 26.2).

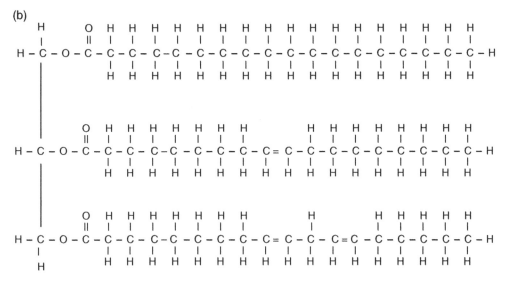

FIG. 26.2   Chemical structure of triglycerides: (a) generic formula of glycerides, where R_1, R_2, and R_3 are alkyl groups; (b) a specific example, where $R_1 = -C_{17}H_{35}$; $R_2 = -C_{17}H_{33}$; and $R_3 = -C_{17}H_{31}$.

Therefore, the molecular weight of a given triglyceride mixture is not unique. In biodiesel production from lipids, it is important to determine the reactant molar ratio of triglycerides to alcohol (e.g., methanol) for efficient transesterification of the oil to biodiesel. Finding the average molecular weight based on the fatty acid profiles of the oil or animal fat is the first step.

The fatty acid profile (i.e., fatty acid composition [% by wt.]) of a lipid is typically determined by gas chromatography (GC) or other analytical means by following standardized procedures, such as the AOCS method Cc-17-79 from the American Oil Chemists' Society. Once the fatty acid composition is known, the average molecular weights of the fatty acids and the triglycerides can be obtained based on the chemical reaction for triglyceride hydrolysis:

$$(26.1)$$

Table 26.1 Fatty acid compositions of selected plant oils and animal fats

| | Fatty acid profiles of seed oils (% wt.) | | | | | | | | | |
	Lauric $C_{12:0}$	Myristic $C_{14:0}$	Palmitic $C_{16:0}$	Stearic $C_{18:0}$	Oleic $C_{18:1}$	Linoleic $C_{18:2}$	Linolenic $C_{18:3}$	Arachidic $C_{20:0}$	Eicosenoic $C_{20:1}$	Erucic $C_{22:1}$
Plant oils										
Microalgal		12–15	10–20		4–19	1–2	5–8			35–48*
Camelina					12–15	15–20	30–40		12–15	2–3
Canola			1–3	2–3	50–60	15–25	8–12			
Coconut oil	45–53	16–21	7–10	2–4	5–10	1–2.5				
Corn oil		1–2	8–16	1–3	20–45	34–65	1–2			
Cottonseed		0–2	20–25	1–2	23–35	40–50				
Jatropha			11–16	6–15	34–45	30–50				
Flax/linseed			4–7	2–4	25–40	35–40	25–60	3–5		
Mustard			1–2	1–2	8–23	10–24	6–18		5–13	20–50
Olive			9–10	2–3	72–85	10–12	0–1			
Palm oil		0.5–2	39–48	3–6	36–44	9–12				
Peanut			8–9	2–3	50–65	20–30				
Rapeseed, high erucic			1–3	0–1	10–15	12–15	8–12	7–10		45–60
Safflower, high linoleic			3–6	1–3	7–10	80–85				
Safflower, high oleic			1–5	1–2	70–75	12–18	0–1			
Sesame oil			8–12	4–7	35–45	37–48				
Soybean oil			6–10	2–5	20–30	50–60	5–11			
Soybean, high oleic			2–3	2–3	80–85	3–4	3–5			
Sunflower			5–8	2–6	15–40	30–70	0–1			
Animal fats										
Chicken fat		1–2	20–30	5–8	35–45	12–18	2–5			
Fish oil			8–12	2–3	11–15	6–12	4–6			
Lard		1–2	25–30	10–20	40–50	6–12	0–1			20–25*
Tallow		3–6	22–32	10–25	35–45	1–3				

Note: *Contains all C22 and above.

The average molecular weight of fatty acids in the lipid, $MW_{ave,FA}$, is calculated by:

$$\frac{1}{MW_{ave,FA}} = \sum \frac{C_{i,FA}}{MW_{i,FA}} \qquad (26.2)$$

where $C_{i,FA}$ is the composition of fatty acid i (% by wt.) and $MW_{i,FA}$ is the molecular weight of fatty acid i.

The average molecular weight of the triglycerides in the lipid, $MW_{ave,trigly}$, is then calculated by:

$$MW_{ave,trigly} = 3MW_{ave,FA} + MW_{gly} - 3\,MW_{water} \qquad (26.3)$$

where MW_{gly} is the molecular weight of glycerol (92.09 kg/kmol) and MW_{water} is the molecular weight of water (18.02 kg/kmol).

Example 26.1: A soybean oil has the following fatty acid profile (i.e., fatty acid composition on a mass basis after being transesterified off the glycerol backbone):

Palmitic	Stearic	Oleic	Linoleic	Linolenic
(C16:0)	(C18:0)	(C18:1)	(C18:2)	(C18:3)
9%	5%	17%	62%	7%

Calculate the average molecular weight of this soybean oil.

Solution: Calculate the average molecular weight of the fatty acids first. The molecular weights of the individual fatty acids are:

FA	C16:0	C18:0	C18:1	C18:2	C18:3
$MW_{i,FA}$ (kg/kmol)	256.5	284.5	282.5	280.5	278.5

According to Equation 26.2:

$$\frac{1}{MW_{ave,FA}} = \frac{9\%}{MW_{C16:0}} + \frac{5\%}{MW_{C18:0}} + \frac{17\%}{MW_{C18:1}} + \frac{62\%}{MW_{C18:2}} + \frac{7\%}{MW_{C18:3}}$$

$$= \frac{0.09}{256.5} + \frac{0.05}{284.5} + \frac{0.17}{282.5} + \frac{0.62}{280.5} + \frac{0.07}{278.5} = 0.0359$$

Therefore, the average molecular weight of the fatty acids is

$$MW_{ave,FA} = \frac{1}{0.0359} = 278.5\,kg/kmol$$

According to Equation 26.3, the average molecular weight of the triglycerides is then:

$$MW_{ave,trigly} = 3MW_{ave,FA} + MW_{gly} - 3\,MW_{water}$$

$$= 3 \times 278.5 + 92.09 - 3 \times 18.02$$

$$= \mathbf{873.6\,kg/kmol}$$

26.3.3 Seed Oil Extraction

The processes for lipid extraction from oilseeds, plant kernels, and other sources are different due to their physical and chemical properties. Mechanical pressing and solvent extraction are typical lipid extraction methods. In this section, lipid extraction from oilseeds is discussed as an example. Detailed discussion on this topic can be found elsewhere (Gupta, 2010).

Mechanical pressing is the oldest and the most commonly used extraction method (Figure 26.3). Two types of mechanical presses are available: hydraulic presses and screw presses. Hydraulic extraction involves pressing batches of oilseeds between two plates, and is less popular due to its low productivity and efficiency. Mechanical screw presses extract oil by squeezing the seeds between a worm shaft and a slotted housing, with a gap that can be adjusted to accommodate the different sizes of oilseed kernels. The pressure that the worm shaft and housing apply to the seeds can reach 140 MPa (or 9,500 psi). Mechanical pressing can extract most of the oil from oilseeds, leaving as little as 10% in the processed meal. The oil yield typically ranges between 60% and 80%, depending on the press type and the operating conditions, such as the moisture content of the oilseeds. Preheating the oilseeds makes the pressing easier and increases the oil yield. Screw presses are simple and relatively inexpensive; however, they require high power and have high maintenance costs. The capacity of screw-type seed presses varies widely from 1–50 tons/day.

Figure 26.4 illustrates the steps in oilseed pressing. Before pressing, oilseeds have to be prepared by removing foreign materials (such as straw debris and dirt), reducing the moisture content to an appropriate level (typically 10–12%), and preheating (typically to 60–70 °C) if desirable. Once the oil is extracted, postextraction treatment is needed to remove gums (phospholipids) from the oil. Then the vegetable oil is ready for use as a feedstock for biodiesel production. However, if the oil is for human consumption, further treatments are needed, including deodorization and bleaching. Also, if the seed oil is for a special application, such as frying, an additional hydrogenation step may be needed. Hydrogenation is becoming less popular nowadays due to its tendency to produce trans-fats, which have negative health effects.

FIG. 26.3   A 45 kg/hr capacity oilseed press with seed preheater. Source: Photo courtesy of B. Brian He, University of Idaho.

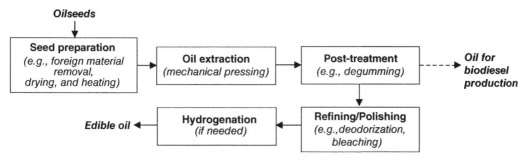

FIG. 26.4   Schematic flow diagram of oilseed pressing.

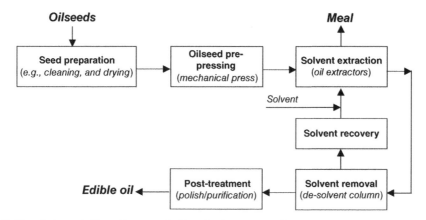

FIG. 26.5   Schematic flow diagram of oilseed solvent extraction.

Solvent extraction is an efficient method of oil extraction with a very high oil yield, commonly leaving less than 1% lipids in the seed meal. As shown in Figure 26.5, it uses a non-toxic solvent, such as n-hexane, to extract oil from the pretreated seeds. Pretreatment may include prepressing for high oil-content seeds, such as canola or sunflower, or just flaking to increase the surface area for low oil content seeds like soybeans. Precrushing of the oilseeds will improve the oil extraction efficiency. Specially designed oil extractors (also known as contactors) are used. The n-hexane solvent is recovered from the oil by evaporation. Two major drawbacks of solvent extraction that make the process ill suited to small producers are the complicated controls and the high initial capital cost.

Mechanical pressing vs. solvent extraction

Mechanical pressing is a simple yet effective method to extract oil from oilseeds. It has a low operating cost and it does not require highly skilled manpower. Mechanical presses can be operated in continuous or in batch mode, which makes them the preferred choice for small facilities. The oil yield varies depending on the equipment and how it is operated, but typically it ranges from 60–80%. Solvent extraction involves application of a solvent, such as n-hexane.

It has a high efficiency with an oil yield up to 99% from oilseeds. It is widely used in large oilseed processing facilities. The disadvantages of solvent extraction include high capital and operating costs, and the safety concerns associated with the use of a highly flammable material.

Example 26.2: A farmer is to crush 1,000 kg of canola seeds to obtain meal for use as livestock feed and oil for biodiesel production in his small facility. The canola seed that he harvested from the field contains (wt. basis) 42% oil and 12% moisture. After the oilseed is pressed, the meal still contains 9.8% oil. The crude oil contains about 1% gum that has to be removed before the oil is used for biodiesel production. Assume that a negligible level of moisture gets into the crude oil phase and the oil loss in degumming is 1.5 times of the quantity of gums, find the following:

(a) How much seed meal is produced and what is its moisture content (MC)?
(b) How much oil is produced after degumming?

Solution: This problem can be solved based on mass balances.

(a) Conduct a mass balance around the seed press, as illustrated in Figure 26.6:

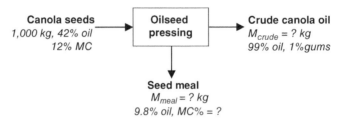

FIG. 26.6   Mass balance around oilseed press.

Overall mass balance : $1,000 = M_{meal} + M_{crude}$

Mass balance on oil : $1,000 \times 42\% = M_{meal} \times 9.8\% + M_{crude} \times 99\%$

Mass balance on MC : $1,000 \times 12\% = M_{meal} \times MC\% + M_{crude} \times 0\%$

Solving first two equations simultaneously gives:

$M_{meal} = 639.0\,kg$ and $M_{crude} = 361.0\,kg$

Substituting $M_{meal} = 639.0$ kg, the third equation gives:

Moisture content = 18.8% (wt. basis)

(b) It is given that the crude oil contains 1% gums (wt. basis) that takes oil of 1.5 times its weight with it when degummed. Therefore, the oil in the gums is

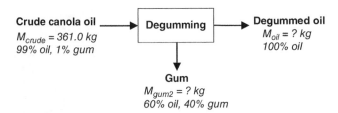

FIG. 26.7   Mass balance around degumming process.

1.5/(1 + 1.5) = 60%. Conduct a mass balance around the degumming process, as illustrated in Figure 26.7.

Overall mass balance: $361.0 = M_{gum2} + M_{oil}$

Mass balance on oil : $361.0 \times 99\% = M_{gum2} \times 60\% + M_{oil} \times 100\%$

Solving these equations simultaneously gives:

$M_{gum2} = 9.0\,kg$ *and* $M_{oil} = 352.0\,kg$

Therefore, the overall mass balances of the oilseed pressing system can be summarized in Figure 26.8.

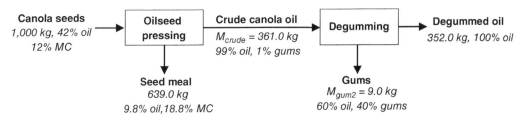

FIG. 26.8   Overall mass balances of oilseed pressing system.

26.4 Biodiesel from Lipids

26.4.1 Biodiesel Production via Transesterification

Vegetable oils have long been used for heating and lighting. They have also been used as engine fuels since internal combustion engines were invented in the late 1800s, but they have usually been found to cause engine deposits, resulting in higher emissions and shorter engine life. These undesirable characteristics are attributed to the high viscosity of the seed oils. This problem has been addressed commercially by converting the oils to *fatty acid methyl* or *ethyl esters*, abbreviated as FAME or FAEE,

FIG. 26.9   Batch biodiesel reactors: 1,000 L and 2,000 L capacity (left); flash evaporation unit for recovery of methanol (right). Source: Photos courtesy of B. Brian He, University of Idaho.

respectively, and commonly known by their trade name as biodiesel. Biodiesel is produced from lipid-based feedstocks by splitting the fatty acid chains from the glycerol backbone and producing methyl, or sometimes ethyl esters.

The chemical reaction to produce the esters (biodiesel) is commonly known as *transesterification*. It involves reacting the triglycerides with a simple alcohol such as methanol, to produce FAME or the biodiesel, and the byproduct, glycerol (also known as glycerin). A batch biodiesel reactor is illustrated in Figure 26.9. The reaction is shown in Equation 26.4 for a representative case of triolein, a triglyceride with all three of its fatty acid chains as C18:1.

$$\underset{\textit{Triolein}}{C_{57}H_{104}O_6} + \underset{\textit{Methanol}}{3\,CH_3OH} \rightarrow \underset{\textit{Methyl ester}}{3\,C_{19}H_{36}O_2} + \underset{\textit{Glycerol}}{C_3H_8O_3} \tag{26.4}$$

In order to have the reaction proceed at a reasonable rate, a catalyst is required. The most commonly used catalysts are strong bases such as sodium methoxide ($NaOCH_3$) or potassium methoxide ($KOCH_3$), although sodium hydroxide ($NaOH$) and potassium hydroxide (KOH; also known as *lye*) are also used. Heterogeneous catalysts have been developed and tested by many researchers, but homogeneous catalysts of strong base solutions are currently preferred in the industry.

It is worth noting that not all lipids are composed of pure triglycerides. When triglycerides are exposed to water and high temperatures, they undergo decomposition reactions that separate the fatty acid chains from the triglyceride, which then becomes a diglyceride. The reaction (Equation (26.5)) produces free fatty acids (FFAs) in the oil.

$$\underset{\textit{Triolein}}{C_{57}H_{104}O_6} + \underset{\textit{Water}}{H_2O} \rightarrow \underset{\textit{Oleic acid}}{C_{18}H_{34}O_2} + \underset{\textit{Diolein}}{C_{39}H_{72}O_5} \tag{26.5}$$

The FFAs in the oil cause a problem for transesterification because they will react with the strong base catalyst to create soap through a process known as *saponification*. This makes the catalyst unavailable for the reaction.

$$\underset{\substack{Oleic\ acid}}{C_{18}H_{34}O_2} + \underset{\substack{Potassium\\hydroxide\,(catalyst)}}{KOH} \rightarrow \underset{\substack{Potassium\ oleate\\(soap)}}{C_{17}H_{34}COOK} + \underset{\substack{Water}}{H_2O} \tag{26.6}$$

When the FFA level is low (<3%), it is common simply to add extra catalyst and allow the soap to be produced, and then removed from the biodiesel at the end of the production process. However, an alternative process is to convert the FFAs to methyl esters by an *esterification* reaction before transesterification. Esterification involves reacting the FFA with a simple alcohol to form an ester. A strong acid, such as sulfuric acid (H_2SO_4), is used as the catalyst and no soap is produced.

$$\underset{\substack{Oleic\ acid}}{C_{18}H_{34}O_2} + \underset{\substack{Methanol}}{CH_3OH} \rightarrow \underset{\substack{Methyl\ ester}}{C_{19}H_{36}O_2} + \underset{\substack{Water}}{H_2O} \tag{26.7}$$

This reaction, typically used as a pretreatment for high-FFA feedstocks (such as waste cooking oils) before transesterification, converts the FFAs to methyl esters so the overall yield of biodiesel is increased compared to that when FFAs are simply removed.

Biodiesel production from lipid-based feedstocks involves several steps. Figure 26.10 shows a schematic diagram of the biodiesel production process. Oil, methanol, and catalyst are added to the reactor in a continuous flow. The reactor may be a system of stirred tanks in series, or may use a high-shear mixer or ultrasound for agitation. After sufficient time has elapsed to ensure a complete reaction, the flow passes to a separator, which consists of a decanter or a centrifuge. The crude glycerol is removed at this point. The biodiesel passes to a methanol recovery unit, in which the residual methanol is evaporated and recirculated for reuse. Then, the biodiesel is either washed by softened water or "dry washed" by using absorbents (such as magnesium silicate) to remove residual methanol and glycerol, as well as soaps and catalyst, leaving the finished biodiesel. Residual water may need to be dried off if water washing is used.

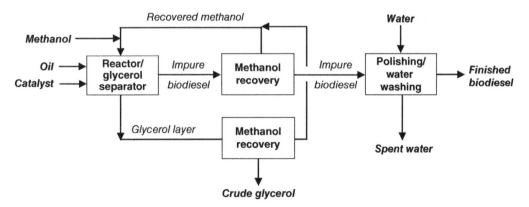

FIG. 26.10   Schematic diagram of a biodiesel production process.

Transesterification

Transesterification is a chemical reaction in which the alkyl group (R_i) in a carboxylic acid ester is replaced by another alkyl group (R_j) from an alcohol to form a new ester and a new alcohol. In biodiesel production from triglycerides (i.e., fatty acid esters of glycerol), the three fatty acid alkyl groups (R_1, R_2, and R_3) on the glycerol are exchanged with the alkyl groups from the alcohol, e.g., methanol or methyl alcohol (R_1 = -CH_3), in three steps:

$$CH_2-OOCR_1$$
$$|$$
$$CH-OOCR_2 \quad + \quad CH_3-OH \quad \Leftrightarrow \quad CH_3-O-\overset{\overset{O}{\|}}{C}-R_1 \quad + \quad \begin{array}{l} CH_2-OH \\ | \\ CH-OOCR_2 \\ | \\ CH_2-OOCR_3 \end{array} \quad \text{(Step 1)}$$

tri-glyceride   methanol   methyl ester   di-glyceride

di-glyceride + methanol ⇔ methyl ester + mono-glyceride (Step 2)

mono-glyceride + methanol ⇔ methyl ester + glycerol (Step 3)

Depending on the types of feedstock and the process applied, the biodiesel yield varies. As an example, the mass balances for biodiesel production from soybean oil and methanol catalyzed by potassium hydroxide are shown in the box.

Mass balances in biodiesel production from soybean oil

Reactants

Soybean oil	1,000 kg	(MW_{ave} = 873.6 kg/kmol)
Methanol	220 kg	(methanol:oil molar ratio 6:1 or 100% excess)
KOH	10 kg	(potassium hydroxide as catalyst; 1% of oil)

Transesterification, separation, and methanol recovery

Biodiesel	995 kg	(98.5% mol conversion rate; 1% impurities)
Recovered methanol	77 kg	(70% recovery/reuse)
Crude glycerol	158 kg	(including spent catalyst, unreacted triglycerides, etc.)

Polishing/water washing

Biodiesel	985 kg	(with minimal impurities; within specification)
Fresh water	1,120 kg	(1 L water per every 1 L of biodiesel)
Spent water	1,130 kg	(with impurities)

More details on biodiesel production and related topics can be found elsewhere (Van Gerpen, 2005; Knothe, 2010; Van Gerpen and Knothe, 2010; Van Gerpen and He, 2010).

26.4.2 Parameters Affecting Biodiesel Production

Reaction Temperature

Transesterification of triglycerides to biodiesel is a reversible chemical reaction involving a number of consecutive reactions. Process parameters, particularly the reaction temperature, strongly affect the reaction rate. At room temperature, transesterification of triglycerides with methanol takes at least 8 hr to reach equilibrium. At 40 °C, the reaction rate increases significantly and the time for a complete transesterification is reduced to 2–4 hr. The time further reduces to 1–2 hr if the reaction temperature is increased to 60 °C. However, the low boiling temperature of the alcohol (64.5 °C for methanol) limits the highest reaction temperature that can be used under atmospheric pressure, unless a pressurized reactor is employed.

Alcohol to Triglyceride Ratio

Due to the reversible nature of the transesterification, the addition of excess alcohol causes the reaction equilibrium to move toward completion. The stoichiometric molar ratio of methanol to triglycerides is 3:1. In commercial facilities, 100% more alcohol than the stoichiometric requirement is typically used to ensure the complete conversion of triglycerides to biodiesel. This excess alcohol is recovered and reused after purification.

Free Fatty Acids

Waste vegetable oils and some microalgal lipids contain significant quantities of FFAs. Soap formation is a concern when these feedstocks contain more than 3–5% free fatty acids. Soap forms when the FFAs react with the basic catalysts, for example KOH or $NaOCH_3$ in the reaction mixture. Soap formation not only consumes catalyst but also potentially causes the product mixture to emulsify, making it difficult to separate. As described earlier, a two-step transesterification process is typically used when a high concentration of FFA is present. A strong acid, such as H_2SO_4, is employed to catalyze the FFA esterification to biodiesel, and then a basic catalyst is used to convert the triglycerides to esters via transesterification.

Reactor Systems for Biodiesel Production

Transesterification can be conducted in batch, semi-continuous, and continuous-flow reactor systems. Batch reactors are inexpensive and flexible in allowing different feedstocks, and production capacities. The drawbacks of batch reactor systems include low productivity, intensive labor involvement, and inconsistent product quality. Continuous-flow reactor systems are commonly used in large-scale facilities. Such reactor systems provide consistent product quality with low operating cost and high throughput. However, continuous-flow reactor systems require a high initial capital cost and sophisticated process control systems, and they are inflexible for accommodating diverse feedstocks. In biodiesel production from lipid-based feedstocks, continuous stirred-tank reactors (CSTR) are most commonly used. Other novel types of reactor, such as tubular flow reactors integrated with an ultrasound unit, are also used by some biodiesel producers.

Example 26.3: In a biodiesel production process, the feedstock oil (average molecular weight 880 kg/kmol) contains 1.2% (wt. basis) free fatty acid (FFA). To ensure complete lipid conversion to biodiesel, additional catalyst potassium hydroxide (KOH) is typically added to compensate for the need to neutralize the FFA. For a batch of 100 kg of oil, find the following:

1. How much KOH is needed for the overall reactions if the catalyst must not be less than 1% (wt. basis) of the quantity of the oil?
2. How much soap will be formed?
3. What is the biodiesel yield if the lipid conversion is 98.5% (wt. basis) and if the finished biodiesel contains 0.4% (wt. basis) of incompletely reacted monoglycerides?

Solution:

1. The total KOH needed includes that for use as catalyst (1% oil) and that for neutralizing the FFA. It is given that 1.2% of the oil is FFA. The total FFA is then:

$$100 \, \text{kg} \times 1.2\% = 1.2 \, \text{kg FFA}$$

Since the average molecular weight of the oil is 880 kg/kmol, according to Equation 26.2 the average molecular weight of the fatty acids is then:

$$MW_{ave, FA} = \left(MW_{ave, trigly} - MW_{gly} + 3 \, MW_{water} \right)/3$$
$$= (880 - 92.1 + 3 \times 18.0)/3 = 280.6 \, \text{kg/kmol}$$

Once mixed, the KOH reacts with the FFA and forms the undesirable byproduct, soap, based on the following reaction:

$$\underset{FFA}{R - COOH} + \underset{Catalyst}{KOH} \rightarrow \underset{Soap}{R - COOK} + \underset{Water}{H_2O}$$

Since 1 kmol of FFA (280.6 kg) reacts with 1 kmol of KOH (56.1 kg), then the amount of KOH required for neutralizing 1.2 kg of FFA is:

$$x = \frac{56.1 \times 1.2}{280.6} = 0.24 \, \text{kg}$$

Therefore, the total required KOH is: 100 kg × 1% + 0.24 kg = **1.24 kg**

2. Since the average molecular weight of FFA is 280.6 kg/kmol, and the atomic weights of hydrogen and potassium are $AW_H = 1.0$ kg/kmol and $AW_K = 39.1$ kg/kmol, respectively, the average molecular weight of the soap can be calculated from the same reaction:

$$MW_{ave,soap} = MW_{ave,FFA} - AW_{H2} + AW_K$$

$$= 280.6 - 1.0 + 39.1$$

$$= 318.7\,kg/kmol$$

Use the same approach as in part 1 to solve for soap formation.

Since 1 kmol of soap (318.7 kg) reacts with 1 kmol of FFA (280.6 kg), the amount of soap produced by 1.2 kg of FFA will be:

$$y = \frac{1.2 \times 318.7}{280.6} = \mathbf{1.36\,kg}$$

3. The total triglycerides of the oil after 1.2% FFA are then $100\,kg \times (1 - 1.2\%) = 98.8\,kg$; of that, 98.5% is converted to biodiesel, or $98.8\,kg \times 98.5\% = 97.3\,kg$.

Since it is known that the average molecular weight of FFA is 280.6 kg/kmol, and the molecular weight of the methyl group is $MW_{methyl} = 15.0\,kg/kmol$, the average molecular weight of the FAMEs, $MW_{ave,esters}$, is then:

$$MW_{ave,esters} = MW_{ave,FFA} - AW_H + MW_{methyl}$$

$$= 280.6 - 1.0 + 15.0$$

$$= 294.6\,kg/kmol$$

The transesterification of the oil to fatty acid methyl esters is based on the following reaction:

$$\underset{\text{Triglycerides}}{(R\text{-}COO)_3\text{-}C_3H_5} + \underset{\text{Methanol}}{3CH_3OH} \rightarrow \underset{\text{FAMEs}}{3R\text{-}COOCH_3} + \underset{\text{Glycerol}}{C_3H_8O_3}$$

Since 1 kmol of triglycerides (880 kg) reacts to produce 3 kmol of FAMEs ($3 \times 294.6\,kg$), with 97.3 kg of triglycerides, the biodiesel yield is:

$$z = \frac{97.3 \times (3 \times 294.6)}{880} = 97.7\,kg$$

Also, if the biodiesel contains 0.4% (wt. basis) monoglycerides, the final biodiesel yield with the minor impurity is $z' = 97.7/(1 - 0.4\%) = \mathbf{98.1\,kg}$.

26.4.3 Quality of Biodiesel

Fuels that will be sold to consumers must be of high quality to ensure that the consumer's vehicle is protected from damage and that its performance meets their expectations. This high level of fuel quality is achieved by strict adherence to fuel standards that are developed by industry experts and are frequently included in government regulations and laws. In the USA, fuel specifications are usually developed through the American Society for Testing and Materials (ASTM). ASTM D 6751 Standard Specification for Biodiesel Fuel Blend Stock (B100) for Middle Distillate Fuels is the quality standard for biodiesel in North America. It is similar to EN14214, the standard used in Europe. These standards list the test procedures and have the acceptable values for many of the fuel properties such as viscosity, sulfur content, ash, and other contaminants. When biodiesel is blended with petroleum diesel fuel, the blend must meet the requirements defined in ASTM D 7467 Standard Specification for Diesel Fuel Oil, Biodiesel Blends (B6 to B20).

26.5 Lipid-Based Biorefinery

In a biorefinery, other high-value products and byproducts besides lipids play an equally important role in generating revenue for the biodiesel industry. Seed oils and other types of lipids can be converted into a wide variety of high-value products other than biodiesel. In this section, some of the important high-value products and byproducts of a lipid-based biorefinery are discussed. Detailed discussion on this subject can be found elsewhere (Meier *et al.*, 2007; Hill, 2000).

26.5.1 High-Value Biobased Products from Seed Oils

The types of fatty acids (saturated and unsaturated) present in the triglyceride will govern the types of product that can be produced, for instance paints, adhesives, or lubricants.

Stains and paints contain lipid-based materials called drying oils. Drying oils consist of highly unsaturated oils that can polymerize (cure or harden) by oxidation in air over time. The principle is that a drying oil reacts with oxygen to form hydroperoxides, which are prone to crosslinking reactions. The reaction can be accelerated by the addition of metal salt catalysts (e.g., Fe, Co, Mn). Linseed oil, which is the common name for oil derived from the flax plant, has been traditionally used in oil-based paints and stains, and in linoleum, a rubbery floor covering product made from linseed oil, cork, sawdust, and a cloth backing.

Seed oils and their derivatives are also used as lubricants. In the 1800s, palm oil (a mixture of C_{14} to C_{18} saturated and unsaturated fatty acids) was a major lubricant used in machinery during the industrial revolution. The most common fatty acid–based lubricant is zinc stearate, which has a complex carboxylate structure $[Zn_4O(O_2C_{18}H_{35})_6]$. Zinc stearate is widely used as a release agent for plastic, rubber, polyurethane, and metal molding applications, and as a lubricant in rubber and plastic compounds. Another example is ethylene bis-stearamide $[CH_2NHC(OC_{17}H_{35})_2]$, a low-toxicity, waxy solid formed from the reaction of ethylenediamine and stearic acid. It is used as a mold release agent as well as a lubricant in plastic compounds.

Surfactants can also be produced from naturally formed lipids and such products are called biosurfactants. Fatty acid salts from palm oil were commercialized in the late 1800s to form soap products (Sunlight and Palmolive), and are still used in detergents, bath soaps, and washing powders. Sodium stearate is the major component (~50%) of most soaps and is formed by saponification of hydrogenated triglycerides. Cocomonoglyceride sulfate (CMGS) is a biosurfactant from coconut oil (or lauric/myristic acid–based oils). It is commonly used in shampoo, shower gel, and cosmetics.

Unsaturated triglycerides are currently being used as monomer building blocks for thermoset resins. The triglyceride can be functionalized to form epoxy and acrylic groups that can react with other resin monomers. The resin properties including tensile strength and the glass transition temperature can be tailored based on the triglyceride composition, functionalization, and resin formulation. These resins can be used in thick coatings or as a matrix for natural (flax, jute, etc.) and synthetic (glass, carbon, etc.) fiber composites.

Printing oils can also be produced by polymerization of an unsaturated oil, such as linseed oil, with heat to produce a gel state, a process that is similar to drying oils. The oils are then blended with additives such as carbon black, pigment thickeners, tackifiers, and brighteners to form finished products for printing.

26.5.2 Seed Meals and Their Applications

Critical to the production of biofuels is the need to consider value-added co-products that increase the economic viability of biofuel production. Seed meals are protein-rich residues obtained following oil recovery that represent valuable co-products for food/feed applications. For high-protein soybean meal, there is a well-established market. However, traditional varieties of rapeseed and mustard seed meals are not suitable for animal feed due to their high glucosinolate and erucic acid contents. Innovative processing methods and product development are required to enhance their value. A good understanding of seed meal chemistry is necessary to optimize the use of seed meals.

Soybean meal contains 43–49% protein (dry wt.), with the highest protein concentrations in solvent-extracted meals obtained from dehulled seed (Table 26.2). Nearly all of the soybean meal produced in the U.S. ends up as an animal feed. On a worldwide basis, soybean meal supplies approximately 69% of the protein used in animal feeds, with canola (rapeseed) providing the second highest protein source at 13%. The dominance of soybean meal as an animal feed is due to its highly digestible total protein content that releases amino acids in proportions matching the nutritional requirements of poultry and swine better than other oilseed meals.

Currently, besides soybean meal the only mass-produced seed meal is that of rapeseed or canola. Rapeseed and canola are essentially the same crop, with canola referring more specifically to a crop in which the oil contains less than 2% erucic acid and the meal contains less than 30.0 µmol of alkenyl glucosinolates per gram of air-dried, oil-free solid. Some varieties of rapeseed fall outside these limits, and thus cannot be termed canola.

The nutritional quality of canola meal as an animal feed is best compared to soybean meal. The amount of protein in canola meal ranges from 36–39% (dry wt.). Although the lysine content of canola meal is only 2.0% compared to 2.7% in soybean meal, canola meal has slightly higher methionine (0.74%) and cysteine (0.86%) contents. The essential amino acids in canola meal are about

Table 26.2 Selected nutritional components of soybean meal

Components	Composition (% dry wt.)			
	Solvent	Dehulled solvent	Expeller	Full-fat
Crude protein	44.0	47.8	42.0	38.0
Lysine	2.70	3.02	2.70	2.40
Tryptophan	0.60	0.70	0.58	0.52
Threonine	1.70	2.00	1.70	1.69
Isoleucine	2.50	2.60	2.80	2.18
Valine	2.40	2.70	2.20	2.02
Methionine	0.65	0.70	0.60	0.54
Cysteine	0.67	0.71	0.62	0.55

Source: Iowa Soybean Association, http://www.soymeal.org/composition.html.

FIG. 26.11 Phytotoxic activity of mustard meal against redroot pigweed and wild oat. Seed meal was added to soil (2,000 kg/ha) prior to seed germination. Source: Photo courtesy of Matthew J. Morra, University of Idaho.

10% less digestible by pigs and poultry as compared to the same amino acids in soybean meal, decreasing their absorption potential in the small intestine.

Rapeseed and mustards contain glucosinolates. Glucosinolates are present in all parts of the plant, but are most concentrated in the seed. Plant tissues also contain an enzyme called myrosinase that catalyzes the hydrolysis of glucosinolates to form a variety of biologically active molecules including isothiocyanate and thiocyanate, which are toxic. Water addition to the meals results in enzymatic hydrolysis of the glucosinolates to form these biologically active molecules. The production of such compounds limits the use of rapeseed and mustard meals as animal feeds. However, hydrolyzed products from glucosinolates can be used as natural pesticides (biopesticides), as evidenced by the comparison of soil treated with glucosinolate-containing mustard meal with soil not treated with the meal shown in Figure 26.11. The presence of different glucosinolates with a spectrum of biological activities creates an opportunity for the extraction of bioactives from meals, increasing the palatability of the residual meals as animal feed and producing value-added biopesticides.

Further reading on this topic can be found elsewhere (Herman and Price, 2013; SzydłOwska-Czerniak, 2013; Newkirk, 2009; Mailer *et al.*, 2008).

26.5.3 Utilization of Glycerol from Biodiesel Production

Glycerol, also known as glycerin, is the principal byproduct of biodiesel production. In principle, for each metric ton of biodiesel produced approximately 105 kg of crude glycerol is produced as a byproduct. High-purity glycerol is a very important industrial feedstock with diverse applications (Silva and Ferreira, 2012; Gu and Jerome, 2010; Pagliaro and Rossi, 2010; Choi, 2008; Pagliaro *et al.*, 2007). Historically, glycerol is produced for traditional applications, such as food, pharmaceuticals, and personal care products, in specialized facilities according to US Food and Drug Administration (FDA) regulations. The crude glycerol from biodiesel production contains impurities and therefore has a very low market value. Further refining of the crude glycerol depends on the economies of scale or the availability of external glycerol refining facilities. Large-scale biodiesel producers refine their crude glycerol and market it to other industries. It is generally treated, and refined through filtration, chemical additions, and vacuum distillation to yield various commercial grades.

For most applications, glycerol is required to have a purity of 99.5% or higher. The impurities in the glycerol include the catalysts, salts from neutralization treatment of high FFA feedstocks, and gums. Typical processes for purifying crude glycerol include vacuum distillation to remove salts and other high-boiling-point compounds from the bottom of the distillation column.

Glycerol, when used in combination with other compounds, yields many useful products. For example, glycerol and ethylene glycol together can be used as a solvent for the alkaline treatment of polyester fabrics. Glycerol can also be used as a dielectric medium for compact, pulse power systems. It acts as a medium in the electro-deposition of indium-antimony alloys from chloride tartrate solutions.

Although crude glycerol can be used as a feedstock in wastewater treatment plants and anaerobic digesters as well as being disposed of through composting, other value-added applications are possible. Such value-added chemicals produced from glycerol include 1,3-propanediol, dihydroxy-acetones, glyceric acids, hydrogen, polyglycerols, succinic acid, and polyesters. All of these can be used as raw materials for a large number of commodity products.

References

Choi, W. (2008). Glycerol-based biorefinery for fuels and chemicals. *Recent Patents on Biotechnology* 2(3): 173–80. doi:10.2174/187220808786241006.

Gu, Y., and Jerome, F. (2010). Glycerol as a sustainable solvent for green chemistry. *Green Chemistry* 12(7): 1127–38. doi:10.1039/c001628d.

Gupta, M. (2010). *Practical Guide to Vegetable Oil Processing*. Urbana, IL: AOCS Publishing.

Herman, R.A., and Price, W.D. (2013). Unintended compositional changes in genetically modified (GM) crops: 20 years of research. *Journal of Agricultural Food Chemistry* 61(48): 11695–701. doi:10.1021/jf400135r.

Hill, K. (2000). Fats and oils as oleochemical raw materials. *Pure and Applied Chemistry* 72(7): 1255–64.

Knothe, G. (2010). Biodiesel and renewable diesel: A comparison. *Progress in Energy and Combustion Science* 36: 364–73.

Maglinao, R., and He, B. (2012). Verification of propylene glycol preparation from glycerol via acetol pathway by *in situ* hydrogenolysis. *Biofuels* 3(6): 675–82. doi:10.4155/bfs.12.65.

Mailer, R.J., McFadden, A., Ayton, J., and Redden, B. (2008). Anti-nutritional components, fibre, sinapine and glucosinolate content, in Australian canola (*Brassica napus* L.) meal. *Journal of the American Oil Chemists' Society* 85: 937–44.

Meier, M.A.R., Metzger, J.O., and Schubert, U.S. (2007). Plant oil renewable resources as green alternatives in polymer science. *Chemical Society Reviews* 36: 1788–1802.

Newkirk, R. (ed.) (2009). *Canola Meal Feed Industry Guide*, 4th edn. Winnipeg: Canola Council of Canada. https://cigi.ca/wp-content/uploads/2011/12/2009-Canola_Guide.pdf, accessed April 2016.

Pagliaro, M., and Rossi, M. (2010). *The Future of Glycerol*, 2nd edn. Cambridge: RSC Publishing.

Pagliaro, M., Ciriminna, R., Kimura, H., Rossi, M., and Della Pina, C. (2007). From glycerol to value-added products. *Advanced Chemie-International Edition* 46(24): 4434–40. doi:10.1002/anie.200604694.

Silva, M., and Ferreira, P. (eds.) (2012). *Glycerol: Production, Structure and Applications*. New York: Nova Science Publishing.

SzydłOwska-Czerniak, A. (2013). Rapeseed and its products: Sources of bioactive compounds. A review of their characteristics and analysis. *Critical Reviews in Food Science and Nutrition* 53: 307–30.

Van Gerpen, J. (2005). Biodiesel processing and production. *Fuel Processing Technology* 86: 1097–107.

Van Gerpen, J., and Knothe, G. (2010). Basics of the transesterification reaction. In: G. Knothe, J. Krahl, and J. Van Gerpen (eds.), *The Biodiesel Handbook*. Champaign, IL: Oil Chemists Society, Chapter 4.1.

Van Gerpen, J.H., and He, B. (2010). Biodiesel production and properties. In: M. Crocker (ed.), *Thermochemical Conversion of Biomass to Liquid Fuels and Chemicals*. London: Royal Society of Chemistry Publishing, pp. 382–415.

Zheng, Y., Chen, X., and Shen, Y. (2008). Commodity chemicals derived from glycerol, an important biorefinery feedstock. *Chemical Reviews* 108(12): 5253–77. doi:10.1021/cr068216s.

Exercise Problems

26.1. A feed of 10,000 kg of canola seed is processed in a sequence of three steps. The seed contains (wt. basis) 35% protein, 10.5% moisture, 40% oil, and other constituents. In the first step, the seeds are crushed and pressed to remove the major portion of the oil content, giving a pure oil stream and a pressed seed stream that still contains 12% oil. Assume no loss of other constituents with the oil stream. In the second step, the pressed seeds are extracted with hexane to produce an extracted meal stream, which still contains 0.5% oil, and a hexane-extracted oil stream. Assume no hexane loss in the extracted meal. Finally, in the last step the extracted meal is dried to give a dried meal of 8% moisture content. Calculate:

 (a) Amount (in kg) of the extracted meal from Step 2.
 (b) Amount (in kg) of oil produced from Steps 1 and 2 combined.
 (c) Amount (in kg) of the final dried meal.
 (d) Percentage (wt. basis) of protein in the dried meal.

26.2. In the USA, biodiesel is mainly produced from soybean oil. In an analysis of a soybean oil sample, the fatty acid profile in Table 26.3 is obtained.

Table 26.3 Fatty acid profile of a soybean oil sample

Fatty acids	Short notation	% wt.
Palmitic	$C_{16:0}$	17
Stearic	$C_{18:0}$	5
Oleic	$C_{18:1}$	16
Linoleic	$C_{18:2}$	48
Linolenic	$C_{18:3}$	12
Eicosenoic	$C_{20:1}$	1
Behenic	$C_{22:0}$	1

 (a) What is the average molecular weight of the soybean oil?
 (b) If 1 gallon of biodiesel is produced from this soybean oil, how much byproduct glycerol (in *kg*) is produced?
 (c) In 2012, 1.2 billion gallons of biodiesel were produced in the USA. How much byproduct glycerol (in metric *tons*) was also produced?

26.3. A seed oil with the fatty acid profile of 8% C16:0, 4% C18:0, 26% C18:1, 54% C18:2, and 8% C18:3 is to be converted to biodiesel using a transesterification reaction.

 (a) What is the average molecular weight for this seed oil?
 (b) If you want to use a 100% excess of the stoichiometric amount of methanol for the transesterification reaction, how many kg of methanol should be added per kg of oil?

26.4. How does the composition of fatty acids in the oil influence the types of bioproducts that can be produced?
26.5. Why is the double bond important in the crosslinking reactions during (a) curing of drying oils and (b) forming of epoxides for making resins?

26.6. Why is it critical to consider seed meals as valuable feedstocks in liquid fuels production from oilseeds?

26.7. How much protein is present in 1 kg of soybeans?

26.8. Why is soybean meal considered a highly desirable animal feed?

26.9. Describe the positive and negative attributes of glucosinolates and how those attributes relate to the potential development of value-added products from rapeseed and mustard seed meals.

26.10. Conduct a literature search and summarize in a tabular form the products that can be produced from glycerol. Provide at least five products that are not mentioned in this textbook.

26.11. Propylene glycol (1,2-propanediol) can be produced from glycerol catalytically via the acetol pathway (Magliano and He, 2012) in an aqueous solution. The hydrogen needed is provided internally by reformation reactions. Considering only the reformation of acetol results in the simplified reactions involved in glycerol conversion to propylene glycol, as shown:

$$\text{Glycerol} \rightleftharpoons \text{Acetol} + H_2O$$

$$\text{Acetol} + H_2 \rightleftharpoons \text{Propylene glycol}$$

$$\text{Acetol} + 4H_2O \rightleftharpoons 3CO_2 + 7H_2$$

What is the theoretical molar yield of propylene glycol per mole of glycerol feed? Theoretically, how much carbon (in molar percentage) has been expensed in order to produce an adequate amount of hydrogen for the process of making propylene glycol?

SECTION VI
Bioenergy System Analysis

Techno-Economic Assessment

Ganti S. Murthy

What is included in this chapter?

This chapter covers the basics of techno-economic assessment for bioenergy systems. The steps and tools to perform the analysis are discussed. Applications of this approach to corn and cellulosic ethanol, and biodiesel production processes, are presented.

27.1 Introduction

Bioenergy is proposed as a sustainable alternative to fossil fuels that addresses the issues of greenhouse gas emissions (GHGs) and energy insecurity. Currently, first-generation biofuels such as ethanol from corn and sugarcane, and biodiesel from soybean, rapeseed, and other oil seeds, are produced in significant quantities in response to various government mandates, incentives, and market demands. Many emerging technologies for the production of bioenergy from biomass feedstocks are currently being tested at a pilot scale and are in various stages of commercialization. It is important to evaluate different technologies based on their technical, economic, and environmental performance metrics in order to compare and select the best technologies. Often the choice of the "*best technology*" is dependent on the definition of "the best." In general, the technical, economic, and environmental aspects of a process technology or bioenergy production pathway are considered critical for the successful production of bioenergy. The question in a broad context when assessing a technology can be framed as: *What are the energetic, economic, and environmental benefits associated with this technology?*

The energetic and economic performance of a technology can be assessed using techno-economic analysis (TEA), while the environmental impacts can be quantified using life-cycle assessment (LCA) methods. This chapter covers TEA with a focus on assessing the technical feasibility, economic viability, and tradeoffs of a process technology for bioenergy systems.

Bioenergy: Principles and Applications, First Edition. Edited by Yebo Li and Samir Kumar Khanal.
© 2017 John Wiley & Sons, Inc. Published 2017 by John Wiley & Sons, Inc.
Companion website: www.wiley.com/go/Li/Bioenergy

27.2 What Is Techno-Economic Analysis?

TEA is used to assess the technical and economic viability of a process, and to identify the optimal unit processes and performance conditions considering *both* technical and economic factors. TEA is an integral part of research, development, and commercialization, as such analysis provides impetus to technology development. Researchers, engineers, plant managers, and investors need information about performance measures and tradeoffs to suggest/implement/operationalize/finance various technologies for bioenergy production. TEA can be performed at multiple levels of detail depending on the accuracy needed in the cost and performance estimates.

One of the first steps in TEA is to define the scope of the analysis and identify the types of estimate, which can be classified as:

- Order of magnitude estimate: ±10 to 50% accuracy
- Study estimate (factored estimate): ±30%
- Preliminary estimate (budget authorization estimate): ±20%
- Definitive estimate (project control estimate): ±10%
- Detailed estimate (firm or contractor's estimate): ±5%

Which estimate?

Consider an investor who wants to build a lignocellulosic-based ethanol facility. The investor would use an *order of magnitude* estimate to see if the venture could make a profit.

If the results are encouraging, the investor would perform a *study estimate (factored estimate)* to identify the best technology for each of the unit processes (dilute acid pretreatment vs. hot water pretreatment vs. steam explosion pretreatment; separate saccharification and fermentation vs. simultaneous saccharification and fermentation; anaerobic digestion vs. multiple-effect evaporator to treat process water) for producing cellulosic ethanol. By the end of this estimate, all major equipment and its specifications (number of units, throughputs, capacities, and capital costs) used in the plant are identified.

The study estimate results are used to inform the *preliminary/budget authorization estimate* for approaching banks/other financial institutions to obtain capital. Once the capital for building the plant is obtained, a *definitive estimate* is commissioned inviting bids from contractors. Contractors will present the *detailed estimates* and build the physical plant for commissioning.

While the accuracy and precision of the estimates improve as we move from order of magnitude to detailed estimate, this is achieved at the expense of higher cost and resources. For example, to develop an order of magnitude estimate for a lignocellulosic ethanol plant about $3,000–13,000 may be required, while the cost for producing a detailed estimate could be over $200,000–1,000,000 (Westney, 1997). Therefore, order of magnitude estimates are used early on in the design process, while the detailed estimate is used by a contracting firm to build the physical plant. Study/factored estimates are the most commonly used types of estimate for identifying the optimal technologies and process conditions, and are the focus of this chapter.

27.3 Basic Steps in TEA

A preliminary evaluation of technical and economic factors such as technology (feasibility of overall process, freedom to operate as defined by absence of restrictions due to competing patents and other intellectual property protection, safety considerations), raw materials (availability, quantity, quality, and cost), product (current and future market demand, expected price trends), current and future competition for the product and raw materials, plant location, and market saturation must be made before starting a detailed TEA.

> **Example 27.1:** A startup company proposes to use corn stover as a feedstock (39% cellulose, 26% hemicellulose, and 23% lignin, 12% ash) to produce ethanol. The company claims to have discovered a new process that can produce 492 L ethanol/dry metric ton (130 gal ethanol/dry ton) of feedstock. Please scrutinize the feasibility of the estimates provided by the company.
>
> **Solution:** In order to scrutinize the claim of 492 L/dry metric ton, it is essential to identify the maximum theoretical yield of ethanol from the feedstock. We can proceed by making a simplifying assumption that hemicellulose is equivalent to cellulose (which is not true in reality) to estimate the theoretical maximum amount of ethanol that can be produced from this feedstock.
>
> It is assumed that hemicellulose is equivalent to cellulose in sugar and ethanol yields and the hydrolysis and fermentation efficiency is 100%.
>
> For 1.0 dry metric ton of corn stover:

Hydrolysable carbohydrates $= 1000 \times (0.39 + 0.26) = 650 \, \text{kg cellulose}$

Sugar yield $= 650 \, \text{kg cellulose} \times 1.11 \, (\text{hydrolysis}) \times 1.0 \, (\text{hydrolysis efficiency})$

$\qquad = 721.5 \, \text{kg glucose}$

Ethanol yield $= 721.5 \, \text{kg glucose} \times 0.51 \, (\text{stoichiometric ethanol yield from glucose})$

$\qquad \times 1.0 \, (\text{fermentation efficiency}) = 367.97 \, \text{kg ethanol} = 367.97 \, \text{kg}$

$\qquad \times 1.267 \, \text{L/Kg} = 466.2 \, \text{L} \, (123.0 \, \text{gal}) \, \text{ethanol}$

The maximum amount of ethanol that can be produced is 466.2 L (123.0 gal)/metric ton of biomass. Since the maximum theoretical ethanol yield is less than the claimed yield, it may not be a wise investment.

The following steps are the most commonly followed in TEA, regardless of the tools used.

Step 1: Base Design

All unit processes and the major equipment involved are specified. This forms the base design on which the plant is constructed. A process flow diagram connecting all the process flows between equipment in the unit operations is developed. The process conditions (temperature, pressure, reaction rates, yields, and residence time in various reactors) of all unit operations are specified.

Step 2: Material and Energy Balances

Material and energy balances are performed under the specified process conditions. This step is often performed in multiple iterations, where the first iteration is carried out without any process recycle streams to obtain a feasible process. Incorporating recycle streams results in process loops that require iterative solution procedures and could result in numerical convergence issues. Further design iterations are performed to optimize both materials (such as minimizing water and other inputs through recycling process streams) and energy (such as heat recovery strategies and a heat exchanger).

Step 3: Optimization Requirements in the Process

Process optimization during TEA is typically performed using computer software (see Example 27.3).

Step 4: Direct Costs Estimation

Direct costs refer to costs related to the actual purchase and installation of equipment to establish a functional production line. These costs are estimated primarily based on the cost of purchased equipment. After the successful calculation of material and energy balances (and optionally optimization of the material and energy inputs), the next step is to input the parameters for estimation of equipment costs. Equipment costs are obtained from various sources such as past data for similar equipment, vendor quotations, industry magazines, and trade journals. Often, the cost for a particular size of equipment is not available and has to be estimated based on a different size of equipment, as illustrated in the box.

Estimation of equipment costs

Capacity vs. capital cost information is very important, but often the cost information may not exist for the exact size of equipment required. In such cases, the new cost can be calculated from the existing information using the following approach:

$$\text{New cost} = \text{Original cost} \times \left(\frac{\text{New size}^*}{\text{Original size}^*} \right)^{\text{exp}}$$

*or characteristic linearly related to the size

The exponent (exp) ranges from 0.6–0.7 depending on the type of equipment. For a list of specific exponents see *Perry's Chemical Engineers Handbook* (Perry and Green, 1998). Similarly, inflation-adjusted costs can be used to estimate the current price of equipment based on historical data.

The purchased equipment cost, however, is only a part of the total cost of the plant. Total costs for piping, electrical equipment, instrumentation, insulation, and installation often equal or exceed the original capital cost of the equipment. These costs are estimated based on factors specific for the process technology and the equipment under consideration. For example, a petroleum refinery may have much higher piping costs compared to a steel manufacturing facility. The factors can

Table 27.1 Factors for estimating fixed capital expenses

Direct costs	
Fixed equipment costs (FEC)	15–40% of fixed capital expenses
Installation	25–55% of FEC
Instrumentation	6–30% of FEC
Piping	10–80% of FEC
Electrical	10–40% of FEC
Land and buildings (new site)	10–80% of FEC
Site preparation and yard improvement	8–20% of FEC
Utilities and other services	30–80% of FEC
Indirect costs	
Engineering and supervision	15–30% of FEC
Contractor's fee	2–8% of FEC
Contingency	5–15% of FEC
Startup expenses	8–10% of FEC

Source: Westney, 1997.

be obtained from literature sources such as handbooks (Perry and Green, 1998; RSMeans, 2009; Westney, 1997), textbooks (Peters and Timmerhaus, 1991), or expert opinion. In addition to factors directly related to the installed equipment, other costs such as site preparation, buildings, utilities, and service facilities are also estimated based on the total equipment cost (Table 27.1).

Example 27. 2: A hammer mill with a rated grind rate of 10,000 kg/hr cost $500,000 in 2004. What would a hammer mill with a grind rate of 15,000 kg/hr cost in 2013? Use a power law coefficient of 0.6 for sizing.

Solution: Given the data and assumptions:

- A hammer mill with a rated grind rate of 10,000 kg/hr cost $500,000 in 2004
- Power law coefficient: 0.6
- $1 in 2004 = $1.24 in 2013 (You can calculate this value using the Consumer Price Index (CPI) Inflation Calculator: http://www.bls.gov/data/inflation_calculator.htm)

$$\text{Hammer mill}\_15000 \text{ (in 2004 \$)} = \$500,000\,(15,000/10,000)^{0.6} = \$637,712$$

$$\text{Hammer mill}\_15000 \text{ (in 2013 \$)} = \$637,712 \times 1.24 = \$790,763$$

Therefore, a 15,000 kg/hr hammer mill will cost an estimated $790,763 in 2013.

Note: The CPI used in this example is just one of the many indices used to account for increases in equipment costs over time due to inflation. While CPI accounts for increases in price for general goods, other indices such as the Chemical Engineering Plant Cost Index (CEPCI) and the Marshall–Swift index (MSI) are also used. The MSI is a composite index for accounting inflation related to industrial chemical equipment costs. It should be noted that the MSI is lower compared to the CPI and hence any estimation using CPI would result in overestimation of actual equipment costs, thus resulting in a conservative estimate. A comparison of MSI, CEPCI, and CPI is given in Table 27.2.

Table 27.2 Comparison of various cost indices

Year	MSI (1926 = 100)	CEPCI (1958–59 = 100)	CPI (1982 = 100)
1990	915.1	357.6	135.44
1995	1,027.5	381.1	157.93
2000	1,089.0	394.1	178.45
2005	1,260.9	468.2	202.38
2010	1,457.4	550.8	225.96

Example 27.3: A cellulosic ethanol plant has the following operational parameters, data, and assumptions for ethanol production from wheat straw:

- Wheat straw composition: uronic acids 2.24%, extractives 12.95%, lignin 16.85%, ash 10.22%, galactan 0.75%, glucans 32.64%, mannan 0.31%, arabinan 2.35%, xylan 19.22% (adapted from NREL data). Combine uronic acid, extractives, lignin, and ash into one stream; galactans, glucans, and mannans into C6 polymers; and arabinan and xylans into C5 polymers for process calculations
- Straw feed rate: 30,000 kg/hr
- Pretreatment method: dilute acid
- Pretreatment efficiency: 90%, conditioning efficiency: 100%, hydrolysis efficiency: 85%, hexose fermentation efficiency: 98%, pentose fermentation efficiency: 60%, and distillation efficiency: 99%
- Inhibitor formation during pretreatment: 2% each from hexoses and pentoses, respectively
- Moisture content of wheat straw: 10%, mash solids content: 90%, Lignin coproduct yield: 14%, pretreatment liquid: 90%, and stillage stream: 95%
- Assume relative volatility of all components except ethanol and CO_2 to be 0
- Azeotrope of ethanol and water (95.63% ethanol, 4.37% water, i.e., ratio of water to ethanol at azeotrope condition = 0.0457)
- Assume a recycle of 0 and 80% of the process water produced in the process
- Plant operates for 330 days/year continuously

Determine the process flows. What is the annual ethanol production capacity? What is the water consumption in this plant? How does that change with water recycling?

Solution: See Table 27.3.

Table 27.3 Capacity and consumption in a cellulosic ethanol plant

	Wheat straw (metric tons)	Ethanol (L)	Lignin co-product (metric tons)	Water use (L)
Recycle 0%				
Hourly	30	7435.1	22.5	260,078
Annual	237,600	58.9 million L	177,953	2.06 billion L
Recycle 80%				
Hourly	30	7694.1	18.5	142,956
Annual	237,600	60.9 million L	146,465	1.13 billion L

Step 5: Indirect Cost Estimation

Indirect costs refer to costs such as contractor's fees, engineering costs, and a contingency cost. These costs are typically specified as a percentage of the direct costs.

Step 6: Fixed Capital Expenses

Fixed capital expenses are calculated as the sum of the direct and indirect costs for the plant. Often it is convenient to refer to total fixed capital expenses in terms of investment per installed capacity (CAPEX) when comparing similar-size plants with different installed technologies. It must be emphasized that unit capital costs vary based on the size of the plant and therefore the unit capital costs must be compared based on same/similar-sized plants. For example, a 75 million L per year corn ethanol plant with fixed capital expenses of $20 million would be specified as a plant with a capital expenditure of $0.267/L ethanol ($20 million/75 million L ethanol).

Step 7: Operational Expenses

Operational costs of plants are divided into facility-dependent and variable costs. Facility-dependent costs (also called fixed operational expenses) are costs that must be paid regardless of the production levels in the plant. Interest on loans, rents, local taxes, insurance, and depreciation costs are all examples of facility-dependent costs. Variable operational costs, on the other hand, include expenses for purchasing raw materials, utilities, operating labor, consumables, royalties, waste product disposal, maintenance, and product sales/advertising. These costs are also often specified on a unit product basis as fixed capital expenses and are generally referred to as OPEX.

Step 8: Cash-Flow Analysis

Establishment of a commercial facility is with a profit motive and therefore it is critical to assess the economic viability of the plant during its lifetime. The internal rate of return, net present value, and payback period for the plant are commonly used metrics to assess its economic viability. Cash-flow analysis (revenues and expenses) is set up based on the expected production schedule of the plant during its life. This is important, as the production facility will require upfront investment for its construction and the design production levels are achieved after an initial startup time. The operational expenditures are calculated for each year since the start of the project and the cash-flow analysis is performed. This is used to calculate the internal rate of return, net present value, and payback period.

Net cash flow for a particular year is the difference between the total revenue and the total expenses for a plant. Various parameters are used to assess the economic viability of a project. The internal rate of return (IRR) is equal to the rate of return when the net present value (NPV) is equal to zero. The IRR indicates the efficiency of the enterprise to return profits and if the IRR is greater than the cost of capital (i.e., cost to invest the money elsewhere in another profitable venture with a similar financial risk profile), the project is accepted.

$$NPV = \sum_{n=0}^{N} \frac{C_n}{(1+r)^n} = 0 \tag{27.1}$$

where C_n refers to cash flow during the year n, for a total of N years.

When the value of IRR is set to a defined discount rate in Equation 27.1, the resulting value is the NPV for the project. The discount rate (i) is the opportunity cost of the capital:

$$NPV = \sum_{n=0}^{N} \frac{C_n}{(1+i)^n} \tag{27.2}$$

A profitable venture must have a positive NPV value and, if all other factors are equal, projects with a higher NPV should be selected. However, note that projects with large upfront costs have lower IRR values compared to projects with lower upfront capital investment. The NPV values can be higher or lower depending on the cash-flow profiles for the two projects.

The payback period refers to the time required to recover the investment from the project. It is important to note that the payback period only indicates how quickly the initial investment can be recovered (i.e., how quickly the project pays for itself) and does not reflect on the future accumulation of profits from the venture. The payback period is always less than the project lifetime for economically viable projects. If all other factors are equal, projects with a shorter payback period are selected.

$$\text{Payback period} = n^- + \frac{S_{n^-}}{S_{n^- + 1}} \tag{27.3}$$

where n^- is the last period with a negative cumulative cash flow, S_{n^-} is the cumulative cash flow at the end of the time period n^-, and $S_{n^- + 1}$ is the cumulative cash flow for the time period $n^- + 1$.

> **Example 27.4:** A 50,000 ton/year biodiesel plant has an estimated fixed capital investment of \$9.3 million. The simplified reaction for biodiesel production can be represented by:
>
> $$C_{57}H_{104}O_6 (\text{Triolein}) + 3CH_3OH (\text{Methanol}) \rightarrow 3C_{19}H_{36}O_2 (\text{Biodiesel})$$
> $$+ C_3H_8O_3 (\text{Glycerol})$$
> $$885.43 \, kg \, (\text{Triolein}) + 96.12 \, kg \, (\text{Methanol}) \rightarrow \, 889.5 \, kg \, (\text{Biodiesel})$$
> $$+ 92.1 \, kg \, (\text{Glycerol})$$
>
> Calculate the unit production cost of biodiesel at 100% efficiency of the overall process. Data for this problem were obtained from Apostolakou *et al.*, 2009.
>
> ### Given data
>
> - Inputs: Rapeseed oil @ \$1,100/ton; methanol @ \$300/ton, NaOH and HCL (catalysts) @\$8.26/ton biodiesel. Assume crude glycerol prices at \$100/ton.
> - Utilities: Heating (as high pressure and low pressure steam) 1.4 GJ/ton biodiesel @ \$10/GJ and electricity 30kWh/ton biodiesel @ \$0.15/kWh.
> - Labor costs: 15 operators @ \$40,000/year·person + overhead costs (insurance, supervision, laboratory costs) of \$36,000/year·person.
> - Other costs (miscellaneous materials, maintenance, capital charges, insurance, taxes, and others): 20% of the fixed capital expenses (FCE). Assume overheads as 5% of the direct costs.
>
> **Solution:** For one metric ton (t) of biodiesel: 0.995 t Rapeseed oil, 0.1081 t methanol are required and 0.1035 t of glycerol is produced as a coproduct.

Yearly cost calculation

Variable costs = Raw materials + Utilities

$$= 50,000(0.995 \times \$1,100 + 0.1081 \times \$300 + \$8.26)$$
$$+ 50,000(1.4 \times \$10 + 30 \times \$0.15)$$
$$= \$\frac{57,684,500}{\text{yr}}$$

Fixed costs = Operating labor costs + Other fixed costs

$$= 15 \times (\$40,000 + \$36,000) + \$9,400,000 \times 0.2 = \frac{\$3,020,000}{\text{yr}}$$

Direct production costs = Variable costs + Fixed costs = $\dfrac{\$60,704,500}{\text{yr}}$

Annual production cost = Direct costs + General overheads = $\$60,704,500 \times 0.15$
$$= \$63,739,725$$

Example 27.5: For the plant specified in Example 27.4, the cash flow is provided in the first three columns of Table 27.4. Calculate the payback period, IRR and NPV assuming a discount rate of 10%.

Table 27.4 Cash-flow analysis for ethanol plant (all numbers in millions $)

Given data			Calculated data			
Year	Expenses	Revenues	Net cash flow	Cumulative cash flow	IRR calculation	NPV calculation
0	23.4	0	−23.4	−23.4	−23.40	−23.40
1	31.2	0	−31.2	−54.6	−27.48	−28.36
2	23.4	0	−23.4	−78	−18.16	−19.34
3	109	119	10	−68	6.83	7.51
4	109	119	10	−58	6.02	6.83
5	109	119	10	−48	5.30	6.21
6	109	119	10	−38	4.67	5.64
7	109	119	10	−28	4.11	5.13
8	109	119	10	−18	3.62	4.67
9	109	119	10	−8	3.19	4.24
10	109	119	10	2	2.81	3.86
11	109	119	10	12	2.48	3.50
12	109	119	10	22	2.18	3.19
13	95	119	24	46	4.61	6.95
14	95	119	24	70	4.06	6.32
15	95	119	24	94	3.58	5.75
16	95	119	24	118	3.15	5.22
17	95	119	24	142	2.78	4.75
18	95	119	24	166	2.44	4.32
19	95	119	24	190	2.15	3.92
20	95	119	24	214	1.90	3.57
21	95	119	24	238	1.67	3.24
22	95	119	24	262	1.47	2.95
				Total NPV	0.00	26.67
				Rate	0.135	0.1

Solution:

Payback period, P = 9 + (|-8|/10) =9.8 yr
IRR = 0.135 corresponding to the total NPV of 0
Total NPV with a discount rate of 10% is $26.67 million

These numbers indicate that the plant will pay for itself in 9.8 years after the start of construction and will have an internal rate of return of 13.5%. The investment of $78 million in this ethanol plant venture (barring any unforeseen financial risks and costs) would result in returns that would be equal to $26.67 million in current dollars over its lifetime. Since the IRR of 13.5% is better than the discount rate (10%) used in the NPV calculation, this indicates that it is a good investment opportunity as returns are higher compared to other investment opportunities.

Example 27.6: You are an engineer in a corn ethanol plant producing 200 million L ethanol/year and 239,180 metric tons DDGS/year and you have to make a decision about installing a new dryer for DDGS. There are two alternatives, both of which have a life of 15 years: a first-generation natural gas–based dryer with an overall efficiency of 80% and an improved design with 85% efficiency. However, the improved-design dryer costs $1.5 million more than the first-generation dryer. Assume that 2 kg water/kg DDGS is evaporated and natural gas prices are $0.003788/MJ. The latent heat of evaporation of water is 2.26 MJ/kg. Which of the two dryers would you recommend?

Solution: Given data and assumptions:

- Annual DDGS production: 239,180 ton (dry wt.)
- Water evaporation requirements and latent heat: 2 kg/kg DDGS and 2.26 MJ/kg water, respectively
- Efficiencies of conventional and improved dryers: 80% and 85%, respectively
- Capital cost difference between dryers: $1.5 million
- Life of dryers: 15 years
- Natural gas price: $0.003788/MJ ($4/MMBtu)
- Annul loan interest and discount rates: 8 and 5%, respectively

Quantity of water to be evaporated $= 239,180$ metric ton DDGS $\times 2 \, ^{kg}/_{kg}$ DDGS

$$= 478,360$$

Heat energy required to evaporate above quantity of water $= 478,360 \times 1000 \times 2.26$ MJ

Savings in energy $= 478,360 \times 1000 \times 2.26 \times \left(^1/_{0.8} - ^1/_{0.85}\right)$ MJ $= 79,492,176$ MJ

Economic value of energy savings $= 79,492,176$ MJ $\times 0.003788/$MJ $= 301,144$

Annual interest for the loan $= 1,500,000 \times 0.08 = 120,000$

Total annual savings $= \$301,144 - \$120,000 = \$181,144$

Net present value (NPV) of the annual savings ($181,144/yr over 15 yr $= \$1,880,211$)

NPV of total investment $= \$1,880,211 - \$1,500,00 = \$380,211 > 0$

Since NPV > 0, this implies that an investment of an additional $1.5 million in the new dryer with 5% higher efficiency has better economic returns, therefore must be selected over the conventional dryer.

27.4 Tools, Software, and Data Sources for Performing TEA

27.4.1 Tools Available for Performing TEA

TEA can be performed using various tools, such as spreadsheets or dedicated process modeling software, based on the level of detail needed for analysis. TEA at the level of study/factored study estimates is performed using process modeling software such as Aspen Plus® and SuperPro Designer®, among others. Process models are developed using such software to visualize the process, to model the process flows, to perform mass and energy balances, to evaluate process economics, to investigate scaleup scenarios, and to investigate the effects of process conditions on process economics.

27.4.2 Procedure for TEA Using Commercial Software

The process for TEA at factored estimate level begins with the specification of all the unit processes and the major equipment used in the process. Commercial modeling software often has built-in data for major equipment used in different unit operations. After specification of all major equipment used in the process and connecting the unit processes, additional details such as specification of the process conditions and equipment capacities are added to the process flowsheet. A process flowsheet can be considered as a computer representation of the actual process. Mass and energy balances are performed using the built-in software routines. Details of calculations such as the inclusion of thermodynamic properties of process flows and error tolerances for mass and energy balance are included and are user customizable to various degrees in commercial software.

After completion of the mass and energy balances for the process, information related to process economics (such as the capacity, capital cost, and number of standby units) for each piece of equipment is an input using built-in/user-defined models. Specifying the capital costs of major capital equipment is the first step in the economic analysis. This information is used to obtain the fixed equipment cost (FEC) or equipment purchase cost (EPC) for all the major equipment specified in the process flowsheet. Other fixed capital costs such as site preparation, installation costs, uninstalled equipment, piping, instrumentation, buildings, auxiliary facilities, engineering consultation, construction, contractor's fees, and contingency funds are estimated as a factor of the EPC. Annual operating costs are calculated using the prices for various raw materials, chemicals, utilities, waste treatment, labor, and facility-dependent costs (such as debt servicing, taxes, and insurance). After completion of the economic inputs, economic analysis is performed using inbuilt software routines.

All process modeling software has integrated reporting functions that will generate reports for mass and energy balances, and the economic analysis. Typical reports include mass, volume, composition, temperature, and pressure of various process flows, details of utilities (heating agents such as high- and low-pressure steam, and cooling agents such as chilled water and refrigerants) used in various processes, and the number and size of all equipment. The economic analysis reports, on the other hand, include total capital investment, operating costs, revenues from main product and co-products, unit production cost, return on investment, payback period, and net present value.

27.4.3 Data Sources for Performing TEA

As with any data-intensive analysis, the availability of accurate data sets is the key requirement for good analysis. Basic process outline information is obtained from process designers and standard peer-reviewed literature. Data for individual process equipment and different scaling factors can be obtained from engineering handbooks such as *Perry's Chemical Engineering Handbook* (Perry and Green, 1998), *The Engineer's Cost Handbook* (Westney, 1997), *RSMeans Estimating Handboook* (RSMeans, 2009), and online databases (Matches, 2013).

27.4.4 Process Optimization Using TEA

The TEA is not limited to performing mass and energy balances and economic analysis for a specified process. In fact, the strength of TEA tools lies in the fact that they can be used for process optimization. Most process modeling software also has the ability to perform analysis of process bottlenecks and scheduling of unit processes. These tools can be used to optimize process times, which are especially critical if the process has a mixture of batch and continuous unit operations, as in the ethanol production process (continuous distillation and batch fermentation). Additionally, process models are also used to investigate different scaleup scenarios to identify the optimum size of the plants. Additional details of these applications can be found in the literature (Eggeman and Elander, 2005; Gnanasounou and Dauriat, 2010; Kazi *et al.* 2010; Kumar and Murthy, 2011).

References

Apostolakou, A.A., Kookos, I.K., Marazioti, C., and Angelopoulos, K.C. (2009). Techno-economic analysis of a biodiesel production process from vegetable oils. *Fuel Processing Technology* 90: 1023–31.

Eggeman, T., and Elander, R.T. (2005). Process and economic analysis of pretreatment technologies. *Bioresource Technology* 96: 2019–25.

Gnansounou, E., and Dauriat, A. (2010). Techno-economic analysis of lignocellulosic ethanol: A review. *Bioresource Technology* 101: 4980–91.

Kazi, F.K., Fortman, J.A., Anex, R.P., *et al.* (2010). Techno-economic comparison of process technologies for biochemical ethanol production from corn stover. *Fuel* 89: S20–S28.

Kumar, D., and Murthy, G.S. (2011). Impact of pretreatment and downstream processing technologies on economics and energy in cellulosic ethanol production. *Biotechnology for Biofuels* 4: 27.

Matches (2013). *Matches' Process Equipment Cost Estimates.* http://www.matche.com/equipcost/Default.html, accessed April 2016.

Perry, R.H., and Green, D.W. (eds.) (1998). Process economics: Fixed capital cost estimation. In: *Perry's Chemical Engineers' Handbook*, 7th edn. New York: McGraw-Hill, Section 9.

Peters, M.S., and Timmerhaus, K.D. (1991). *Plant Design and Economics*, 4th edn. New York: McGraw-Hill.

RSMeans (ed.) (2009). *RSMeans Estimating Handbook*, 3rd edn. Chichester: John Wiley & Sons Ltd.

Westney, R.E. (1997). *The Engineer's Cost Handbook: Tools for Managing Project Costs.* Boca Raton, FL: CRC Press.

Exercise Problems

27.1. A startup company proposes to use a cellulosic feedstock (35% cellulose, 23% hemicellulose, 20% lignin, 11% ash, and 11% extractives) for ethanol production. The company claims to have discovered a new process that can produce 416 L/dry ton (109.8 gal ethanol/dry ton) of feedstock. Is this a feasible technology? Assume any missing data.

27.2. A fermenter of 2,525 m^3 volume cost \$450,000 in 2007. What would a fermenter of 3,500 m^3 volume cost in 2013? Use a power law coefficient of 0.66 for sizing.

27.3. Hydrolysis, fermentation, and distillation are three main unit operations in the corn ethanol process. Assume that corn (10% moisture, 68% starch, 6% protein, 6% oil, 6% fiber, and 4% ash on a wet basis) is processed in a corn dry-grind ethanol process with 90% hydrolysis (liquefaction + saccharification) efficiency, 98% fermentation efficiency, and 99.5% distillation efficiency. Assume a resistant starch content of 3% of total starch and DDGS moisture content as 12% dry basis. Assume a corn price of \$0.12/kg, ethanol selling price of \$2.0/gal, and DDGS price of \$187/t. Calculate the ethanol and DDGS yields per ton of corn processed. Calculate the maximum allowable production cost assuming a 15% profit margin. Assume any missing data.

27.4. A startup company proposes to use switchgrass as its feedstock. The composition of switchgrass on a dry basis is extractives 16.83, lignin 17.96, ash 5.63, C6 polymers 34.63, C5 polymers 24.94 (adapted from NREL data). Efficiencies for hydrolysis, hexose fermentation, pentose fementation, and distillation are 80%, 98%, 65%, and 99%, respectively. Inhibitors formed during the pretreatment process are 0.7% and 1.2% from hexoses and pentoses, respectively. Make any suitable assumptions for any missing data. Based on this information, estimate the following. What is the ethanol yield for the above process conditions? What is the yield if only hexoses are fermented to produce ethanol? What would be the yield if pentose fermentation efficiency is 98%? If the spot price for ethanol is \$2.3/gal and the cost of switchgrass as \$50/Mt (dry basis), comment on the maximum allowable production cost.

27.5. Using the data given in Problem 27.4, determine the ethanol production rate for the following cases:

(a) Pentose fermentation efficiency is 40% instead of 60%.
(b) Water recycle is 90%. Discuss the challenges associated with increasing the water recycle rates in the performance of enzymatic hydrolysis and fermentation.
(c) Hydrolysis efficiency is 90%.

27.6. You are an engineer in a corn ethanol plant (producing 200 million L ethanol/year and 239,180 metric tons DDGS/year) who has to make a decision about installing a new dryer for DDGS. There are two alternatives, both of which have a life of 15 years: a first-generation natural gas–based dryer with an overall efficiency of 83% and an improved design with 85% efficiency. However, the improved-design dryer costs \$1.5 million more than the first-generation dryer. Assume that 2 kg water/kg DDGS are evaporated, and natural gas prices are \$0.003788/MJ. The latent heat of evaporation of water is 2.27 MJ/kg. Which of the two dryers would you recommend?

27.7. An anaerobic digestion plant with a 15-year design life for production of 250 KW of renewable electricity and 2340 t/yr of compost consists of three components: biogas plant, microturbine, and composting system, with a capital cost of \$5,000/KW electricity, \$1,100/KW electricity, and \$110/KW electricity, respectively. Assume annual operation and maintenance costs as 2.5% of the capital costs, annual labor costs as \$14,261, and other variable costs as \$0.015/KW electricity. Assuming an electricity and compost price of \$0.09/KWh and \$140/t, respectively, calculate the payback period for the plant.

27.8. For Problem 27.7, calculate the payback period if the developer obtains a loan with an interest rate of 8% per annum for financing the initial installation of the plant.

27.9.   Given the cash-flow analysis in Table 27.5 for a biofuel plant with 3 years of construction and 17 years of design life, calculate the payback period, IRR, and NPV using a discount rate of 0.09.

Table 27.5

Year	Expenses	Revenues	Year	Expenses	Revenues
0	49.9	0	11	95	115
1	37.4	0	12	95	115
2	37.4	0	13	95	115
3	95	115	14	90	115
4	95	115	15	90	115
5	95	115	16	90	115
6	95	115	17	90	115
7	95	115	18	90	115
8	95	115	19	90	115
9	95	115	20	90	115
10	95	115			

27.10.   A bioethanol plant has a total fixed equipment cost of $50 million. Various factors for direct costs are piping = 0.4, instrumentation = 0.3, insulation = 0.1, electrical facilities = 0.05, buildings = 0.2, yard improvement = 0.1, other auxiliary facilities = 0.2, and unlisted equipment = 0.075 of fixed equipment cost. Assume indirect costs factors as engineering = 0.25, construction = 0.35, contractor's fee = 0.05, and contingency = 0.1 of direct costs. Calculate the fixed capital expenses for this plant. A bioethanol plant uses 13 million tons/year sugarcane purchased at $20/ton and producing 83 L ethanol/ton sugarcane. The plant obtains all its utility needs through sugarcane bagasse combustion. The excess energy is used to produce electricity (173.2 kWh/ton sugarcane), which is sold to a power company at $0.1/kWh. Materials required for the fermentation of the sugarcane juice are estimated to cost $0.02/ton sugarcane. The plant requires 10 operators who have yearly benefits of $35,000 and the overhead costs (insurance, supervision, and laboratory costs) are estimated at 85% of the yearly salary for an operator. Other costs (miscellaneous materials, maintenance, capital charges, insurance, taxes, and others) are estimated to be 25% of the fixed capital expenses. Assume general overheads as 5% of the direct costs. Calculate the variable costs, fixed costs, direct costs, co-product revenue, and the unit production cost for the plant.

Life-Cycle Assessment

Ganti S. Murthy

What is included in this chapter?

This chapter covers the basics of life-cycle assessments for the analysis of bioenergy systems. The procedures to perform these analyses are discussed. Applications of this approach to soybean oil biodiesel, corn ethanol, and cellulosic ethanol production processes are also presented.

28.1 Introduction

Sustainability is proposed as one of the positive aspects of bioenergy. It is necessary to evaluate the positive or, in some cases, negative impacts of increased bioenergy generation objectively, using quantitative metrics to compare different bioenergy resources and their respective technologies.

Two key questions that are central to this discussion are: What exactly is sustainability? How should we assess it? The Brundlandt Commission of the United Nations (UN) defined sustainable development as "development which meets the needs of current generations without compromising the ability of future generations to meet their own needs." While there are various definitions of sustainability, it is essential to acknowledge that sustainability encompasses multiple dimensions of environmental, economic, and social factors. There are different techniques to assess the environmental aspects of sustainability, and life-cycle assessment (or analysis) is one of the widely used methods today. This chapter focuses on the implementation of life-cycle assessment methods as a means to compare the environmental impacts of current and future (bio)process technologies for bioenergy production.

The term life-cycle assessment/analysis (LCA) was used for the first time in the United States in 1990. However, the analytical framework (called resource environmental profile analysis, REPA) that ultimately turned into LCA was developed by Harry E. Teasley, Jr. while working for Coca-Cola in 1969 (Hunt and Franklin, 1996). At that time, the framework was used to analyze the

---

Bioenergy: Principles and Applications, First Edition. Edited by Yebo Li and Samir Kumar Khanal.
© 2017 John Wiley & Sons, Inc. Published 2017 by John Wiley & Sons, Inc.
Companion website: www.wiley.com/go/Li/Bioenergy

material, energy, and environmental consequences (impacts) of different beverage packaging material alternatives such as aluminum cans, glass, and plastic bottles. Similar efforts were also going on in Europe in the 1970s and were known as "Ecobalance" methods. These methods were formalized and published into the *Handbook of Industrial Energy Analysis* in 1979. The LCA methodologies received significant attention in the 1980s and 1990s, and were the subject of active development and practice, with several developments in inventory and impact-assessment methodologies. The methods were formalized by the ISO 14040–14044 (2006) standards. The method has evolved from passive applications for reporting the environmental impacts to active use in process optimization, business decisions, and policy formulations among various stakeholders. The LCA standards and practices are still evolving to address the many challenges that come with applying the methodology to new processes and systems.

28.2 What Is Life-Cycle Assessment (LCA)?

While a techno-economic analysis helps in making technically feasible and economically optimal choices, LCA techniques are widely used to compare quantitatively the impacts of various processes, products, and services on the environment. Comparing the environmental impacts of different products/processes/services is essential to make environmentally responsible choices. Methods to compare environmental impacts can be divided into process-oriented and environmentally emphasized metrics. Process-oriented metrics such as a typical LCA are useful in assessing competing technologies. On the other hand, environmentally emphasized metrics, such as the sustainability process index (SPI), are focused on the depletion of resources.

LCA is used to assess the potential environmental impacts and resources used throughout a product's entire life cycle: from raw material acquisition, via production and use phases, to waste management (ISO, 2006). LCA is comprehensive and considers all attributes or aspects of the natural environment, human health, and resources (ISO, 2006). The unique scope of LCA is useful in order to avoid problem shifting, for example from one phase of the life cycle to another, from one region to another, or from one environmental problem to another. It is important to note that LCA is not a risk assessment; that is, it does not provide any information on the inherent risk associated with a particular product/technology. LCA also does not provide any information on the profitability of the process/product. The information on profitability is obtained through life-cycle costing methods.

Shifting of the environmental burden: Importance of system boundaries

The concept of shifting the environmental burden can be shown using the example of a hydrogen car. If only the hydrogen car using H_2 as the sole fuel source is considered (depicted by the inner rectangle in Figure 28.1), there are no CO_2 emissions. However, when the fossil fuel–based production processes for H_2 are considered (outer rectangle), the net emissions of CO_2 are significant. In this case, the analysis limited to the hydrogen car only (inner box) shifted the environmental burden to an upstream process (hydrogen production). LCA is designed to avoid such shifting.

Elementary flows are defined as flows to/from the environment that will not be subjected to any further human intervention. *Intermediate flows* occur between the unit processes within the system boundary. *Product flows* are products of some other product system. In

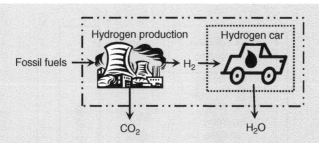

FIG. 28.1   Shifting of environmental burden.

Figure 28.1, H_2O and CO_2 would be considered elementary flows, while H_2 would be considered an intermediate flow. Fossil fuels would be considered as a product flow, as it is a product of another process (e.g., petroleum refining, coal mining, or natural gas extraction in this example).

There are two variants of LCA that answer different questions: *attributional LCA* and *consequential LCA*. Attributional LCA is the most widely used form of LCA and is the focus of this chapter. In LCAs, all material and energy inputs to the life cycle of a product/process, including the acquisition of raw materials, production, distribution, and disposal, are considered. This information is used to estimate various outputs such as atmospheric emissions, liquid wastes, solid wastes, co-products, and other releases to the environment.

Attributional vs. consequential LCA

Attributional life-cycle assessment (aLCA) is defined by its focus on describing the environmentally relevant physical flows to and from a life cycle and its subsystems. Consequential life-cycle assessment (cLCA) is defined by its aim to describe how environmentally relevant flows will change in response to possible decisions (Zamagni *et al.*, 2012).

The two types of LCAs answer different questions:

- aLCA: What are the *total* emissions from the process during the life cycle of the product?
- cLCA: What is the *change* in total emissions from the process during the life cycle of the product?

The different focuses of aLCA and cLCA are reflected in several methodological choices in LCA (Tillman, 2000). One is the choice between average and marginal data in the modeling of subsystems of the life cycle. Average data for a system are those representing the average environmental burdens for producing a unit of the good and/or service in the system. Marginal data represent the effects of a small change in the output of goods and/or services from a system on the environmental burdens of the system. In aLCA, only average data reflecting the actual physical flows are used. On the other hand, in cLCA, marginal data are used for the purpose of assessing the consequences (Ekvall and Weidema, 2004).

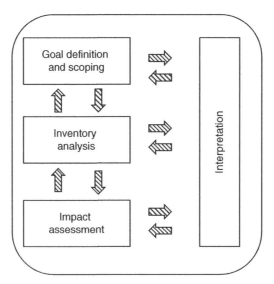

FIG. 28.2   Steps in LCA.

28.3 Procedure for LCA

LCA consists of four distinct steps: goal and scope definition; inventory; impact assessment; and interpretation. Often these steps are interlinked and are iterative (Figure 28.2). The methodology for LCA is standardized in the International Organization for Standardization (ISO) 14040–14044 (ISO, 2006). Note that the abbreviation (i.e., ISO) is not reflective of the English translation of the organization's name.

28.3.1 Goal Definition and Scoping

Formulation of the goal and scope of the study is one of the critical aspects of LCA. The goals of the LCA are stated in terms of the intended application, purpose, intended audience, and whether the results will be disclosed to the public. The scope definition for the LCA consists of a description of the products, processes, and systems, system boundaries, functional unit definition, allocation procedures, impact categories, data requirements, and assumptions and limitations of the study.

> **Example 28.1:** Consider a very simplified case of two processes. Process 1 produces A as the main product and B as the co-product, and has total emissions of 2 kg CO_2 equivalent. Process 2, on the other hand, produces product C and has total emissions of 3 kg CO_2 equivalent. Also assume that co-product B from Process 1 can be used as a substitute for product C (Figure 28.3).
>
> *Solution:*
>
> aLCA for product A+B: Total emissions are 2 kg CO_2 equivalent
> aLCA for product C: Total emissions are 3 kg CO_2 equivalent
> cLCA for product A+B: Total emissions are −1 kg CO_2
>
> cLCA total emissions = 2 kg CO_2 (for Process 1 that produces A+B) − 3 kg CO_2 (credit for replacing C by B that is produced in Process 1).

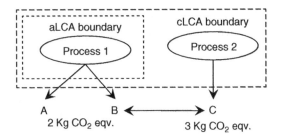

FIG. 28.3   Difference between aLCA and cLCA.

Note that aLCA results can never have negative emissions, while it is possible for a process to have negative CO_2 emissions in the case of cLCA. In this example we have considered marginal values (for cLCA) to be equivalent to the average values (aLCA). This is only true, if the production technology, scale, and time horizon for the processes are held constant for both aLCA and cLCA analyses.

Also note that the example provided here is only a very simplistic case to demonstrate the differences between the approaches of aLCA and cLCA. In reality, the cLCA analysis will consist of a detailed analysis using partial/general equilibrium models such as the Global Trade Analysis Project (GTAP) model to determine the sectors that are affected by the product/process under consideration. Then aLCA is performed for each of those sectors and the results are analyzed to obtain the cLCA analysis.

The ISO 14040:2006 definition of a functional unit states: "Functional unit defines the quantification of the identified functions of the product. The primary purpose of a functional unit is to provide a reference to which inputs and outputs are related." For example, the functional unit in the case of LCAs to compare different transportation fuels could be 1 MJ of energy available to the customer at a gas station. In the case of an LCA for plastic bottles vs. aluminum cans for beverages, the functional unit could be defined as 1 L of beverage delivered to customers in either plastic bottles or aluminum cans.

System boundary definitions are also a critical aspect of LCA. The choice of different system boundaries can lead to dramatically different results and conclusions from an LCA. Ideally the system boundary should be expanded until all inputs and outputs across the system boundary are elementary flows. However, in reality such a system boundary is too difficult to analyze for most products, and thus system boundaries are drawn using cut-off criteria. A formal method that is used for an objective definition of the system boundary is the relative mass, energy, and economic value (RMEE) method proposed by Raynolds *et al.* (2000). In this method, a cut-off percentage is defined by the LCA practitioner based on available resources and data. Relative mass, energy, and the economic values of all the inputs to the system are calculated as a fraction of the mass, energy, and economic value of the functional unit. For all inputs with relative mass, energy, or economic values above the cut-off percentage, the upstream processes are included. The process is repeated until all the relative mass, energy, and economic values of all inputs are below the cutoff percentage. This is similar to ISO-14044, which also suggests the use of mass, energy, and environmental significance as cut-off criteria to define system boundaries.

Based on differences in the goals and scope of LCAs, they can be classified as cradle-to-cradle, cradle-to-grave, or cradle-to-gate LCA. In a cradle-to-cradle LCA, all processess starting from raw material extraction, manufacturing, maintenance and repair, waste disposal, and recycling are considered for calculating the environmental impacts. On the other hand, a cradle-to-grave LCA terminates at the waste disposal step and does not include recycling step. Producers of manufactured goods often conduct LCAs that are limited to assessing the impacts of products and processes up to the factory gate and are called cradle-to-gate LCAs. In the context of transportation fuels, well-to-wheel LCA refers to an LCA that considers all the processes from the extraction of raw materials from a petroleum well to the combustion of the fuels in the vehicle delivering energy to the wheel. Similarly, well-to-pump refers to a limited LCA that terminates at the fuel pump station.

28.3.2 Life-Cycle Inventory

The life-cycle inventory (LCI) step is the most time-consuming step in an LCA, and typically accounts for over 70% of overall time and resources. Key steps in LCI include development of a process flow diagram, listing all flows in the process, data collection plans, and data collection. Some of the factors that must be considered are the amount of resources available for the study, type of information needed, data sources, data quality goals, and quality measures. Some of the data quality indicators specified in the ISO 14044 standard include age of data, geographical area from where the data was collected, specific technology or technology mix, precision, completeness, representativeness, consistency, reproducibility, source, and uncertainty in the data. Since an LCA is driven by the data input into the model, the quality of the data is of immense importance.

Input data collected during the LCI process may be classified as follows (ISO 14044):

- Energy, raw materials, and other inputs to the process
- All products, co-products, and waste streams
- All releases to air, water, and soil
- Other environmental aspects

LCI data can be obtained from a variety of open and commercial databases maintained around the world. Some of the important LCI databases are:

- US NREL LCI
- US Department of Agriculture LCA Digital Commons
- European reference Life Cycle Database (ELCD)
- EcoInvent database
- GaBi database
- UNEP/SETAC Life Cycle Initiative Database Registry
- New Energy Externalities Development for Sustainability (NEEDS).
- ProBas
- Okubau.dat

Although these databases are comprehensive and include LCI data for thousands of products, it is important to recognize that they do not always contain all data suitable for a particular LCA. In such cases, datasets have to be built based on actual field literature and experimental data.

Co-Product Allocation

In real-world systems/processes, it is often the case that multiple co-products, in addition to the main product of interest, are also produced. For example, in the corn ethanol process, distiller's grains are produced as a co-product while wet milling produces starch as the main product, and co-products like gluten meal and corn germ. Similarly, anaerobic digestion of energy crops produces electricity as the main product, and heat and digestate as co-products. Partitioning the life-cycle impacts becomes challenging in such cases. The ISO 14044 standard recommends the following strategies for allocations:

- *System partition or system expansion.* Wherever possible, allocation should be avoided by dividing the unit processes into two or more subprocesses. Additionally, the system boundaries could be expanded to include additional processes related to co-products in a manner consistent with the system boundary delineation criteria defined earlier in the process. System expansion is also known as a displacement/replacement/substitution method.
- *Mass/energy-based allocation.* When the previous approach is not feasible, the allocation must be based on physical relationships between the main product and the co-product. For example, the allocations could be based on the mass or energy content distribution among the products.
- *Economic value–based allocation.* If the physical relationships cannot be established, the allocation procedure must be based on the economic value of the products.

It is to be noted that waste streams are not considered in allocation procedures. A consistency check should be performed to verify that the inputs and outputs of the process are exactly the same before and after an allocation procedure.

Example 28.2: Consider the hypothetical process in Figure 28.4 producing one main (generic) product as an output. The numbers in square brackets indicate the absolute values of mass, energy, and economic values of various streams, respectively. The RMEE boundary cut-off value is set at 5%. Determine the system boundary.

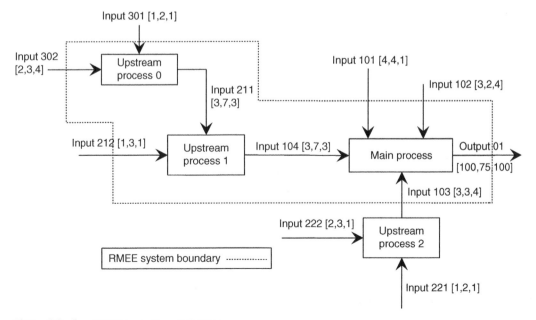

FIG. 28.4   RMEE boundary definition.

Solution: Table 28.1 indicates the procedure to identify the system boundary. For any input stream whose relative percentage of mass, energy, or economic value exceeds the RMEE cut-off, previous processes are included. For Input 103, none of the values exceeds the RMEE cut-off, so the upstream process (Process 2) is outside the RMEE boundary. For Input 104, energy is 13% (>5% of the RMEE cut-off) and hence the upstream process (Process 0) is included and the calculations are performed for input streams (Inputs 301, 302) of process 0. Note that all the inputs coming into the system boundary identified by the RMEE method are below the RMEE cut-off value.

Table 28.1 Procedure to identify the system boundary

RMEE cut-off: 5%

Input	Absolute values			As a percentage of output		
	Mass	Energy	Economic value	Mass	Energy	Economic value
101	4	3	1	4%	4%	1%
102	3	2	4	3%	3%	4%
103	3	3	4	3%	4%	4%
104	4	10	4	4%	**13%**	4%
211	3	7	3	3%	**9%**	3%
212	1	3	1	1%	4%	1%
221	1	2	1	1%	3%	1%
222	2	3	1	2%	4%	1%
301	1	2	1	1%	3%	1%
302	2	3	4	2%	4%	4%
Output	100	75	100			

Example 28.3: Consider the two datasets for corn yields and nitrogen use in Nebraska in Table 28.2. Analyze possible reasons for these dramatic differences and discuss the implications of using one dataset versus the other in LCA.

Table 28.2 Corn yields and nitrogen use in Nebraska

	Corn yield (kg/ha)	Nitrogen use (kg/ha)	Year of data collection
Dataset A	4100	110	1965
Dataset B	9500	140	2005

Ref: http://cropwatch.unl.edu/web/cropwatch/archive?articleID=4585476

Analysis: The yields of corn grain are almost double for Dataset A compared to Dataset B, although both use similar amounts of nitrogen. A closer look at the underlying geographic source and the time when these data were collected yields some clues.

Dataset A was collected in 1965, while Dataset B was collected in 2005. Over the years, farmers developed strategies that increased nitrogen use efficiency for corn production. A direct implication of the use of these datasets for LCA is that Dataset A, which has lower nitrogen use efficiency (37.3 kg corn/kg nitrogen) compared to Dataset B (67.9 kg corn/kg nitrogen), will have higher nitrogen use per functional unit. Nitrogen fertilizers are often the most significant contributors to GHG emissions for two reasons:

- Synthetic nitrogen fertilizers are made from fossil fuels and hence have a large GHG footprint (4.148 and 3.335 kg GHG/kg of NH_3 and urea, respectively).
- Application of nitrogen fertilizers to soil contributes significantly to the GHG footprint via N_2O emissions (N_2O has 233 times more GHG potential compared to CO_2).

Hence an LCA that uses Dataset A would report higher nitrogen use (and hence higher emissions of GHG) per unit of grain produced compared to Dataset B. Additionally, these data were collected specifically for Nebraska and hence an LCA practitioner must be careful when extrapolating these data to other geographic regions of the world where agricultural practices may be completely different.

Example 28.4: Consider a corn dry-grind ethanol process that produces ethanol as the main product and distiller's dry grains (DDGS) as the only co-product (Figure 28.5). In this process, 9.0 kg of corn grain are used to produce 3.78 L (1 gal) of ethanol and 2.26 kg of DDGS. DDGS is a protein-rich co-product that can be used as a replacement for soybean meal or corn grain in animal feeds. Every 0.782 kg of corn grain can be substituted with 1 kg of DDGS in animal feeds. Determine the GHG emissions allocation based on mass, energy, price, and system expansion methods.

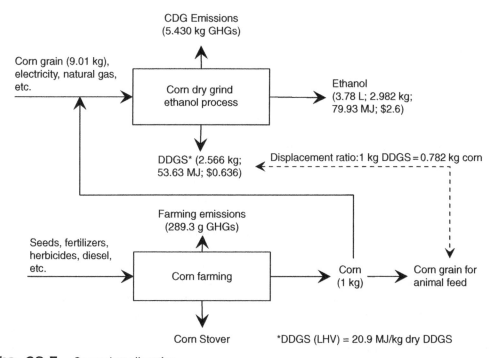

FIG. 28.5   Co-product allocation.

Solution:

Mass-based allocation

Fraction of DDGS produced = DDGS/(Ethanol + DDGS)

$$= 2.566/(2.982 + 2.566) = 0.4625$$

DDGS share of emissions = 0.4625 × 5.430 = 2.511 kg GHGs

Ethanol share of emissions = 5.430 − 2.511 = 2.920 kg GHGs

Similar calculations can be performed for energy and economic value-based allocations using the data provided here.

System expansion

1 kg DDGS ≈ 0.782 kg corn; 1 kg of corn has 289.3 g GHGs

⇨   2.566 kg DDGS = 0.782 × 2.566 × (289.3/1,000) = 0.5860 kg GHGs

⇨   Ethanol share of emissions = 5.430 − 0.5860 = 4.844 kg GHGs

Table 28.3 Share of GHG emissions based on different allocation strategies

Allocation method →	Mass	Energy	Price	Displacement
Ethanol	2.920	3.250	4.363	4.844
DDGS	2.511	2.180	1.067	0.586

As can be seen from Table 28.3, allocation based on economic value is often the best approximation for the allocation based on the displacement method.

28.3.3 Life-Cycle Impact Assessment

Completion of the LCI step results in an exhaustive list of all inputs and outputs crossing the system boundary. Often there are hundreds of different inputs and outputs for a process, and it is very difficult to understand the significance and meaningfully to compare LCI results for different processes/ products. The life-cycle impact assessment (LCIA) step reduces the LCI results to meaningful indicators by calculating different impact categories to various output flows. In this process, several assumptions and models may be used to make this assignment to particular impact categories.

However, the use of additional assumptions and models to assign LCI results to different meaningful impact categories leads to increased uncertainties in the LCIA results. To address this issue, impact categories are classified into midpoint and endpoint indicators. *Midpoint indicators* reduce the complexity of the LCI data by calculating the impact of LCI outputs on the environment through various environmental mechanisms with less uncertainty. *Endpoint indicators* include additional characterization factors to link the midpoint indicators through additional environmental mechanisms of greater uncertainty. An example of this mapping for the ReCiPe-2008 method (Figure 28.6) is shown in Figure 28.7. ReCiPe-2008 is one of numerous LCIA methods that are used. As can be seen from this mapping, the relationship between LCI and the midpoint/endpoint indicators is strongly dependent on environmental mechanisms, which, in turn, are dependent on regional factors

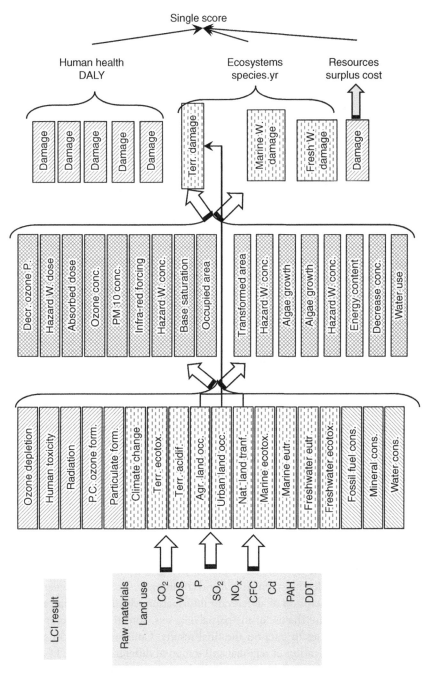

FIG. 28.6 A mapping of life-cycle inventory (LCI) results to midpoint and endpoint indicators in the ReCiPe-2008 life-cycle impact assessment method.
Source: Adapted from Goedkoop *et al.,* 2009.

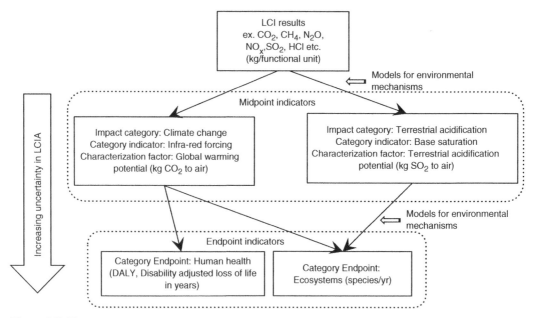

FIG. 28.7 Example of LCI results mapped to midpoint and endpoint indicators in the ReCiPe-2008 life-cycle impact assessment method.

such as climate, soil, vegetation, population density, etc. Many LCIA methods have been developed with an emphasis on different environmental mechanisms and regions around the world.

These elements of LCIA – namely, selection of impact categories, classification or assignment of LCI results to impact categories, and characterization defined as the calculation of category indicator results – are mandatory for any LCIA. In addition to these mandatory elements, optional elements for LCIA include normalization (with respect to a reference scenario), grouping (ranking of impact categories), weighting (converting the impact categories into a single score by assigning weights to the impact categories), and data quality analysis (assessing the reliability of the LCIA results), as per the ISO 14044 standards.

28.3.4 Life-Cycle Interpretation

In the life-cycle interpretation phase of LCA, the results of LCI and LCIA are evaluated to identify any significant issues such as completeness, sensitivity, and consistency. The completeness of the LCI and LCIA data is assured by the inclusion of all relevant information, clearly specifying any missing/incomplete data, and justification for the use of any partial data sets. Sensitivity checks are performed to assess the impact of uncertainties in data on the final results. Consistency checks are also performed to assure a consistent application of regional and temporal differences in the LCI and LCIA phases. Based on this assessment, the goal and scope of the LCA may be revised, and LCI and LCIA may be performed again until all the issues identified in the interpretations stage are mitigated. For example, the RMEE cut-off boundary may be revised and new system boundaries specified based on the LCI results. Finally, conclusions, limitations, and recommendations for the LCA study are presented in the interpretation phase of the LCA.

28.4 Tools Available to Perform LCA

Software development for performing rigorous LCAs is a very active area and is continually evolving. Currently, commonly available software includes OpenLCA, Simapro, GaBi, Sustainable Minds, Enviance System, Economic Input-Output LCA (EIOLCA), Eco-LCA, and Greenhouse gas and Regulated Emissions for Transportation (GREET). While Simapro is among the most commonly used LCA software, OpenLCA is open-source software with comparable features. GREET, on the other hand, was specifically developed by Argonne National Laboratory for the transportation sector. It is important to recognize that while most of the LCA software has internal databases that may be limited in scope of coverage of unit processes, although it often has the ability to import/link to external LCI databases, as discussed previously. Economic input–output models such as EIOLCA and Eco-LCA are freely available tools to perform aLCA using input–output models for regional and national economies.

28.5 Advanced Topics

28.5.1 Sensitivity Analysis

Sensitivity analysis is performed to evaluate the impact of scope of the problem, system boundaries, co-product allocation, and other assumptions and data inputs on the LCA results. Sensitivity analysis is performed by systematically varying the inputs within a given range (usually ± 50%, but dependent on the problem) and analyzing the impacts of the LCA results. Results of sensitivity analysis provide a clear quantitative measure of the importance of various inputs to the LCA. The inputs identified to be the most significant for LCA results should be considered more carefully and any assumptions must be rigorously justified and validated.

28.5.2 Process Optimization Using LCA

One of the key strengths of the LCA framework is that it can be used to identify the unit processes with the highest contribution to various impact categories. Once such unit processes are identified, strategies to reduce the environmental impacts of that particular process can be devised based on process improvements or replacements with better technologies. Examples of LCA as a tool for process selection and optimization can be found elsewhere (Azapagic, 1999; Jacquemin *et al.*, 2012; Pieragostini *et al.*, 2012).

28.5.3 Consequential LCA

Despite their suitability for comparing environmental impacts at individual farm and firm levels, attributional LCAs are not suitable for understanding the indirect effects of the large-scale production of biofuels as influenced by interaction effects and policy initiatives. Consequential LCA (cLCAs), on the other hand, are designed to account for these interactions and thus are suitable for investigating the environmental impacts of different policy choices, the technology-adoption behaviors of farmers, and interaction effects. As discussed earlier, cLCA is focused on quantifying marginal changes in environmental impacts due to a marginal increase in product/processes. Since marginal changes in product are dependent on various economic factors, economic analyses are an integral part of the cLCA. Computable general equilibrium (CGE) models are often used to generate the marginal economic and production data. However, CGE models are not always available and in such cases, simple/multi-region partial equilibrium models may be used to obtain marginal input data for cLCA.

Attributional LCA, together with the marginal data obtained from the CGE models, serves as an input to the cLCA. Similar to aLCA, two issues are critical in consequential LCA: systematic boundary definition; and realistic data with consideration for interaction effects. However, in contrast to aLCA, scale of production and time horizon for technologies are critical in cLCA, as these factors influence the marginal data. Additional information on cLCA can be found in Earles and Halog (2011) and Zamagni *et al.* (2012).

Rebitzer *et al.* (2004) and Pennington *et al.* (2004) provide detailed explanations of various steps involved in the LCA process. The *International Reference Life Cycle Data System Handbook* (JRC, 2013) provides comprehensive guidance to current practices in LCA, various issues that arise in LCAs, and strategies to address them. For a review of recent advances in LCA methodology, refer to Finnveden *et al.* (2009).

Example 28.5: Consider the LCA results for grass straw production in Figure 28.8 and suggest some process improvements that could result in reduced GHG emissions.

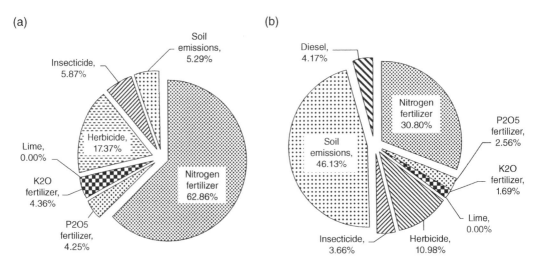

FIG. 28.8 Fossil energy used and GHG emissions contribution from different inputs during grass straw production: (a) fossil energy used; (b) GHG emissions. Source: Kumar and Murthy 2012. Reproduced with permission of Springer.

Solution: It can be readily identified that nitrogen fertilizers contribute 62.86% of fossil energy and 30.80% of GHG emissions. The largest contribution to GHG emissions is associated with soil emissions. As discussed previously in this chapter, application of nitrogen fertilizers releases N_2O, a potent GHG gas. Hence it is reasonable to associate soil emissions (46.13%) to direct emissions associated with nitrogen fertilizer (30.80%) to fertilizer use.

Possible strategies to reduce nitrogen fertilizer use are:

- Application of fertilizers in smaller doses, and matching application timing with plant physiological needs.
- Precision application of fertilizers to plants to minimize dispersion.
- Consider both climatic factors (temperature and soil moisture) and crop physiological needs in deciding the timing of fertilizer applications to reduce the volatility of fertilizers.

Example 28.6: A process flow for a corn stover to ethanol (CSE) process is given in Figure 28.9 with the emissions information for various stages (GREET, 2015). Estimate the LCA emissions for ethanol. The emissions associated with electricity are calculated based on a displacement ratio of 1.0 (i.e., electricity produced in the CSE process is exactly the same as electricity from the grid).

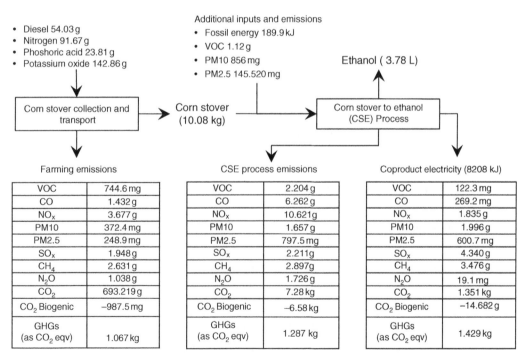

FIG. 28.9   Process flow for a corn stover to ethanol process.

Solution: The results for the LCA are presented in Table 28.4. It can observed that most of the credits for electricity generation are as a co-product.

Table 28.4 Results for LCA

	Farming F	Process C	Electricity E	Ethanol Emissions Ethanol=F+C–E
VOC (g)	0.745	1.459	0.122	2.082
CO (g)	1.432	4.830	0.269	5.993
NO_x (g)	.677	6.944	1.835	8.786
PM10 (g)	0.372	1.657	1.996	0.033
PM2.5 (g)	0.248	0.549	0.601	0.197
SO_x (g)	1.948	0.263	4.340	–2.129
CH_4 (g)	2.631	0.266	3.476	–0.579
N_2O(g)	1.038	0.688	0.019	1.707
CO_2 (kg)	0.693	6.587	1.351	5.929
CO_2 Biogenic (kg)	–0.001	–6.579	–0.015	–6.565
GHGs (kg)	1.067	0.220	1.429	–0.142
(as CO_2 eq.)				

Example 28.7: Consider a biodiesel production process using soybeans that produces soybean meal and crude glycerin as the two co-products in addition to the biodiesel. The process can be divided into soybean farming, soybean oil extraction, and biodiesel production steps. The simplified process here does not consider the transportation of the soybeans, soybean oil, co-products, or biodiesel. Various inputs for these stages are given in Table 28.4. Allocation factors are provided in Table B. The GHG emission factors for GHG emissions and fossil energy inputs are also provided in Table C. Calculate the GHG emissions and fossil energy use of the soybean biodiesel using different allocation methods. Use a functional unit of 1L biodiesel. Soybean yields are 2,811 kg/ha with an oil content of 21.29% and soybean oil to biodiesel conversion is accomplished with a yield of 1.083L biodiesel/kg soybean oil. The data in this example is modified from the data reported by Krohn and Fripp (2012).

Solution:

Soybean oil yield = 2,811 kg soybean/ha × (21.29 kg/100 kg) oil/soybean
= 598.46 kg soybean oil/ha

Biodiesel yield = 1.083 L biodiesel/kg soybean oil × 598.46 kg soybean oil/ha
= 648.13 L/ha

Application of fertilizers leads to emissions of N_2O and CO_2 from the field. These emissions can be calculated as follows:

Field N_2O emissions = 4.8 kg N fertilizer × 1.33 × 0.01 (% emitted as N_2O) = 63.84 g N_2O/ha = 63.84 g N_2O/ha × 233 gCO_2e/g N_2O =14.87 kgCO_2e/ha

Field CO_2 (urea) emissions = 4.8 kg N fertilizer/ha × 0.211 kg urea/kg N fertilizer applied × 0.3316 kg CO_2/kg urea = 0.3 kgCO_2e/ha

Field CO_2 (lime) emissions= 401 kg/ha × 0.44 kg CO_2/kg lime = 176.4 kgCO_2e/ha

Total field emissions = 14.87 + 0.3 + 176.4 = 191.7 kgCO_2e/ha

The life-cycle inventory for the complete process is provided in Table 28.8.

Sample calculation: GHG emissions for gasoline: From Table 28.5, gasoline use is 11.8 L/ha, and from Table 28.7, GHG factors for gasoline are 92.6 gCO_2e/MJ and energy density is 39.52 MJ/L.

Table 28.5 Life-cycle inventory data

	Unit	Value
Nitrogen	Kg/ha	4.8
Phosphorous	Kg/ha	14.2
Potassium	Kg/ha	28.6
Lime	Kg/ha	401
Herbicide	Kg/ha	1.4
Insecticide	Kg/ha	0.02
Seed	Kg/ha	76.1
Gasoline	L/ha	11.8
Diesel	L/ha	38
LPG	L/ha	6.8
Natural gas	m3/ha	4.1
Electricity	kWh/ha	61.9
N-hexane	Kg/kg oil	0
Methanol	g/kg oil	25.73
Sodium hydroxide	g/kg oil	3.58
Sodium hydroxide	g/kg oil	70.73
Hydrochloric acid	g/kg oil	67.84

Table 28.6 Allocation factors

Allocation Method →	Mass	Energy (MJ/kg)	Economic value ($/kg)
Biodiesel	1.0	37.25	1.08
Soy meal	3.85	9.87	0.60
Glycerin	0.214	18.54	0.33

Therefore, GHG emissions related to gasoline use are 11.8 L/ha × 39.52 MJ/L × 92.6 gCO_2e/MJ = 43.2 $kgCO_2e/ha$.

Mass-based allocation

Using the mass allocation factors from Table 28.7:

Fraction of soybean meal produced = 3.85/(1+3.85+0.214) = 0.76
Fraction of glycerin produced = 0.214/(1+3.85+0.214) = 0.04
Fraction of biodiesel produced = 1.0/(1+3.85+0.214) = 0.20

Emissions without allocation = 1.79 $kgCO_2e/L$ biodiesel (Table 28.8)
Share of GHG emissions for soybean meal = 1.79 $kgCO_2e/L$ biodiesel × 0.76
$\qquad\qquad\qquad\qquad$ = 1,363.8 gCO_2e/L biodiesel
Share of GHG emissions for glycerin = 1.79 $kgCO_2e/L$ biodiesel × 0.04
$\qquad\qquad\qquad\qquad$ = 75.8 gCO_2e/L biodiesel
Share of GHG emissions for biodiesel = 1.79 $kgCO_2e/L$ biodiesel × 0.20
$\qquad\qquad\qquad\qquad$ = 354.6 gCO_2e/L biodiesel

Table 28.7 GHG and fossil energy factors

Input	GHG Emissions		Fossil Energy		Input	GHC Emissions		Fossil Energy	
	Unit	Value	Unit	Value	Field Emissions	Unit	Value	Unit	Value
Agricultural emissions									
Nitrogen	gCO_2e/g	3.08	MJ/kg	47.7	Field N_2O emissions	%N emitted as N_2O	1.33		
Phosphorous	gCO_2e/g	1.12	MJ/kg	13.35	Field CO_2 (Lime)	kg CO_2/kg lime	0.44		
Potassium	gCO_2e/g	0.78	MJ/kg	8.09	Urea fraction of N	% of applied N	21.1		
Lime	gCO_2e/g	0.04	MJ/kg	0.42	Field CO_2 (urea)	kg CO_2/kg N fertilizer	0.332		
Herbicide	gCO_2e/g	23.3	MJ/kg	274.63					
Insecticide	gCO_2e/g	27.15	MJ/kg	313.46					
Seed	gCO_2e/g	0.39	MJ/kg	2.35					
Fuel Emissions	Unit	Value	Unit	Value	Chemical Emissions	Unit	Value	Unit	Value
Gasoline	gCO_2e/MJ	92.6	MJ/L	39.52	N-hexane	gCO_2e/g	3.56	MJ/kg	44.41
Diesel	gCO_2e/MJ	93.1	MJ/L	45.3	Methanol	gCO_2e/g	0.61	MJ/kg	33.67
LPG	gCO_2e/MJ	76.9	MJ/L	26.62	Sodium hydroxide	gCO_2e/g	2.84	MJ/kg	19.87
Natural gas	gCO_2e/MJ	66.3	MJ/m^3	39.24	Sodium methoxide	gCO_2e/g	2.43	MJ/kg	39
Electricity	gCO_2e/kWh	780	MJ/Kwh	7.94	Hydrochloric acid	gCO_2e/g	2.7	MJ/kg	20.9

The share of emissions and fossil energy use using different allocation strategies is shown in Tables 28.9 and 28.10.

Table 28.8 Life-cycle inventory

	GHC Emissions ($kgCO_2e$/ha)	Fossil Energy use (MJ/ha)
Nitrogen	14.8	229.0
Phosphorous	15.9	189.6
Potassium	22.3	231.4
Lime	16.0	168.4
Herbicide	32.6	384.5
Insecticide	0.5	6.3
Seed	29.7	178.8
Total Field Emissions	191.7	—
Gasoline	43.2	466.3
Diesel	160.3	1721.4
LPG	13.9	181.0
Natural gas	10.7	160.9
Electricity	383.4	491.5
N-hexane	0.0	0.0
Methanol	9.4	518.5
Sodium hydroxide	6.1	42.6
Sodium methoxide	102.9	1650.8
Hydrochloric acid	109.6	848.5
Total	1162.9	7469.4
Per functional unit (1 L biodiesel)	1.79	11.52

Table 28.9 GHG emissions (gCO_2e/L biodiesel)

Allocation Method →	Mass	Energy	Economic value
Biodiesel	2.3	5.4	3.9
Soy meal	0.5	0.6	0.2
Glycerin	8.8	5.5	7.4

Table 28.10 Fossil energy use (MJ/L biodiesel)

Allocation Method →	Mass	Energy	Economic value
Biodiesel	354.6	844.2	607.3
Soy meal	1363.8	89.8	35.2
Glycerin	75.8	860.2	1151.7

References

Azapagic, A. (1999). Life cycle assessment and its application to process selection, design and optimization. *Chemical Engineering Journal* 73: 1–21.

Chaffee, C., and Yaros, B.R. (2007). Life Cycle Assessments for Three Types of Grocery Bags: Recyclable Plastic; Compostable, Biodegradable Plastic; and Recycled, Recyclable Paper. Philadelphia, PA: Boustead Consulting & Associates. http://heartland.org/sites/default/files/threetypeofgrocerybags.pdf, accessed March 2014.

Earles, J.M., and Halog, A. (2011). Consequential life cycle assessment: A review. *International Journal of Life Cycle Assessment* 6: 445–53.

Eggeman, T., and Elander, R.T. (2005). Process and economic analysis of pretreatment technologies. *Bioresource Technology* 96: 2019–25.

Ekvall, T., and Weidema, B.P. (2004). System boundaries and input data in consequential life cycle inventory analysis. *International Journal of Life Cycle Assessment* 9: 161–71.

Finnveden, G., Hauschild, M.Z., Ekvall, T., *et al.* (2009). Recent developments in life cycle assessment. *Journal of Environmental Management* 91: 1–21.

Gnansounou, E., and Dauriat, A. (2010). Techno-economic analysis of lignocellulosic ethanol: A review. *Bioresource Technology* 101: 4980–91.

Goedkoop, M., Heijungs, R., Huijbregts, M.A.J., de Schryver, A., Struijs, J., and Van Zelm, R. (2009). *ReCiPE 2008: A Life-Cycle Impact Assessment Method Which Comprises Harmonized Category Indicators of the Midpoint and the Endpoint Level. Report I: Characterisation.* The Hague: VROM.

GREET (2015) *The Greenhouse Gases, Regulated Emissions, and Energy Use in Transportation Model.* https://greet.es.anl.gov/, accessed March 2016.

Hunt, R.G., and Franklin, W.E. (1996). LCA – How it came about: Personal reflections on the origin and development of LCA in the USA. *International Journal of Life Cycle Assessment* 1: 4–7.

ISO (2006). *ISO 14044:2006 – Environmental Management. Life Cycle Assessment. Requirements and Guidelines.* Geneva: International Standards Organization.

Jacquemin, L., Pontalier, P., and Sablayrolles, C. (2012). Life cycle assessment (LCA) applied to the process industry: A review. *International Journal of Life Cycle Assessment* 17: 1028–41.

JRC (2013). *International Reference Life Cycle Data System (ILCD) Handbook: General Guide for Life Cycle Assessment: Detailed Guidance.* Brussels: European Commission Joint Research Centre. http://bookshop.europa.eu/en/international-reference-life-cycle-data-system-ilcd-handbook-general-guide-for-life-cycle-assessment-detailed-guidance-pbLBNA24708/, accessed August 2013.

Kazi, F.K., Fortman, J.A., Anex, R.P., *et al.* (2010). Techno-economic comparison of process technologies for biochemical ethanol production from corn stover. *Fuel.* 89: S20–S28.

Krohn, B.J., and Fripp, M. (2012). A life cycle assessment of biodiesel derived from the "niche filling" energy crop camelina in the USA. *Applied Energy.* 92: 92–8.

Kumar, D., and Murthy, G.S. (2011). Impact of pretreatment and downstream processing technologies on economics and energy in cellulosic ethanol production. *Biotechnology for Biofuels* 4: 27.

Kumar, D., and Murthy, G.S. (2012). Life cycle assessment of energy and GHG emissions during ethanol production from grass straws using various pretreatment processes. *International Journal of Life Cycle Assessment* 17: 388–401.

Li, C., Narayanan, V., and Harriss, R.C. (1996). Model estimates of nitrous oxide emissions from agricultural lands in the United States. *Global Biogeochemical Cycles* 10: 297–306.

Matches (2013). *Matches' Process Equipment Cost Estimates.* http://www.matche.com/equipcost/Default.html, accessed April 2016.

Pennington, D.W., Potting, J., Finnveden, G., *et al.* (2004). Life cycle assessment. Part 2: Current impact assessment practice. *Environment International* 30: 721–39.

Perry, R.H., and Green, D.W. (eds.) (1998). Process economics: Fixed capital cost estimation. In: *Perry's Chemical Engineers' Handbook*, 7th edn. New York: McGraw-Hill, Section 9.

Pieragostini, C., Mussati, M.C., and Aguirre, P. (2012). On process optimization considering LCA methodology. *Journal of Environmental Management* 96: 43–54.

Raynolds, M., Fraser, R., and Checkel, D. (2000). The relative mass–energy economic (RMEE) method for system boundary selection. Part 1: A means to systematically and quantitatively select LCA boundaries. *International Journal of Life Cycle Assessment* 5: 37–46.

Rebitzer, G., Ekvall, T., Frischknecht, R., *et al.* (2004). Life cycle assessment. Part 1: Framework, goal and scope definition, inventory analysis, and applications. *Environment International* 30: 701–20.

RSMeans (ed.) (2009). *RSMeans Estimating Handbook*, 3rd edn. Chichester: John Wiley & Sons Ltd.

Tillman, A.-M. (2000). Significance of decision-making for LCA methodology. *Environmental Impact Assessment Review* 20: 113–123. doi:10.1016/S0195-9255(99)00035-9

Westney, R.E. (1997). *The Engineer's Cost Handbook: Tools for Managing Project Costs.* Boca Raton, FL: CRC Press.

Zamagni, A., Guinée, J., Heijungs, R., Masoni, P., and Raggi, A. (2012). Lights and shadows in consequential LCA. *International Journal of Life Cycle Assessment* 17: 904–18.

Exercise Problems

28.1. What is life-cycle assessment? What information does LCA provide? What distinguishes it from techno-economic analysis or risk analysis? What are the essential steps in an LCA?

28.2. A biofuels researcher wants to perform an environmental impact comparison of biofuels with fossil fuels. The researcher defines the functional unit as "1 L of fuel available to the customer at the gas station." The researcher's colleague defines the functional unit as "1,000 MJ of energy available to the customer at the gas station." Which of these is an appropriate functional unit for comparison of biofuels with fossil fuels, and why?

28.3. Discuss the importance of system boundaries and objective delineation of the system boundaries in LCA. What happens if the system boundaries are not consistent within the two systems being compared?

28.4. What is the difference between attributional and consequential LCA?

28.5. What is the importance of data quality in LCA? How does the state of technology influence the data sources? What are the different LCA databases?

28.6. What are the different methods of allocation used in LCA? What is the recommended method of allocation as per ISO-14044?

28.7. What is the purpose of life-cycle impact assessment? What are the mechanisms that are used to reduce uncertainty in the results?

28.8. Consider the flow chart in Figure 28.10. Indicate which of the inputs will be inside a RMEE boundary with 5% cut-off. Although in a full RMEE procedure you will have to use mass, energy, and economic value, use mass/energy values for boundary delineation in this problem.

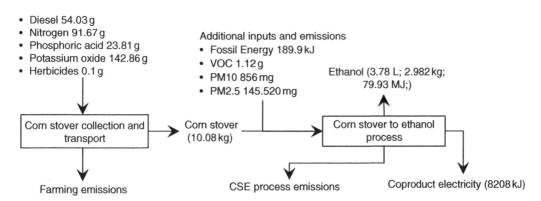

FIG. 28.10   RMEE boundary selection.

28.9. N_2O emissions vary based on environmental conditions such as temperature, field moisture content, and tillage practices. The N_2O emissions at different annual temperatures for a sample field in Iowa are shown in Table 28.11.

Assuming N_2O to be 233 time more potent a greenhouse gas compared to CO_2, calculate the variability (as a % of variation from the base case at 9.8 °C annual temperature) in the overall greenhouse gas emissions for different annual temperatures due to differences in N_2O emissions. Assume that the base case total emissions are 10,000 kg CO_2/ha-year.

Table 28.11 Sensitivity of N₂O emissions to annual average temperatures

Annual average temperature (°C)	7.8	8.8	*9.8*	10.8	11.8
Soil N mineralization (kg N/ha-yr)	171.4	180.8	*190.1*	199.6	211.9
N₂O emissions (kg N/ha-yr)	3.7	4.2	*4.7*	5.3	6.3
N₂ emissions (kg N/ha-yr)	14.4	14.9	*15.4*	16.0	16.6

Source: Data from Li *et al.*, 1996.

Note: The exercise here is limited to variations in N_2O emissions due to annual temperature differences. In the source paper (Li *et al.*, 1996), variations in rainfall, atmospheric deposition of nitrogen, and tillage practices are included.

28.10. Consider Example 28.8. What would be the emissions from ethanol if electricity were allocated based on the energy content? Would this be significantly different from ethanol emissions allocated based on the economic value of electricity? Assume a $0.05/KWh price for electricity. Discuss the implications of the ethanol emissions resulting from different allocation procedures for a company producing cellulosic ethanol.

28.11. A simple process flow diagram for the anaerobic process used to produce renewable natural gas (methane) from animal wastes is given in Figure 28.11. Factors for the calculation of GHG CO_2 equivalents for various emissions are CO_2 –1, CO –1.57143, N_2O –265, CH_4 –30. The emissions at various stages are provided in Table 28.12. Assuming that it takes 1,438.931 kg of animal wastes to produce 1,055 MJ of renewable natural gas (RNG), calculate the emissions per MJ of RNG.

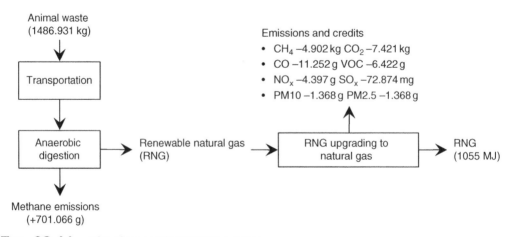

FIG. 28.11   LCA of anaerobic digestion process.

Table 28.12 Emissions associated with processing 1,438.91 kg of animal waste to 1,055 MJ RNG

Emissions	Transportation	Anaerobic digestion	RNG upgrading	Units
VOC	0.369	33.050	–0.164	g
CO	1.273	144.163	27.295	g
NOₓ	4.432	10.614	2.01	g
PM10	0.235	4.478	0.848	g
PM2.5	0.199	4.478	0.848	g

Table 28.12   (*Continued*)

Emissions	Transportation	Anaerobic digestion	RNG upgrading	Units
SO_x	219.702	141.793	26.846	mg
CH_4	1.363	235.069	0.045	kg
N_2O	2.227	0.457	0.087	g
CO_2	0.001	31.183	20.587	kg
CO_2 biogenic	−252.789	0.000	0	mg
Black carbon	30.061	−0.001	0	mg
Primary organic carbon	38.287	0.000	0	mg

Government Policy and Standards for Bioenergy

Sami Kumar Khanal, Gal Hochman, Ajay Shah, and Jeffrey M. Bielicki

---

What is included in this chapter?

This chapter provides an overview of bioenergy markets and presents the importance of government interventions for their success. Existing biofuels policies and instruments, and their implications for various social and environmental sectors, are also introduced.

29.1 Overview of the Bioenergy Market

Energy policies should consider the availability of conventional energy resources and their prices when trying to meet present and future energy demands. World energy demands are projected to increase by 56%, from 524 quadrillion Btu in 2010 to 820 quadrillion Btu in 2040. More than 85% of this increase is expected to occur in developing nations, like China and India, where strong economic growth and rapidly growing populations should increase demand for energy (EIA, 2013). As a result, there is concern that in future the supply of petroleum fuels (i.e., gasoline and diesel) may no longer meet global demands (Suplee, 2008).

A number of energy technologies (e.g., renewable sources, like wind turbines and solar photovoltaics, and carbon-free sources, like nuclear power) can substitute for, or displace, coal and natural gas in the electricity sector, but the transportation sector has fewer options. At present, there are no alternatives that can economically compete with the high-energy densities of fossil fuels like petroleum and natural gas (Rajagopal and Zilberman, 2007; Boyle, 2012). Decades ago, a great deal of effort was put into developing biomass-derived fuels (i.e., biofuels), mostly in the aftermath of the oil crises in the 1970s. For example, after oil prices peaked in 1979 and concerns about dependence on foreign oil became widespread, Brazil accelerated its national sugarcane-based ethanol program and the USA began a corn-based ethanol program. However, the US biofuels program was short-lived because oil

---

Bioenergy: Principles and Applications, First Edition. Edited by Yebo Li and Samir Kumar Khanal.
© 2017 John Wiley & Sons, Inc. Published 2017 by John Wiley & Sons, Inc.
Companion website: www.wiley.com/go/Li/Bioenergy

prices decreased substantially during the late 1980s and into the 1990s. As a result, much of the attention in the USA shifted away from biofuels.

Interest in biofuels was renewed in the early 2000s in part as a result of political strife in the Middle East, unstable oil prices, concerns about environmental degradation, and growing acceptance that carbon dioxide (CO_2) emissions from burning fossil fuels exacerbate global climate change. As a result, various government-enacted biofuels policies have been implemented and efforts were renewed to develop cost-effective commercial-scale production of biofuels. Between 2007 and 2012, global ethanol production increased from 13 billion to 21 billion gallons (Licht, 2013), and global biodiesel production increased from 2.7 billion to 5.6 billion gallons (Licht, 2012; Figure 29.1).

The global biofuels market is presently dominated by technologies that use *first-generation feedstocks* that compete with use for food or animal feed. The USA and Brazil produce most of the world's ethanol, mainly from corn (USA) and sugarcane (Brazil). Historically, the European Union (EU) has been the main producer of biodiesel. The USA was the second largest biodiesel producer until 2008, but production decreased by over 240 million gallons in 2009 after the tax credit lapsed in December 2009. As a result, Central and South America and Asia surpassed the USA in biodiesel production in 2010. Then US biodiesel production increased substantially after the tax credit was reinstated in 2011. This example demonstrates the importance of government policy support for the biodiesel industry.

In contrast to the development of first-generation biofuels (i.e., biofuels produced from food/feed crops), *second-generation biofuels* (i.e., biofuels produced from lignocellulosic biomass) industries have been slow to produce large quantities of biofuels, and are far from being commercially viable. Second-generation biofuels are typically derived from agricultural residues (e.g., corn stover) and energy crops (e.g., switchgrass, energy cane, miscanthus, poplar) that are not used for food or feed, and may emit fewer greenhouse gases (GHGs) than first-generation biofuels (Davis *et al.*, 2011). Additionally, these dedicated energy crops can be grown on marginal land, and are thus less likely to compete for space with food crops. As a consequence, second-generation biofuels have the potential to create new opportunities with advances in technologies and innovations in bioprocessing.

The costs of cultivating second-generation feedstocks comprise more than half of the overall production costs. Despite the remarkable improvements in technologies, second-generation biofuels have not been economical, even under the most favorable conditions of high oil prices and low

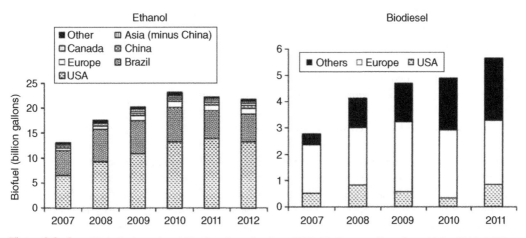

FIG. 29.1   Global ethanol and biodiesel production, 2007–12. Source: Data from Licht, 2012, 2013.

feedstock prices (Kojima *et al.*, 2007). In contrast, fossil fuels have been the cheapest and most ubiquitous source of energy, but the economics of these fuels to date have generally not included the environmental and social implications of using them. Government policies have thus been put in place to help biofuels compete with fossil fuels. In the absence of mandates and government support, ethanol production in places like the USA has been sensitive to feedstock prices. For example, ethanol production decreased substantially in mid-1996 because corn prices were high. The spectrum of policies that are currently in place around the world to promote biofuels include tax exemptions, mandatory blending requirements, and renewable standards, as well as other indirect policies, such as carbon taxes, and agricultural and trade policies.

29.2 Rationale for Government Intervention

It is necessary to present some basic elements of microeconomics in order to understand the basis for the government to create and enact policy. Microeconomics is the study of the behavior of individual consumers, workers, investors, businesses, and so on. Some of these entities are producers who make or market goods and services, and some of them are consumers who buy or consume these goods and services. The goods and services are bought and sold in the market. These market transactions are based on the prices that buyers, or consumers, are willing to pay for goods and services, and the prices at which suppliers, or producers, are willing to sell goods and services. In general, there is high demand for a good when it is cheap, but little demand when it is expensive. A *demand curve* represents the relationship between the price of a good or service and the quantity that will be demanded. One explanation for this relationship is that only a few people can afford the good when the price is high, so only a small quantity will be demanded. However, many more people can afford it when prices are low, so a much larger quantity will be demanded. The supply of the good tends to function in the opposite way. Producers are willing to supply a large amount of the good at a high price, but only a small number of producers are willing, or able, to provide the good when the price is low. Similar to the demand curve, there is a *supply curve* for each good and service on the market. One part of this explanation is that only some companies have production costs that are low enough to be able to sell the good for a low price, whereas more producers are able cost-effectively to produce and sell the good when the price is high.

These relationships for supply and demand are shown in Figure 29.2. The *demand curve* can also be referred to as the *marginal benefit curve*, because when consumers are deciding between multiple goods, they are implicitly weighing up the incremental or marginal benefit that they would receive from consuming one more unit of this good instead of one more unit of another one. Similarly, the *supply curve* can also be referred to as the *marginal cost curve*, because a producer chooses how much of a good it can supply based on the cost of producing one more unit of the good. Market transactions will converge at *market equilibrium*, where the supply curve and the demand curve intersect, and the price will be P^E for quantity Q^E. Since there is more demand at prices that are lower than P^E, those consumers who would have been willing to purchase the good at a price higher than P^E receive a benefit because they are paying less than they were willing to pay. The total amount of this benefit is the area under the demand curve to the left of (P^E, Q^E). This aggregate benefit to consumers is called the *consumer surplus*, and is indicated in Figure 29.2. Similarly, producers that are willing to sell at prices higher than P^E yield a benefit. This benefit is shown in Figure 29.2 as the *producer surplus*.

In Figure 29.2, the supply curve is termed the *marginal private cost* (MPC) curve, because producing the good creates *externalities* that can negatively affect other entities but are not included in the cost of production. An externality is the cost or benefit of an activity that does not directly accrue

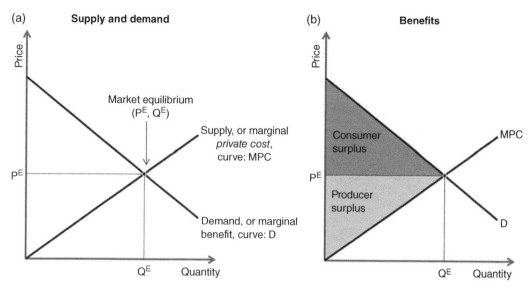

FIG. 29.2   Basic elements of microeconomics: (a) market equilibrium at the intersection of supply and demand curves; and (b) the benefits to consumers and producers of goods in the market.

to the person or institution engaging in that activity. Externalities can be positive or negative. For example, the negative impact on human health due to particulates that are emitted from burning coal is a negative externality. People who get asthma from breathing these particulates, for example, did not burn the coal, but they are the ones who are negatively affected. Other negative externalities include decreases in air and water quality that result from the expansion and intensification of agriculture to produce bioenergy crops. In contrast, public education is a positive externality that results from research and development. The public did not engage in the research, but they benefit from it. Classical economic principles state that the social implications of externalities are not automatically embedded in market prices, and thus these ramifications are not directly considered when making decisions. That is, the market equilibrium (P^E, Q^E) in Figure 29.2 is not the socially best outcome, because producing the good results in a negative externality that costs society.

Figure 29.3 shows the negative externality from producing the good in an unregulated free market, and how the market should incorporate that externality into transactions. Assuming that the externality is linearly related to the quantity of the good that is produced, the *marginal externality cost* (MEC) is linear and increasing. The full cost of production to society is the *marginal social cost* (MSC), which is the sum of MPC and MEC. The *socially optimal outcome* occurs when decisions in the market consider all of the costs, not just those that are incurred by the producer. This occurs where Q^O units of the good are supplied at a price of P^O. With a negative externality, the optimal allocation of a good is where it is supplied at a price higher than that obtained under competitive equilibrium: the market oversupplies the good ($Q^E > Q^O$) for too low a price ($P^E < P^O$). This is referred to as a *market failure*, because the unregulated free market does not allocate goods in a socially optimal way.

Welfare maximization theory states that governments should attempt to correct market failures by policies that encourage the efficient allocation of resources. This approach provides a strong rationale for government-supported bioenergy policies. The British economist Arthur C. Pigou argued in 1920 that firms seek to maximize their private interests, and that they do not have an incentive to consider social interests when their private costs and benefits do not include social costs and

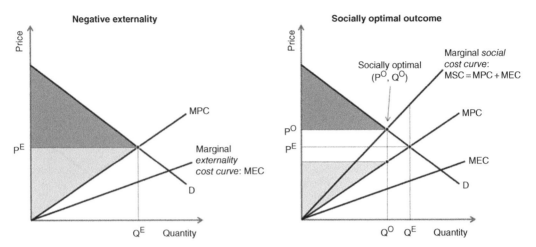

FIG. 29.3   The negative externality from producing the good and the socially optimal outcome when the negative externality is internalized in market transactions.

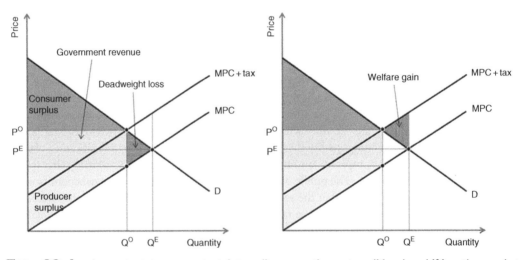

FIG. 29.4   A constant tax on output internalizes negative externalities by shifting the market equilibrium to a higher price and lower quantity (P°, Q°) than in an unregulated free market (Pᵉ, Qᵉ). The lefthand panel shows that the tax reduces consumer and producer surpluses, generates revenue for the government, and incurs a deadweight loss. The righthand panel shows the welfare gain by how the tax shifts the market equilibrium.

benefits. To deal with the divergence between social and private costs, Pigou recommended levying a tax to align private and social incentives. Such a *Pigouvian tax* could produce a market equilibrium outcome that is the same as the socially optimal outcome, and provide a logical rationale for taxes or subsidies to internalize externalities via market prices to various parties. For example, environmental regulations are often designed to encourage firms to internalize externalities by considering the external costs of producing a good.

Figure 29.4 shows the effects of a Pigouvian tax. If firms face a constant pollution tax, their marginal private cost will be shifted upward by an amount equal to the tax. The amount of this tax can be set so that the new marginal private cost curve (MPC + tax) intersects the demand curve at the same point as the marginal social cost curve. As a result, the new market equilibrium will then be (P^O, Q^O). The regulated market will have a higher price and lower quantity than the unregulated market. At this new equilibrium, consumer surplus is reduced and producer surplus is also reduced, as shown in Figure 29.4.

Example 29.1 illustrates the distribution of revenues between cotton and dairy farmers when this tax is imposed. The government receives revenue equal to the amount of the tax multiplied by the quantity that is exchanged: Government revenue = Tax • Q^O. Welfare maximization seeks to maximize total benefits which include the consumer surplus, the producer surplus, the gain from controlling the externality, and the government revenue regardless of how that welfare is distributed. The Pigouvian tax corrects the market failure by internalizing the externality, but it changes and redistributes the total benefits. These changes include a *deadweight loss* and the *welfare gain*, as indicated in Figure 29.4.

Example 29.1: Consider a cotton farmer and a neighboring dairy farmer. The cotton farmer wants to spray an insecticide on his crops to reduce the damage by pests. The cotton farmer will earn higher profits if he sprays the insecticide four times per year. Each spray, however, contaminates water in the local aquifer and causes damage that has a negative impact on the dairy farmer. The impact of the number of sprays on the cotton farmer's profits and dairy farmer's damages/losses are depicted in Table 29.1.

Table 29.1 Benefits and losses of insecticide spray

Sprays per year	Cotton farm profits	Cotton farm marginal profit	Dairy farm damage	Cotton farm profit minus damage
0	40	0	0	40
1	100	60	10	90
2	150	50	15	135
3	180	30	30	150
4	185	5	45	140
5	180	–5	60	120

How can a tax be used to correct the competitive market and internalize the negative externality into the cotton farmer's decision about how many times per year he will spray his cotton plants? When the government imposes a tax of 30 units per spray, the cotton farmer reduces the number of times per year he sprays insecticide to three. With the introduction of this tax, the private and social incentives align and the socially optimal solution is achieved. Although the cotton farmer's profits decreased by 5 units, the damage to the dairy farmer decreased by 15 units: a 10 unit gain in total. With three sprays, the cotton farmer's profits minus damages to the dairy farmer are maximized. At the optimum, the tax equals the marginal damage (30 units in this example).

Externalities can also be positive. For example, government-funded research and development (R&D) often produces *knowledge spillovers*, which occur when knowledge generated in one domain is applied to other domains. Knowledge spillovers are also *public goods*, a term that refers to goods for which it is difficult to restrict consumption once it is provided to some consumers. A lighthouse, for example, is a public good; once it is built, the marginal cost of another ship using the light from the lighthouse is zero, and one ship cannot exclude other ships from using the light. From a societal standpoint, it is desirable to produce as much general and widespread knowledge as possible. However, when public goods are produced, individual firms are not able to reap the full return on their investment because others may benefit as well. As a result, the private sector will likely under-invest in research and development that can spill over, and there is little support from risk-averse investors in relatively new industries, such as bioenergy. Consequently, government support may be necessary to address the drawbacks of positive externalities in order to stimulate the bioenergy industry. In this manner, government intervention helps to provide a stable foundation to build skills and capacities for the timely emergence and development of renewable energy technologies. An example of intervention by the German government to reduce the uncertainty of economic returns to investment in order to stimulate the development of the renewable electricity sector is highlighted in the box.

The Renewable Sources Act

In response to investor uncertainty and sluggish investments, Germany amended its feed-in tariffs (FIT) and replaced the Feed-in Act with the Renewable Sources Act (*Erneuerbare-Energien-Gesetz*, or EEG). The EEG tied remuneration rates to the generation costs of renewable energy technologies, rather than average electricity prices. For biogas producers, this rate increased to an average of €9.5 ct/kWh. The EEG also required utilities to connect renewable electricity operators to the electricity grid and guaranteed that the length of the FIT would be 20 years. This policy change reduced the uncertainty and risk around investing in anaerobic digestion technologies by providing higher remuneration rates and guaranteeing revenues for a long timeframe.

29.3 Government Intervention through Policy Tools

Externalities are often diverse and difficult to monetize. As a result, designing and analyzing policies can be a challenge. Several policy tools are available to correct market failures, but the efficiency of any policy tool depends on numerous factors that are difficult to quantify and can change over time (e.g., transaction costs, availability of information, budgets, and resources). A Pigouvian tax that is equal to the marginal social cost of the externality is a "first-best" instrument. Establishing a tax to internalize the externalities associated with oil production, its conversion to gasoline, and the use of that gasoline is difficult in part because identifying and monetizing the environmental impacts are not straightforward. In addition, the public may not be receptive to taxes, especially those that may be high in order to internalize the externalities sufficiently. Other approaches and policy tools include subsidies for new technologies and direct controls through quotas, targets, and standards. As an example of direct control, the US Environmental Protection Agency requires pollution control devices, such as catalytic converters on cars and flue-gas desulphurization equipment on the smokestacks of power plants. For instance, power plants in 27 states must install scrubbers to reduce emissions of dioxins, arsenic, mercury, or lead, in an effort to minimize and/or eliminate risks to human health. These states have forests, farms, lakes, and streams that have been polluted for many decades as a result of emissions from power plants.

29.4 Biofuels Policy Implementations: Existing Policy Instruments

At present, ethanol and biodiesel are the biofuels that can most effectively displace fuels derived from petroleum – namely, gasoline and diesel – for the transportation sector. Policies supporting these two biofuels fall into two categories: those that replace the consumption of petroleum fuels (e.g., mandates for biofuels use, comparative reductions in fuel taxes for biofuels); and those that stimulate domestic biofuels production (e.g., producer subsidies, import tariffs to protect domestic producers, and direct government support and funding for research and development of new or improved technologies). This section highlights some of the policies for biofuels around the world, and their implications for economic welfare and the environment.

29.4.1 Tax Credit/Subsidy

Tax credits or subsidies for biofuels are the most direct and widely used financial support instrument to promote biofuels. Subsidies or tax credits lower the cost of production while also increasing output. Tax credits shift the supply curve down and to the right (Figure 29.5), lowering the price from P^E to P^S and increasing the amount produced from Q^E to Q^S. In the USA, the 1978 Energy Tax Act established a subsidy of $0.40 per gallon of ethanol blended with gasoline. This subsidy was the first major support for the modern corn ethanol industry. In 2004, the policy was revised and expanded under the Volumetric Ethanol Excise Tax Credit (VEETC), which provided a fixed tax credit of $0.51 per gallon of ethanol, $1.00 per gallon of biodiesel, and $0.50 per gallon of recycled oils when blended with gasoline. To promote cellulosic ethanol production, the 2007 Farm Bill increased the blender's tax credit to $1.01 per gallon of cellulosic ethanol, and reduced the tax credit for corn ethanol to $0.45 per gallon. The Energy Independence and Security Act (EISA) of 2007 stated that each gallon of cellulosic ethanol can be counted as 2.5 gallons toward the renewable fuel standard (Carriquiry *et al.*, 2011). In the EU, tax credits for ethanol have been as high as $3.18 per gallon. For biodiesel, tax credits have reached $2.27 per gallon in the EU, $1.05 per gallon in Australia, and $1.00 per gallon in the USA. In the EU, biogas and liquid biofuels are 100% tax exempt until 2015. In China, ethanol producers are exempted from the 5% consumption tax and 17% value-added tax (VAT; Sorda *et al.*, 2010). Tax credits primarily benefit biofuels industries. While some benefits are

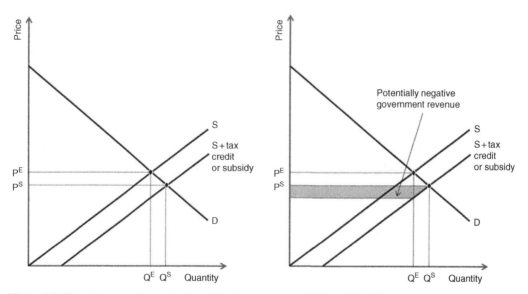

FIG. 29.5   A tax credit or subsidy shifts the market equilibrium (P^e, Q^e) to a lower price and higher quantity (P^s, Q^s) by shifting the supply curve(s) downward by the amount of the tax credit or subsidy.

passed on to the farmer, government revenue can be negative by the amount shown in Figure 29.5. Tax credits that do not vary with the price of oil and do not cap the amount of production to which they are applicable, or do not terminate at a fixed period of time (a "sunset clause"), can substantially increase the subsidy burden if oil prices decrease and/or the production of biofuels increases (Rajagopal and Zilberman, 2007).

29.4.2 Carbon Tax

A *carbon tax* is a Pigouvian tax that could be levied on the carbon content of the fuel, or on the emissions from combusting the fuel. The difference is important. The carbon content of a fuel such as coal, gasoline, or natural gas is relatively constant. That is, the chemical structure has a fixed number of carbon atoms relative to the hydrogen, oxygen, and other potential atoms. A tax applied to the carbon content of the fuel would then be a tax directly applied to the fuel as a means to directly discourage the use of that fuel, in part by encouraging a shift toward fuels that have a lower carbon content. In contrast, a carbon tax applied to the emissions from using that resource discourage its use. Despite actually being applied to CO_2 and not C, the term "carbon tax" is usually used when referring to CO_2 emissions. Like the incentive when the carbon content of the fuel is taxed, the first effect of a tax on CO_2 emissions is the "fuel-switching" incentive to reduce the amount of CO_2 that is emitted per unit of energy produced. The second effect provides an incentive to use the fuel more efficiently by, for example, refining and improving combustion and manufacturing processes as well as incentivizing technological development that improves the efficiency by which the fuel is converted into energy.

Although economists advocate this tax, politicians are often reluctant to use it. There are numerous possible explanations for this reluctance. An economic explanation is that producers sell fewer units, the price of these units increases, and the total surplus decreases. These effects of a carbon tax can be seen in Figure 29.4, where the market equilibrium would shift from (P^E, Q^E) to (P^O, Q^O), and the government would earn the revenue indicated by the shaded area.

Carbon taxes were first implemented in northern European countries. Finland was the first country to introduce a carbon-based tax in 1990. Other countries implemented their own carbon taxes soon thereafter: The Netherlands (1990), Norway (1991), Sweden (1991), and Denmark (1992). In 2001, the UK and Australia also implemented carbon taxes. Such taxes were introduced in some Canadian provinces, including Alberta and Quebec in 2007 and British Columbia in 2008, and in some regions of the USA, including Boulder, Colorado in 2006 and Montgomery, Maryland in 2010. Table 29.2 summarizes the fuels and sources subject to carbon taxes as of late 2014.

Carbon taxes in Europe are high compared to other countries. In Sweden, the standard tax rate is equivalent to $105 per metric ton of CO_2 emissions, but the rate for industry is $23 per metric ton of CO_2 emissions. Norway's tax on gasoline is $62 per metric ton of CO_2 emissions. In contrast, the Bay Area Air Quality Management District (BAAQMD) in California has one of the lowest tax rates: $0.045 per metric ton of CO_2 emissions. These taxes have raised substantial revenue in some of the European countries that implement them. In 2007, the revenue from carbon taxes equaled about 0.3% of GDP in Finland and in Denmark, and about 0.8% of GDP in Sweden.

Carbon taxes are implemented with the goal of reducing CO_2 emissions, but some applications aggressively encourage changes in behavior and consumption patterns, while others are more modest and generate revenue to support the development and deployment of cost-effective technologies that have minimal CO_2 emissions (Sumner *et al.*, 2009). Higher carbon taxes may provide a stronger message to consumers to change their consumption behavior, but if the demand is *inelastic* (i.e., demand does not change with price), the amount of the tax will be directly passed on to

Table 29.2 Major taxed sectors in existing carbon tax systems

Sectors	Finland	Netherland	Norway	Sweden	Denmark	UK	France	Quebec	British Columbia	Boulder, CO	BAAQMD, CA*	CARB*
Natural gas	X	X	X	X	X	X	X	X	X			
Gasoline	X		X	X		X	X	X	X			
Coal	X			X			X	X				
Electricity	X	X			X	X		X		X		
Diesel	X				X			X				
Light and heavy fuel oil	X	Light only	X	X	X			X	X			
Liquefied petroleum gas (LPG)			X	X		X		X				
Home heating oil		X		X				X				
Permitted facilities											X	X

Note: *Bay Area Air Quality Management District (BAAQMD) in California; California Air Resource Board (CARB).
Source: Adapted from Sumner et al., 2009.

the consumer through an increase in price (Figure 29.4). Some programs as in British Columbia in Canada and in the UK return tax revenues to customers through different means (e.g., income tax reductions, credits to low-income households) that can reduce the burden of the tax on low-income households.

Carbon taxes increase the costs of "dirty" technologies i.e., those that have a high carbon intensity (CO_2 emitted per MJ of energy produced), and provide a comparative benefit to "clean" technologies with lower carbon intensities. Carbon taxes are an effective and efficient way to stimulate innovation, but the transition phase from dirty to clean technologies can be economically challenging. For example, Penrice Soda, a Australian soda ash company that makes glass and detergents, was unable to pay the AUD$8 million annual carbon tax bill in 2013 and petitioned for a decrease to AUD$1 million.

29.4.3 Feed-In Tariff

A *feed-in tariff* (FIT) is a policy intended to accelerate investment in renewable energy technologies. Feed-in tariffs guarantee prices for fixed periods of time for electricity that is produced from renewable energy sources. These prices vary by the technology, the installed capacity, the quality of the resource, the location of the project, and a number of other project-specific variables. Feed-in tariffs enable numerous stakeholders (e.g., homeowners, landowners, farmers, municipalities, and small business owners) to invest in renewable energy technologies and thus stimulate their broader deployment. Along with stable policies, feed-in tariffs for on-farm anaerobic digesters are the primary factors that account for the difference in adoption of on-farm anaerobic digestion between Europe and the USA. A fixed remuneration rate and guaranteed revenues for an extended timeframe under FIT help reduce uncertainty and risk to investment. In Germany, feed-in tariffs for biogas have promoted the development of commercial anaerobic digesters, resulting in an increase in the number of biogas plants from about 140 in 1992 to about 7,720 by the end of 2013. Biogas that is upgraded so that it contains at least 98% methane can be substituted for natural gas in compressed natural gas (CNG) vehicles. Sweden has nearly 11,500 CNG vehicles, and it is estimated that biogas supplies half of the demand for transportation fuels in the country (IEA Bioenergy, 2011).

> **Example 29.2:** How much money can you earn from a feed-in tariff? Feed-in tariffs offset upfront costs, but the amount of that offset depends on the specifics of the technology.
> How much can you earn from the installation of a 3.5 kW solar PV system?
> Assume:
>
> - Generation tariff is 14.38 cents per kWh.
> - Export tariff is 4.77 cents per kWh.
> - The rates are guaranteed for 20 years.
>
>
> If the system generates at its peak capacity of 3.5 kW for 1 hour and sells 0.5 kWh to the grid, it receives 14.38 cents/kWh × 3.5 kWh + 4.77 cents/kWh × 0.5 kWh = 52.72 cents for that hour. However, this amount is not guaranteed for every hour of every day. For example, the system may only generate 1.5 kWh on a cloudy day and not sell electricity to the grid. For that hour, the unit earns 14.38 cents/kWh × 1.5 kWh = 21.57 cents from the feed-in tariff.

29.4.4 Biofuels Regulations and Standards

Mandatory blending requirements and biofuels standards are two examples of direct control of governments over the biofuels market. When implementing a *biofuels mandate*, a country usually specifies a quantity of biofuels to be produced each year. For example, in the USA, the revised Renewable Fuel Standard (RFS2) mandates that 36 billion gallons of biofuels be produced annually by 2022, with a cap of 15 billion gallons for corn ethanol. Out of the remaining 21 billion gallons, at least 16 billion gallons must be cellulosic biofuels, and the rest in the form of biomass-based diesel and other advanced biofuels, including those derived from renewable feedstocks, such as sugarcane. The Brazilian government mandates 20–25% ethanol blends in all gasoline sales. In India, China, and Thailand, mandatory blend ratios range from 5–10%. The 2003 EU Biofuels Directive requires member countries to blend into transportation fuels 5.75% of biofuels in 2010 and 10% in 2020. Germany and Austria mandate 20% biomethane in their CNG vehicles.

Figure 29.6 depicts the market for gasoline, assuming a mandate for M gallons of biofuels in gasoline equivalent units. Let D be the demand curve for gasoline before biofuels are introduced, and let S be the supply curve for gasoline. The market must purchase M units of biofuels, suggesting that the residual demand for gasoline is $DR = D - M$. The mandate causes the equilibrium to move from (P^E, Q^E) to (P^{DR}, Q^{DR}), which is a decrease in both price and quantity. The amount of gasohol (a blend of gasoline and biofuels) consumed is $Q^{DR} + M$, but the price to consumers who purchase gasohol may increase if the cost of producing a gallon of biofuels is more than the cost of producing a gallon of gasoline. When a quantity of biofuels is specified (M), the demand for gasoline decreases (assuming that the mandate is binding). A blending standard is revenue neutral from the perspective of the government, and usually increases producer surplus but decreases consumer surplus. This decrease in consumer surplus is because the energy content of ethanol is about

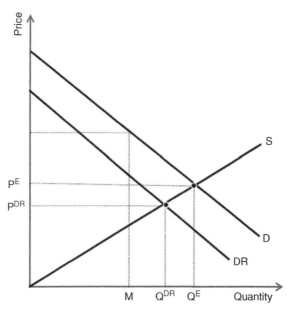

FIG. 29.6   Biofuels mandates (M) influence the supply and demand of gasoline by effectively shifting the demand curve for gasoline.

two-thirds the energy content of gasoline; a car using ethanol will not go as far as it would if it used the same amount of gasoline, and thus a gallon of high-blend ethanol will not provide the same vehicle mileage as a gallon of gasoline. If the price of a gallon of gasoline is the same as the price of a gallon of ethanol, consumers would pay more overall while driving with an ethanol–gasoline blend than with gasoline alone.

29.4.5 Emissions Trading or "Cap-and-Trade"

Emissions trading, or a "*cap-and-trade*" system, is a market-based approach that caps emissions at some level and establishes a system where emitters are provided with permits that they can trade on the market. The total amount of emissions that is allowed by the permits is capped below the present amount of emissions, and individual emitters can choose to buy or sell permits on the market. Presumably, these decisions are based on their individual ability to reduce their emissions, and the cost of doing so relative to the price of the permit. If the cost to an emitter of reducing emissions is less than the price of the permit, the emitter should do so by, for example, improving the efficiency of processes or installing new technologies. If the emitter can reduce emissions below the amount that is allowable by its permits, the emitter should sell those permits on the market. Conversely, if an emitter cannot cost-effectively reduce its emissions below the amount allowable by the permits it holds, it should buy additional permits on the market. As a consequence, the price of the emissions permit is established on the market by the marginal costs to each emitter. If the costs are higher, they buy permits; if the costs are lower, they sell permits. The overall effect is to reduce emissions as economically efficiently as possible while providing flexibility to how the sources that generate those emissions do so.

Regulatory frameworks, such as the Low Carbon Fuel Standard (LCFS) implemented in California, British Columbia, and the EU, use market-based cap-and-trade approaches to reduce emissions from petroleum-based transportation fuels. The LCFS requires transportation fuels to compete based on their carbon intensity rather than their price. It means that substantially reducing the emissions associated with petroleum fuels by carbon dioxide capture and sequestration technologies is as beneficial as bringing biofuels to the market. In California, the LCFS was implemented by an Executive Order pursuant to AB 32 and requires petroleum-based fuel producers to reduce the carbon intensity of their products by 10% by 2020. Petroleum importers, refiners, and wholesalers can either develop low-carbon alternative fuels, or buy carbon permits from other companies that develop and sell low-carbon alternative fuels such as biofuels, natural gas, or hydrogen.

Cap-and-trade systems encourage emissions to be reduced to the amount that is capped, as economically as possible. Policy-makers can change the amount that is capped, if it turns out that the permit prices are low (i.e., it is easy for emitters to reduce emissions) or if the prices are high (i.e., it is difficult for emitters to reduce emissions). In contrast, carbon taxes do not ensure that emissions will be reduced to a specific level, but rather that the marginal cost of these emissions reductions will equal the carbon tax. As a consequence, cap-and-trade systems provide more certainty about the quantity of emissions reductions than they do about the cost of those reductions, whereas carbon taxes provide more certainty about the cost of the reductions than they do about the quantity that will be reduced. From the standpoint of the individual emitters, this certainty of costs is helpful for decision-making and long-term planning of investments. From the perspective of the climate, however, the certainty over the amount of emissions that will be reduced is more important than the cost of those reductions.

Carbon credit

A carbon credit is a permit that allows facilities to emit a specified amount of CO_2 (usually a metric ton of CO_2 or its equivalent gases, CO_{2eq}) and to trade that permit if the full amount is not emitted. For carbon trading, CO_2 emissions are capped and individual emitters are given an allowance of credits, or quotas, where each credit gives the owner the right, but not the obligation, to emit the specified amount of CO_2. Emitters that have not used their quotas can sell their unused allowances, whereas facilities that may exceed their emissions quotas can buy the extra allowances on the market.

29.4.6 Flex-Fuel Vehicles

Governments have promoted biofuels by giving preferential treatment to alternative-fuel vehicles, both directly and indirectly. Direct support has been manifested by tax credits (subsidies), whereas indirect support has resulted from credits given to automobile manufacturers for improving energy efficiency. Flex-fuel vehicles can use more than one fuel, usually gasoline blended with either ethanol or biodiesel, and have been supported by numerous countries, including the USA and Brazil. In the USA, for example, the 1992 Energy Policy Act mandated that 75% of the new light-duty vehicles acquired by federal agencies be alternative-fuel vehicles. In addition, in 2007, Executive Order 13423 required federal agencies with 20 vehicles or more in their US fleet to reduce their petroleum consumption by 2% per year, relative to their fiscal year (FY) 2005 baseline, through 2015. Agencies must also increase their alternative fuel consumption by at least 10% per year over the FY 2005 baseline. In Brazil, by the late 1980s, almost all new cars were built with the capacity to run on ethanol. The subsequent demand for ethanol resulted in a substantial shortage in 1990, and the policies were modified to align the supply of vehicles with the supply of ethanol. In 2007, flex-fuel vehicles accounted for 85% of new auto sales in Brazil (Embassy of Brazil, 2007).

29.4.7 Farm Policies

At present, biofuels are produced primarily from agricultural crops and residues, and the costs of these feedstocks comprise more than half the cost of the production of biofuels. Agricultural and trade policies affect the supply, demand, and prices of agricultural commodities, and thus affect the cost of biofuels. Agricultural policies in the USA typically seek to control the agricultural supply, using price supports, land-use regulations, and trade regulations that affect imports and exports (Rajagopal and Zilberman, 2007). Historically, agricultural policies have focused on protecting domestic farmers from being undercut by imports from lower-cost producers.

Producer support programs in the USA, such as Loan Deficiency Payments (LDPs) and Counter-Cyclical Payments (CCPs), provide benefits to farmers when market prices for certain crops fall below designated levels. Other programs support the production of energy crops. For example, the Biomass Crop Assistance program under the 2008 Farm Bill financially supported the establishment of energy crops, including switchgrass, hybrid poplar, and miscanthus, in 11 project areas across the USA. Many of these projects were associated with major efforts to develop "second-generation" lignocellulosic biofuels. The 2014 Farm Bill provided protection against production and/or revenue losses for crops, including sweet sorghum, biomass sorghum, sugarcane, and penny-cress, that are research and development priorities.

The EU's Common Agricultural Policy supports the biofuels sector by allowing farmers to grow biofuels crops on land that is set aside, and grants direct payments to energy crops that are grown on such land. In Brazil, a regional producer subsidy seeks to balance differences in the costs of production across the country. The north–northeastern part of Brazil includes the poorest states in the country, and the growers in this region receive a subsidy of $5.0 per metric ton of sugarcane produced for up to 10,000 metric tons per grower (USDA, 2010). In China, the government provides direct support for cultivating second-generation jatropha and cassava feedstocks by providing allowances for up to $438 per hectare of jatropha and $394 per hectare of cassava (Sorda *et al.*, 2010).

29.4.8 Trade Policies

The global trade in biofuels is modest relative to production levels: approximately 10% of the total volume of biofuels is traded internationally, with Brazil accounting for about half of exports (Kojima *et al.*, 2007). This is in part due to several trade restrictions imposed by most countries on both feed-stocks and biofuels. For example, import tariffs and quotas in most biofuels-producing countries protect the interests of domestic producers and restrict benefits to certain countries. Export taxes create another barrier to trade in biofuels, but export subsidies could motivate high-cost domestic producers to adopt efficient technologies in order to compete with low-cost producers in international markets (Rajagopal and Zilberman, 2007). In order to protect domestic corn ethanol producers, the USA used to have an import tariff of $0.14 per liter and an ad valorem tariff of 2.5% on Brazilian sugarcane. Argentina has export taxes of 27.5% on soybean seeds and 24% on soybean oil, but only 5% on biodiesel. This policy seeks to promote the export of value-added finished products rather than the raw materials. Likewise, Malaysia does not place export taxes on processed palm oil or biodiesel, but it does have export charges on crude palm oil. Malaysia sets export charges based on conditions in the global markets. For example, in September 2014, Malaysia exempted crude palm oil from a 4.5% tax that was in place to boost export demand. Global palm prices are based on those for Malaysian palm, which plunged 14.5% in August 2014 and reduced exports of palm oil.

Brazil eliminated its 20% tariff on ethanol imports in 2010 in order to transform ethanol into a globally traded commodity. In the USA, import tariffs expired in 2011 and the new policies require imported ethanol to meet the EISA advanced biofuels mandate. Trade in biofuels is also limited by regulatory measures such as the EU's sustainability criteria for palm oil imports from Malaysia and Indonesia, and Thailand's ban on palm oil imports (USAID, 2009). Given the concerns that biofuels production may lead to the deforestation of the rainforest, barriers to trade may be appropriate (Kojima *et al.*, 2007).

29.4.9 Funding for Research and Development

Investment in research and development (R&D) for biofuels has the potential to increase productivity and reduce the associated costs and CO_2 emissions. However, private investment in the development of biofuels technologies is likely to be limited because of the positive externalities of R&D: knowledge is a public good. As a consequence, economists generally agree that governments should conduct and fund R&D to develop these technologies (Klette *et al.*, 2000). For example, in 2011, the US Department of Agriculture and the US Department of Energy jointly announced $47 million in funding for R&D projects on bioenergy and bioproducts. The projects, funded by the Biomass

Research and Development Initiative, seek to develop advanced biofuels that emit less than 50% of the CO_2 from fossil fuels as well as increase the availability of alternative renewable fuels and biobased products (DOE, 2011). These R&D activities have been very effective at reducing the cost of producing enzymes for cellulosic ethanol production thirtyfold (Hunt, 2007). In the EU, Germany, France, and Sweden have programs (e.g., Horizon 2020, Intelligent Energy Europe, European Industrial Bioenergy Initiative) that support biofuels R&D.

29.5 Implications of Biofuels Policies

Biofuels markets around the world are promoted through a number of policies. Promoting biofuels has a positive impact on agricultural incomes, but the benefits of biofuels processing facilities depend on the fluctuations in biomass supply and price. If the feedstock price is low, biofuels facilities are likely to benefit from various forms of tax credits.

Biofuels production could affect the price of food and impose other costs on consumers. Biofuels policies promote the production of energy crops, which could lead to the conversion of land that is presently used to grow staple crops. This conversion could increase agricultural prices, and affect food industries and customers. The livestock industry that uses corn and other feedstocks could also be hampered by high commodity prices. While taxes and mandates increase the overall cost of energy, price support for biofuels crops may increase the consumer surplus by decreasing the cost of producing feedstocks and helping to lower the cost of energy. Subsidies are usually effective in stimulating an increase in the supply of biofuels, but unconditional subsidies could potentially transfer too much income to producers, especially when the price of oil is high.

In general, most biofuels policies increase biofuels supply and reduce oil imports. As a result of these policies and the introduction of efficient light vehicle fleets and improved ethanol technologies, a fair amount of gasoline consumption has been displaced by ethanol. This displacement has improved the balance of trade in the USA. For example, in 2011 the USA consumed 134 billion gallons of finished motor gasoline, which is 7 billion gallons less than the 141 billion gallons consumed in 2005. In 2011, consumption of ethanol was responsible for 67.25% of the decrease in the consumption of finished motor gasoline (Hochman *et al.*, 2013), and resulted in a decrease of financial outflows to other countries.

Biofuels policies create incentives for and constraints on the import and export of ethanol, and result in biofuels price volatility. For example, the cost to produce sugarcane ethanol in Brazil is lower than the cost to produce corn ethanol in the USA. However, factors like the transportation costs of sugarcane ethanol, mandates for blending biofuels, and tax credits for US corn ethanol alter the relative competitiveness. In addition, the cost of ethanol in the USA and Brazil is highly sensitive to the exchange rate and the price of feedstocks.

Various measures are used by governments to promote and stimulate demand for biofuels, as well as to increase their cost-competitiveness. If biofuels production costs remain high despite large-scale production, policies that promote their production and use could result in large financial burdens on governments. Even if low biofuels production costs are to be realized, the promotion of biofuels would be justified only if national benefits (e.g., rural development, energy security) were substantial. Biofuels must also have fewer environmental impacts than gasoline. Whether these benefits can be realized or not depends on the success of policies and strategies for biofuels production.

Biofuels policies (see Table 29.3) are designed to promote the use of biofuels. Biofuels are perceived to have fewer CO_2 emissions than oil, but a clear consensus on the climate benefits of biofuels

Table 29.3 Biofuels politics

Country	Major biofuels policies	Highlights
USA	Revised Renewable Fuel Standard (RFS2) authorized under Energy Independence and Security Act (EISA)	Production of 36 billion gallons of biofuels annually by 2022
	Tax credit for biofuels blend	Blender's tax credit of $1.01 per gallon of cellulosic ethanol, and $0.45 per gallon of corn ethanol
	Carbon tax	California has the lowest carbon tax per metric ton of CO_2 emissions
Brazil	Biofuels regulation	20–25% ethanol blend in all regular gasoline sales
	Tax credit for flex-fuel vehicles	Flex-fuel vehicles represent more than 80% of new auto sales in Brazil
European Union	Biofuels regulation under the EU Biofuels Directive	10% blend of renewable fuels content in transportation fuels by 2020
	Tax credit for biofuels blends	Tax credit for ethanol as high as $3.18 per gallon
	Feed-in tariff	Rapid adoption of commercial anaerobic digesters with the introduction of the feed-in tariff
	Carbon tax	Countries in the EU have high carbon tax rates
Other countries	Biofuels regulation	Biofuels blend ratio varies across countries, e.g., 5–10% in India, China, and Thailand
	Tax credit	In China, biofuels producers are exempted from consumption tax and value-added tax
	Trade policies	Argentina has high export taxes on soybean seeds and soybean oil; Malaysia sets export charges on biofuels feedstock (palm oil) on a monthly basis

is presently lacking (Fargione *et al.*, 2008; Searchinger *et al.*, 2008). At present, it is also unclear whether policies for renewable fuels (e.g., carbon taxes, fuel efficiency standards, and cap-and-trade systems) result in reductions in CO_2 emissions. There are several ways in which biofuels production can result in CO_2 emissions (e.g., using fossil fuels to produce and process feedstocks). Using corn as the feedstock for first-generation biofuels has other environmental issues, including the loss of soil carbon and erosion due to tilling, eutrophication due to fertilizer runoffs, and habitat loss due to land-use change. Similarly, the expansion of soybean and oil palm plantations has resulted in the deforestation of Brazilian and Malaysian rainforests. As a result, there is a need to coordinate policies worldwide if biofuels are to be effective in contributing to the mitigation of climate change and environmental deterioration while also meeting energy demands.

Concerns about the sustainability of biofuels have resulted in a number of certification schemes, which are mostly the result of regulations in the EU. The Roundtable on Sustainable Biofuels is one of the most noticeable international multi-stakeholder initiatives. It was established in 2006 and seeks to produce a consensus on a set of principles and criteria for sustainable liquid biofuels feedstock production and processing, and for biofuels transportation and distribution. International Sustainability and Carbon Certification (ISCC) is another government-supported, privately run organization that focuses on reducing CO_2 emissions, sustainable land use, ecosystem protection, and selected social concerns (Elbehri *et al.*, 2013). Following the sustainability criteria, in 2006 the Committee on Industry, Research, and Energy of the European Parliament called for a ban in the EU on the use of biofuels that are derived from palm oil imported from Malaysia and Indonesia (Kojima *et al.*, 2007).

References

Boyle, G. (2012). *Renewable Energy: Power for a Sustainable Future*, 3rd edn. Oxford: Oxford University Press.

Carriquiry, M.A., Du, X., & Timilsina, G.R. (2011). Second generation biofuels: Economics and policies. *Energy Policy* 39(7), 4222–34.

Davis, S.C., Parton, W.J., Grosso, S.J.D., *et al.* (2011). Impact of second-generation biofuel agriculture on greenhouse-gas emissions in the corn-growing regions of the US. *Frontiers in Ecology and the Environment* 10(2), 69–74.

DOE (2011). *Bioenergy Technologies Office Solicitations*. Washington, DC: U.S. Department of Energy. http://energy.gov/eere/bioenergy/past-solicitations, accessed April 2016.

EIA (2013). *International Energy Outlook 2013*. Washington, DC: U.S. Energy Information Administration. http://www.eia.gov/forecasts/ieo/pdf/0484(2013).pdf, accessed April 2016.

Elbehri, A., Segerstedt, A., and Liu, P. (2013). *Biofuels and the Sustainability Challenge: A Global Assessment of Sustainability Issues, Trends and Policies for Biofuels and Related Feedstocks*. Rome: Food and Agriculture Organization of the United Nations.

Embassy of Brazil (2007). *Clean Energy: The Brazilian Ethanol Experience*. London: Embassy of Brazil.

Fargione, J., Hill, J., Tilman, D., Polasky, S., and Hawthorne, P. (2008). Land clearing and the biofuel carbon debt. *Science* 319(5867): 1235–8.

Hochman, G., Barrows, G., and Zilberman, D. (2013). U.S. biofuels policy: Few environmental benefits but large trade gains. *Agricultural and Resource Economics Update* 17(2): 1–3.

Hunt, S. (ed.) (2007). *Biofuels for Transport: Global Potential and Implications for Sustainable Energy and Agriculture*. Sterling, VA: World Watch Institute.

IEA Bioenergy (2011). *Country Reports of Member Countries, Istanbul*. Paris: IEA Bioenergy. http://www.iea-biogas.net/country-reports.html, accessed April 2016.

Klette, T.J., Møen, J., and Griliches, Z. (2000). Do subsidies to commercial R&D reduce market failures? Microeconometric evaluation studies. *Research Policy* 29(4): 471–95.

Kojima, M., Mitchell, D., and Ward, W. (2007). *Considering Trade Policies for Liquid Biofuels*. Energy Sector Management Assistance Program, Special Report 004/07. Washington, DC: World Bank.

Licht, F.O. (2012). *World Ethanol and Biofuels Report*, 7(18): 323. Ratzeburg: F.O. Licht.

Licht, F.O. (2013). *World Ethanol and Biofuels Report*, 11(9): 365. Ratzeburg: F.O. Licht.

Rajagopal, D., and Zilberman, D. (2007). *Review of Environmental, Economic and Policy Aspects of Biofuels*. Publication no. 4341. Washington, DC: World Bank.

Searchinger, T., Heimlich, R., Houghton, R.A., *et al.* (2008). Use of U.S. croplands for biofuels increases greenhouse gases through emissions from land-use change. *Science* 319(5867): 1238–40.

Sorda, G., Banse, M., and Kemfert, C. (2010). An overview of biofuel policies across the world. *Energy Policy* 38(11): 6977–88.

Sumner, J., Bird, L., and Smith, H. (2009). *Carbon Taxes: A Review of Experience and Policy Design Considerations*. Technical Report, NREL/TP-6A2-47312. Golden, CO: National Renewable Energy Laboratory.

Suplee, C. (2008). *What You Need to Know about Energy*. Washington, DC: National Academies Press.

USAID (2009). *Biofuels in Asia: An Analysis of Sustainability Options*. Bangkok: United States Agency for International Development.

USDA (2010). *Brazil Biofuels Annual*. Report BR10006. Washington, DC: USDA Foreign Agricultural Service.

Exercise Problems

29.1. Suppose that the demand and supply schedule for gas is as in Table 29.4.

    (a) Draw the demand and supply curve for the gas.

    (b) What are the equilibrium price and quantity of gas consumption? (Hint: equilibrium is reached when the price and quantity the consumer demand equal the amount that the suppliers are willing to supply.)

Table 29.4 Demand and supply schedule for gas

Buyer demand per consumer		Gas supply per consumer	
Price per liter	Quantity (liters) demanded per week	Price per liter	Quantity (liters) supplied per week
$3.00	40	$2.20	40
$2.75	50	$2.30	50
$2.50	65	$2.50	65
$2.25	90	$2.75	90
$2.00	120	$3.15	120

29.2. A car gets 30 miles per gallon (mpg) on regular gasoline but would get 22.2 mpg with 15% of ethanol blended in gasoline (E15). Assume that E15 costs $2.5 and regular gasoline costs $3.0 gallon. How would the cost compare between driving on E15 and regular gasoline?

29.3. Discuss some of the areas that need to be assessed before promoting the production of feedstocks for biofuels.

29.4. Discuss the biofuels policies that might result in negative revenue to the feedstock producer.

29.5. Discuss the biofuels policies that might result in positive revenue to the feedstock producer.

29.6. Discuss the biofuels production targets set by the Energy Independence and Security Act of 2007.

29.7. How does the carbon tax compare with cap-and-trade policy tools? Discuss, summarizing the comparative merits of each.

29.8. Why is government intervention necessary for the success of the overall biofuels industry?

29.9. Which is the most effective and practical biofuels policy tool? Provide your opinion, discussing the near- and long-term implications of the selected policy tool.

29.10. Summarize Brazilian biofuels-related policies, mandates, and programs.

Index

acclimation, 259, 329
acetic acid or acetate, 59, 90–92, 98, 173, 207,
    219, 257, 264, 279–283, 297, 308,
    316–320, 333, 370, 423, 433
acetogenesis or acetogens, 92, 316–319, 332
    see also anaerobic digestion
acetone-butanol-ethanol (ABE) fermentation or
    butanol fermentation, 92, 277–279
  biochemical pathway, 280
  butanol recovery, 288–292
  co-factors, 285
  medium composition, 286
  microbiology, 284
  product inhibition, 287
  stoichiometry, 282
  substrates, 287
acetotrophic/aceticlastic methanogens, 316, 319,
    320 *see also* anaerobic digestion
acetyl-CoA, 90, 280–283, 299, 300, 409
acidogenesis or acidogens, 279, 281, 282, 285,
    317, 325 *see also* anaerobic digestion
acid pretreatment, 207 *see also* pretreatment
activation energy, 47, 227, 376, 429
adenosine triphosphate (ATP), 71, 89, 90, 131, 172,
    176, 226, 253–257, 280, 285, 299, 365
adiabatic flame temperature, 395
advanced biofuel, 48, 89–91, 467, 482, 555, 558, 559
air/oxygen ratio, 425
alcohols, 56–60, 89, 91, 92, 96, 193, 213, 220, 297,
    298 *see also* ethanol
alcohol to triglyceride ratio, 495
aldehydes, 57, 58, 434
algae or algal cells/biomass, 14, 89, 170, 483
  biodiesel, 176
  cultivation, 178
  growth, 17, 172, 174
  harvesting, 183
  modeling, 182
aliphatic compounds, 52, 61
alkaline pretreatment, 209 *see also* pretreatment

alkalinity, 325, 326, 333
alkanes, 52
alkenes, 54
alkynes, 55
alpha-amylase or α-amylase, 225, 226, 242, 243,
    246, 454, 456, 457
amino acids, 66, 90, 98, 131, 225, 226, 267,
    315–317, 499
ammonium/ammonia, 131, 176, 209–211, 267, 319,
    326, 328, 333, 349, 358, 380, 405, 433
amylase, 224, 225, 242 *see also* alpha-amylase or
    α-amylase; glucoamylase
anaerobic digester, 313, 324, 330, 339, 345 *see also*
    bioreactor configurations
  anaerobic baffled reactor, 345
  anaerobic contact reactor, 345
  anaerobic filter, 315, 330, 347
  anaerobic sequencing batch reactor, 345
anaerobic digestion (AD), 14, 109, 170, 313, 328,
    338, 463, 527, 550, 554
  alkalinity, 325
  attached growth system, 346
  household digester, 351
  hydraulic retention time, 330
  hydrolysis, 315
  Mouras Automatic Scavenger, 314
  nutrients, 327
  solids retention time, 330
  solid-state anaerobic digestion (SS-AD), 349
  start-up, 331
  stoichiometry, 320
  suspended growth system, 339
  toxic compounds, 328
  volumetric organic loading rate, 330
Anaerobic Digestion Model No.1 (ADM1), 320, 331
anaerobic microorganisms/anaerobes, 244, 305,
    317, 323, 330, 346
animal fats, 484
anode, 184, 363–372
anode chamber, 363, 379, 380

Bioenergy: Principles and Applications, First Edition. Edited by Yebo Li and Samir Kumar Khanal.
© 2017 John Wiley & Sons, Inc. Published 2017 by John Wiley & Sons, Inc.
Companion website: www.wiley.com/go/Li/Bioenergy

Printed and bound by CPI Group (UK) Ltd, Croydon, CR0 4YY

16/04/2025

14658354-0004